Fire Service Pump Operator
Principles and Practice

JONES AND BARTLETT PUBLISHERS
Sudbury, Massachusetts
BOSTON   TORONTO   LONDON   SINGAPORE

Jones and Bartlett Publishers, LLC
World Headquarters
40 Tall Pine Drive
Sudbury, MA 01776
978-443-5000
info@jbpub.com
www.jbpub.com

National Fire Protection Association
1 Batterymarch Park
Quincy, MA 02169-7471
www.NFPA.org

International Association of Fire Chiefs
4025 Fair Ridge Drive
Fairfax, VA 22033
www.IAFC.org

Jones and Bartlett's books and products are available through most bookstores and online booksellers. To contact Jones and Bartlett Publishers directly, call 800-832-0034, fax 978-443-8000, or visit our website, www.jbpub.com.

Substantial discounts on bulk quantities of Jones and Bartlett's publications are available to corporations, professional associations, and other qualified organizations. For details and specific discount information, contact the special sales department at Jones and Bartlett via the above contact information or send an email to specialsales@jbpub.com.

Production Credits

Chief Executive Officer: Clayton Jones
Chief Operating Officer: Don W. Jones, Jr.
President, Higher Education and Professional Publishing: Robert W. Holland, Jr.
V.P., Design and Production: Anne Spencer
V.P., Manufacturing and Inventory Control: Therese Connell
Publisher, Public Safety Group: Kimberly Brophy
Senior Acquisitions Editor, Fire: William Larkin
Editor: Jennifer S. Kling
Production Manager: Jenny L. Corriveau
Associate Photo Researcher: Jessica Elias
Director of Sales: Matthew Maniscalco
Director of Marketing: Alisha Weisman
Marketing Manager, Fire: Brian Rooney
Cover Image: Courtesy of Thomas O'Connell, Division Chief, Sunrise Fire Rescue, FL
Text Design: Anne Spencer
Cover Design: Kristin E. Parker
Composition: Publishers' Design and Production Services
Text Printing and Binding: Courier Companies
Cover Printing: Courier Companies

Copyright © 2011 by Jones and Bartlett Publishers, LLC, an Ascend Learning Company, and the National Fire Protection Association®.

All rights reserved. No part of the material protected by this copyright notice may be reproduced or utilized in any form, electronic or mechanical, including photocopying, recording, or by any information storage and retrieval system, without written permission from the copyright owner.

The procedures and protocols in this book are based on the most current recommendations of responsible sources. The International Association of Fire Chiefs (IAFC), National Fire Protection Association (NFPA®), and the publisher, however, make no guarantee as to, and assume no responsibility for, the correctness, sufficiency, or completeness of such information or recommendations. Other or additional safety measures may be required under particular circumstances.

Additional photographic and illustration credits appear on page 394, which constitutes a continuation of the copyright page.

Notice: The individuals described in "You Are the Driver/Operator" and "Driver/Operator in Action" throughout the text are fictitious.

Library of Congress Cataloging-in-Publication Data
Fire service pump operator : principles and practice.
    p. cm.
  Includes index.
  ISBN-13: 978-0-7637-3908-9 (pbk.)
  ISBN-10: 0-7637-3908-1 (pbk.)
  1. Fire pumps. 2. Fire engines.
  TH9363.F58 2009
  628.9'25—dc22
                        2009024205

6048
Printed in the United States of America
16 15 14 13 12   10 9 8 7 6 5 4 3

Brief Contents

1. The History of the Driver/Operator 1
2. Types of Fire Apparatus 16
3. Water 32
4. Mathematics for the Driver/Operator 56
5. Pumper Apparatus Overview 108
6. Performing Apparatus Check-Out and Maintenance 126
7. Pumper Operations 152
8. Approaching the Fire Ground 190
9. Responding on the Fire Ground 218
10. Water Supply 254
11. Foam 296
12. Performance Testing 328

Appendix A: An Extract From: NFPA® 1002, *Fire Apparatus Driver/Operator Professional Qualifications*, 2009 Edition 362
Appendix B: NFPA® 1002 Correlation Guide 366
Appendix C: Fire Protection Hydraulics and Water Supply (FESHE) Correlation Guide 367
Glossary 368
Index 378
Credits 394

Contents

1 The History of the Driver/Operator ... 1
Introduction ... 2
Evolution of Fire Apparatus and Equipment ... 2
   Fire Protection in Early American Cities ... 3
   Hand Pumps and Hose Carts ... 3
   Fire Apparatus Evolution ... 4
   Ladder Wagons Pulled by Horses ... 4
   Elevated Streams ... 5
   Gasoline-Powered Fire Apparatus ... 5
   Adding a Pump ... 5
   Triple-Combination Pumper ... 6
   Improvements in Aerial Devices ... 6
   Other Uses for Motorized Apparatus ... 6
   Diesel-Powered Fire Apparatus ... 9
Safety Considerations ... 9
   Protecting the Driver/Operator at the Scene ... 9
   Visibility for the Driver/Operator ... 9
   The Driver/Operator's Role in Safety ... 10
Modern Fire Apparatus ... 10
Driver/Operator Selection ... 11

2 Types of Fire Apparatus ... 16
Introduction ... 18
Fire Apparatus Requirements ... 18
   Working with a Manufacturer ... 18
Water on the Fire Apparatus ... 19
Pumper ... 20
Initial Attack Fire Apparatus ... 21
Mobile Water Supply Apparatus ... 24
Aerial Fire Apparatus ... 25
Quint Fire Apparatus ... 27
Special Service Fire Apparatus ... 27
Mobile Foam Fire Apparatus ... 28

3 Water ... 32
Introduction ... 34
Chemical Properties of Water ... 35
   Physical Properties of Water ... 35
   Harmful Characteristics of Water ... 36
Municipal Water Systems ... 36
   Water Sources ... 37
   Water Treatment Facilities ... 37
   Water Distribution System ... 37
Flow and Pressure ... 38
Fire Hydrants ... 41
   Dry-Barrel Hydrants ... 41
   Wet-Barrel Hydrants ... 41
   Operation of Fire Hydrants ... 41
   Shutting Down a Hydrant ... 42
   Locations of Fire Hydrants ... 42
   Inspecting and Maintaining Fire Hydrants ... 44
   Testing Fire Hydrants ... 46
      Fire Hydrant Testing Procedure ... 47
Rural Water Supplies ... 49
   Static Sources of Water ... 49

4 Mathematics for the Driver/Operator ... 56
Introduction ... 59
Pump Discharge Pressure ... 60
   Nozzle Pressures ... 61
   Determining Nozzle Flow ... 62
Friction Loss ... 63
Calculating Friction Loss ... 65
   Multiple Hose Lines of Different Sizes and Lengths ... 66
   Elevation Pressure ... 70
   Appliance Loss ... 72
   Total Pressure Loss ... 76
Wyed Hose Lines ... 77
   Siamese Hose Lines ... 78
   Calculating Elevated Master Streams ... 83
   Standpipe Systems ... 87
   Pressure-Regulating Valves ... 92
   Net Pump Discharge Pressure ... 93
Fire Service Hydraulic Calculations ... 93
   Charts ... 97
   Hydraulic Calculators ... 98
   Hand Method ... 98
Subtract 10 Method (GPM Flowing Method) ... 100
Condensed Q Method ... 100
   The Condensed Q Method in a Relay Operation ... 101
   Flow Meters and Electronic Pump Controllers ... 101
   Electronic Pump Controllers and Pressure Governors ... 102
   Preincident Plan ... 102

5 Pumper Apparatus Overview ... 108
Introduction ... 110
Exterior of the Pumper ... 110
Interior of the Pumper: The Cab ... 111

Pumps 111
Types of Fire Pumps 112
   Positive-Displacement Pumps 112
   Centrifugal Pump 114
   Power Supplies for Pumps. 120

6 Performing Apparatus Check-Out and Maintenance 126

Introduction 129
Inspection. 130
Inspection Process. 133
Fire Apparatus Sections. 134
   Exterior Inspection 134
   Engine Compartment .136
   Cab Interior. 137
   Brake Inspection 139
   General Tools/Equipment Inspection 140
   Pump Inspection. 141
   Aerial Device Inspection 145
Safety 148
   Weekly, Monthly, or Other Periodic
     Inspection Items 148
   Completing Forms 148

7 Pumper Operations 152

Introduction 155
The Many Roles of the Driver/Operator 155
   Promoting Safety. 156
   Educating Crew Members 157
   Trust and Team Building 158
Maintaining a Safe Work Environment 158
   Lead by Example 159
Fire Apparatus and Equipment:
   Functions and Limitations. 161
   Fire Apparatus and Equipment Inspections 162
   Safety Across the Board 162
Emergency Vehicle Response 163
   Dispatch 163
   Maps 164
   360-Degree Inspection 166
Starting the Fire Apparatus 166
   Seat Belt Safety 168
   Getting Underway 169
Driving Exercises 170
   Performing a Serpentine Maneuver 175
   Performing a Confined-Space Turnaround. 175
   Performing a Diminishing Clearance Exercise . .177
Returning to the Station 180

Reserving the Fire Apparatus 180
Shutting Down the Fire Apparatus. 183

8 Approaching the Fire Ground 190

Introduction 193
Emergency Vehicle Laws 193
   Safe Driving Practices .194
Approaching the Scene of an Emergency. 199
   Slow Down 199
   Identify the Address/Location 199
   Recognize Potential Hazards 200
Fire Scene Positioning 200
   Nothing Showing 200
   Working Fire. 200
   Staging 201
   Engines and Ladders 203
   Positioning of Other Fire Scene Apparatus 205
Positioning at an Intersection or
   on a Highway 206
   Never Trust Traffic. 206
   Engage in Proper Protective Parking 206
   Reduce Motorist Vision Impairment 206
   Wear High-Visibility Reflective Vests 208
Manual on Uniform Traffic Control Devices 208
   Motor Vehicle Accidents 210
Positioning at the Emergency
   Medical Scene 212
Special Emergency Scene Positioning 213

9 Responding on the Fire Ground 218

Fire Hose, Appliances, and
   Nozzles Overview 220
   Functions of Fire Hose 220
   Sizes of Hose 220
   Attack Hose. 222
   Supply Hose 222
   Fire Hose Appliances. .223
   Nozzles 228
   Other Types of Nozzles 232
   Nozzle Maintenance and Inspection 232
Fire Hose Evolutions 232
   Supply Line Evolutions 233
   Connecting a Fire Department Engine to a
     Water Supply 236
Cab Procedures 236
   Exiting the Cab 237
Securing a Water Source 239
   Hand Lays. 239
Standpipe/Sprinkler Connecting 241

Connecting Supply Hose Lines to Standpipe
  and Sprinkler Systems. 241
Performing a Changeover 245
Duties on Scene 249

10 Water Supply 254
Introduction 257
The Mechanics of Drafting. 257
Inspections, Routine Maintenance, and
  Operational Testing. 258
  Inspecting the Priming System. 258
  Performing a Vacuum Test 260
  Finding a Vacuum Leak 261
Water Supply Management in the Incident
  Management System 261
Selecting a Drafting Site 263
  Determining the Reliability of Static
    Water Sources. 263
  Estimating the Quantity of Water Available 263
  Accessibility to the Static Water Source 264
  Special Accessibility Considerations 265
  Operational Considerations for
    Site Selection. 266
Pumping from a Draft. 266
  Making the Connection 267
Preparing to Operate at Draft by
  Priming the Pump 270
Drafting Operations 270
Producing the Flow of Water 271
Complications during Drafting Operations 273
Uninterrupted Water Supply 275
Relay Pumping Operations 275
  Components of a Relay Pumping Operation . . . 275
  Equipment for Relay Pumping Operations. 276
  Personnel for Relay Pumping Operations. 277
  Preparing for a Relay Pumping Operation 277
  Calculating Friction Loss. 278
  Types of Relay Pumping Operations. 279
  Operating the Source Pumper 279
  Operating the Attack Pumper 279
  Operating the Relay Pumper 280
  Water Relay Delivery Options 281
  Joining an Existing Relay
    Pumping Operation 281
  Pressure Fluctuations in a Relay
    Pumping Operation 282
  Shutting Down a Relay Pumping Operation. . . 282
  Safety for Relay Pumping Operations. 283
Water Shuttle 283
  Fill Sites 283
  Filling Tankers. 286

Safety for Water Shuttle Operations 287
Establishing Dump Site Operations 287
  Using Portable Tanks 289
  Offloading Tankers 289
  Use of Multiple Portable Tanks 290
  Source Pumper Considerations for Portable
    Tank Operations 290
  Traffic Flow within a Dump Site 291
Nurse Tanker Operations. 292
  Communication for Tanker Operations 292
Water Shuttle Operations in the Incident
  Management System 292

11 Foam 296
Introduction 298
History. 298
Overview. 299
  What Is Foam? 299
  Foam Tetrahedron 299
  Foam Characteristics 300
Foam Classifications 301
  Class A Foams. 301
  Class B Foams 302
  Synthetic Foams 303
  Alcohol-Resistant Aqueous
    Film-Forming Foams 303
  Synthetic Detergent Foams
    (High-Expansion Foams) 304
Foam Concentrates 305
Foam Expansion Rates. 305
  Low-Expansion Foam 305
  Medium-Expansion Foam 305
  High-Expansion Foam. 305
Foam Proportioning 306
  Proportioning Foam Concentrate. 306
  Foam Proportioning Systems. 306
  Around-the-Pump Proportioning System. 309
  Balanced-Pressure Proportioning Systems 311
  Injection Systems 313
  Compressed-Air Foam System. 314
Nozzles 317
  Medium- and High-Expansion
    Foam Generators. 317
  Master Stream Foam Nozzles. 317
  Air-Aspirating Foam Nozzles 317
  Smooth-Bore Nozzles 317
  Fog Nozzles. 317
Foam Supplies 317
Foam Application 319
  Applying Foam 319
Foam Compatibility 322

12 Performance Testing............ 328

Introduction 330
Fire Apparatus Requirements 331
   Environmental Requirements 331
   Test Site................................331
   Equipment Requirements 333
No-Load Governed Engine Speed Test 334
Intake Relief Valve System Test.............. 335
Pump Shift Indicator Test 335
Pump Engine Control Interlock Test 336
Gauge and Flow Meter Test 338
Tank-to-Pump Flow Test.................... 341
Vacuum Test 341
Priming System Test....................... 344
Pumping Test Requirements 344
Pump Performance Test.................... 348
   Capacity Test/150 psi Test
   (100 Percent Test)......................350
   Overload Test/165 psi Test...............350
   200 psi Test (70 Percent Test) 352
   250 psi Test (50 Percent Test) 352
Pressure Control Test...................... 354
Post Performance Testing 356
   Final Test Results......................356
   Problem Solving 357
Re-rating Fire Pumps 357

Appendix A: An Extract From: NFPA® 1002, *Fire Apparatus Driver/Operator Professional Qualifications*, 2009 Edition 362

Appendix B: NFPA® 1002 Correlation Guide......... 366

Appendix C: Fire Protection Hydraulics and Water Supply (FESHE) Correlation Guide 367

Glossary 368

Index 378

Credits 394

Skill Drills

Chapter 3
Skill Drill 3-1	Operating a Fire Hydrant	43
Skill Drill 3-2	Shutting Down a Hydrant	44
Skill Drill 3-3	Obtaining the Static Pressure	48
Skill Drill 3-4	Operating a Pitot Gauge	50

Chapter 4
Skill Drill 4-1	**[Standard]** Calculating the Flow of a 1¼″ Smooth-Bore Nozzle on a 2½″ Handline	64
Skill Drill 4-1	**[Metric]** Calculating the Flow of a 32-mm Smooth-Bore Nozzle on a 64-mm Handline	64
Skill Drill 4-2	**[Standard]** Calculating Friction Loss in Single Hose Lines	66
Skill Drill 4-2	**[Metric]** Calculating Friction Loss in Single Hose Lines	67
Skill Drill 4-3	**[Standard]** Calculating Friction Loss in Multiple Hose Lines of Different Sizes and Lengths	68
Skill Drill 4-3	**[Metric]** Calculating Friction Loss in Multiple Hose Lines of Different Sizes and Lengths	69
Skill Drill 4-4	**[Standard]** Calculating the Elevation Pressure (Loss and Gain)	71
Skill Drill 4-4	**[Metric]** Calculating the Elevation Pressure (Loss and Gain)	72
Skill Drill 4-5	Determining the Friction Loss in Appliances	75
Skill Drill 4-6	Using a Pitot Gauge	76
Skill Drill 4-7	Using In-Line Gauges to Test Friction Loss	77
Skill Drill 4-8	**[Standard]** Determining the Pump Discharge Pressure in a Wye Scenario with Equal Lines	78
Skill Drill 4-8	**[Metric]** Determining the Pump Discharge Pressure in a Wye Scenario with Equal Lines	79
Skill Drill 4-9	**[Standard]** Determining the Pump Discharge Pressure in a Wye Scenario with Unequal Lines	80
Skill Drill 4-9	**[Metric]** Determining the Pump Discharge Pressure in a Wye Scenario with Unequal Lines	81
Skill Drill 4-10	**[Standard]** Determining the Pump Pressure for a Siamese Line by the Split Flow Method	82
Skill Drill 4-10	**[Metric]** Determining the Pump Pressure for a Siamese Line by the Split Flow Method	82
Skill Drill 4-11	**[Standard]** Calculating Friction Loss in Siamese Lines by the Coefficient Method	83
Skill Drill 4-11	**[Metric]** Calculating Friction Loss in Siamese Lines by the Coefficient Method	83
Skill Drill 4-12	**[Standard]** Calculating Friction Loss in Siamese Lines by the Percentage Method	84
Skill Drill 4-12	**[Metric]** Calculating Friction Loss in Siamese Lines by the Percentage Method	85
Skill Drill 4-13	**[Standard]** Calculating the Pump Discharge Pressure for a Prepiped Elevated Master Stream Device	88
Skill Drill 4-13	**[Metric]** Calculating the Pump Discharge Pressure for a Prepiped Elevated Master Stream Device	89
Skill Drill 4-14	**[Standard]** Calculating the Pump Discharge Pressure for an Elevated Master Stream	90
Skill Drill 4-14	**[Metric]** Calculating the Pump Discharge Pressure for an Elevated Master Stream	91
Skill Drill 4-15	**[Standard]** Calculating the Discharge Pressure for a Standpipe During Preplanning	92
Skill Drill 4-15	**[Metric]** Calculating the Discharge Pressure for a Standpipe During Preplanning	93
Skill Drill 4-16	**[Standard]** Calculating Pump Discharge Pressure for a Standpipe	94
Skill Drill 4-16	**[Metric]** Calculating Pump Discharge Pressure for a Standpipe	95
Skill Drill 4-17	Finding the Flow Rate with a Fire Stream Calculator	98
Skill Drill 4-18	Using a Manual Friction Loss Calculator	98
Skill Drill 4-19	Performing the Hand Method of Calculation	99
Skill Drill 4-20	Performing the Hand Method of Calculation for 2½″ Hose	99
Skill Drill 4-21	Using the Hand Method Calculation for 1¾″ Hose	100

Skill Drill 4-22	Using the Subtract 10 Method (GPM Flowing Method).	100
Skill Drill 4-23	Using the Condensed Q Method	101

Chapter 6

Skill Drill 6-1	Performing an Apparatus Inspection	147

Chapter 7

Skill Drill 7-1	Identifying the Critical Information Received from the Dispatch	164
Skill Drill 7-2	Performing a 360-Degree Inspection	167
Skill Drill 7-3	Starting a Fire Apparatus	171
Skill Drill 7-4	Performing the Serpentine Exercise	176
Skill Drill 7-5	Performing a Confined-Space Turnaround	178
Skill Drill 7-6	Performing a Diminishing Clearance Exercise	179
Skill Drill 7-7	Backing a Fire Apparatus into a Fire Station Bay	184
Skill Drill 7-8	Shutting Down and Securing a Fire Apparatus	185

Chapter 8

Skill Drill 8-1	Performing the Alley Dock Exercise	214

Chapter 9

Skill Drill 9-1	Connecting a Storz Coupling Soft Suction Hose from a Hydrant to a Pump	224
Skill Drill 9-2	Inspecting a Solid-Stream Nozzle	229
Skill Drill 9-3	Inspecting a Fog-Stream Nozzle	231
Skill Drill 9-4	Engaging the Fire Pump	238
Skill Drill 9-5	Hand Laying a Supply Line	242
Skill Drill 9-6	Connecting Hose to a Fire Department Connection	246
Skill Drill 9-7	Performing a Changeover Operation	248
Skill Drill 9-8	Operating an Auxiliary Cooling System	249
Skill Drill 9-9	Disengaging the Fire Pump	249

Chapter 10

Skill Drill 10-1	Performing a Vacuum Test	262
Skill Drill 10-2	Positioning the Fire Apparatus for Drafting Operations	269
Skill Drill 10-3	Drafting from a Static Water Source	272
Skill Drill 10-4	Providing Water Flow for Handlines and Master Streams	273

Chapter 11

Skill Drill 11-1	Batch Mixing Foam	307
Skill Drill 11-2	Operating an In-line Eductor	310
Skill Drill 11-3	Operating an Around-the-Pump Proportioning System	312
Skill Drill 11-4	Operating a Balanced-Pressure Proportioning System	313
Skill Drill 11-5	Operating an Injection Foam System	313
Skill Drill 11-6	Operating a Compressed-Air Foam System	315
Skill Drill 11-7	Applying Class A Foam on a Fire	320
Skill Drill 11-8	Applying Foam with the Roll-on Method	321
Skill Drill 11-9	Applying Foam with the Bankdown Method (FF2)	322
Skill Drill 11-10	Applying Foam with the Raindown Method	323

Chapter 12

Skill Drill 12-1	Conducting the No-Load Governed Engine Speed Test	336
Skill Drill 12-2	Performing the Pump Shift Indicator Test	337
Skill Drill 12-3	Performing the Pump Engine Control Interlock Test	339
Skill Drill 12-4	Performing a Gauge Meter Test	340
Skill Drill 12-5	Performing a Flow Meter Test	342
Skill Drill 12-6	Testing the Tank-to-Pump Rate	343
Skill Drill 12-7	Performing a Vacuum Test	345
Skill Drill 12-8	Performing a Priming System Test	346
Skill Drill 12-9	Performing a Capacity Test (150 psi Test)	351
Skill Drill 12-10	Performing an Overload Test	352
Skill Drill 12-11	Performing the 200 psi Test	352
Skill Drill 12-12	Performing the 250 psi Test	352
Skill Drill 12-13	Performing a Pressure Control Test	355

Resource Preview

Fire Service Pump Operator: Principles and Practice

Jones and Bartlett Publishers, the National Fire Protection Association®, and the International Association of Fire Chiefs have joined forces to raise the bar for the fire service once again with the release of *Fire Service Pump Operator: Principles and Practice*.

Safety Fundamentals

Fire Service Pump Operator: Principles and Practice features a laser-like focus on driver operator safety and responsibility. Dedicated chapters on Performing Apparatus Check-Out and Maintenance (Chapter 6) and Performance Testing (Chapter 12) stress the importance of safety whether that is in the station or behind the wheel. Actual Near-Miss Reporting System cases are discussed to drive home important points about safety and the lessons learned from these real-life incidents. It is our profound hope that this textbook will contribute to the goal of reducing line-of-duty deaths by 25 percent in the next 5 years.

Chapter Resources

Fire Service Pump Operator: Principles and Practice thoroughly supports instructors and prepares future driver/operators for the job. This text meets and exceeds the driver/operator requirements as outlined in Chapters 4, 5, and 10 of NFPA 1002, *Fire Apparatus Driver/Operator Professional Qualifications*, 2009 Edition. It also addresses all of the course outcomes from the National Fire Academy's Fire and Emergency Services Higher Education (FESHE) Associates (Core) *Fire Protection Hydraulics and Water Supply* course.

Fire Service Pump Operator: Principles and Practice serves as the core of a highly effective teaching and learning system. Its features reinforce and expand on essential information and make information retrieval a snap. These features include:

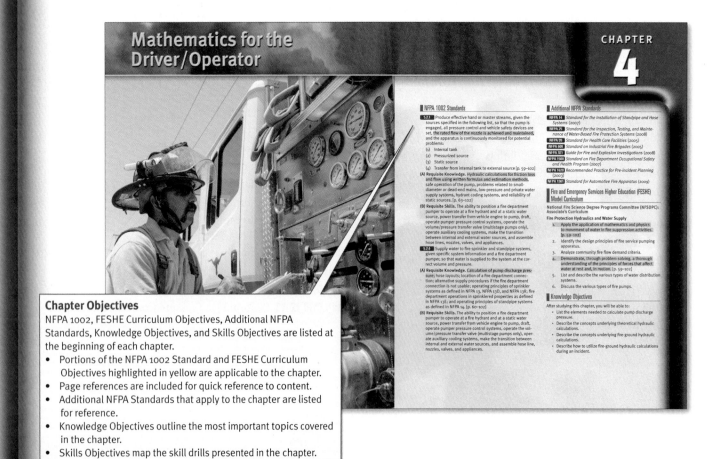

Chapter Objectives
NFPA 1002, FESHE Curriculum Objectives, Additional NFPA Standards, Knowledge Objectives, and Skills Objectives are listed at the beginning of each chapter.
- Portions of the NFPA 1002 Standard and FESHE Curriculum Objectives highlighted in yellow are applicable to the chapter.
- Page references are included for quick reference to content.
- Additional NFPA Standards that apply to the chapter are listed for reference.
- Knowledge Objectives outline the most important topics covered in the chapter.
- Skills Objectives map the skill drills presented in the chapter.

Resource Preview

You Are the Driver/Operator

Each chapter opens with a case study intended to stimulate classroom discussion, capture students' attention, and provide an overview for the chapter. An additional case study is provided in the end-of-chapter Wrap-Up material.

You Are the Driver/Operator

Your crew has entered a house, searching for the victims who have been reported to be trapped there by fire. The attack lines are pulled and advanced into the house by the attack crew. You charge the attack line and locate a fire hydrant within 100 feet of your engine. The second engine arrives and takes a second line from your engine to back up the first crew. You are asked to charge that engine's hose line before you can complete making the hydrant connection. You know an adequate water supply is essential for the crews on the inside.

1. What would you do if there was no water available from the hydrant you located right by the engine?
2. What is the pressure of your incoming water supply?
3. How many more attack lines can you support with the water supply you have available?

Introduction

The importance of a dependable and adequate water supply for fire suppression operations is self-evident. The hose line is not just the primary weapon for fighting a fire; it is also the fire fighter's primary defense against being burned or driven out of a burning building. The basic plan for fighting most fires depends on water to confine, control, [...] supply is interrupted while [...] then fire fighters could [...] ters entering a burning [...] eir water supply is both [...] ines for their protection

[...]ly is a critical fire-ground [...] as soon as possible. A [...] the same time as other [...]ce. At many fire scenes, [...] raising ladders, search [...]g a water supply all oc-

[...] sources. First, munici-[...]h water under pressure [...]ond, rural areas may de-[...]akes and streams. These [...]department apparatus to [...]ne. Often, the water car-[...]cles is used in the initial

Figure 3-1 The water that comes from a hydrant is provided by a municipal or private water system.

Voices of Experience

In the middle of the night, sometime around 0130 hours, a complement of two engines, a squad, a ladder, an ambulance, and a command officer were dispatched to a reported structure fire in an annex building of the Historic Arlington Park Race Track. Upon their arrival, the fire companies found a working fire in the wood-framed paddock building, which was attached to the five-story wood-frame grandstand. Over the years, these buildings had been renovated multiple times, which created multiple void spaces. The fire spread quickly from the annex to the grandstands and continued to burn throughout the night.

This fire occurred early in my career. I was assigned to the relief crew on the snorkel for shift change at 0800 hours. At that time, the fire had spread to the grandstands but was still being aggressively fought from interior positions. Heavy-timber building construction was on our side, but all of the remodeling work was working against us. The fire grew to eight alarms throughout the morning as more water was being used to fight the fire.

The property had a private water system that was immediately overcome. Units were used to pump water from the municipal water system into the private system and through extremely long relay operations. Those long hose lays challenged driver/operators to maintain hose line pressures. Engines were driven out across the dirt track to a large pond on the infield of the race track. Drafting operations were established as an additional water supply to the south of the building.

Millions of gallons of water from the three types of water sources were used on this fire. Despite the determined efforts of everyone on scene, the historic landmark building eventually collapsed onto itself and smoldered for days. The community, property owners, and the area fire service knew that they had done everything possible to save it. A strong reliable water source allowed crews to operate for hours in their attempt to save this vital piece of local history.

Bill Stipp
Goodyear Fire Department
Goodyear, Arizona

> "Despite the determined efforts of everyone on scene, the historic landmark building eventually collapsed onto itself and smoldered for days."

Voices of Experience

In the Voices of Experience essays, veteran driver/operators share their accounts of memorable incidents while offering advice and encouragement. These essays highlight what it is truly like to be a driver/operator.

Chapter 2 Types of Fire Apparatus

Near Miss REPORT

Report Number: 05-0000495
Report Date: 08/30/2005 18:54

Synopsis: Fire fighters nearly struck by broken coupling from overpressurized supply line.

Event Description: Our engine and truck company were participating in probationary fire fighter training for our newest fire fighter on the shift. The drill entailed establishing and flowing an elevated master stream through our 105' pre-piped aerial. Also being trained is a relatively new fire fighter who is being qualified as a back-up engine driver/operator. Because this was a familiarization-only drill, a single 20' pony sleeve of 4-inch hose was used to connect from the engine to the aerial's waterway intake. The gate was closed on the aerial side of the hose.

What followed was a series of communication errors and human errors. Due to a miscommunication from earlier in the morning, the fire fighter operating the pump panel mistakenly thought that the starting pressure for the operation was to be 225 psi, our starting pressure for aerial master streams is 150 psi. In addition, the aerial turntable operator had the intake gate closed. The fire fighter on the pump panel thought that the truck crew had stated they were "ready for water." In fact, they were not. The pump operator opened the gate valve on the pump outlet, and began to increase pressure on the line. Within 5 seconds, a loud "pop" was heard. The engine officer and two truck company fire fighters were standing within 20 feet of the apparatus. None had their helmets on. With the "pop" noise, the 4-inch hose failed at the coupling immediately behind the (unisex) connection on the truck's waterway intake gate. The metal collar that holds the hose onto the (unisex) coupling flew off the hose as it initially whipped when it came off of the truck, and it narrowly missed striking the engine captain who was standing nearby. Luckily, no one was hurt in this event and property damage was limited to the 4-inch hose and its (unisex) coupling.

Lessons Learned:
1. Whenever pumping apparatus and/or aerial apparatus are being used, all participants in the immediate area of the equipment should at least be wearing helmets and leather gloves.
2. Assure before starting a training evolution that all participants fully and completely understand their roles and responsibilities. For new trainees, ensure that an experienced member is overseeing their activity.
3. Consider a "challenge and response" means of communication. In this case, if the pump o— truck operator was truly ready for water, the event may have been avoided.
4. Know the limitations of all equipment you are using either in operations or training scena— recommended specifications. The hose in this case had stamped on it "Test to 200 psi" yet sure should be 225 psi.
5. Establish a "safety perimeter" around training evolutions and keep non-active participants

Demographics
Department type: Combination, Mostly paid
Job or rank: Captain
Department shift: 24 hours on—48 hours off
Age: 43–51
Years of fire service experience: 11–13
Region: FEMA Region III
Service Area: Suburban

Event Informati
Event type: Train
in-station drills, n
Event date and
Hours into the s
Event participat
Do you think thi
What were the
- Communication
- Situational Awa
- Human Error
What do you be
- Property damag
- Life-threatenin

Near-Miss Report
Utilizing incident data, National Fire Fighter Near-Miss Reporting System cases are discussed to highlight important points about safety and the lessons learned from real-life incidents.

Fire Service Pump Operator: Principles and Practice

Skill Drill 8-1

Performing the Alley Dock Exercise

1. Position the rear of the fire apparatus past the dock's opening and at a 90-degree angle to the marker cones.

2. Ensure that a spotter is correctly positioned behind the fire apparatus.

3. Activate the emergency lights.

4. Roll down the windows.

5. Turn off any mounted stereo equipment.

6. Disengage the parking brake, if set.

7. Shift the transmission into reverse.

8. Proceed in a reverse mode and turn the fire apparatus to align it with the objective.

9. Continue backing the fire apparatus until it has reached the desired objective or the spotter signals "stop."

Skill Drills
Skill Drills provide written step-by-step explanations and visual summaries of important skills and procedures. This clear, concise format enhances student comprehension of sometimes complex procedures. In addition, each Skill Drill identifies the corresponding NFPA job performance requirement.

Resource Preview

Comprehensive Measurements
Both U.S. Imperial units and Metric units are used throughout the text.

Safety Tip
Safety Tips reinforce safety-related concerns for driver/operators and crews.

Driver/Operator Tip
This series of tips provide advice from masters of the trade.

Hot Terms
Hot Terms are easily identifiable within the chapter and define key terms that the student must know. A comprehensive glossary of Hot Terms also appears in the Wrap-Up.

Fire Service Pump Operator: Principles and Practice

Wrap-Up
End-of-chapter activities reinforce important concepts and improve students' comprehension. Additional instructor support and answers for all questions are available on the Instructor's ToolKit CD-ROM.

Chief Concepts
Chief Concepts highlight critical information from the chapter in a bulleted format to help students prepare for exams.

Driver/Operator in Action
This feature promotes critical thinking through the use of case studies and provides instructors with discussion points for the classroom presentation.

Wrap-Up

Chief Concepts

- To obtain a fire apparatus that meets the needs of your fire department, you must work closely with the manufacturer.
- The fire pump is the key component in getting water from the fire apparatus, through the hose, and onto the fire.
- The pumper is the most common fire apparatus and is a part of almost every fire department.
  - The pumper is used on small fires such as dumpster fires, vehicle fires, and brush fires in urban settings.
  - It is also the main source of fire attack for larger fires that involve structures.
- The initial attack fire apparatus is utilized much like the pumper, but its specifications are much different. This type of apparatus is designed to be more maneuverable than the pumper, especially on off-road terrain.
- Mobile water supply fire apparatus (commonly referred to as tenders) provide enough onboard water to extinguish most fires that are held to only one or two bedrooms of a home.
- An aerial fire apparatus is a vehicle equipped with an aerial ladder, elevating platform, or water tower that is designed and equipped to support firefighting and rescue operations by positioning personnel, handling materials, providing continuous egress, or discharging water at positions elevated from the ground.
- A quint has five functions associated with it: pump, water tank, fire hose storage, aerial, and ground ladders.
- The special service fire apparatus is designed for a particular purpose and does not quite fit into the other categories. For example, a hazardous materials apparatus or a heavy technical rescue apparatus would fall into this category.
- The mobile foam fire apparatus is a fire apparatus with a permanently mounted fire pump, foam proportioning system, and foam concentrate tank(s) whose primary purpose is for use in the control and extinguishment of flammable and combustible liquid fires in storage tanks and other flammable liquid spills.

Hot Terms

aerial fire apparatus A vehicle equipped with an aerial ladder, elevating platform, or water tower that is designed and equipped to support firefighting and rescue operations by positioning personnel, handling materials, providing continuous egress, or discharging water at positions elevated from the ground. (NFPA 1901)

aerial ladder A self-supporting, turntable-mounted, power-operated ladder of two or more sections permanently attached to a self-propelled automotive fire apparatus and designed to provide a continuous egress route from an elevated position to the ground. (NFPA 1904)

chassis The basic operating motor vehicle, including the engine, frame, and other essential structural and mechanical parts, but exclusive of the body and all appurtenances for the accommodation of driver, property, passengers, appliances, or equipment related to other than control. Common usage might, but need not, include a cab (or cowl). (NFPA 1901)

initial attack fire apparatus Fire apparatus with a fire pump of at least 250 GPM (1000 L/min) capacity, water tank, and hose body, whose primary purpose is to initiate a fire suppression attack on structural, vehicular, or vegetation fires, and to support associated fire department operations. (NFPA 1901)

miscellaneous equipment Portable tools and equipment carried on a fire apparatus, not including suction hose, fire hose, ground ladders, fixed power sources, hose reels, cord reels, breathing air systems, or other major equipment or components specified by the purchaser to be permanently mounted on the apparatus as received from the apparatus manufacturer. (NFPA 1901)

mobile foam fire apparatus Fire apparatus with a permanently mounted fire pump, foam proportioning system, and foam concentrate tank(s) whose primary purpose is for use in the control and extinguishment of flammable and combustible liquid fires in storage tanks and other flammable liquid spills. (NFPA 1901)

mobile water supply apparatus A vehicle designed primarily for transporting (pickup, transporting, and delivering) water to fire emergency scenes to be applied by other vehicles or pumping equipment. Also known as a tanker or tender. (NFPA 1901)

pumper Fire apparatus with a permanently mounted fire pump of at least 750 GPM (3000 L/min) capacity, water tank, and hose body, whose primary purpose is to combat structural and associated fires. (NFPA 1901)

quint Fire apparatus with a permanently mounted fire pump, a water tank, a hose storage area, an aerial ladder or elevating platform with a permanently mounted waterway, and a complement of ground ladders. (NFPA 1901)

Driver/Operator in Action

The station captain and some other fire fighters in the station are very helpful, answering any questions you have about the fire apparatus and the equipment it carries. After you complete your inspection of the aerial apparatus, you sit down and talk with the rest of the station's crew. Some of the fire fighters have not worked on a pumper apparatus in years, which sparks a conversation about the different types of fire apparatus that the fire department now has. During the discussion, you start to recall some of the different features of those fire apparatus and the equipment that they are required to carry.

1. Which type of apparatus has a pump, carries water, fire hose, ground ladders and has an aerial device?
   A. Mobile water supply apparatus
   B. Quint
   C. Aerial apparatus
   D. Pumper

2. What would be an example of a special service fire apparatus?
   A. Pumper
   B. Hazardous materials apparatus
   C. Mobile foam apparatus
   D. Initial attack fire apparatus

3. What is the minimum amount of tank water a pumper should have?
   A. 200 gallons
   B. 1000 gallons
   C. 500 gallons
   D. 300 gallons

4. What is the total length of fire service ground ladders an aerial apparatus should carry?
   A. 80′
   B. 100′
   C. 115′
   D. 85′

5. What is the NFPA standard that covers standards for automotive fire apparatus?
   A. NFPA 1002
   B. NFPA 1021
   C. NFPA 1901
   D. NFPA 1931

Hot Terms
Hot Terms provide key terms and definitions from the chapter.

Instructor Resources

A complete teaching and learning system developed by educators with an intimate knowledge of the obstacles that instructors face each day supports *Fire Service Pump Operator: Principles and Practice*. These resources provide practical, hands-on, time-saving tools such as PowerPoint presentations, customizable lesson plans, test banks, skill sheets, and image/table banks to better support instructors and students.

Instructor's ToolKit CD-ROM

Preparing for class is easy with the resources on this CD-ROM. The CD-ROM includes the following resources:

- **Adaptable PowerPoint Presentations.** Provides instructors with a powerful way to create presentations that are educational and engaging to their students. These slides can be modified and edited to meet instructors' specific needs.
- **Detailed Lesson Plans.** The lesson plans are keyed to the PowerPoint presentations with sample lectures, lesson quizzes, and teaching strategies. Complete, ready-to-use lecture outlines include all of the topics covered in the text. The lecture outlines can be modified and customized to fit any course.
- **Image and Table Bank.** Offers a selection of the most important images and tables found in the text. Instructors can use these graphics to incorporate more images into the PowerPoint presentations, make handouts, or enlarge a specific image for further discussion.
- **Skill Sheets.** Provides you with a resource to track students' skills and conduct skill proficiency exams. The skill sheets are customizable.

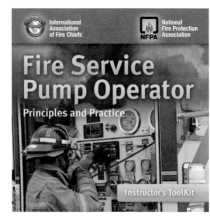

ISBN: 978-0-7637-6058-8

Instructor's Test Bank CD-ROM

This electronic test bank contains over 300 multiple-choice questions and allows instructors to create tailor-made classroom tests and quizzes quickly and easily by selecting, editing, organizing, and printing a test along with an answer key, including page references to the text.

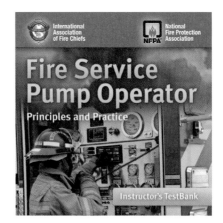

ISBN: 978-0-7637-8543-7

JBCourseManager

Combining our robust teaching and learning materials with an intuitive and customizable learning platform, JBCourseManager enables instructors to create an online course quickly and easily. The system allows instructors to readily complete the following tasks:

- Customize preloaded content or easily import new content
- Provide online testing
- Offer discussion forums, real-time chat, group projects, and assignments
- Organize course curricula and schedules
- Track student progress, generate reports, and manage training and compliance activities

JBCourseManager is free to adopters of *Fire Service Pump Operator: Principles and Practice*. Contact your sales specialist at 800-832-0034 for details today.

Student Resources

Student Workbook

This resource is designed to encourage critical thinking and aid comprehension of the course material through use of the following activities:
- Case studies and corresponding questions
- Figure-labeling exercises
- Crossword puzzles
- Matching, fill-in-the-blank, short-answer, and multiple-choice questions
- Skill drill activities
- Mathematical calculation exercises
- Answer key with page references

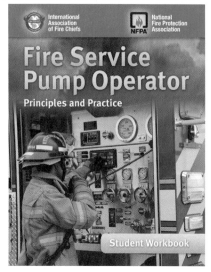

ISBN: 978-0-7637-6057-1

JBTest Prep: Driver/Operator Success

ISBN: 978-0-7637-8349-5

JBTest Prep: Driver/Operator Success is a dynamic program designed to prepare students to sit for Driver/Operator certification examinations by including the same type of questions they will likely see on the actual examination.

It provides a series of self-study modules, organized by chapter and level, offering practice examinations and simulated certification examinations using multiple-choice questions. All questions are page referenced to *Fire Service Pump Operator: Principles and Practice* for remediation to help students hone their knowledge of the subject matter.

Students can begin the task of studying for Driver/Operator certification examinations by concentrating on those subject areas where they need the most help. Upon completion, students will feel confident and prepared to complete the final step in the certification process—passing the examination.

Technology Resources

www.Fire.jbpub.com

This site has been specifically designed to complement *Fire Service Pump Operator: Principles and Practice* and is regularly updated. Resources include:

- Chapter Pretests that prepare students for training. Each chapter has a pretest and provides instant results, feedback on incorrect answers, and page references for further study.
- Interactivities that allow students to reinforce their understanding of the most important concepts in each chapter.
- Hot Term Explorer, a virtual dictionary, allowing students to review key terms, test their knowledge of key terms through quizzes and flashcards, and complete crossword puzzles.

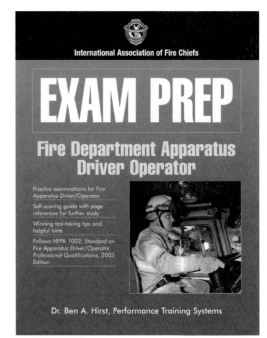

Also Available:

Exam Prep: Fire Apparatus Driver/Operator

Exam Prep: Fire Apparatus Driver/Operator is designed to thoroughly prepare you for Fire Apparatus Driver/Operator certification, promotion, or training examinations by including the same type of multiple-choice questions you are likely to encounter on the actual exam.

To help improve examination scores, this prep guide follows the Performance Training Systems, Inc Systematic Approach to Examination Preparation. *Exam Prep: Fire Apparatus Driver/Operator* was written by fire personnel explicitly for fire personnel, and all content was verified with the latest reference materials and by a technical review committee.

Your exam performance will improve after using this system!

ISBN: 978-0-7637-2845-4

Acknowledgments

Editorial Board

Alan Caldwell
International Association of Fire Chiefs
Fairfax, Virginia

Paul Dow
Albuquerque Fire Department
Albuquerque, New Mexico

Ken Holland
National Fire Protection Association
Quincy, Massachusetts

James B. Hylton
Roanoke Fire EMS
Roanoke, Virginia

Christopher R Niebling
Sunrise Fire Rescue
Sunrise, Florida

Contributors

Raul A. Angulo
Seattle Fire Department
Seattle, Washington

Jerry Clark (Retired)
NYS Office of Fire Prevention and Control
Albany, New York

Joseph Cordeiro
Pawtucket Fire Department
Pawtucket, Rhode Island

Brian Davis
Fairbanks Fire Department
Fairbanks, Alaska

Todd Dettman
Rocky Mount Fire Department
Rocky Mount, North Carolina

Robert H. Grunmeier II
Bucks County Community College
Department of Public Safety Training & Certification
Newtown, Pennsylvania

Joseph P. Guarnera
Revere Fire Department
Revere, Massachusetts

Daniel G. Klein
Cologne Fire and Rescue Department
Cologne, Minnesota

Walter James Knapp
Broward Sheriff's Office Department of Fire Rescue
Ft. Lauderdale, Florida

Michael A. Noah
Newport News Fire Department
Newport News, Virginia

Todd Poole
South Burlington Fire Department
South Burlington, Vermont

Scott Raygor
Fairbanks Fire Department
Fairbanks, Alaska

Jerry Schroeder
Emergency Services Training
Idaho Division of Professional-Technical Education
Boise, Idaho

R. Peter Sells
Medteq Solutions CA, Ltd.
Mississauga, Ontario
Canada

Daryl W. Songer
Roanoke Fire EMS
Roanoke, Virginia

William Stipp, EFO
Goodyear Fire Department
Goodyear, Arizona

Jessie D. Vaca
Monroe Township Fire Department
Henryville, Indiana Mike Warriner
Phoenix Fire Department
Phoenix, Arizona

Michael Washington
Fire Department City of Cincinnati
Cincinnati, Ohio

Roger Westhoff
Houston Fire Department
Houston, Texas

Brent Willis
Martinez-Columbia Fire Rescue
Martinez, Georgia

Reviewers

William Alexander
Pierre Fire Department
Pierre, South Dakota

W. Chris Allan
City of Albemarle Fire and Rescue
Fire Training Coordinator
Stanly Community College
Albemarle, North Carolina

Harold "Lou" Brungard III
Centre County Public Safety Training Facility and
Pleasant Gap Fire Department
Bellefonte, Pennsylvania

Billy Carter
Ozark Fire Department
Ozark, Arkansas

Greg Chatham
North Carolina Department of Insurance
Office of the State Fire Marshal
Raleigh, North Carolina

Jesse Chesser
New Mexico Firefighter's Training Academy
Soccorro, New Mexico

Richard Connelly
Boston Fire Department
Boston, Massachusetts

Gary Cooper
Houston Community College Fire Training Academy
Houston, Texas

Jason D'Eliso
City of Scottsdale Fire Department
Scottsdale, Arizona

William B. Eisenhardt
Fire Academy Drill Instructor and Fire Science Degree Lecturer
Cuyahoga Community College
Cleveland, Ohio

Frank H. Hammond, Jr
Maine Fire Training & Education
Bangor, Maine

Skip Heflin
Hall County Fire Services
Gainesville, Georgia

Rick Hilinski
Community College of Allegheny County
Public Safety Institute
Pittsburgh, Pennsylvania

Timothy Hongo
Shelton Fire Department
Shelton, Connecticut

Kevin Kalmus
Austin Fire Department
Austin, Texas

Richard Kosmoski
Middlesex County Fire Academy
Sayreville, New Jersey

Michael Lant
Kent Fire Department
Kent, Washington

Kenneth J. Loo
Contra Costa County Fire Protection District
Pleasant Hill, California

Dale Mashuga
Fire Instruction Rescue Education Inc. F.I.R.E.
Motley, Minnesota

Philip J. Oakes
Laramie County Fire District #4
Laramie, Wyoming

Douglas Ott
Akron Fire Department
Akron, Ohio

Randy Pearson
Killeen Fire Department Training Academy
Killeen, Texas

Tracy E. Rickman
Rio Hondo College
Santa Fe Springs, California

William Rowley
Palm Beach County Fire Rescue
Indian River State College
Port St. Lucie, Florida

Heath Ryan
Calloway County Fire-Rescue
Murray, Kentucky

Joe Schulz
Assistant Chief of Training
Minneapolis Fire Department Training Division
Minneapolis, Minnesota

Jay Sikes
San Antonio Fire Department
San Antonio, Texas

Gary Suan
Las Vegas Fire and Rescue
Las Vegas, Nevada

George L. Thomas IV
Frederick County Division of Fire and Rescue Service
Frederick, Maryland

Acknowledgments

Photographic Contributors

We would like to extend a huge thank you to Glen E. Ellman, the photographer for this project. Glen is a commercial photographer and fire fighter based in Fort Worth, Texas. His expertise and professionalism are unmatched!

We would also like to thank the following departments, which opened up their facilities for these photo shoots.

Burleson Fire Department
Burleson, Texas
- Gary A. Wisdom—Fire Chief
- Tom E. Foster—Battalion Chief
- Mike Jones—Lieutenant
- Shane Mobley—Fire Fighter
- Casey Davis—Fire Fighter
- Shelby Stone—Fire Fighter
- Juliet Knight—Fire Fighter
- Rob Moore—Fire Fighter
- Jake Hopps—Fire Fighter

Crowley Fire Department
Crowley, Texas
- Robert Loftin—Fire Chief
- Larry Swartz—Division Chief
- Christopher Young—Lieutenant
- Scott Calhoun—Driver Operator
- Bud Hardin—Fire Fighter

Fort Worth Fire Department
Fort Worth, Texas
- Rudy Jackson—Fire Chief
- Homer Robertson—Captain, Apparatus Officer
- Frank Becerra—Fire Fighter
- Larry Manasco—Captain

A special thanks to the Training Facility, Maintenance Shop, and Stations #1, #2, #5, #10, and #26.

Granbury Volunteer Fire Department
Granbury, Texas
- Darrell Grober—Fire Chief
- Kurt Brown—Fire Fighter

The History of the Driver/Operator

CHAPTER 1

NFPA 1002 Standard

4.1 General. Prior to operating fire department vehicles, the fire apparatus driver/operator shall meet the job performance requirements defined in Sections 4.2 and 4.3. [p. 1–360]

5.1 General. The requirements of Fire Fighter 1 as specified in NFPA 1001 (or the requirements of Advanced Exterior Industrial Fire Brigade Member or Interior Structural Fire Brigade Member as specified in NFPA 1081) and the job performance requirements defined in Sections 5.1 and 5.2 shall be met prior to qualifying as a fire department driver/operator—pumper. [p. 1–360]

Additional NFPA Standards

NFPA 1500 *Standard on Fire Department Occupational Safety and Health Program* (2007)

NFPA 1521 *Standard for Fire Department Safety Officer* (2008)

NFPA 1901 *Standards for Automotive Fire Apparatus* (2009)

Knowledge Objectives

After studying this chapter, you will be able to:
- Describe how the fire service in America has progressed from the colonial period to the present day.
- Describe the major changes in transport equipment and personnel from the colonial period to the present day.
- Identify the role of the driver/operator in the safe operation of fire apparatus.

Skill Objectives

There are no skill objectives for this chapter.

You Are the Driver/Operator

As a driver/operator candidate, you will learn how to maintain and operate all of the fire apparatus and equipment that your department currently uses. The instructors will teach you the steps for inspecting the equipment and completing the proper documentation. They will also show you how to troubleshoot problems that you may encounter while working with the front-line equipment that the department uses.

The information that you will learn focuses on the equipment and fire apparatus that the department is currently using. Today, your lead instructor asks that each candidate complete a research paper on the history of the driver/operator. While researching this topic, you come up with the following questions to discuss with the other members of your class and your instructor:

1. In the past, which types of equipment and fire apparatus was the driver/operator responsible for?
2. Which features in modern fire apparatus have made firefighting safer and more efficient?
3. What is the driver/operator's role in ensuring safety at an incident scene?

Introduction

According to NFPA 1002, *Standard for Fire Apparatus Driver/Operator Professional Qualifications*, the **driver/operator** is responsible for getting the fire apparatus to the scene safely, setting up, and running the pump or operating the aerial ladder once the vehicle arrives on the scene. In the fire service, a driver/operator is also called an engineer or technician. This function is a full-time role in some fire departments; in other fire departments, it is rotated among the fire fighters.

The operation of fire apparatus and equipment is a critical life-safety issue for all fire fighters. Too many lives are lost every year as a result of fire apparatus accidents. In fact, driving to and from the scene can be as dangerous as operating on the fire ground itself. In addition, if the equipment fails on the fire ground, the consequences can be disastrous. A sudden loss of water could jeopardize the safety of a line crew, an overextended aerial device could result in a sudden failure with fire fighters on the device, or difficulty in troubleshooting a mechanical problem could result in the inability to prime a pump. For all these reasons, the driver/operator has a tremendous responsibility for the success and safety of the entire company.

As the driver/operator, you will be responsible for all aspects of the call, including the following issues:

- Preparation of the fire apparatus and equipment for a safe response
- Driving the fire apparatus in an emergency response mode to a call
- Placing the fire apparatus at the scene to ensure the maximum effectiveness of the equipment
- Safely and properly operating the equipment to support all operations on the fire ground
- Securing the equipment and safely returning the fire apparatus and members of the company to the fire station

Evolution of Fire Apparatus and Equipment

Today's fire apparatus and equipment have evolved over a number of centuries as new inventions have emerged and been adapted to the needs of the fire service. Colonial-era fire fighters had only buckets, ladders, and fire hooks at their disposal. Because early American settlements featured buildings with thatch roofs and wooden chimneys, firefighting with these limited tools posed a serious challenge. Fire protection was limited to the use of water relays with leather buckets passed from person to person, a system called a **bucket brigade** Figure 1-1 . These buckets were made from leather, which was either secured by

Chapter 1 The History of the Driver/Operator

Figure 1-1 A leather fire bucket from colonial times.

rivets or sewn together. In many communities, residents were required to place a bucket filled with water on their front steps at night in case of a fire in the community. Any fire would require all residents to help stop the progress of the fire and save the settlement. Water would be dipped from a well and passed from one person to another until the bucket reached the fire. A second line would return the bucket to the water supply. Compared to today's fire suppression capabilities, this method of firefighting seems archaic and time-consuming.

Some towns also required that ladders be available so that fire fighters could access the roof to extinguish small fires. If all else failed, the **fire hook** would be used to pull down a burning building and prevent the fire from spreading to nearby structures. The "hook-and-ladder truck" evolved from this early equipment.

Driver/Operator Tip

Buckets made of leather could not hold water for any length of time, so in some parts of the country, fire buckets were filled with sand. After the sand was thrown on the fire, the bucket would be refilled with water and used in a bucket brigade.

Fire Protection in Early American Cities

In the early years of American history, many cities employed watchmen to walk the streets at night looking for fires. If a fire was spotted, the watchman would run through the streets with a rattle device alerting residents to turn out and help fight the fire—thus the term "rattle watch." In some cities, a drum was used to notify fire companies of a fire. In Cincinnati, Ohio, the beating of the drum was a duty of the night watchman. Fire companies were formed, and small storage buildings were erected to store one or more ladders specifically for firefighting. The first organized companies would have one or two straight ladders, a few leather buckets, a selection of axes, and a small selection of hand tools.

Hand Pumps and Hose Carts

The evolution of firefighting tools was heavily influenced by fire suppression teams in Europe, which had developed hand-operated pumping devices. The first equipment to replace the bucket brigade was the hand pump. In 1720, Richard Newsham developed the first such pumper in London. Several strong men powered the pump, making it possible to propel a steady stream of water from a safe distance.

In the early 1800s, the city of Philadelphia developed a municipal water supply using wooden water mains that could be tapped for firefighting. To get the water from the **fire plugs** (early hydrants), fire fighters used hose made from strips of leather held together with rivets. To access water, the fire fighters would drill into the main water pipes, access the water spilling out of the leak, and then plug the leak in the pipe when the fire was suppressed. The leather hose was stored on hose carts that had a large reel. Often these hose cart wheels were ornately decorated to reflect the name of the fire company **Figure 1-2**.

The couplings for the fire plugs were made by local blacksmiths. Each community had its own coupling threads based on the capabilities of the local blacksmith. The Great Baltimore Fire in 1904 resulted in the development of standardizing coupling threads after mutual aid forces supplied by Philadelphia, New York City, and other locales were stymied when their equipment could not hook up to the local Baltimore hydrants.

Driver/Operator Tip

Many communities today maintain a unique thread size for their fire hydrants. As the driver/operator, you should be familiar with any adapters that will allow you to connect to hydrants outside your normal response area.

Figure 1-2 Horses once supplied the energy to transport the crew to the scene.

Figure 1-3 Larger-capacity hand pumps required more personnel to operate them.

Hand pumps were a type of piston-driven, positive-displacement pump that was pushed up and down by fire fighters manning poles at the side of the pump. Over time, these hand pumps became larger to accommodate the new, more plentiful water supplies available in municipal areas; thus they were placed on wagon wheels to allow them to be more easily transported to the fire ground. The larger-capacity hand pumps also required more personnel to operate **Figure 1-3**; in fact, some required six to eight fire fighters to operate the pump. Many of these units maintained a fixed straight-stream nozzle, while others used a leather hose with a long pipe nozzle that was held by a crew of fire fighters.

Fire Apparatus Evolution

With the advent of the steam-powered pumps called steamers in the mid-1800s, horses replaced fire fighters in moving equipment to the fire ground **Figure 1-4**. The first mechanized fire pumps were generally fueled by coal. A coal fire in the fire box of the steamer would be ignited, thereby creating the steam used to power the pump. Drivers of the horses, called teamsters, were hired with their horses to stay at the firehouse and transport the steamer to the fire ground. While en route to the scene, a fire fighter stokes the fire in the fire box, leading to production of the needed steam pressure. One steam pumper could deliver the same volume of water as six or more hand pumpers with eight or more fire fighters working each pump handle in 15- to 20-minute shifts. The term "engineer" was used to identify the fire fighter charged with operating the steam engine. This task required knowledge of hydraulics, pressure regulators, and use of a unique source of energy to propel a pump.

Driver/Operator Tip

In early firefighting companies, a lot of excitement was caused in the streets owing to the horses racing through the streets, bells clanging, smoke from the steamer stack, and teamsters snapping whips. The Dalmatian dog became a friend of the teamster and fixture of the fire station. The Dalmatian had a good temperament around horses, was a good companion, and was effective in racing ahead of the belching and noisy steamer, thereby helping to clear the streets of errant animals—and people—blocking the path of the fire wagon.

Ladder Wagons Pulled by Horses

Wagons loaded with ladders and fire hooks were once drawn by horses and driven by teamsters **Figure 1-5**. Ladders were constructed with strong materials. One-person ladders with a hook could be placed over a window sill and used to climb from floor to floor. A fire fighter would get from the first floor to the second floor by throwing the hook over the second floor window sill and climbing the ladder. Once on the second-floor window sill, the fire fighter would repeat this process, allowing a single fire fighter with a hook to climb several stories. This activity was

Figure 1-4 With the advent of the steam-powered pumps (called steamers) in the mid-1800s, horses replaced fire fighters in moving equipment to the fire ground.

Figure 1-5 A ladder hook truck.

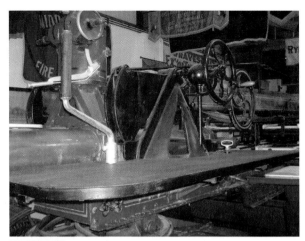

Figure 1-6 A hand cranked wooden aerial ladder.

Figure 1-7 The first gasoline-powered fire apparatus emerged around 1900.

the origin of the term "hook and ladder," which is still used today to identify ladder companies.

As extension ladders were developed, wooden aerial ladders were built to serve fire fighters. These ladders were operated by hand and carried to the scene by horse-drawn cart. The wooden aerial ladder used cranks to rotate the ladder on a turntable and had a large wheel to raise, extend, or retract the wooden aerial ladder. Improvements to the wooden aerial ladder included heavy springs that caused the ladder to spring from its resting position **Figure 1-6**. Fire fighters would be positioned at the raising wheels when the release was pulled and used the momentum caused by the compressed springs to raise the wooden aerial ladder. When the ladder was lowered, fire fighters would compress the springs so that the aerial portion of the ladder was locked in place.

As the length of the ladders increased even further, a steering wheel was placed on the rear axle so that a tiller person could help maneuver the long ladders through the streets. These ladders became the first horse-drawn fire apparatus to be motorized when the front axle was replaced with a motorized single-axle tractor.

Elevated Streams

The first elevated streams were water towers mounted on wagons and pulled by horses. When raised, these elevated streams could raise 30 feet (9 meters) in the air. Devices evolved when pipes were attached to ladders to provide elevated water streams.

Gasoline-Powered Fire Apparatus

The first gasoline-powered fire apparatus emerged around 1900 **Figure 1-7**. These units were initially small cars used by chief officers to drive to the fire ground. The next gasoline-powered unit to evolve within the fire service was the ladder wagon, in which the front axle and horses were replaced by a single-axle tractor used to propel and steer the fire apparatus.

The first gasoline-powered fire apparatus to discharge water was a unit called a <u>chemical wagon</u>. This small truck was constructed with a large soda acid extinguisher and 50 feet (15 meters) or more of small hose (booster line size). It also carried a supply of double-jacked cotton hose, a short ladder, and a running board for the fire fighters to ride. Small fires were extinguished with the soda acid device, while larger fires used the hose connected directly to the hydrant.

> **Driver/Operator Tip**
>
> In 1911, the Savannah, Georgia, fire department became one of the first fully motorized departments.

Adding a Pump

As trucks became larger and stronger, a small pump, called a <u>booster pump</u>, was added; a small hose line, called a <u>booster line</u>, was attached to this pump **Figure 1-8**. Many of these

Figure 1-8 A booster pump.

booster pumps were rotary gear pumps and could generate as much as 200 gallons (757 liters) of water per minute. Because they could not deliver as much water as a fire hydrant, they were used mainly for small fires and to supply the booster line with water.

■ Triple-Combination Pumper

The real breakthrough for today's modern fire apparatus came in 1906, when the <u>triple-combination pumper</u> was introduced Figure 1-9 ▼ . This truck carried water in a tank generally used for the booster line and commonly called the booster tank. The triple-combination pumper also had a pump capacity of 250 gallons (946 liters) per minute (GPM) or greater. Additionally 2½" (6.35-cm) hose was carried for water supply purposes, as well as firefighting hose, commonly 1½" (3.81 cm) in diameter. A driver/operator and a fire officer rode in the cab, and others members of the company rode on the tailboard.

■ Improvements in Aerial Devices

With the gasoline-powered engine, a power supply was now available to turn a hydraulic pump. Hydraulic pressure could be used to perform the tasks of lifting, rotating, and extending the aerial device. This development allowed the aerial ladders to be made of steel, reach farther, and carry greater loads—and today's modern ladder truck was born Figure 1-10 ▶ . In the new devices, ladder fly and bed pipes were added to improve elevated streams. With the stronger metal ladders, larger-caliber nozzles could be used to flow more water onto the fire.

Later additional aerial devices were developed that included <u>articulated booms</u> and <u>tower ladders</u> Figure 1-11 ▶ . The articulated booms provided greater flexibility in operating in areas filled with electrical wires and other obstructions. The tower ladder provided a solid platform at the end of a boom. This gave fire fighters a platform to work from, a device to use for rescue that eliminated the need for the victims to navigate down a ladder, and a platform to provide water tower equipment.

Figure 1-10 An example of a modern aerial device in action.

■ Other Uses for Motorized Apparatus

The gasoline engine eventually became the sole power source to drive fire apparatus. Additional uses for apparatus also emerged. A wide variety of specialty units were developed, such as salvage companies; heavy rescue squads began to use large enclosed vehicles to carry their tools and personnel hose tenders, which had the capacity to carry more hose than a standard apparatus. Air supply units carrying cascade units and compressors were now available at the fire ground.

Figure 1-9 A major breakthrough for today's modern fire apparatus came in the early 1900s when the triple-combination pumper was introduced.

Figure 1-11 Additional aerial devices were developed that included articulated booms and tower ladders.

Voices of Experience

The word "SAFETY" is consistently reinforced in every fire department throughout our nation. From the east coast to the west, whether professional or volunteer, it's the tie that binds us together. As a fire service instructor and company officer with over twenty-six years of experience, I have encountered numerous situations where safety was only verbalized. Accomplishing an objective doesn't necessarily account for safe operations. It is important that all fire fighters understand this concept.

There are three general components to safety that can be instituted at any incident, which must be known in order to accomplish specific tasks in a safe manner.

1. Take your physical fitness seriously. The number one cause of death in the fire service is stress. Being fit will assist you in performing the tasks of your job safely. Your life and the safety of those around you depend on your physical fitness.

2. Learn, know, and understand your job. Your actions throughout your career will depend upon you acquiring and maintaining a working knowledge of the job. Your safety and the safety of your crew depend upon your skill set.

3. Most importantly, always maintain situational awareness. In stressful situations, it is common to acquire tunnel vision. Be aware of your situation and the actions of those around you. This could save your life.

> *"Understanding how to operate safely is the key to your success and longevity as a driver/operator."*

Every driver/operator is responsible for his or her own safety first. Utilizing the three components of "SAFETY" has enhanced not only my own safety, but the safety of my crews and my students. Understanding how to operate safely is the key to your success and longevity as a driver/operator. Remember these components to "SAFETY" during your journey.

Dave Winkler
El Camino Fire Academy
Inglewood, California

Near Miss REPORT

Report Number: 06-0000341
Report Date: 06/25/2006 13:20

Synopsis: Apparatus lurches forward during training exercise.

Event Description: While preparing for an acquired structure live fire training exercise, apparatus were being brought into position to allow for a NFPA 1403 layout. The apparatus operator of one of the engines had not driven nor operated the unit in quite some time. The operator dismounted the fire apparatus to attach a suction line to the engine to draft from a portable tank. Once the suction was in place and the tank was filled with water, the operator entered the cab of the truck to engage the pump. He dismounted the unit and raised the RPM's of the motor to prime the pump. As he pulled the primer, the truck violently lurched forward and the rear wheels were turning with such momentum that the engine dug holes in the ground with its spinning wheels. Immediately ahead of the apparatus was the personnel staging area for the live burn. There were approximately 18 fire fighters in this location who were donning their protective equipment on a large tarp. The engine moved forward approximately 12 feet, onto the tarp, crushing several SCBA cases and running over the hoselines that were deployed. The engine stopped because of the soft ground that it encountered where it became stuck. Fortunately, all of the fire fighters in the path of the apparatus were able to run to safety and no one was injured. The apparatus operator upon recognizing that the engine was moving forward climbed from the top mount pump operator's position into the cab where he discovered that he had only placed the air-operated pump shift into the middle or "neutral" position and the truck was in drive. This event caused many of the fire fighters on scene to reevaluate the capabilities and competence of the engine operator. It is because of the large volume of rain that the area had received, causing the apparatus to become stuck, that no fire fighters' lives were lost in this event.

Lessons Learned: The lessons that were learned from this event were recognizing minimum qualifications for apparatus operators and mandatory refresher training. The member who was operating this fire engine had recently been given a warning for lack of attendance at training. He was using this training burn to try to counter his warning. In the event that a member has had initial training but has become "rusty" one recommendation would be to appoint a preceptor to shadow the member to afford some inherent protection against these types of situations.

[Reviewers Note: The reporter of this incident stated that there were no wheel chocks used by the operator. "The apparatus operator did not have wheel chocks in place. This would have more than likely prevented most of the drama associated with this event."]

Demographics

Department type: Volunteer
Job or rank: Fire Fighter
Department shift: Respond from home
Age: 25–33
Years of fire service experience: 17–20
Region: FEMA Region I
Service Area: Rural

Event Information

Event type: Vehicle event: responding to, returning from, routine driving, etc.
Event date and time: 06/19/2006 06:30
Hours into the shift: 0–4
Event participation: Witnessed event but not directly involved in the event
Do you think this will happen again? Yes
What were the contributing factors?
- Training Issue
- Individual Action
- Human Error

What do you believe is the loss potential?
- Life-threatening injury

Figure 1-12 Diesel engines quietly replaced the gasoline engine as the power system of choice for the fire service.

Diesel-Powered Fire Apparatus

With the development of larger-capacity pumps that required greater horsepower, larger vehicles carrying more equipment, and greater loads to carry, the diesel engine started to replace the gasoline engine. At the same time that greater horsepower was needed, gasoline engines were being equipped with emissions-reduction equipment that reduced their horsepower. Diesel engines were more reliable, burned fuel more efficiently, and provided the desired horsepower. Over time, diesel engines quietly replaced the gasoline engine as the power system of choice for the fire service **Figure 1-12**.

Additionally, diesel engines could be used to assist with braking of the fire apparatus. Devices were available that allowed the cylinders to be turned into compressors, creating resistance and slowing the fire apparatus quickly, and reducing the need to use the vehicle's brakes. Some fire apparatus also used transmission retarders (hydraulic) in an attempt to assist with the braking component. Transmission retarders use the transmission fluid located in the vehicle transmission to create backpressure, thereby slowing the vehicle down. With these devices, the driver/operator needed to keep a watchful eye on the transmission oil temperature, as it could quickly overheat, potentially damaging the transmission.

Today's fire apparatus may be equipped with an anti-lock braking system (ABS), which monitors all wheels on the vehicle to determine if one or more wheels are starting to skid or actually skidding during braking. Many of today's transmissions are also equipped with an automatic gear shifter. Another device commonly found on modern-day fire apparatus is the automatic traction control system (ATC), an electronic sensor that controls wheel slippage. Regardless of which system is used, it is the driver/operator's responsibility to be familiar with the driving and braking functionality on the apparatus.

Safety Considerations

Growing concerns for fire fighter safety during the movement of fire apparatus resulted in the adoption of standard NFPA 1500, *Standard on Fire Department Occupational Safety and Health Program*. This standard required that new fire apparatus be designed so that all members riding on the fire apparatus would be provided a seated area inside an enclosed cab. Adoption of this standard has resulted in the near elimination of rear steps on fire apparatus. Now apparatus includes large enclosed cabs, which ensure that each fire fighter can be seated and wear a seat belt. NFPA 1901, *Standards for Automotive Fire Apparatus*, requires that fire apparatus manufacturers install seat belt warning devices that verify that a fire fighter is seated and belted.

Some fire apparatus are monitored by black boxes, which are recording systems that note what the fire apparatus was actually doing prior to a collision. Additionally, fire apparatus specifications have been developed to provide a safe riding area in case of a fire apparatus rollover. The corner posts of the cab should be strengthened so that in case of a rollover, the fire fighters inside the cab will not be crushed.

Protecting the Driver/Operator at the Scene

Because the driver/operator of the fire apparatus is likely to be operating in the vicinity of moving traffic, protective efforts must be taken to guard against collisions and other traffic-related incidents. As a fire fighter, you can assist in this effort by wearing equipment with reflective material, such as a structural turnout coat. If the coat is removed, then you should don a reflective vest designed for use in or near traffic, and place traffic cones appropriately to alert drivers to the incident and control traffic. You should take all necessary measures to protect your safety when operating in traffic, including using all appropriate emergency lighting on the fire apparatus.

Some fire apparatus have been designed with mid-ship pump panels that provide a better vantage point for viewing the entire scene **Figure 1-13**. These panels also remove you from the street. Fire apparatus have also been built with pump panels on the fire officer's side (curb side) or at the rear of the fire apparatus.

Additionally, the fire apparatus may be placed on arrival in such a manner as to protect the pump panel from oncoming traffic. Many of today's firefighting vehicles are equipped with traffic directional arrows. Traffic cones may be placed to divert traffic, and some fire departments have policies of replacing red lights at the scene with yellow lights.

Visibility for the Driver/Operator

Fire apparatus are designed with larger windshields to provide greater visibility and with heated mirrors to prevent icing or frosting. Today many devices are equipped with cameras mounted on the rear of the fire apparatus so that you can view the back of the fire apparatus and reverse it more safely.

Unfortunately, fire fighter injuries and fatalities have occurred during the reversal of fire apparatus—both at fire stations and on the fire ground. Many fire departments now have policies of using spotters before the fire apparatus backs up **Figure 1-14**. Fire apparatus are also equipped with alarms to warn persons behind the fire apparatus that the vehicle is in reverse and may be backing up. Even when such an alarm is present, the driver/operator must be cautious when backing up: Some people, such as small children, do not recognize the sound and do not heed the warning.

Figure 1-13 Mid-mounted panels also provide a better vantage point for viewing the entire scene.

Figure 1-15 The second leading cause of fire fighter fatalities is injuries sustained while responding or returning from the fire ground.

The Driver/Operator's Role in Safety

The second leading cause of fire fighter fatalities is injuries sustained while responding or returning from the fire ground **Figure 1-15**. According to the U.S. Fire Administration (USFA), 19 fire fighters died while responding to or returning from a call in 2008. In 2007, 26 fire fighters died while responding to or returning from a call.

Figure 1-14 Spotters assist in the backing up of fire apparatus.

As the driver of the fire apparatus, you are a critical factor in ensuring a safe response. You are responsible for operating the vehicle in a safe manner and driving in a defensive manner to ensure both a safe arrival at the scene and a safe return to the fire station. A critical component of this effort is the proper maintenance of the fire apparatus, which will be discussed later in this book.

Every fire department should develop, train members on, and enforce standard operating procedures (SOP) regarding how the fire apparatus should be driven and operated. Such SOPs should include how to follow at traffic control devices (traffic lights and stop signs), how fast the vehicle should be operated, and other safe operational procedures. Of particular importance is the process of proceeding safely through traffic control devices—this is where a significant number of accidents occur.

As the operator of the fire apparatus, you play a critical role in ensuring the safety of the fire fighters operating at the incident scene. Fire fighters advancing a hose line are dependent on the constant flow of water at an adequate pressure not only to control the fire, but also to provide a safe retreat if needed. The sudden loss of water or pressure could result in serious injury or even death. The incorrect positioning of an aerial device adjacent to electrical power lines could result in fire fighters being electrocuted. A ladder placed at an improper angle could result in structural failure and collapse.

Modern Fire Apparatus

The fire apparatus being used by many fire departments today is larger, heavier, taller with a higher center of gravity, and generally more difficult to maneuver through traffic than the devices used by earlier-era fire fighters. Additionally, traffic continues to increase, cars are better insulated and often have stereos blaring, and drivers are generally less patient and often distracted by devices such as cell phones. This provides an ever-increasing challenge for you to maneuver the fire apparatus safely through traffic.

In addition, the capacities of fire apparatus continue to increase. Only a few years ago, a 750 GPM [2838 liters] pump was common. Today, however, the water delivery rate of many pumps exceeds 1500 GPM (5677 liters). Water capacities have also increased from 100 gallons on a pumper to 1000 gallons (3785 liters) or more today. This increased water capacity, along with enclosed cabs, makes the fire apparatus heavier, larger, and more challenging to drive.

Finally, the computer age has made its way into the fire apparatus. Today's fire apparatus have computer systems to prevent the vehicle from skidding (antilocking brake system), electrical load management systems to make sure the electrical load does not exceed the ability of the fire apparatus to produce electricity, and computer systems that monitor the engine's performance. In addition, some vehicles are able to provide monitoring information on the fire apparatus's speed and engine performance (rpm) to the fire officer's monitor.

Figure 1-16 One of the best methods for understanding apparatus systems is to reference the manufacturer's operating manual.

Driver/Operator Selection

What makes a good driver/operator candidate? Knowledge, skill, and attitude are some key attributes. Driver/operators must be knowledgeable in the services provided by their fire departments. They must understand how a consistent delivery of services provides for a safe and successful outcome. Driver/operators must be educated and trained in operating at emergency scenes. They must possess the required skills to complete critical life-preserving actions. These skills must be developed through a continuing process of performance and repetition. Reviewing the job performance requirements (JPRs) identified in NFPA 1002 will give potential candidates a better understanding of their fire department's expectations regarding their performance.

An effective driver/operator is a good problem solver. When fire apparatus or equipment malfunctions arise, he or she must be able to quickly evaluate all possible solutions to rectify the problem and seamlessly implement a backup plan. To attain this level of efficiency, driver/operators must understand the basic and complex systems of the fire apparatus and the equipment that it carries. Just as driver/operators need a working knowledge of fire-ground hydraulics to achieve positive results during fire suppression efforts, they must also know all of the onboard devices and systems carried on the fire apparatus. Today, the fire apparatus provides many functions to support fire departmental services to the community. The modern fire apparatus is multitasking, and many of the functions it performs require many support and power systems. Hydraulic, electrical, and pneumatic energy sources are present on many modern fire apparatus. Driver/operators must be proficient in correcting the problems that are related to all of these onboard systems. One of the best sources for understanding these systems is the manufacturer's operating manual, many of which are available in an electronic format **Figure 1-16**.

Selecting a driver/operator candidate can be difficult, so the selection process is based on a candidate's knowledge, ability, and willingness to pursue the challenge. To start this selection process, fire officers may consider a few basic qualities: thorough written and verbal communication skills, physical and mental

> **Driver/Operator Tip**
>
> To gain and maintain proficiency in the role of the driver/operator, you should take every opportunity to engage in skill exercises or training sessions requiring the use of the onboard systems and equipment. The more familiar you are with the related processes, the more effective you will be in solving the problems that may arise during emergency responses.
>
> Always attempt to train as if you were on an assignment, using the same crew sizes, same fire apparatus, and same equipment resources as you would on a real call. During these training sessions, you may discover better methods to make you and your crew more efficient. Be sure to pass on the lessons that you learn to other driver/operators and other crew members.
>
> Remember—do not just train for the big one. It is usually the basic skill requirements needed at your bread-and-butter responses that cause the most problems.

fitness, basic mechanical ability, and attitude. Driver/operators must be thorough when completing reporting forms, inspections forms, and documentation of deficiencies or corrective actions. Reporting is a major job function of the driver/operator. Do you remember being told sometime in your career, "If it's not documented, it didn't happen"? This holds true for the driver/operator. Undocumented events, conditions, or actions may lead to disaster. Crews cannot be expected to perform their assignments without the required equipment that is in proper working condition. Driver/operators must explain the importance of reporting to their respective crews. Educating the other crew members will make your job easier and begin to shift the paradigm away from the notion that "reporting is complaining." Explain to your crew that reporting is preventing a problem or dangerous situation.

Another desirable quality is for the driver/operator to be physically capable of performing the assigned tasks. Good vision, hearing, and physical ability are all characteristics that a driver/operator will need to be safe and successful. Fire apparatus and equipment can be very unforgiving; you will be challenged both mentally and physically by the malfunctions and errors that occur on scene. Your patience will be tested and your reactions watched. By remaining focused on your assignment and being acutely aware of the ongoing operation, you will be better positioned to handle the various stressors headed your way. Remember this lesson: "Everything under pressure is worse." This includes the driver/operator.

Next on this list of preferred driver/operator qualities is mechanical ability. Driver/operators should have a basic understanding of mechanics. Familiarity with basic preventive maintenance processes such as lubricating components, adding fluids, and inspecting equipment and onboard systems for proper operation forms an essential skill set for driver/operators. Driver/operators are charged with handling minor system repairs and preventive maintenance servicing only; major repairs or adjustments should be performed by qualified emergency vehicle technicians.

Finally, attitude is a key consideration for the driver/operator. The right attitude of the driver/operator is critical for building a successful team, instilling confidence in crew members, and supporting positive response outcomes. Driver/operators must take their roles and responsibilities seriously. Many variables hinge on the ready availability of a well-maintained apparatus and service-ready equipment. Having a positive attitude toward the job will breed additional support and confidence from your crew members. There will be operational setbacks; these come with the job. However, a driver/operator who is committed to the mission of the fire department and dedicated to providing service to the community will be able to meet these challenges and make a difference when it counts.

Wrap-Up

Chief Concepts

- Early fire suppression efforts relied on buckets filled with water thrown at the fire. With buildings constructed of wood and roofs of thatch, fire was a constant threat in the colonial era.
- Formal fire protection efforts began with the adoption of response equipment such as a hand-operated pumper, a reel of hose, and a ladder.
- Fire protection has continued to improve and enhance its level of sophistication based on the improvement of fire apparatus and equipment. The key to its success is the person who will drive the fire apparatus and operate the fire apparatus at the fire incident scene.

Hot Terms

<u>articulated booms</u> An aerial device consisting of two or more folding boom sections whose extension and retraction modes are accomplished by adjusting the angle of the knuckle joints.

<u>booster line</u> A rigid hose that is ¾″ (2 cm) or 1″ (2.5 cm) in diameter. Such a hose delivers water at a rate of only 30 to 60 GPM (113 to 227 liters), but can do so at high pressures.

<u>booster pump</u> A water pump mounted on the fire apparatus in addition to a fire pump and used for firefighting either in conjunction with or independent of the fire pump.

bucket brigade An early effort at fire protection that used leather buckets filled with water to combat fires. In many communities, residents were required to place a bucket filled with water typically on their front steps at night in case of a fire in the community.

chemical wagon A small truck constructed with a large soda acid extinguisher and 50 feet (15 meters) or more of small hose (booster line size).

driver/operator A person having satisfactorily completed the requirements of driver/operator as specified in NFPA 1002, *Standard for Fire Apparatus Driver/Operator Professional Qualifications*. (NFPA 1521)

fire hook A tool used to pull down burning structures.

fire plug A valve installed to control water accessed from wooden pipes.

fire pump A water pump with a rated capacity of 250 GPM (1000 L/min) or greater at 150 psi (10 bar) net pump pressure that is mounted on a fire apparatus and used for firefighting. (NFPA 1901)

hand pump A piston-driven, positive-displacement pump that is pushed up and down by fire fighters manning poles at the side of the pump.

triple-combination pumper A truck that carries water in a tank generally used for the booster line and commonly called the booster tank.

tower ladders A device that provides fire fighters with a solid platform at the end of a boom.

Driver/Operator in Action

During training, you and several other fire fighters in the class form a study group. While studying for the final exam of the class, the group decides that each fire fighter should make up several questions and bring the questions to the next study group session. All of the questions will then be compiled, and each of you will have a practice test to take before the final. Here are some of the questions that your study group created:

1. In the early 1800s, the city of _____ developed a municipal water supply using wooden water mains that could be tapped for firefighting.
   A. Chicago
   B. Cincinnati
   C. Philadelphia
   D. New York

2. What were the drivers of the first steam-powered pumps called?
   A. Teamsters
   B. Chauffeurs
   C. Minutemen
   D. Operators

3. How far in the air could the first elevated streams reach?
   A. 15′
   B. 25′
   C. 30′
   D. 35′

4. When did the first gasoline-powered apparatus start to emerge?
   A. 1900
   B. 1800
   C. 1850
   D. 1940

Types of Fire Apparatus

CHAPTER 2

NFPA 1002 Standards

There are no NFPA 1002 standards covered in this chapter.

Additional NFPA Standards

NFPA 1901 *Standard for Automotive Fire Apparatus* (2009)
NFPA 1931 *Standard for Manufacturer's Design of Fire Department Ground Ladders*
NFPA 1963 *Standard for Fire Hose Connections* (2009)
NFPA 1981 *Standard on Open-Circuit Self-Contained Breathing Apparatus for Fire and Emergency Services* (2007)
NFPA 1983 *Standard on Life Safety Rope and Equipment for Emergency Services* (2006)

Fire and Emergency Services Higher Education (FESHE) Model Curriculum

National Fire Science Degree Programs Committee (NFSDPC): Associate's Curriculum

Fire Protection Hydraulics and Water Supply

1. Apply the application of mathematics and physics to movement of water in fire suppression activities.
2. Identify the design principles of fire service pumping apparatus. [p. 18–29]
3. Analyze community fire flow demand criteria.
4. Demonstrate, through problem solving, a thorough understanding of the principles of forces that affect water at rest and, in motion.
5. List and describe the various types of water distribution systems.
6. Discuss the various types of fire pumps.

Knowledge Objectives

After studying this chapter, you will be able to:
- Describe which components are needed to classify a piece of fire apparatus as a pumper.
- Describe which components are needed to classify a piece of fire apparatus as an initial attack fire apparatus.
- Describe which components are needed to classify a piece of fire apparatus as a mobile water supply.
- Describe which components are needed to classify a piece of fire apparatus as an aerial apparatus.
- Describe which components are needed to classify a piece of fire apparatus as a quint.
- Describe which components are needed to classify a piece of fire apparatus as a special service fire apparatus.
- Describe which components are needed to classify a piece of fire apparatus as a mobile foam apparatus.

Skill Objectives

There are no skill objectives for this chapter.

You Are the Driver/Operator

You are a new driver/operator assigned to a single-engine company fire station. Over the past few months, you have become very familiar with the equipment that is carried on your fire apparatus—a pumper. You know the amount of water it has for fighting fires, the types of hand tools found in its compartments, and every section, size, and length of hose that it carries.

During an overtime shift, you are assigned to a ladder company for the first time. Although you were trained on how to operate the fire apparatus for the ladder company, it has been a while since you last drove it. While you are completing the morning inspection of this fire apparatus, you ask yourself the following questions:

1. Which equipment is carried on the ladder company's fire apparatus that is not on the engine company's fire apparatus?
2. Does the ladder company's fire apparatus carry water?
3. How different is the ladder fire company's fire apparatus from the engine company's fire apparatus?

Introduction

The nature of the calls that today's fire fighters respond to vary from day to day. The need to respond solely to wood-structure building fires is a thing of the past. Instead, the modern-day fire service's duties include responding to high-rise building fires, hazardous chemicals spills, brush fires, confined-space rescues, and many other emergency incidents, all of which may occur in large cities, small towns, and rural areas.

Given all of these variables, no single fire apparatus is adequate for responding to all calls. Rather, specialized equipment has been developed for particular situations. These types of fire apparatus are distinguished based on their function and their capabilities at the emergency scene. You should know what each piece of fire apparatus in your department is capable of doing and what equipment each brings to the emergency scene. For example, you should know the difference between an aerial apparatus and a quint: Although both of these fire apparatus may look the same, they are not identical. You should understand that all fire apparatus are not created equal. In this chapter, you will learn about various fire apparatus and what they are designed to do.

This chapter discusses the various types of fire apparatus and explores what is required of each type. It covers pump sizes, tank sizes, ladders, storage, and more. Finally, it discusses the equipment that *must* be on a fire apparatus if it is to be in accordance with NFPA 1901, *Standard for Automotive Fire Apparatus*.

Fire Apparatus Requirements

Working with a Manufacturer

With so many types of fire apparatus available for performing a wide variety of jobs, the fire apparatus used by different departments will inevitably vary. Nevertheless, there are many general requirements for all fire apparatus. With any fire apparatus, the jurisdiction purchasing the fire apparatus must convey the following specifications to the manufacturer:

- The specific performance requirements
- The maximum number of fire fighters to ride on the fire apparatus
- The specific electrical loads required

The jurisdiction must also inform the manufacturer of any additional equipment that will be on the fire apparatus above and beyond what is required Figure 2-1 ▶. For example, if the jurisdiction needs room on the fire apparatus for swiftwater rescue equipment to be purchased in the future, the department should inform the manufacturer. This way the fire apparatus can be designed for this type of equipment and space will be provided to store it. Waiting until the fire apparatus is complete and then attempting to cram additional equipment into it may not work.

It is the responsibility of the jurisdiction purchasing the fire apparatus to conduct ongoing training to ensure its personnel's proficiency with regard to the safe and proper use of the fire apparatus and its equipment. If a fire department has never had an

Figure 2-1 The jurisdiction must inform the manufacturer of any additional equipment that will be on the fire apparatus above and beyond what is required.

aerial apparatus, then it should ensure that all of the fire fighters operating it receive proper training. Many times, the manufacturer is available to provide this initial operational training for the department members.

The manufacturer must also describe the fire apparatus, including the estimated weight, wheel base, turning radius, principal dimensions, transmission, axle ratios, and capacity of the aerial platform, if applicable. While designing a new fire apparatus, it is easy to keep adding features and equipment to the vehicle without realizing what problems may occur with their inclusion. Too many added features may make the vehicle grossly overweight; also, some of the equipment may not be designed to work together. The manufacturer should ensure that all of the components will interact correctly and that the vehicle does not exceed the required weight limits.

The jurisdiction and the manufacturer must work together cohesively to construct the fire apparatus. When a city or town purchases a piece of firefighting equipment whose price could exceed $500,000, it is incumbent upon the two entities to produce an excellent piece of equipment that will do the job. Once the fire apparatus is delivered to the purchasing jurisdiction, the manufacturer will typically supply a qualified representative to demonstrate the fire apparatus and provide training. This training should include all aspects from operation, maintenance, and

Driver/Operator Tip

Before investing in a new fire apparatus, it is well worth the time and effort to seek out the feedback of other fire departments. You should make every effort to speak with representatives from fire departments that have purchased fire apparatus from any manufacturer that you are considering working with. Do not just take the salesperson's recommendations at face value: What are the experts in the field saying about this vehicle? The negative experience of another fire department could save your department hundreds of thousands of dollars and untold hours of frustration!

Driver/Operator Tip

Bear in mind that your new fire apparatus will probably be sitting at the fire station for a good month while training on the new vehicle and its features occurs. During this month, your old fire apparatus will continue to be used to respond to incidents.

equipment. NFPA 1901 defines all of the documents and components that should be incorporated for a piece of equipment to be considered an NFPA-compliant fire apparatus. For example, all fire service apparatus must be able to attain a speed of 35 mph (55 kmph) from a standing start within 25 seconds.

Water on the Fire Apparatus

A major component of fire attack is water. To get the water on the fire, or "put the wet stuff on the red stuff," you must first pressurize it. To do so, you need a fire pump. NFPA 1901 defines a fire pump as a water pump that is mounted on the fire apparatus and used for firefighting. The fire pump must be capable of delivering a minimum capacity of 250 gallons per minute (GPM) (1000 L/min) at 150 pounds per square inch (psi) (1000 kilopascals [kPa]) net pump pressure. When the fire apparatus is designed for pump-and-roll operations, the fire apparatus engine and drive train must be arranged so that it can pump at least 20 gpm (76 L/min) at a gauge pressure of 80 psi (550 kPa) while the fire apparatus is moving at a speed of 2 mi/h (3.2 km/h) or less **Figure 2-2**.

Some fire pumps are designed for responding to small vegetation fires and have relatively small capacities because they do not carry large amounts of water. These fire pumps are usually mounted on the fire apparatus and have a separate engine to power them, other than the one that drives the fire apparatus. Other fire pumps are operated in a stationary position; they tend to be much larger and have a greater capacity. These fire pumps, which are typically used to combat fires that involve structures such as homes and commercial buildings, are powered through the transmission in the fire apparatus's engine. Still other fire pumps may be equipped with pump-and-roll capabilities.

Figure 2-2 A fire pump.

No matter what size the fire pump, it must meet certain requirements. For example, NFPA 1901 states that if the pumping system is rated at 3000 GPM (12,000 L/min) or less, it shall be able to deliver water at the following rates:

- 100 percent of rate capacity at 150 psi (1000 kPa) net pump pressure
- 70 percent of rated capacity at 200 psi (1400 kPa) net pressure
- 50 percent of rated capacity at 250 psi (1700 kPa) net pressure

This tells us that the maximum amount of water (GPM) the fire pump can deliver is at 150 psi (1000 kPa) net pump pressure. As the fire pump's psi rate is increased, the amount of water that can be delivered decreases. After the fire pump reaches capacity, the only thing to gain by increasing the revolutions per minute (rpm) rate is an increase in pressure. This information may be useful if you are supplying attack lines for a high-rise fire and need the added pressure but not the entire capacity of the fire pump.

While drafting, if the fire pump is rated at less than 1500 GPM (5700 L/min), it is capable of taking suction through 20 feet (6 m) of suction hose and can discharge water in less than 30 seconds. If the fire pump is rated at greater than 1500 GPM, it is capable of taking suction through 20 feet (6 m) of suction hose and can discharge water in less than 45 seconds. Larger fire pumps are allowed more time to operate from a draft because they have more space inside that must be exhausted of air before water can enter the pump.

The fire apparatus manufacturer should certify that the fire pump has the capability to pump water at 100 percent of rated capacity at 150 psi (1000 kPa) with a net pump pressure from draft through 20 feet (6 m) of suction hose with a strainer at an atmosphere of 2000 feet (600 m) above sea level, an atmospheric pressure of 29.9 in. Hg (101 kPa), and a water temperature of 60°F (15.6°C). The size and number of suction lines required vary according to the size of the fire pump. All fire pumps are rated from a draft, so that no added water pressure from an external source is used to offset the capacity of the fire pump. The pump test is usually conducted at the manufacturer by a third party—for example, to determine Underwriters Laboratory (UL) certification.

Once you engage the fire pump, you need to have a mechanism to ensure that it is running. When a separate engine is used to drive the fire pump, an indicator light is provided in the driving compartment that is illuminated when the fire pump is running. The indicator light is marked with a label such as "Pump Engine Running." If a fire apparatus is equipped with an automatic **chassis** transmission, the fire pump is driven by the transmission. This type of fire pump is used for stationary pumping only. A "Pump Engaged" indicator in the driver/operator's compartment signals when the pump shift process is complete, and an "OK to Pump" indicator in the driver/operator's compartment identifies when the fire pump is engaged. A separate "Throttle Ready" indicator is provided at the pump panel to signal when all necessary steps have been taken to ensure proper safe pumping. Other indicators of pump engagement include that the speedometer in the cab is operating and that the pressure is indicated on the master discharge gauge of the pump panel.

What good would a fire pump be without water? NFPA 1901 defines the requirements of fire apparatus equipped with water

Figure 2-3 A fire apparatus with a water tank.

tanks **Figure 2-3**. Depending on the type of fire apparatus, the water tank may hold several hundred or several thousand gallons of water. Each fire department should carefully evaluate its water supply needs and the available water delivery systems when considering the size of water tank on the fire apparatus. Regardless of the tank's size, NFPA 1901 states that all water tanks shall be constructed with noncorrosive materials or other materials that are treated against corrosion and deterioration. All water tanks should be equipped with a baffling system with a minimum of two transverse or longitudinal vertical baffles that cover at least 75 percent of the area of the plane that contains the baffle. Each tank includes one or more cleanout sumps with 3" (75 mm) or larger removable pipe plugs. These plugs are often used to clean out the silt and debris that may be trapped in the water tank over time. NFPA 1901 covers many other aspects of the water tank, including water-level indicators, tank-to-pump intakes, filling, and venting, as well as external filling issues.

Pumper

The **pumper** fire apparatus is the bread-and-butter of the fire service **Figure 2-4**. The most common type of fire apparatus, it is a part of almost every fire department. This equipment is used to respond to small incidents such as dumpster fires, vehicle fires, and brush fires in urban settings. It is also the main source

Figure 2-4 The pumper secures the water source and extinguishes the fire.

of fire attack for larger fires that involve structures. The pumper is *the* piece of fire apparatus that is critical for the initial extinguishment of the fire because it brings the initial water supply as well as tools to the fire scene.

Pumpers can reach lengths of 30 feet (9 m) (or longer) and are capable of holding 1250 gallons (4710 liters) of water. Per NFPA 1901, a pumper shall be equipped with a permanently mounted pump with a minimum rating of 750 GPM (2800 L/min). Each fire department should determine which types of fires the fire apparatus will respond to and, therefore, what the appropriate pump capacity is. Because the pumper is usually the first piece on scene, it needs its own water supply to sustain operation. Pumpers can be designed with tanks that carry more than 1000 gal (3784 L) of water, but must carry, at a minimum, 300 gal (1135 L). Each fire department should determine the water supply needs for its jurisdiction and adjust the tank size accordingly.

The pumper does more than just carry water; it carries a multitude of tools and equipment. At a minimum, the fire apparatus must carry one straight ladder (with roof hooks), one extension ladder, and one attic ladder. The fire apparatus must carry several types of hose as well. At a minimum, it must have 15′ (4.5 m) of soft suction hose or 20′ (6 m) of hard suction hose. The following types of fire hose and nozzles are required by NFPA 1901:

- 800′ (240 m) of 2½″ (65 mm) or larger supply/attack hose
- 400′ (120 m) of 1½″ (38 mm), 1¾″ (45 mm), or 2″ (52 mm) attack hose
- One combination spray nozzle capable of delivering 200 GPM (750 L/min) minimum
- Two combination spray nozzles capable of delivering 95 GPM (360 L/min) minimum
- One playpipe, with shut-off and 1″ (25 mm), 1⅛″ (29 mm), and 1¼″ (32 mm) tips

Each pumper must be designed with a minimum of 40 ft³ (1.1 m³) of enclosed weather-resistant compartments to store miscellaneous equipment. As stated in NFPA 1901, the **miscellaneous equipment** listed in Table 2-1 ▶ is the minimum carried on the fire apparatus that is required by NFPA 1901. Most fire departments exceed this list and will add equipment that best suits the needs for their jurisdiction.

Table 2-1 Miscellaneous Equipment on a Pumper

Amount	Equipment
1	6 lb (2.7 kg) or greater flathead axe mounted in a bracket fastened to the pumper
1	6 lb (2.7 kg) pickhead axe mounted in a bracket fastened to the pumper
1	6′ (2 m) pike pole or plaster hook mounted in a bracket fastened to the pumper
1	8′ (2.4 m) or longer pike pole mounted in a bracket fastened to the pumper
2	Portable hand lights mounted in a bracket fastened to the pumper
1	Approved dry chemical portable fire extinguisher with a minimum 80-B:C rating mounted in a bracket fastened to the pumper
1	2½ gal (9.5 L) or larger water extinguisher mounted in a bracket fastened to the pumper
1	Self-contained breathing apparatus (SCBA) complying with NFPA 1981, *Standard on Open-Circuit Self-Contained Breathing Apparatus for Emergency Services*, for each assigned seating position, but not fewer than four units, mounted in a bracket fastened to the apparatus or containers supplied by the SCBA manufacturer
1	Spare cylinder for each SCBA unit carried, mounted in a bracket fastened to the fire pumper or stored in a specially designed storage space
1	First-aid kit
4	Combination spanner wrenches mounted in a bracket fastened to the pumper
2	Hydrant wrenches mounted in a bracket fastened to the fire pumper
1	Double female 2½″ (65 mm) adapter with National Hose threads, mounted in a bracket fastened to the pumper
1	Double male 2½″ (65 mm) adapter with National Hose threads, mounted in a bracket fastened to the pumper
1	Rubber mallet, suitable for use on suction hose connections, mounted in a bracket fastened to the pumper
2	Salvage covers each a minimum size of 12′ × 14′ (3.7 m × 4.3 m)
2	Wheel chocks, mounted in readily accessible locations, each designed to hold the pumper, when loaded to its maximum in-service weight, on a 10 percent grade with the transmission in neutral and the parking brake released

Initial Attack Fire Apparatus

The **initial attack fire apparatus** is used much like the pumper, but its specifications are much different. This type of fire apparatus is not as commonly encountered as the pumper, but is nevertheless used by many fire departments. The initial attack fire apparatus is a smaller version of the pumper that is designed to be more maneuverable than the pumper, especially on off-road terrain. This apparatus is usually equipped with four-wheel drive and used to fight fires in urban and rural settings. For example, a fire department may have an area in its jurisdiction that is difficult for larger pumpers to access. Because of narrow roads and small bridges, a pumper may not be capable of accessing the buildings in this area. In such circumstances, the initial attack fire apparatus provides a scaled-down version of a pumper for responding to hard-to-reach areas.

Many initial attack fire apparatus are built on a commercial chassis platform with a custom-built body. This construction makes many parts of the initial attack fire apparatus easier to maintain and repair because the parts can be found at most local auto parts stores. Although many of these fire apparatus units are custom built by a private company, some are not. Fire fighters should be aware of the potential for overloading the chassis with too much weight, thereby creating a safety hazard.

NFPA 1901 states that if a fire apparatus is to be used as an initial attack fire apparatus, then it *shall* conform to these

Voices of Experience

A number of years ago, I was the chief of a fire department in rural Vermont. One February evening, the air temperature was 20°F [–7°C] and falling. The call came in for a fire in a large two-story wood frame hardware store. The hardware store was approximately 40 × 30 feet (12m × 9m) above-the-ground with inventory in the basement and the attic. The hardware store abutted the owner's residence of the same size and construction type. The building was built in the 1940s.

The first arriving company found fire on the first floor of the hardware store and proceeded with search and rescue operations. Multiple additional companies were immediately requested. The initial water supply comprised of a 1000 gallons per minute (GPM) (3800 L/min) pump drafting from a large local river. This drafting pump supplied three tankers, all carrying 1200 gallons (4500 L) of water. Due to the short travel distance between the fill site and the fire scene, the tankers were able to initially support the attack engines and four hand lines.

> *"Thanks to the uninterrupted water supply, the defensive operations contained the fire to the single structure."*

Because of the building construction type and the fuels within the structure, the operation was declared defensive. All companies were evacuated from the hardware store and the adjoining residence. Master stream devices were required to suppress the fire and protect the one wooden 80 × 40 foot (24m × 12m) exposure to the East of the store and a 20 × 30 foot (6m × 9m) wooden storage shed on the North side. Additional companies laid a large diameter hose (LDH) line from the initial fill site to the fire scene. A second water supply site was set-up to support an additional elevated master stream device. Large flows from the LDH lines enabled constant, uninterrupted water for the defensive operation.

The investigation after the fire found two origins, one in the hardware store and one in the residence. In the end the fire was confined to the building(s) of origin, with no loss in the exposed structures. Thanks to the water supply, this fire was suppressed safely.

Todd Poole
Tunbridge Volunteer Fire Department
Tunbridge, Vermont

Near Miss REPORT

Report Number: 05-0000495
Report Date: 08/30/2005 18:54

Synopsis: Fire fighters nearly struck by broken coupling from overpressurized supply line.

Event Description: Our engine and truck company were participating in probationary fire fighter training for our newest fire fighter on the shift. The drill entailed establishing and flowing an elevated master stream through our 105' pre-piped aerial. Also being trained is a relatively new fire fighter who is being qualified as a back-up engine driver/operator. Because this was a familiarization-only drill, a single 20' pony sleeve of 4-inch hose was used to connect from the engine to the aerial's waterway intake. The gate was closed on the aerial side of the hose.

What followed was a series of communication errors and human errors. Due to a miscommunication from earlier in the morning, the fire fighter operating the pump panel mistakenly thought that the starting pressure for the operation was to be 225 psi, our starting pressure for aerial master streams is 150 psi. In addition, the aerial turntable operator had the intake gate closed. The fire fighter on the pump panel thought that the truck crew had stated they were "ready for water." In fact, they were not. The pump operator opened the gate valve on the pump outlet, and began to increase pressure on the line. Within 5 seconds, a loud "pop" was heard. The engine officer and two truck company fire fighters were standing within 20 feet of the apparatus. None had their helmets on. With the "pop" noise, the 4-inch hose failed at the coupling immediately behind the (unisex) connection on the truck's waterway intake gate. The metal collar that holds the hose onto the (unisex) coupling flew off the hose as it initially whipped when it came off of the truck, and it narrowly missed striking the engine captain who was standing nearby. Luckily, no one was hurt in this event and property damage was limited to the 4-inch hose and its (unisex) coupling.

Lessons Learned:

1. Whenever pumping apparatus and/or aerial apparatus are being used, all participants in the immediate area of the equipment should at least be wearing helmets and leather gloves.

2. Assure before starting a training evolution that all participants fully and completely understand their roles and responsibilities. For new trainees, ensure that an experienced member is overseeing their activity.

3. Consider a "challenge and response" means of communication. In this case, if the pump operator had taken a second to ensure that the truck operator was truly ready for water, the event may have been avoided.

4. Know the limitations of all equipment you are using either in operations or training scenarios and never exceed manufacturer-recommended specifications. The hose in this case had stamped on it "Test to 200 psi" yet the pump operator thought the pump pressure should be 225 psi.

5. Establish a "safety perimeter" around training evolutions and keep non-active participants outside of the perimeter.

Demographics

Department type: Combination, Mostly paid
Job or rank: Captain
Department shift: 24 hours on—48 hours off
Age: 43–51
Years of fire service experience: 11–13
Region: FEMA Region III
Service Area: Suburban

Event Information

Event type: Training activities: formal training classes, in-station drills, multi-company drills, etc.
Event date and time: 08/30/2005 13:30
Hours into the shift: 5–8
Event participation: Involved in the event
Do you think this will happen again? Uncertain
What were the contributing factors?
- Communication
- Situational Awareness
- Human Error

What do you believe is the loss potential?
- Property damage
- Life-threatening injury

guidelines. The initial attack fire apparatus is equipped with a fire pump as defined in NFPA 1901. The fire pump must have the minimum rated capacity of 250 GPM (1000 L/min). The apparatus's water tank must have, at a minimum, a certified capacity of 200 gal (750 L).

Because an initial attack fire apparatus has pumping capabilities, it requires a hose to expel the water. Hose bed areas, compartments, and reels must meet the specifications in NFPA 1901. A minimum of 15′ (4.5 m) of soft suction hose or 20′ (6 m) of hard suction hose with a strainer must be carried. The purchaser of the initial attack fire apparatus specifies whether hard or soft suction is needed. The following fire hose and nozzles are required by NFPA 1901:

- 300′ (90 m) of 2½″ (65 mm) or larger fire supply hose
- 400′ (120 m) of 1½″ (38 mm), 1¾″ (45 mm), or 2″ (52 mm) attack fire hose
- Two combination spray nozzles with a minimum capacity of 95 GPM (360 L/min)

Like the pumper, the initial attack fire apparatus needs storage compartments. NFPA 1901 mandates a minimum of 22 ft^3 (0.62 m^3) of enclosed weather-resistant compartments, which are to be used for storage of equipment. The compartment space is usually a custom body that is added to a commercial cab and chassis.

The initial attack fire apparatus is also used to bring equipment to the scene of the fire. Most of these tools are the same ones found on the larger pumper. Remember—the initial attack fire apparatus responds to basically the same types of fires as the larger pumpers, but is capable of accessing hard-to-reach areas.

A 12′ (3.7 m) or longer ground ladder is carried on the initial attack fire apparatus. All ground ladders must conform to the requirements in NFPA 1931, *Standard for Manufacturer's Design of Fire Department Ground Ladders*. Hand tools and other miscellaneous equipment must also be carried on the initial fire attack apparatus.

As stated in NFPA 1901, the initial attack fire apparatus must carry the miscellaneous equipment listed in Table 2-2. This equipment is the minimum that is required by NFPA 1901. Most fire departments exceed this list and add equipment that best suits the needs for their jurisdiction.

Table 2-2 Miscellaneous Equipment on an Initial Attack Fire Apparatus

Amount	Equipment
1	6 lb (2.7 kg) pickhead axe mounted in a bracket fastened to the initial attack fire apparatus
1	6′ (2 m) pike pole or plaster hook mounted in a bracket fastened to the initial attack fire apparatus
2	Portable hand lights mounted in a bracket fastened to the initial attack fire apparatus
1	Approved dry chemical portable fire extinguisher with a minimum 80-B:C rating mounted in a bracket fastened to the initial attack fire apparatus
1	2½ gal (9.5 L) or larger water extinguisher mounted in a bracket fastened to the initial attack fire apparatus
1	SCBA complying with NFPA 1981, for each assigned seating position, but no fewer than two units, mounted in a bracket fastened to the apparatus or stored in containers supplied by the SCBA manufacturer
1	Spare SCBA cylinder for each SCBA unit carried, each mounted in a bracket fastened to the initial attack fire apparatus or in a specially designed storage space(s)
1	First-aid kit
2	Combination spanner wrenches mounted in a bracket fastened to the initial attack fire apparatus
1	Hydrant wrench mounted in a bracket fastened to the initial attack fire apparatus
1	Double female adapter, sized to fit 2½″ (65 mm) or larger fire hose, mounted in a bracket fastened to the initial attack fire apparatus
1	Double male adapter, sized to fit 2½″ (65 mm) or larger fire hose, mounted in a bracket fastened to the initial attack fire apparatus
1	Rubber mallet, for use on suction hose connections, mounted in a bracket fastened to the initial attack fire apparatus
2	Wheel chocks, mounted in readily accessible locations, each designed to hold the initial attack fire apparatus, when loaded to its maximum in-service weight, on a 10 percent grade with the transmission in neutral and the parking brake released

Mobile Water Supply Apparatus

Many rural communities do not have hydrants or readily accessible water for use at the fire scene. For this reasons, fire fighters need fire apparatus with large-capacity water tanks at their disposal. These mobile water supply fire apparatus are commonly referred to as tenders and can provide enough onboard water to extinguish most fires that are held to only one or two bedrooms of a home. When fires are larger and more water is needed, a water shuttle operation may be used by these types of fire apparatus to establish a sustained water supply for fire attack.

Mobile water supply apparatus are defined in NFPA 1901 Figure 2-5. Such an apparatus may be designed with or without a fire pump. If the mobile water supply apparatus has a fire pump, then that fire pump must meet the criteria stated in NFPA 1901. This type of fire apparatus is designed to carry a large ca-

Figure 2-5 A mobile water supply apparatus are commonly referred to as tenders.

pacity of water to the fire scene. It is equipped with one or more water tanks that meet the requirements of NFPA 1901 and have a minimum certified capacity of 1000 gal (400 L).

If the mobile water supply apparatus is equipped with a fire pump, then a minimum of 15′ (4.5 m) of soft suction hose or 20′ (6 m) of hard suction hose with a strainer is carried. Soft suction hose includes couplings compatible with local hydrants. The fire department specifies to the maker whether hard or soft suction is needed.

Fire hose and nozzles are needed as well. At least 200′ (60 m) of 2½″ (65 mm) or larger supply hose must be available on the mobile water supply apparatus. If the mobile water supply apparatus is equipped with a fire pump, 400′ (120 m) of 1½″ (45 mm) or 2″ (52 mm) attack hose and two combination spray nozzles with a capacity of 95 GPM (360 L/min) minimum are required.

Equipment storage is very important. A minimum of 20 ft³ (0.57 m³) of enclosed weather-resistant compartment space must be provided for the storage of equipment. Minor equipment should be organized and mounted in brackets or compartments. At a minimum, to comply with NFPA 1901 the mobile water supply fire apparatus must carry the equipment listed in **Table 2-3**. Most fire departments exceed this list and will add equipment that best suits the needs for their jurisdiction.

Aerial Fire Apparatus

With so many multistory buildings in their jurisdictions, what would fire fighters do without ladders? With the growth of city populations, the need for multilevel dwellings increased. Fire fighters relied on ladders to enter the upper levels to fight fires and rescue people. In 1830, Abraham Wivell, an English fire fighter, created the first fly ladder. Wivell's ladder could reach the second story with the main ladder and extend the next two flies to the upper floors. Known as "escapes" in Europe, these ladders became a necessity. All too soon, however, cities began outgrowing this ladder's reach.

In the late nineteenth century, companies such as Scott-Uda began to design height extension ladders. The Scott-Uda design used eight extensions countered by weights and balances. These aerials were short-lived, as many people died due to their collapse. In 1868, Daniel Hayes successfully patterned an 85′ hand-cranked aerial ladder. As mechanics, pneumatics, and hydraulics became more popular, the race for the perfect aerial ladder began. Seagraves introduced the first aerial truck, and in 1916 American LaFrance used air compression to raise the ladders. Today's ladder trucks are equipped with computerized aerial ladders that can extend for an average distance of 110′.

<u>Aerial fire apparatus</u> are defined in NFPA 1901 **Figure 2-6**. Such apparatus can be configured in many different ways—for example, with an <u>aerial ladder</u>, elevating platform, water tower, or water pump **Figure 2-7**. If the aerial fire apparatus is designed with any of these configurations, they are subject to the requirements outlined in specific chapters of NFPA 1901. This chapter discusses the basic fire apparatus without specialties.

To be considered an aerial fire apparatus as defined by NFPA 1901, the apparatus must include, at a minimum, a 115′ (35 m) total complement of ground ladders that are supplied and

Table 2-3 Miscellaneous Equipment on Mobile Water Supply Apparatus

Amount	Equipment
1	6 lb (2.7 kg) pickhead axe mounted in a bracket fastened to the mobile water supply apparatus
1	6′ (2 m) pike pole or plaster hook mounted in a bracket fastened to the mobile water supply apparatus
2	Portable hand lights mounted in a bracket fastened to the mobile water supply apparatus
1	Approved dry chemical portable fire extinguisher with a minimum 80-B:C rating mounted in a bracket fastened to the mobile water supply apparatus
1	2½ gal (9.5 L) or larger water extinguisher mounted in a bracket fastened to the mobile water supply apparatus
1	SCBA complying with NFPA 1981 for each assigned seating position, but no fewer than two units, mounted in a bracket fastened to the mobile water supply apparatus or stored in containers supplied by the SCBA manufacturer
1	Spare SCBA cylinder for each SCBA unit carried, each mounted in a bracket fastened to the mobile water supply apparatus or in a specially designed storage space(s)
1	First-aid kit
2	Combination spanner wrenches mounted in a bracket fastened to the mobile water supply apparatus
1	Hydrant wrench mounted in a bracket fastened to the mobile water supply apparatus
1	Double female adapter, sized to fit 2½″ (65 mm) or larger fire hose, mounted in a bracket fastened to the mobile water supply apparatus
1	Double male adapter, sized to fit 2½″ (65 mm) or larger fire hose, mounted in a bracket fastened to the mobile water supply apparatus
1	Rubber mallet, for use on suction hose connections, mounted in a bracket fastened to the mobile water supply apparatus
2	Wheel chocks, mounted in readily accessible locations, each designed to hold the mobile water supply apparatus, when loaded to its maximum in-service weight, on a 10 percent grade with the transmission in neutral and the parking brake released

Figure 2-6 An aerial apparatus can be configured in many different ways.

Figure 2-7 A tiller is a type of aerial apparatus and is also known as a hook and ladder truck.

installed by the manufacturer. NFPA 1901 states that as a minimum, the following ladders shall be provided:

- One attic ladder
- Two straight ladders (with folding roof hooks)
- Two extension ladders

All ground ladders shall be mounted with brackets provided by the manufacturer and shall meet all requirements stated in NFPA 1931. In addition, NFPA 1901 states the apparatus must include a minimum of 40 ft³ (1.1 m³) of enclosed weather-resistant compartment space for equipment storage. Aerial fire apparatus are commonly known in the fire service as "toolboxes on wheels." They usually carry more equipment to the fire scene than any other type of fire apparatus. These vehicles are designed to assist fire fighters more with support functions than with extinguishment functions while on the fire ground. Aerial fire apparatus do not usually carry hose or nozzles because they are designed for search and rescue, forcible entry, and ventilation—all of which are support functions.

The items listed in **Table 2-4** are found on the aerial fire apparatus. The equipment listed in the table is the minimum

Table 2-4 Equipment on an Aerial Fire Apparatus

Amount	Equipment
2	6 lb (2.7 kg) or larger flathead axes mounted in brackets fastened to the aerial fire apparatus
3	6 lb (2.7 kg) pickhead axes mounted in brackets fastened to the aerial fire apparatus
4	Pike poles mounted in brackets fastened to the aerial fire apparatus
2	3′ to 4′ (1–1.2 m) plaster hooks with D handles mounted in brackets fastened to the aerial fire apparatus
2	Crowbars mounted in brackets fastened to the aerial fire apparatus
2	Claw tools mounted in brackets fastened to the aerial fire apparatus
2	12 lb (5 kg) sledgehammers mounted in brackets fastened to the aerial fire apparatus
4	Portable hand lights mounted in brackets fastened to the aerial fire apparatus
1	Approved dry chemical portable fire extinguisher with a minimum 80-B:C rating mounted in brackets fastened to the aerial fire apparatus
1	2½ gal (9.5 L) or larger water extinguisher mounted in brackets fastened to the aerial fire apparatus
1	SCBA complying with NFPA 1981, for each assigned seating position, but not fewer than four units, mounted in brackets fastened to the aerial fire apparatus or stored in containers supplied by the SCBA manufacturer
1	Spare SCBA cylinder for each unit carried, each mounted in brackets fastened to the aerial fire apparatus or stored in a specially designed storage space(s)
1	First-aid kit
6	Salvage covers, each a minimum size of 12″ × 18′ (3.6 m × 5.5 m)
4	Combined spanner wrenches mounted in brackets fastened to the aerial fire apparatus
2	Scoop shovels mounted in brackets fastened to the aerial fire apparatus
1	Pair of bolt cutters, 24″ (0.6 m) minimum, mounted in brackets fastened to the aerial fire apparatus
4	Ladder belts meeting the requirements of NFPA 1983, *Standard on Life Safety Rope and Equipment for Emergency Services*
1	150′ (45 m) light-use life safety rope meeting the requirements of NFPA 1983
1	150′ (45 m) general-use life safety rope meeting the requirements of NFPA 1983
2	150′ (45 m) utility ropes having a breaking strength of at least 5000 lb (2300 kg)
1	Box of tools including the following: • One hacksaw with three blades • One keyhole saw • One 12″ (0.3 m) pipe wrench • One 24″ (0.6 m) pipe wrench • One ballpeen hammer • One pair of tin snips • One pair of pliers • One pair of lineman's pliers Assorted types and sizes of screwdrivers Assorted adjustable wrenches Assorted combination wrenches
2	Wheel chocks, mounted in readily accessible locations, each designed to hold the aerial fire apparatus, when loaded to its maximum in-service weight, on a 10 percent grade with the transmission in neutral and the parking brake released

Table 2-5 **Additional Equipment for the Aerial Fire Apparatus**

Amount	Equipment
1	Double female 2½" (65 mm) adapter with National Hose threads, mounted in a bracket fastened to the aerial fire apparatus
1	Double male 2½" (65 mm) adapter with National Hose threads, mounted in a bracket fastened to the aerial fire apparatus
1	Rubber mallet, for use on suction hose connections, mounted in a bracket fastened to the aerial fire apparatus
2	Hydrant wrenches mounted in a bracket fastened to the aerial fire apparatus
Note:	If the supply hose carried does not use sexless couplings, an additional double female adapter and double male adapter, size 2½" (65 mm) or larger fire hose, mounted in a bracket fastened to the aerial fire apparatus

Figure 2-8 A quint has five functions associated with it: pump, water tank, fire hose storage, aerial, and ground ladders.

that is required by NFPA 1901. Most fire departments exceed this list and will add equipment that best suits the needs for their jurisdiction. If the fire apparatus is equipped with a fire pump, the equipment in Table 2-5 must be provided to be compliant with NFPA 1901.

Quint Fire Apparatus

What is a quint? Is it a pumper or is it an aerial fire apparatus? Or is it both? This debate began in 1912, when Metz of Germany patented the first quint. In the United States, American LaFrance built its first quint in 1935; Seagraves followed with its own model in 1940. "Quint" is short for "quintuple," meaning five. A quint has five functions associated with it: pump, water tank, fire hose storage, aerial, and ground ladders. The quint was very popular until the 1990s.

Due to service-wide budget cuts, many fire departments reconsidered the use of the multipurpose quint. Many believed that the purchase of a quint would mean the department had available a double-functioning piece of fire apparatus—that is, *both* a pumper *and* a ladder. In reality, the quint works as *either* a pump *or* an aerial ladder truck. Once this apparatus is running as a pump, it cannot be maneuvered for ladder duty.

NFPA 1901 defines what an apparatus needs to be defined a **quint** Figure 2-8 . As previously discussed, the quint is equipped with a fire pump. This standard states that the fire pump shall meet the fire pump requirements and shall have a rated capacity of 1000 GPM (4000 L/m). NFPA 1901 also states that the quint shall be equipped with an aerial ladder or elevating platform with a permanently installed waterway, as well as a water tank that has a minimum certified capacity of 300 gal (1100 L). The quint shall carry a minimum total complement of 85' (26 m) of ground ladders to include at least one extension ladder, one straight ladder equipped with roof hooks, and one attic ladder; all of these ladders must satisfy the requirements in NFPA 1931.

Because the quint has a fire pump, hose is a necessity, consisting of a minimum of 15' (4.5 m) of soft suction with compatible couplings, or 20' (6 m) of hard suction hose with strainer. It is up to the purchaser to specify whether hard or soft suction will be provided. Table 2-6 lists the fire hose and nozzles that must be carried on a quint.

Like the pumper, the quint must have a minimum of 40 ft^3 (1.1 m^3) of enclosed weather-resistant compartment space for storage of equipment. This fire apparatus will need some of the same equipment required for a pumper and an aerial apparatus. NFPA 1901 specifies that the miscellaneous equipment listed in Table 2-7 shall be carried on a quint. This equipment is the minimum that is required by NFPA 1901; most fire departments exceed this list and will add equipment that best suits the needs for their jurisdiction.

Special Service Fire Apparatus

The special service fire apparatus is designed for a particular purpose and does not quite fit into the other categories Figure 2-9 . For example, a hazardous materials apparatus or a heavy technical rescue apparatus would fall into this category. The majority of the special service fire apparatus is devoted to providing compartment space for the unique equipment that

Table 2-6 **Fire Hose and Nozzles for the Quint**

Amount	Hose
	800' (240 m) of 2½" (65 mm) or larger supply hose, in any combination
	400' (120 m) of 1½" (38 mm), 1¾" (45 mm), or 2" (52 mm) attack hose, in any combination
1	Combination spray nozzle, 200 GPM (750 L/min) minimum
2	Combination spray nozzles, 95 GPM (360 L/min) minimum
1	Playpipe with shut-off and 1" (25 mm), 1⅛" (29 mm), and 1¼" (32 mm) tips

Table 2-7 Miscellaneous Equipment on the Quint

Amount	Equipment
1	6 lb (2.7 kg) flathead axe mounted in a bracket fastened to the quint
1	6 lb (2.7 kg) pickhead axe mounted in a bracket fastened to the quint
1	6′ (2 m) pike pole or plaster hook mounted in a bracket fastened to the quint
1	8′ (2.4 m) or longer pike pole or plaster hook mounted in a bracket fastened to the quint
2	Portable hand lights mounted in a bracket fastened to the quint
1	Approved dry chemical portable fire extinguisher with a minimum 80-B:C rating mounted in a bracket fastened to the quint
1	2½ gal (9.5 L) or larger water extinguisher mounted in a bracket fastened to the quint
1	SCBA complying with NFPA 1981, for each assigned seating position, but not fewer than four units, mounted in brackets fastened to the quint or stored in containers supplied by the SCBA manufacturer
1	Spare SCBA cylinder for each unit carried, each mounted in brackets fastened to the quint or stored in a specially designed storage space(s)
1	Spare SCBA cylinder for each SCBA unit carried
1	First-aid kit
4	Combination wrench mounted in a bracket fastened to the quint
2	Hydrant wrenches mounted in a bracket fastened to the quint
2	Double female 2½″ (65 mm) adapter with National Hose threads, mounted in a bracket fastened to the quint
1	Double male 2½″ (65 mm) adapter with National Hose threads, mounted in a bracket fastened to the quint
1	Rubber mallet, for use on suction hose connections, mounted in a bracket fastened to the quint
4	Salvage covers, each a minimum size of 12′ × 14′ (3.7 m × 4.3 m)
4	Ladder belts meeting the requirements of NFPA 1983
1	150′ (45 m) light-use safety rope meeting the requirements of NFPA 1983
1	150′ (45 m) general-use safety rope meeting the requirements of NFPA 1983
2	Wheel chocks, mounted in readily accessible locations, each designed to hold the quint, when loaded to its maximum in-service weight, on a 10 percent grade with the transmission in neutral and the parking brake released

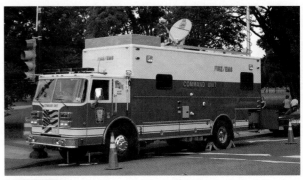

Figure 2-9 The special service fire apparatus is designed for a particular purpose and does not quite fit into the other categories.

it carries. However, if the fire apparatus is to be equipped with a fire pump, then the fire pump shall meet all of the requirements listed in NFPA 1901. If the special service fire apparatus is equipped with ground ladders, it shall meet all of the requirements listed in NFPA 1931.

Because the primary function of this fire apparatus is to supply a certain type of equipment for the incident, it is required to have more compartment space than the other types of fire service apparatus. NFPA 1901 states it must have a minimum of 120 ft³ (3.4 m³) of enclosed, weather-resistant compartment space for storage of equipment. According to NFPA 1901, the minimum equipment that must be carried comprises the items listed in **Table 2-8**. Most fire departments exceed this list and will add equipment that best suits the needs for their jurisdiction.

Mobile Foam Fire Apparatus

The **mobile foam fire apparatus** is a fire apparatus with a permanently mounted fire pump, foam proportioning system, and foam concentrate tank(s) whose primary purpose is for use in the control and extinguishment of flammable and combustible liquid fires in storage tanks and other flammable liquid spills **Figure 2-10**. The mobile foam fire apparatus can be configured with or without an aerial device. Its job is to deliver foam imme-

Table 2-8 Minimum Equipment on a Special Service Fire Apparatus

Amount	Equipment
2	Portable hand lights mounted in a bracket fastened to the special service fire apparatus
1	Approved dry chemical portable fire extinguisher with a minimum 80-B:C rating mounted in a bracket fastened to the special service fire apparatus
1	2½ gal (9.5 L) or larger water extinguisher mounted in a bracket fastened to the special service fire apparatus
1	SCBA complying with NFPA 1981, for each assigned seating position, but not fewer than four units, mounted in brackets fastened to the special service fire apparatus or stored in containers supplied by the SCBA manufacturer
1	Spare SCBA cylinder for each unit carried, each mounted in brackets fastened to the special service fire apparatus or stored in a specially designed storage space(s)
1	First-aid kit
2	Wheel chocks, mounted in readily accessible locations, each designed to hold the special service fire apparatus, when loaded to its maximum in-service weight, on a 10 percent grade with the transmission in neutral and the parking brake released

A minimum of 15′ (4.5 m) of soft suction hose with compatible couplings or 20′ (6 m) of hard suction hose with strainer shall be carried on the mobile foam fire apparatus. NFPA 1901 identifies the fire hose and nozzles to be carried on the mobile foam fire apparatus, which are also listed in **Table 2-9**.

Equipment storage on a mobile foam fire apparatus must include a minimum of 40 ft³ (1.13 m³) of enclosed weather-resistant compartment space. According to NFPA 1901, the minimum equipment that must be carried comprises the items listed in **Table 2-10**. Most fire departments exceed this list and will add equipment that best suits the needs for their jurisdiction.

Figure 2-10 The mobile foam fire apparatus is a fire apparatus with a permanently mounted fire pump, foam proportioning system, and foam concentrate tank(s).

diately at the scene without requiring fire fighters to attach special containers or change nozzles. This task may be accomplished through use of a turret attached to the top of the mobile foam fire apparatus or an aerial device that extends to reach the fire; both can be operated from inside the cab of the fire apparatus.

The mobile foam fire apparatus, as defined in NFPA 1901, shall be equipped with a fire pump that has a minimum rated capacity of 750 GPM (3000 L/min), or an industrial supply pump that meets the requirements of NFPA 1901. Because the mobile foam fire apparatus is designed to produce foam, a foam proportioning system is needed. The mobile foam fire apparatus shall be equipped with one or more foam concentrate tanks that meet the NFPA requirements and have a minimum capacity (combined if applicable) of 500 gal (2000 L).

Table 2-9 Equipment on Mobile Foam Fire Apparatus

Amount	Equipment
	800′ (240 m) of 2½″ (65 mm) or larger supply hose, in any combination
	400′ (120 m) of 1½″ (38 mm), 1¾″ (45 mm), or 2″ (52 mm) attack hose, in any combination
4	Foam or spray nozzles, 200 GPM (750 L/min)
2	Foam or spray nozzles, 95 GPM (360 L/min)
1	Preconnected monitor, rated to discharge a minimum of 1000 GPM (4000 L/min), mounted on top of the fire apparatus with a spray or foam nozzle rated at a minimum of 1000 GPM (4000 L/min)

Table 2-10 Minimum Equipment on Mobile Foam Fire Apparatus

Amount	Equipment
1	6 lb (2.7 kg) pickhead axe mounted in a bracket fastened to the mobile foam fire apparatus
1	6′ (2 m) pike pole or plaster hook mounted in a bracket fastened to the mobile foam fire apparatus
2	Portable hand lights mounted in a bracket fastened to the mobile foam fire apparatus
1	Approved dry chemical portable fire extinguisher with a minimum 80-B:C rating mounted in a bracket fastened to the mobile foam fire apparatus
1	SCBA complying with NFPA 1981, for each assigned seating position, but not fewer than four units, mounted in a bracket fastened to the mobile foam fire apparatus
1	Spare SCBA cylinder for each SCBA unit carried, each mounted in a bracket fastened to the mobile foam fire apparatus or stored in a specially designed storage space
1	First-aid kit
4	Combination wrench mounted in a bracket fastened to the mobile foam fire apparatus
2	Hydrant wrenches mounted in a bracket fastened to the mobile foam fire apparatus
1	Double female 2½″ (65 mm) adapter with National Hose threads, mounted in a bracket fastened to the mobile foam fire apparatus
1	Double male 2½″ (65 mm) adapter with National Hose threads, mounted in a bracket fastened to the mobile foam fire apparatus
1	Rubber mallet, for use on suction hose connections, mounted in a bracket fastened to the mobile foam fire apparatus
2	Wheel chocks, mounted in readily accessible locations, each designed to hold the mobile foam fire apparatus, when loaded to its maximum in-service weight, on a 10 percent grade with the transmission in neutral and the parking brake released

Wrap-Up

■ Chief Concepts

- To obtain a fire apparatus that meets the needs of your fire department, you must work closely with the manufacturer.
- The fire pump is the key component in getting water from the fire apparatus, through the hose, and onto the fire.
- The pumper is the most common fire apparatus and is a part of almost every fire department.
  - The pumper is used on small fires such as dumpster fires, vehicle fires, and brush fires in urban settings.
  - It is also the main source of fire attack for larger fires that involve structures.
- The initial attack fire apparatus is utilized much like the pumper, but its specifications are much different. This type of apparatus is designed to be more maneuverable than the pumper, especially on off-road terrain.
- Mobile water supply fire apparatus (commonly referred to as tenders) provide enough onboard water to extinguish most fires that are held to only one or two bedrooms of a home.
- An aerial fire apparatus is a vehicle equipped with an aerial ladder, elevating platform, or water tower that is designed and equipped to support firefighting and rescue operations by positioning personnel, handling materials, providing continuous egress, or discharging water at positions elevated from the ground.
- A quint has five functions associated with it: pump, water tank, fire hose storage, aerial, and ground ladders.
- The special service fire apparatus is designed for a particular purpose and does not quite fit into the other categories. For example, a hazardous materials apparatus or a heavy technical rescue apparatus would fall into this category.
- The mobile foam fire apparatus is a fire apparatus with a permanently mounted fire pump, foam proportioning system, and foam concentrate tank(s) whose primary purpose is for use in the control and extinguishment of flammable and combustible liquid fires in storage tanks and other flammable liquid spills.

■ Hot Terms

aerial fire apparatus A vehicle equipped with an aerial ladder, elevating platform, or water tower that is designed and equipped to support firefighting and rescue operations by positioning personnel, handling materials, providing continuous egress, or discharging water at positions elevated from the ground. (NFPA 1901)

aerial ladder A self-supporting, turntable-mounted, power-operated ladder of two or more sections permanently attached to a self-propelled automotive fire apparatus and designed to provide a continuous egress route from an elevated position to the ground. (NFPA 1904)

chassis The basic operating motor vehicle, including the engine, frame, and other essential structural and mechanical parts, but exclusive of the body and all appurtenances for the accommodation of driver, property, passengers, appliances, or equipment related to other than control. Common usage might, but need not, include a cab (or cowl). (NFPA 1901)

initial attack fire apparatus Fire apparatus with a fire pump of at least 250 GPM (1000 L/min) capacity, water tank, and hose body, whose primary purpose is to initiate a fire suppression attack on structural, vehicular, or vegetation fires, and to support associated fire department operations. (NFPA 1901)

miscellaneous equipment Portable tools and equipment carried on a fire apparatus, not including suction hose, fire hose, ground ladders, fixed power sources, hose reels, cord reels, breathing air systems, or other major equipment or components specified by the purchaser to be permanently mounted on the apparatus as received from the apparatus manufacturer. (NFPA 1901)

mobile foam fire apparatus Fire apparatus with a permanently mounted fire pump, foam proportioning system, and foam concentrate tank(s) whose primary purpose is for use in the control and extinguishment of flammable and combustible liquid fires in storage tanks and other flammable liquid spills. (NFPA 1901)

mobile water supply apparatus A vehicle designed primarily for transporting (pickup, transporting, and delivering) water to fire emergency scenes to be applied by other vehicles or pumping equipment. Also known as a tanker or tender. (NFPA 1901)

pumper Fire apparatus with a permanently mounted fire pump of at least 750 GPM (3000 L/min) capacity, water tank, and hose body, whose primary purpose is to combat structural and associated fires. (NFPA 1901)

quint Fire apparatus with a permanently mounted fire pump, a water tank, a hose storage area, an aerial ladder or elevating platform with a permanently mounted waterway, and a complement of ground ladders. (NFPA 1901)

Driver/Operator in Action

The station captain and some other fire fighters in the station are very helpful, answering any questions you have about the fire apparatus and the equipment it carries. After you complete your inspection of the aerial apparatus, you sit down and talk with the rest of the station's crew. Some of the fire fighters have not worked on a pumper apparatus in years, which sparks a conversation about the different types of fire apparatus that the fire department now has. During the discussion, you start to recall some of the different features of those fire apparatus and the equipment that they are required to carry.

1. Which type of apparatus has a pump, carries water, fire hose, ground ladders and has an aerial device?
   A. Mobile water supply apparatus
   B. Quint
   C. Aerial apparatus
   D. Pumper

2. What would be an example of a special service fire apparatus?
   A. Pumper
   B. Hazardous materials apparatus
   C. Mobile foam apparatus
   D. Initial attack fire apparatus

3. What is the minimum amount of tank water a pumper should have?
   A. 200 gallons
   B. 1000 gallons
   C. 500 gallons
   D. 300 gallons

4. What is the total length of fire service ground ladders an aerial apparatus should carry?
   A. 80'
   B. 100'
   C. 115'
   D. 85'

5. What is the NFPA standard that covers standards for automotive fire apparatus?
   A. NFPA 1002
   B. NFPA 1021
   C. NFPA 1901
   D. NFPA 1931

Water

CHAPTER 3

NFPA 1002 Standard

There are no NFPA 1002 Standards covered in this chapter.

Additional NFPA Standards

NFPA 25 *Standard for the Inspection, Testing, and Maintenance of Water-Based Fire Protection Systems* (2008)

NFPA 291 *Recommended Practice for Fire Flow Testing and Marking of Hydrants* (2010)

NFPA 1144 *Standard for Reducing Structure Ignition Hazards from Wildland Fire* (2008)

NFPA 1150 *Standard on Foam Chemicals for Fires in Class A Fuels* (2004)

Fire and Emergency Services Higher Education (FESHE) Model Curriculum

National Fire Science Degree Programs Committee (NFSDPC): Associate's Curriculum

Fire Protection Hydraulics and Water Supply

1. Apply the application of mathematics and physics to movement of water in fire suppression activities.
2. Identify the design principles of fire service pumping apparatus.
3. Analyze community fire flow demand criteria. [p. 34–41]
4. Demonstrate, through problem solving, a thorough understanding of the principles of forces that affect water at rest and, in motion.
5. List and describe the various types of water distribution systems. [p. 36–38]
6. Discuss the various types of fire pumps.

Knowledge Objectives

After studying this chapter, you will be able to:
- Describe the properties of water.
- Describe the three sources of water.
- Describe water's role in extinguishing a fire.
- Describe static water pressure.
- Describe residual water pressure.
- Describe flow pressure.

Skills Objectives

After studying this chapter, you will be able to:
- Operate a hydrant.
- Shut down a hydrant.
- Demonstrate how to obtain the static pressure.
- Demonstrate the correct use of a Pitot gauge.

You Are the Driver/Operator

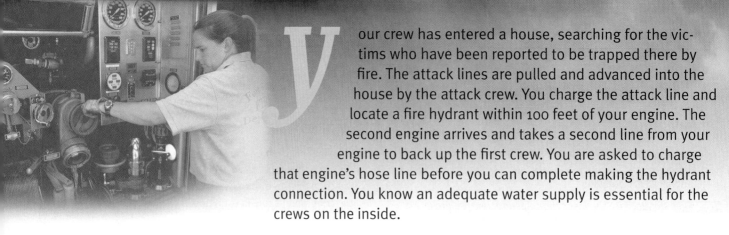

Your crew has entered a house, searching for the victims who have been reported to be trapped there by fire. The attack lines are pulled and advanced into the house by the attack crew. You charge the attack line and locate a fire hydrant within 100 feet of your engine. The second engine arrives and takes a second line from your engine to back up the first crew. You are asked to charge that engine's hose line before you can complete making the hydrant connection. You know an adequate water supply is essential for the crews on the inside.

1. What would you do if there was no water available from the hydrant you located right by the engine?
2. What is the pressure of your incoming water supply?
3. How many more attack lines can you support with the water supply you have available?

Introduction

The importance of a dependable and adequate **water supply** for fire suppression operations is self-evident. The hose line is not just the primary weapon for fighting a fire; it is also the fire fighter's primary defense against being burned or driven out of a burning building. The basic plan for fighting most fires depends on having an adequate supply of water to confine, control, and extinguish the fire. If the water supply is interrupted while crews are working inside a building, then fire fighters could be trapped, injured, or killed. Fire fighters entering a burning building need to be confident that their water supply is both reliable and adequate to operate hose lines for their protection and to extinguish the fire.

Ensuring a dependable water supply is a critical fire-ground operation that must be accomplished as soon as possible. A water supply should be established at the same time as other initial fire-ground operations take place. At many fire scenes, the processes of size-up, forcible entry, raising ladders, search and rescue, ventilation, and establishing a water supply all occur concurrently.

You obtain water from one of two sources. First, **municipal** and **private water systems** furnish water under pressure through fire hydrants Figure 3-1. Second, rural areas may depend on **static water sources** such as lakes and streams. These resources serve as drafting sites for fire department apparatus to obtain and deliver water to the fire scene. Often, the water carried in the tank of the first-arriving vehicles is used in the initial

Figure 3-1 The water that comes from a hydrant is provided by a municipal or private water system.

Figure 3-2 Mobile water supply fire apparatus can deliver large quantities of water to the scene of the fire.

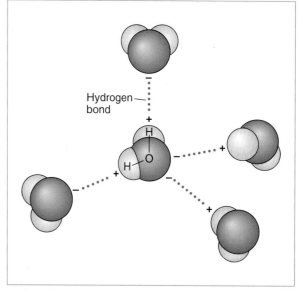

Figure 3-3 Water has many hidden qualities contained in its chemical description: H_2O.

attack. Although many fires are successfully controlled using tank water, this tactic does not ensure the ongoing adequacy and reliability of the water supply throughout the incident. The establishment of an adequate, continuous water supply then becomes the primary objective to support the fire attack **Figure 3-2**. The operational plan must ensure that an adequate and reliable water supply is available before the tank is empty.

Chemical Properties of Water

Approximately 70 to 75 percent of the earth's surface is covered by water. Water is so abundant that it is often easy to take for granted. According to the U.S. Geological Survey, the U.S. population consumes 408 billion gallons of water per day.

Water is a virtually colorless, odorless, and tasteless liquid. It has many hidden qualities contained in its chemical description: H_2O. That is, in water, one atom of oxygen is bound by two atoms of hydrogen attached on the same side. This arrangement gives the water molecule a positive charge at the hydrogen side and a negative charge on the oxygen side **Figure 3-3**. Opposite electrical charges attract each other (hydrogen bonding), making water molecules stick to one another. As a result of this attraction, the bond between the molecules becomes stronger and forms a surface film that makes it more difficult to move an object through the surface than to move it when it is completely submersed. This **surface tension** allows water to flow, puddle, and remain together even after leaving a nozzle. It is not until gravity acts upon the water stream that it starts to break up into smaller droplets. Likewise, as water is heated, the surface tension decreases. The lower surface tension allows the water to soak into the object with which it comes into contact, rather than just skating over its outer surface.

The bond between hydrogen and oxygen in water is so strong that water dissolves more substances than any other liquid. As water passes through or moves by a substance, the positive and negative charges of the molecule take the various chemicals, minerals, and solvents along with it, thus earning water the title of **universal solvent**. For example, pour a teaspoon of salt into a glass of water and note the result: The salt dissolves into the water. The mixing of the water molecules with the sodium chloride causes the sodium and chloride molecules to separate from each other and be carried away in the solution. Whether a substance dissolves into water is determined by whether the substance's components can break the hydrogen bond of water. If they cannot, the substance's molecules are passed among the water molecules but do not dissolve (i.e., interact with the water molecules). Substances dissolved in water are referred to as **aqueous**. For example, Class A foam dissolves in water, which reduces the water's surface tension, and allows it to saturate the fuel.

Water has been used as the primary extinguishing agent in the fire service from the earliest days, mostly because of its widespread availability. Water, however, possesses other properties that make it ideal for extinguishing fires. For example, water is able to absorb heat because of its hydrogen-bonding characteristics. The amount of heat energy required to increase the temperature of a substance is known as its **specific heat index**. The specific heat index of water is one of the highest of any known chemical compound. Consider that sodium bicarbonate—a chemical once used in fire extinguishers—has a specific heat index of 0.22 compared to water's specific heat index of 1. Thus it would take almost five times as much heat to raise the temperature of water as it does to raise the temperature of an equal amount of sodium bicarbonate. Simply stated, water absorbs almost five times more heat than sodium bicarbonate.

Physical Properties of Water

Water exists in all three property states on earth: liquid (oceans, rivers), gas (clouds), and solid (ice) **Figure 3-4**. All three property states affect the use of water in firefighting operations. Water in its liquid form weighs 8.33 pounds (3778.424 g) per gallon,

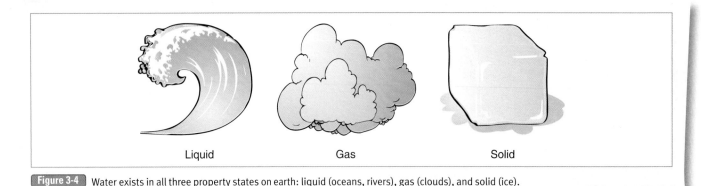

Figure 3-4 Water exists in all three property states on earth: liquid (oceans, rivers), gas (clouds), and solid (ice).

adding 6248 pounds (3.124 tons) of weight to the chassis of an engine with a 750-gallon (2800 L) tank.

Temperature has a significant effect on the weight of water. Water at 32°F (0°C) is almost one-half pound per cubic foot heavier than water at 100°F (38°C). Water begins freezing at 32°F (0°C). When in its solid form, it has a unique characteristic: Water is less dense as a solid than it is as a liquid. The molecular geometry of water when frozen causes ice to float, a very unique property. With a few exceptions, most other substances are denser when in solid form compared to their liquid form. In addition, water freezes from the top down as it gradually cools. Open lakes, for example, freeze from the top down and the ice thickens over time.

On the other end of the heat spectrum, water begins boiling at 212°F (100°C) at sea level (184.4°F [84°C] at an altitude of 14,000 feet [4300 m]). Whereas its high heat absorption allows water to have notable cooling effects, water may also be converted to a gas (steam) through the application of heat, an effect attributable to its high **heat of vaporization**.

In summary, water is the primary substance used to extinguish a fire for the following reasons:

- Water turns to a vapor (steam) when it comes in contact with a fire. The volume of water vapor is 1700 times greater than that of an equal amount of liquid water; as a consequence, water displaces oxygen and effectively smothers the fire.
- Water absorbs heat and cools the smoke, air, and room and its contents, thereby reducing the ignition temperature and decreasing the amount of fuel available for continued combustion.
- The water molecules in steam vapor carry the elements of smoke with them as the steam is ventilated from the area.

Harmful Characteristics of Water

A common misconception is that water is a good conductor of electricity. Technically, pure H_2O has a relatively low ability to conduct electricity. Thus the ability of water to conduct electrical current actually depends on the amount of dissolved solids in the water. Salt water, for example, is highly conductive.

The amount of particulate matter suspended in water is known as **turbidity**. The color of the water is directly related to the level of turbidity present. Limestone turns water turquoise, whereas iron compounds turn water reddish-brown, copper compounds create an intense blue color, and algae commonly color the water green. Turbidity affects the care and maintenance of the fire pump. As the particulates settle in the pump housing and water tank, they start to cause a breakdown of valves, gaskets, seals, and piping, which can lead to leaks and operational deficiencies.

Another effect that water has on pump maintenance derives from its **hardness**. The amount of dissolved calcium and magnesium, which increases the mineral content within water, determines the water's level of hardness. The surface tension of "hard water" is lower than that of pure water, which in turn allows for more minerals to be present. The presence of these minerals may ultimately affect the operation of fire pumps, valves, and piping. As the calcium collects in equipment over time, it starts breaking down the seals; when not regularly exercised, the valves begin to stick.

Municipal Water Systems

Municipal water systems make clean water available to people in populated areas and provide water for fire protection. As the name suggests, most municipal water systems are owned and operated by a local government agency, such as a city, county, or special water district. Some municipal water systems are privately owned; however, the basic design and operation of both private and government systems are very similar.

Driver/Operator Tip

As a driver/operator, you have an obligation to do all that you can to keep your fire apparatus well maintained. How clean the water is and its turbidity play critical roles in determining the level of maintenance required. You may need to flush the pump after each use to keep dirty water out, depending on the characteristics of your local water system. The water conditions of the static water source you use will affect your clean-up efforts following the incident as well. Without the knowledge of how the characteristics of water affect fire apparatus care, you may experience higher than normal out-of-service rates due to costly pump repairs.

Municipal water is supplied to homes, commercial establishments, and industries. Hydrants make the same water supply available to the fire department. In addition, most automatic sprinkler systems and many standpipe systems are connected to a municipal source. There are three parts to a municipal water system: the water source, the treatment plant, and the distribution system.

Water Sources

Municipal water systems can draw water from wells, rivers, streams, lakes, or human-made storage facilities called **reservoirs**. The source will depend on the geographic and hydrologic features of the area. Many municipal water systems draw water from several sources to ensure a sufficient supply. Underground pipelines or open canals supply some cities with water from sources that are many miles away.

The water source for a municipal water system needs to be large enough to meet the total demands of the service area. Most municipal water systems include large storage facilities, which are intended to ensure they will be able to meet the community's water supply demands if the primary water source is interrupted. The backup supply for some systems can provide water for several months or years; in other systems, this supply may last only a few days.

Water Treatment Facilities

Municipal water systems also include a water treatment facility, where impurities are removed **Figure 3-5**. The nature of the water treatment system depends on the quality of the untreated source water. Water that is clean and clear from the source requires little treatment. Other systems must use extensive filtration to remove impurities and foreign substances. Some treatment facilities use chemicals to remove impurities and improve the water's taste. In the end, all of the water in the system must be suitable for drinking. Chemicals also are used to kill bacteria and harmful organisms and to keep the water pure as it moves through the distribution system to individual homes or businesses. After the water has been treated, it enters the distribution system.

Figure 3-5 Impurities are removed at the water treatment facility.

Water Distribution System

The distribution system delivers water from the treatment facility to the end users and fire hydrants through a complex network of underground pipes, known as **water mains**. In most cases, the distribution center also includes pumps, storage tanks, reservoirs, and other necessary components to ensure that the required volume of water can be delivered where and when it is needed, at the required pressure.

Water pressure requirements differ, depending on how the water will be used. Generally, the water pressure ranges from 20 pounds per square inch (psi) [140 kPa] to 80 psi (560 kPa) at the delivery point. The recommended minimum pressure for water coming from a fire hydrant is 20 psi (140 kPa), but it is possible to operate with lower hydrant pressures under some circumstances.

Most water distribution systems rely on an arrangement of pumps to provide the required pressure, either directly or indirectly. Some water distribution systems use pumps to supply direct pressure. If the pumps stop operating, the pressure is lost, and the system will be unable to deliver the adequate water to the end users or to hydrants. Most municipal systems have multiple pumps and backup power supplies to reduce the risk of a service interruption due to a pump failure. The extra pumps can sometimes be used to boost the flow for a major fire or a high-demand period.

In a pure **gravity-feed system**, the water source, treatment plant, and storage facilities are located on high ground while the end users live in lower-lying areas, such as a community in a valley **Figure 3-6**. This type of system may not require any pumps, because gravity, through the elevation differentials, provides the pressure necessary to deliver the water. In some systems, the elevation pressure is so high that pressure-control devices are needed to keep from over-pressurizing parts of the system.

Most municipal water supply systems use both pumps and gravity to deliver water. Pumps may be used to deliver the water from the treatment plant to **elevated water storage towers** or to reservoirs located on hills or high ground areas. The elevated storage facilities maintain the desired water pressure in the distribution system, so that water can be delivered under pressure, even if the pumps are not operating **Figure 3-7**. When the elevated storage facilities need refilling, large supply pumps are used. Additional pumps may be installed to increase the pressure in particular areas, such as a booster pump that provides extra pressure to deliver water to a hilltop neighborhood.

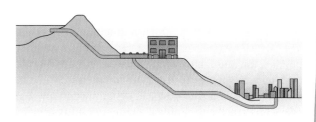

Figure 3-6 A gravity-feed system can deliver water to a low-lying community without the need for pumps.

Figure 3-7 Water that is stored in an elevated tank can be delivered to the end users under pressure.

Figure 3-8 A shut-off valve controls the water supply to an individual user or fire hydrant.

A combination pump-and-gravity-feed system must maintain enough water in the elevated storage tanks and reservoirs to meet the anticipated demands. If more water is being used than the pumps can supply, or if the pumps are out of service, some systems will be able to operate for several days using only their elevated storage reserves, whereas others will be able to function for only a few hours.

The underground water mains that deliver water to the end users come in several different sizes. Large mains, known as **primary feeders**, carry large quantities of water to a section of the town or city. Smaller mains, called **secondary feeders**, distribute the water to a smaller area. The smallest pipes, called **distributors**, carry water to the users and hydrants along individual streets.

The size of the water mains required depends on the amount of water needed both for normal consumption and for fire protection in that location. Most jurisdictions specify the minimum-size main that can be installed in a new municipal water system to ensure an adequate flow. Other municipal water systems, however, may have undersized water mains in older areas of the community. You must know the arrangement and capacity of the water systems in your department's response areas.

Water mains in a well-designed system will follow a grid pattern. A grid arrangement provides water flow to a fire hydrant from two or more directions and establishes multiple paths from the source to each area. This helps to ensure an adequate flow of water for firefighting. In addition, the grid design helps to minimize downtime for the other portions of the system if a water main breaks or needs maintenance work. With a grid, water flow can be diverted around the affected section.

Older water distribution systems may have dead-end water mains, which supply water from only one direction. Such water mains may also be found in the outer reaches of a municipal system. Hydrants on a dead-end water main will have a limited water supply. If two or more hydrants on the same dead-end main are used to fight a fire, the upstream hydrant will have more water and water pressure than the downstream hydrants.

Control valves installed at intervals throughout a water distribution system allow different sections to be turned off or isolated. These valves are used when a water main breaks or when work must be performed on a section of the system.

Shut-off valves are located at the connection points where the underground mains meet the distributor pipes. Shut-off valves control the flow of water to individual customers or to individual fire hydrants **Figure 3-8**. If the water system in a building or to a fire hydrant is damaged, the shut-off valves can be closed to prevent further water flow.

The fire department should notify the water department when the fire operations will require prolonged use of large quantities of water. The water department may be able to increase the normal volume and/or pressure by starting additional pumps. In some systems, the water department can open valves to increase the flow to a certain area in response to fire department operations at major fires.

Flow and Pressure

To understand the procedures for testing the fire hydrants, you must understand the terminology that is used. The flow or quantity of water moving through a pipe, hose, or nozzle is measured by its **volume**, usually given in terms of gallons (or liters) per minute. **Water pressure** refers to an energy level and is measured in psi (or kilopascals [kPa]). Volume and pressure are two different, but mathematically related measurements.

Water that is not moving has potential energy. When the water is moving, it has a combination of **potential energy** and **kinetic energy**. Both the quantity of water flowing and the pressure under a specific set of conditions must be measured as part of testing any water system, including hydrants.

<u>Static pressure</u> is the pressure in a system when the water is not moving. Static pressure is potential energy, because it would cause the water to move if there were some place the water could go. This type of pressure causes the water to flow out of an opened fire hydrant. If there was no static pressure, nothing would happen when the hydrant was opened.

Static pressure is generally created by **elevation pressure** and/or pump pressure. An elevated storage tank, for example, creates elevation pressure in the water mains. Gravity also creates elevation pressure in the water system as the water flows from a hilltop reservoir to the water mains in the valley below. Pumps create pressure by bringing the energy from an external source into the system.

Static pressure in a water distribution system can be measured by placing a pressure gauge on a hydrant port and opening the hydrant valve. There cannot be any water flowing out of the hydrant when the static pressure is measured; static pressure measured in this way assumes that there is no flow in the system. Because municipal water systems deliver water to hundreds or thousands of users, there is almost always water flowing within the system. Thus, in most cases, a static pressure reading is actually measuring the normal operating pressure of the system.

<u>Normal operating pressure</u> refers to the amount of pressure in the water distribution system during a period of normal consumption. In a residential neighborhood, for example, people are constantly using water to care for lawns, wash clothes, bathe, and do other normal household activities. In an industrial or commercial area, normal consumption occurs during a normal business day as water is used for various purposes. The system uses some of the static pressure to deliver this water to residents and businesses. A pressure gauge connected to a hydrant during a period of normal consumption will indicate the normal operating pressure of the system.

As a fire fighter, you need to know how much pressure will be in the system when a fire occurs. Because the regular users of the system will be drawing off a normal amount of water even during firefighting operations, the normal operating pressure is sufficient for measuring available water.

<u>Residual pressure</u> is the amount of pressure that remains in the system when the water is flowing. When you open a hydrant and start to draw large quantities of water out of the system, some of the potential energy of still water is converted into the kinetic energy of moving water. However, not all of the potential energy turns into kinetic energy; some of it is used to overcome friction in the pipes. The pressure remaining while the water is flowing is residual pressure.

Residual pressure is an important measurement because it provides the best indication of how much more water is available in the system. The more water that is flowing, the less residual pressure there is. In theory, when the maximum amount of water is flowing, the residual pressure is zero, and there is no more potential energy to push more water through the system. In reality, 20 psi is considered the minimum usable residual pressure, as this level reduces the risk of damage to underground water mains or pumps.

At the scene of a fire, you can use the difference between static pressure and residual pressure to determine how many more attack lines or appliances can be operated from the available water supply. Typically, a set of tables has been developed to help you calculate the maximum amount of available water; these tables are based on the static and residual pressure readings taken during hydrant testing. **Flow pressure** measures the quantity of water flowing through an opening during a hydrant test. When a stream of water flows out through an opening (known as an orifice), all of the pressure is converted to kinetic energy.

To calculate the volume of water flowing, measure the pressure at the center of the water stream as it passes through the opening, and factor the size and flow characteristics of the orifice. According to Pascal's law, pressure acts in all directions equally. When water is flowing through hose or pipes, the water is slowed by the contact it has with the walls of the hose or pipe, which results in **friction loss**. Bernoulli's equation further defines this friction loss as a drop in pressure while the water maintains its speed (velocity). The result of the interaction between the hose or pipe and the flowing water is a reduction in pressure at the point of discharge. The farther the water must flow through the hose or pipe, the greater the friction loss will be.

Knowing the static pressure, the flow in gallons (liters) per minute, and the residual pressure enables you to calculate the amount of water that can be obtained from a hydrant or a group of hydrants on the same water main.

Gravity continues to exert its effects even on flowing water based on the stream's elevation and altitude. Elevation, for you as a driver/operator, refers to the position of the pump (above or below) as compared to the nozzle. Pressure will increase when the nozzle is below the location of the pump and decrease when the nozzle is above the location of the pump. Atmospheric pressure changes depending on where the location is compared to sea level. Atmospheric pressure drops as altitude increases, and this drop in pressure causes pumps to work harder to generate the same pressures at lower altitudes. Atmospheric pressure decreases approximately 0.5 psi (3.4 kPa) for every 1000 feet (305 m) that the geographic location rises above sea level.

<u>Water hammer</u> is a pressure surge or wave caused by the kinetic energy of a fluid in motion when it is forced to stop or change direction suddenly. Quickly closing valves, nozzles, or

Safety Tip

Water hammer is a serious condition that can cause injuries on the fire ground. Improperly closing valves at the pump panel can result in hose lines bursting. Many driver/operators work right next to the hose lines that discharge from the pump. A burst line may injure the driver/operator as well as endanger crews operating in the fire-ground area. Opening and closing all valves slowly will prevent water hammer.

Voices of Experience

In the middle of the night, sometime around 0130 hours, a complement of two engines, a squad, a ladder, an ambulance, and a command officer were dispatched to a reported structure fire in an annex building of the Historic Arlington Park Race Track. Upon their arrival, the fire companies found a working fire in the wood-framed paddock building, which was attached to the five-story wood-frame grandstand. Over the years, these buildings had been renovated multiple times, which created multiple void spaces. The fire spread quickly from the annex to the grandstands and continued to burn throughout the night.

This fire occurred early in my career. I was assigned to the relief crew on the snorkel for shift change at 0800 hours. At that time, the fire had spread to the grandstands but was still being aggressively fought from interior positions. Heavy-timber building construction was on our side, but all of the remodeling work was working against us. The fire grew to eight alarms throughout the morning as more water was being used to fight the fire.

The property had a private water system that was immediately overcome. Units were used to pump water from the municipal water system into the private system and through extremely long relay operations. Those long hose lays challenged driver/operators to maintain hose line pressures. Engines were driven out across the dirt track to a large pond on the infield of the race track. Drafting operations were established as an additional water supply to the south of the building.

> "Despite the determined efforts of everyone on scene, the historic landmark building eventually collapsed onto itself and smoldered for days."

Millions of gallons of water from the three types of water sources were used on this fire. Despite the determined efforts of everyone on scene, the historic landmark building eventually collapsed onto itself and smoldered for days. The community, property owners, and the area fire service knew that they had done everything possible to save it. A strong reliable water source allowed crews to operate for hours in their attempt to save this vital piece of local history.

Bill Stipp
Goodyear Fire Department
Goodyear, Arizona

hydrants can create damaging pressure spikes, leading to blown diaphragms, seals, and gaskets and destroyed hoses, piping, and gauges. Liquid, for the most part, is not compressible; any energy that is applied to it is instantly transmitted back through it. This energy becomes dynamic in nature when a force such as quickly closing a valve or a nozzle applies velocity to the fluid. This is why nozzles and valves must be closed slowly and all large-diameter discharge gates are required to be slow-open and -close types of valves.

Fire Hydrants

Fire hydrants provide water for firefighting purposes. Public hydrants are part of the municipal water distribution system and draw water directly from the public water mains. Hydrants also can be installed on private water systems supplied by the municipal water system or from a separate source. The water source as well as the adequacy and reliability of the supply to private hydrants must be identified to ensure that they will be sufficient in fighting fires.

Most fire hydrants consist of an upright steel casing (barrel) attached to the underground water distribution system. The two main types of hydrants are the dry-barrel hydrant and the wet-barrel hydrant. Hydrants are equipped with one or more valves to control the flow of water through the hydrant. One or more outlets are provided to connect fire department hoses to the hydrant. These outlets are sized to fit the 2½" (51-mm), or larger fire hoses used by the local fire department.

Dry-Barrel Hydrants

<u>Dry-barrel hydrants</u> are used in climates where temperatures can be expected to fall below the freezing level. The valve that controls the flow of water into the barrel of this type of hydrant is located at the base, below the frost line, to keep the hydrant from freezing. The length of the barrel depends on the climate and the depth of the valve. Water enters the barrel of the hydrant only when it will be used. Turning the nut on the top of the hydrant rotates the operating stem, which opens the valve so that the water flows up into the barrel of the hydrant.

Whenever this kind of hydrant is not in use, the barrel must remain dry. If the barrel contains standing water, it will freeze in cold weather and render the hydrant inoperable. After each use, the water drains out through an opening at the bottom of the barrel. The drain is fully open when the hydrant valve is fully shut. When the hydrant valve is fully opened, the drain closes, which prevents water from being forced out of the drain when the hydrant is under pressure.

A partially opened valve means that the drain is also partially open, and pressurized water can flow out. This kind of drainage can erode (undermine) the soil around the base of the hydrant and may damage the hydrant. For this reason, a dry-barrel hydrant should always be either fully opened or fully closed. A fully opened hydrant also makes maximum flow available to fight a fire.

Most dry-barrel hydrants have only one large valve controlling the flow of water Figure 3-9 ▶ . Each outlet must be con-

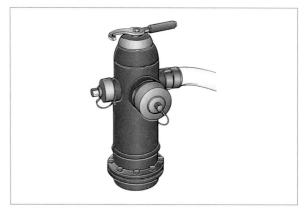

Figure 3-9 Most dry-barrel hydrants have only one large valve that controls the flow of water.

nected to a hose or an outlet valve, or have a hydrant cap firmly in place before the valve is turned on. Many fire departments use special hydrant valves so additional connections can be made after water is flowing through the first hose. If additional outlets are needed later, separate outlet valves can be connected before the hydrant is opened.

In some areas, vandals may dispose of trash or foreign objects in the empty barrels of dry-barrel hydrants. These materials can obstruct the water flow or damage a fire department pumper if they are drawn into the pump. To avoid this problem, many fire departments check the operation and flush out debris before connecting a hose to this type of hydrant. The fire fighter making the connection opens a large outlet cap and then releases the valve just enough to ensure that water flows into the hydrant and flushes out any foreign matter. This step should take just a few seconds. The fire fighter then closes the valve, connects the hose, and reopens the valve all the way. Fire departments should also perform regular inspections and tests to keep hydrants operating smoothly.

Wet-Barrel Hydrants

<u>Wet-barrel hydrants</u> are used in locations where the temperatures do not drop below the freezing mark. These hydrants always have water in the barrel and do not have to be drained after each use. Wet-barrel hydrants usually have separate valves that control the flow to each individual outlet Figure 3-10 ▶ . You can hook up one hose line and begin flowing water, and later attach a second hose line and open the valve for that outlet, without shutting down the hydrant.

Operation of Fire Hydrants

You must be proficient in operating a fire hydrant. Skill Drill 3-1 ▶ outlines the steps in obtaining water from a dry-barrel hydrant efficiently and safely. These same steps, with the modifications noted, apply to wet-barrel hydrants as well.

Figure 3-10 A wet-barrel hydrant has a separate valve for each outlet.

1. Remove the cap from the outlet you will be using. (**Step 1**)
2. Quickly look inside the hydrant opening for any objects that may have been thrown into the hydrant. (**Step 2**) (Omit this step for a wet-barrel hydrant.)
3. Check to ensure that the remaining hydrant caps are snugly attached. (**Step 3**) (Omit this step for a wet-barrel hydrant.)
4. Place the hydrant wrench on the stem nut. Check the top of the hydrant for the arrow indicating which direction to turn the nut to open the hydrant valve. (**Step 4**)
5. Open the hydrant just enough to determine that there is a good flow of water and to flush out any objects that may have been put into the hydrant. (**Step 5**) (Omit this step for a wet-barrel hydrant)
6. Shut off the flow of water. (**Step 6**) (Omit this step for a wet-barrel hydrant)
7. Attach the hose or valve to the hydrant outlet. (**Step 7**)
8. When instructed, start the flow of water. Turn the hydrant wrench to fully open the valve. This may take 12 to 14 turns, depending on the type of hydrant. (**Step 8**)
9. Open the hydrant slowly to avoid a pressure surge. Once the flow of water has begun, you can open the hydrant valve more quickly. Make sure that you open the hydrant valve completely. If the valve is not fully opened, the drain hole will remain open. (**Step 9**)

Note: This skill drill applies to a dry-barrel hydrant. For a wet-barrel hydrant, omit steps 2, 3, 5, and 6, and simply open the valve for the particular outlet that will be used.

Individual fire departments may vary the procedures they specify for opening a hydrant. For example, some fire departments specify that the wrench be left on the hydrant. Other fire departments require that the wrench be removed and returned to the fire apparatus so that an unauthorized person cannot interfere with the operation. Always follow the standard operating procedures (SOP) for your fire department.

Shutting Down a Hydrant

Shutting a hydrant down properly is just as important as opening a hydrant properly. If the hydrant is damaged during shutdown, it cannot be used until it has been repaired. Following the steps shown in **Skill Drill 3-2** will enable you to shut down a hydrant efficiently and safely.

1. Turn the hydrant wrench slowly until the valve is closed. (**Step 1**)
2. Allow the hose to drain by opening a drain valve or disconnecting a hose connection downstream. Slowly disconnect the hose from the hydrant outlet, allowing any remaining pressure to escape. (**Step 2**)
3. On dry-barrel hydrants, leave one outlet open until the water drains from the hydrant. (**Step 3**)
4. Replace the hydrant cap. (**Step 4**)

Note: Do not leave or replace the caps on a dry-barrel hydrant until you are sure that the water has completely drained from the barrel. If you feel suction on your hand when you place it over the opening, the hydrant is still draining. In very cold weather, you may have to use a hydrant pump to remove all the water and prevent the hydrant from freezing.

Driver/Operator Tip

Establishing an uninterruptible water supply is a critical fire-ground factor needed to ensure the safety of the fire company operating in a fire building. This can most efficiently be accomplished through teamwork within the fire company. A fire fighter dropped at a hydrant with the proper tools can begin the process of establishing the water supply while the other crew members are deploying handlines and you complete the connection to the engine. With teamwork, this step can be accomplished before the first handline is charged.

Locations of Fire Hydrants

Fire hydrants are located according to local standards and nationally recommended practices. Fire hydrants may be placed a certain distance apart, about every 500′ (152-m) in residential areas and every 300′ (91-m) in high-value commercial and industrial areas. In many communities, fire hydrants are located at every street intersection, with mid-block fire hydrants being established if the distance between intersections exceeds a specified limit.

Skill Drill 3-1

Operating a Fire Hydrant

1 Remove the cap from the outlet you will be using.

2 Quickly look inside the hydrant opening for foreign objects (dry-barrel hydrant only).

3 Check that the remaining caps are snugly attached (dry-barrel hydrant only).

4 Attach the hydrant wrench to the stem nut. Check for an arrow indicating the direction to turn to open.

5 Open the hydrant enough to verify the flow and flush the hydrant (dry-barrel hydrant only).

6 Shut off the flow of water (dry-barrel hydrant only.)

7 Attach the hose or valve to the hydrant outlet.

8 When instructed, turn the hydrant wrench to fully open the valve.

9 Open the hydrant slowly to avoid a pressure surge.

Skill Drill 3-2

Shutting Down a Hydrant

1. Turn the wrench to slowly close the hydrant valve.

2. Drain the hose line. Slowly disconnect the hose from the hydrant outlet.

3. Leave one hydrant outlet open until the hydrant is fully drained (dry-barrel hydrant only).

4. Replace the hydrant cap.

In some cases, the requirements for locating fire hydrants are based on occupancy, construction, and size of a building. A builder may be required to install additional fire hydrants when a new building is constructed so that no part of the building will be more than a specified distance from the closest fire hydrant.

Knowing the plan for installing fire hydrants makes them easier to find in emergency situations. Fire companies that perform fire inspections or develop preincident plans should identify the locations of nearby fire hydrants for each building or group of buildings as part of their survey.

Inspecting and Maintaining Fire Hydrants

Because fire hydrants are essential to fire suppression efforts, you must understand how to inspect and maintain them. Fire hydrants should be checked on a regular basis—no less than once a year—to ensure that they are in proper operating condition. During inspections, you may encounter some common problems and should know how to correct them.

The first factors to check when inspecting fire hydrants are visibility and accessibility. Fire hydrants should always be

Near Miss REPORT

Report Number: 09-0000676
Report Date: 07/14/2009 21:16

Synopsis: Loss of water during fire attack.

Event Description: In a working structural fire, while fire fighters were inside the house, the operating engine came out of pump gear. Several attempts were made to place the engine back in pump gear without success. A radio report was made for the loss of water to attack lines. With a delay, a second engine was used to continue pump operations.

Lessons Learned: To prevent this, follow the SOP's of having a second engine always pumping in series. This will not delay or decrease time without water. Also ensure regular maintenance on the truck. Regular training regarding SOP's and pump operation are important.

Demographics

Department type: Paid Municipal

Job or rank: Fire Fighter

Department shift: 24 hours on—48 hours off

Age: 25–33

Years of fire service experience: 0–3

Region: FEMA Region V

Service Area: Urban

Event Information

Event type: Fire emergency event: structure fire, vehicle fire, wildland fire, etc.

Event date and time: 11/22/2006 14:00

Event participation: Involved in the event

Weather at time of event: Clear and Dry

What were the contributing factors?
- Equipment
- SOP / SOG

What do you believe is the loss potential?
- Lost time injury
- Life-threatening injury
- Property damage

Figure 3-11 Hydrants should not be hidden or obstructed.

Figure 3-12 All fire hydrants should be checked at least annually.

visible from every direction, so they can be easily spotted. A fire hydrant should not be hidden by tall grass, brush, fences, debris, dumpsters, or any other obstructions Figure 3-11. In winter, fire hydrants must be clear of snow. No vehicles should be allowed to park in front of a fire hydrant.

In many communities, fire hydrants are painted in bright reflective colors for increased visibility. The bonnet (the top of the fire hydrant) may also be color-coded to indicate the available flow rate of a fire hydrant Table 3-1. Colored reflectors are sometimes mounted next to the fire hydrants or placed in the pavement in front of them to make them more readily visible at night.

Fire hydrants should be installed at an appropriate height above the ground. The outlets should not be so high or so low that the fire companies have difficulty connecting hose lines to them. NFPA 291, *Recommended Practice for Fire Flow Testing and Marking of Hydrants*, requires a minimum of 8″ (2-m) from the center of a hose outlet to the finish grade. Fire hydrants should be positioned so that the connections—and especially the large steamer connection—are facing the street.

During a fire hydrant inspection, check the exterior for signs of damage. Open the steamer port of each dry-barrel hydrant to ensure that the barrel is dry and free of debris. Make sure that all caps are present and that the outlet hose threads are in good working order Figure 3-12.

The second part of the inspection ensures that the hydrant is working properly. Open the hydrant valve just enough to ensure that water flows out and flushes any debris out of the barrel. After flushing, shut down the hydrant. A properly draining hydrant will create suction against a hand placed over the outlet opening Figure 3-13. When the hydrant is fully drained, replace the cap on a wet-barrel hydrant; leave the cap off a dry-barrel hydrant to ensure that it drains properly.

If the threads on the discharge ports need cleaning, use a steel brush and a small triangular file to remove any burrs in the threads. Also check the gaskets in the caps to make sure they are not cracked, broken, or missing. Replace worn gaskets with new ones, which should be carried with each apparatus. Always follow the manufacturer's recommendations for any parts that require lubrication.

Testing Fire Hydrants

The amount of water available to fight a fire at a given location is a crucial factor in planning an attack. Will the fire hydrants deliver enough water at the needed pressure to enable fire fighters

Table 3-1 Fire Hydrant Colors

NFPA 291 recommends that fire hydrants be color-coded to indicate the water flow available from each hydrant at 20 psi. It is recommended that the top bonnet and the fire hydrant caps be painted according to the following system, which provides an idea of how much water can be obtained from a fire hydrant during a fire.

Class	Flow Available at 20 psi	Color
Class C	Less than 500 GPM	Red
Class B	500 to 999 GPM	Orange
Class A	1000 to 1499 GPM	Green
Class AA	1500 GPM and higher	Light blue

Figure 3-13 Feel suction to indicate the fire hydrant is draining.

gauge. A <u>Pitot gauge</u> is used to measure flow pressure in psi (or kilopascals), and to calculate the flow in gallons (liters) per minute. As part of testing, fire fighters measure static pressure and residual pressure at one hydrant, and open the other to let water flow out. The two hydrants should be connected to the same water main and preferably at about the same elevation **Figure 3-14**.

The cap gauge is placed on one of the outlets of the first hydrant. The hydrant valve is then opened to allow water to fill the hydrant barrel. The initial pressure reading on this gauge is recorded as the static pressure. To obtain the static pressure, follow the steps in **Skill Drill 3-3**:

1. Remove the cap from the hydrant port, open the hydrant, and allow water to flow until it runs clear. Do your best to avoid property damage with the flowing water. **(Step 1)**
2. Close the hydrant valve. **(Step 2)**
3. Install a cap gauge on the port. **(Step 3)**
4. Open the hydrant valve fully to fill the barrel. No water should be flowing. **(Step 4)**
5. Note the pressure reading on the gauge; this is the static pressure of the system. **(Step 5)**
6. Close the hydrant fully.
7. Bleed down the pressure.
8. Remove the cap gauge and replace it with the hydrant cap. **(Step 6)**

At the second hydrant, fire fighters remove one of the discharge caps and open the hydrant. They put the Pitot gauge into the middle of the stream and take a reading. This value is recorded as the Pitot pressure. At the same time, fire fighters at the first hydrant record the residual reading.

Using the size of the discharge opening (usually 2½″ [51-mm]) and the Pitot pressure, fire fighters can calculate the flow in gallons per minute or look it up in a table; the table usually incorporates factors to adjust for the shape of the discharge opening. Fire fighters can use special graph paper or computer

to control a fire? If not, what can be done to improve the water supply? How can you obtain additional water if a fire does occur?

Fire-suppression companies are often assigned to test the flow from fire hydrants in their districts. The procedures for testing fire hydrants are relatively simple, but a basic understanding of the concepts of hydraulics and careful attention to detail are required. This section explains some of the basic theory and terminology of hydraulics and describes how the tests are conducted and the results are recorded.

■ Fire Hydrant Testing Procedure

The procedure for testing fire hydrant flows requires two adjacent hydrants, a Pitot gauge, and an outlet cap with a pressure

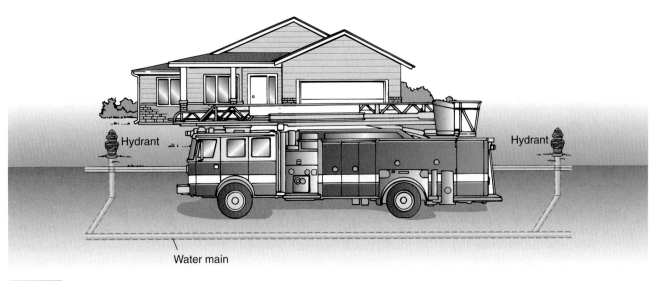

Figure 3-14 Testing hydrant flow requires two hydrants on the same water main.

Skill Drill 3-3

Obtaining the Static Pressure

1. Remove the cap from the hydrant port, open the hydrant, and allow water to flow until it runs clear.

2. Close the hydrant valve.

3. Install a cap gauge on the port.

4. Open the hydrant valve fully to fill the barrel. No water should be flowing.

5. Note the pressure reading on the gauge; this is the static pressure of the system.

6. Close the hydrant fully, bleed down the pressure, remove the cap gauge, and replace it with the hydrant cap.

Figure 3-15 The Pitot gauge.

Figure 3-16 Any accessible body of water can be used as a static source.

software to plot the static pressure and the residual pressure at the test flow rate. The line defined by these two points shows the number of gallons (liters) per minute that is available at any residual pressure. The flow available for fire suppression is usually defined as the number of gallons (liters) per minute available at 20 psi (140 kPa) residual pressure.

Several special devices are available to simplify the process of taking accurate Pitot readings. Some outlet attachments have smooth tips and brackets that hold the Pitot gauge in the exact required position Figure 3-15 ▲. The flow can also be measured with an electronic flow meter instead of a Pitot gauge.

To operate a Pitot gauge, follow the steps in Skill Drill 3-4 ▶:
1. Remove the cap from a hydrant port, preferably the 2½" (51-mm) discharge port. (**Step 1**)
2. Measure the inside diameter of the discharge port, and record the size. (**Step 2**)
3. Fully open the hydrant and allow water to flow. (**Step 3**)
4. Hold the Pitot tube into the center of the flow and place it parallel to the discharge opening at a distance one-half of the inside diameter of the discharge port. (**Step 4**)
5. Record the pressure reading. (**Step 5**)
6. Close the hydrant valve fully and replace the cap. (**Step 6**)

Rural Water Supplies

Many fire departments protect areas that are not serviced by municipal water systems. In these areas, residents usually depend on individual wells or cisterns to supply the water for domestic uses. Because there are no fire hydrants in these areas, fire fighters must depend on water from other sources. In rural areas, you must know how to get water from the sources that are available.

Static Sources of Water

Several potential static water sources can be used for fighting fires in rural areas. Both natural and human-made bodies of water such as rivers, streams, lakes, ponds, oceans, reservoirs, swimming pools, and cisterns can be used to supply water for fire suppression Figure 3-16 ▲. Some areas have many different static sources, whereas others have few or none at all. Water from a static source can be used to fight a fire directly, if it is close enough to the fire scene. Otherwise, it must be transported to the fire using long hose lines, engine relays, or mobile water supply tankers.

Static water sources must be accessible to a fire engine or portable pump. If a road or hard surface is located within 20′ (6-m) of the water source, a fire engine can drive close enough to draft water directly into the pump through a hard suction hose. Rural fire departments should identify these areas and practice establishing drafting operations at all of these locations. Some fire departments construct special access points so engines can approach the water source.

Dry hydrants also provide quick and reliable access to static water sources. A dry hydrant is a pipe with a strainer on one end and a connection for a hard suction hose on the other end. The strainer end should be placed below the water's surface and away from any silt or potential obstructions. The other end of the pipe should be accessible to fire apparatus, with the connection at a convenient height for an engine hook-up Figure 3-17 ▶. When a hard suction hose is connected to the dry hydrant, the engine can draft water from the static source.

Dry hydrants are often installed in lakes and rivers and close to clusters of buildings where there is a recognized need for fire protection. Dry hydrants also may be installed in farm cisterns or connected to swimming pools on private property to make water available for the local fire department. In some areas, dry hydrants are used to enable fire fighters to reach water under the frozen surface of a lake or river.

Skill Drill 3-4

Operating a Pitot Gauge

1. Remove the cap from a hydrant port, preferably the 2½" (51-mm) discharge port.

2. Measure the inside diameter of the discharge port, and record the size.

3. Fully open the hydrant and allow water to flow.

4. Hold the Pitot tube into the center of the flow and place it parallel to the discharge opening at a distance one-half of the inside diameter of the discharge port.

5. Record the pressure reading.

6. Close the hydrant valve fully and replace the cap.

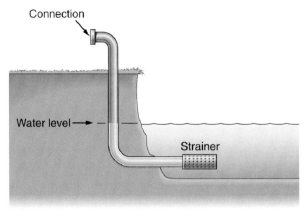

Figure 3-17 A dry hydrant or drafting hydrant can be placed at an accessible location near a static water source.

The portable pump is an alternative means of obtaining water in areas that are inaccessible to fire apparatus . The portable pump can be hand-carried or transported by an off-road vehicle to the water source. Portable pumps can deliver up to 500 gallons (1893 litres) of water per minute.

Figure 3-18 A portable pump can be used if the water source is inaccessible to a fire department engine.

Wrap-Up

Chief Concepts

- The basic plan for fighting most fires depends on having an adequate supply of water to confine, control, and extinguish the fire.
- Three types of fire hydrants are available: wet-barrel, dry-barrel, and dry hydrants.
- Fire hydrants are located according to local standards and nationally recommended practices.
- Fire fighters must be proficient in operating a fire hydrant, including being able to turn on and shut down a fire hydrant.

Hot Terms

<u>aqueous</u> Pertaining to, related to, similar to, or dissolved in water.

<u>distributors</u> Relatively small-diameter underground pipes that deliver water to local users within a neighborhood.

<u>dry-barrel hydrant</u> The most common type of hydrant; it has a control valve below the frost line between the footpiece and the barrel. A drain is located at the bottom of the barrel above the control valve seat for proper drainage after operation. (NFPA 25)

<u>dry hydrant</u> An arrangement of pipes that is permanently connected to a water source other than a piped, pressurized water supply system, provides a ready means of water supply for firefighting purposes, and utilizes the drafting (suction) capability of fire department pumpers. (NFPA 1144)

<u>elevated water storage tower</u> An above-ground water storage tank that is designed to maintain pressure on a water distribution system.

<u>elevation pressure</u> The amount of pressure created by gravity.

<u>flow pressure</u> The amount of pressure created by moving water.

<u>friction loss</u> The result of the interaction between the hose or pipe and the flowing water, which leads to a reduction in flow pressure at the point of discharge.

<u>gravity-feed system</u> A water distribution system that depends on gravity to provide the required pressure. The system storage is usually located at a higher elevation than the end users of the water.

<u>hardness</u> The mineral content of the water; it usually consists of calcium and magnesium, but can also include iron, aluminum, and manganese.

<u>heat of vaporization</u> The energy required to transform a given quantity of a substance into a gas.

<u>kinetic energy</u> The energy possessed by an object as a result of its motion.

<u>municipal water system</u> A water distribution system that is designed to deliver potable water to end users for domestic, industrial, and fire protection purposes.

<u>normal operating pressure</u> The observed static pressure in a water distribution system during a period of normal demand.

<u>Pitot gauge</u> A type of gauge that is used to measure the velocity pressure of water that is being discharged from an opening. It is used to determine the flow of water from a hydrant.

<u>potential energy</u> The energy that an object has stored up as a result of its position or condition. A raised weight and a coiled spring have potential energy.

primary feeder The largest-diameter pipes in a water distribution system, which carry the largest amounts of water.

private water system A privately owned water system that operates separately from the municipal water system.

reservoir A water storage facility.

residual pressure The pressure that exists in the distribution system, measured at the residual hydrant at the time the flow readings are taken at the flow hydrant. (NFPA 291)

secondary feeder A smaller-diameter pipe that connects the primary feeder to the distributors.

shut-off valve Any valve that can be used to shut down water flow to a water user or system.

specific heat index The amount of heat energy required to raise the temperature of a substance. These values are determined experimentally and then made available in tabular form.

static pressure The pressure in a water pipe when there is no water flowing.

static water source The pressure that exists at a given point under normal distribution system conditions measured at the residual hydrant with no hydrants flowing. (NFPA 291)

surface tension The elastic-like force at the surface of a liquid, which tends to minimize the surface area, causing water drops to form. (NFPA 1150)

turbidity The amount of particulate matter suspended in water.

universal solvent A liquid substance capable of dissolving other substances, which does not result in the first substance changing its state when forming the solution.

volume The quantity of water flowing; usually measured in gallons (liters) per minute.

water hammer The surge of pressure caused when a high-velocity flow of water is abruptly shut off. The pressure exerted by the flowing water against the closed system can be seven or more times that of the static pressure. (NFPA 1962)

water main A generic term for any underground water pipe.

water pressure The application of force by one object against another. When water is forced through the distribution system, it creates water pressure.

water supply A source of water that provides the flows (L/min) and pressures (bar) required by a water-based fire protection system. (NFPA 25)

wet-barrel hydrant A hydrant used in areas that are not susceptible to freezing. The barrel of this type of hydrant is normally filled with water.

Driver/Operator in Action

Your engine company is dispatched in response to a reported fire in an auto parts store located in a large strip mall. The first engine reports a working fire in the rear of the store but has set up in the front of the building. Your engine, which is second to arrive, is directed to the rear of the shopping center to establish a water supply and take a 1¾" (44-mm) line to the rear of the store. Lieutenant Bills spots the engine to the rear of the store within 75′ (23-m) of the nearest hydrant. He directs you to make your own hydrant connection after charging his line. Engine 3 is assigned to the rear and is directed to take a second 1¾" (44-mm) handline from your engine as a backup to your crew.

1. When you reach the hydrant to establish your water supply, you should open the hydrant to flow water before making your connections to ensure that
   A. The hydrant works.
   B. All foreign matter is flushed out.
   C. You are turning the hydrant stem in the right direction.
   D. The drain works before using it.

2. Which task, establishing the hydrant connection or charging the second line, should be completed first and why?
   A. The hydrant connection—because that was the first thing you started to do.
   B. Charging the second line—because command ordered it.
   C. Charging the second line—because it is being used to back up your crew.
   D. The hydrant connection—because if you run out of water with a crew on the interior, they might get hurt.

After several hours, it appears the fire is under control. Lieutenant Bills calls you on the radio and tells you to shut down his line immediately without any further details.

3. You should close the discharge valve slowly despite the immediacy of the request
  A. To ensure that you didn't misunderstand the request of the company officer from Engine 3.
  B. To prevent water hammer on the municipal water system.
  C. To prevent water hammer on the pump and hose.
  D. You should shut down the line quickly, just as you were ordered.

4. You can check whether a dry-barrel hydrant is draining water by
  A. Using a flashlight and looking through an open port.
  B. Attaching a Pitot gauge and checking for a vacuum.
  C. Holding your hand over the open port to feel for suction.
  D. Attaching a hydrant pump and creating a vacuum.

Mathematics for the Driver/Operator

CHAPTER 4

NFPA 1002 Standards

5.2.1 Produce effective hand or master streams, given the sources specified in the following list, so that the pump is engaged, all pressure control and vehicle safety devices are set, the rated flow of the nozzle is achieved and maintained, and the apparatus is continuously monitored for potential problems:
(1) Internal tank
(2) Pressurized source
(3) Static source
(4) Transfer from internal tank to external source [p. 59–102]

(A) Requisite Knowledge. Hydraulic calculations for friction loss and flow using written formulas and estimation methods, safe operation of the pump, problems related to small-diameter or dead-end mains, low-pressure and private water supply systems, hydrant coding systems, and reliability of static sources. [p. 63–102]

(B) Requisite Skills. The ability to position a fire department pumper to operate at a fire hydrant and at a static water source, power transfer from vehicle engine to pump, draft, operate pumper pressure control systems, operate the volume/pressure transfer valve (multistage pumps only), operate auxiliary cooling systems, make the transition between internal and external water sources, and assemble hose lines, nozzles, valves, and appliances.

5.2.4 Supply water to fire sprinkler and standpipe systems, given specific system information and a fire department pumper, so that water is supplied to the system at the correct volume and pressure.

(A) Requisite Knowledge. Calculation of pump discharge pressure; hose layouts; location of a fire department connection; alternative supply procedures if the fire department connection is not usable; operating principles of sprinkler systems as defined in NFPA 13, NFPA 13D, and NFPA 13R; fire department operations in sprinklered properties as defined in NFPA 13E; and operating principles of standpipe systems as defined in NFPA 14. [p. 60–102]

(B) Requisite Skills. The ability to position a fire department pumper to operate at a fire hydrant and at a static water source, power transfer from vehicle engine to pump, draft, operate pumper pressure control systems, operate the volume/pressure transfer valve (multistage pumps only), operate auxiliary cooling systems, make the transition between internal and external water sources, and assemble hose line, nozzles, valves, and appliances.

Additional NFPA Standards

NFPA 14 *Standard for the Installation of Standpipe and Hose Systems* (2007)
NFPA 25 *Standard for the Inspection, Testing, and Maintenance of Water-Based Fire Protection Systems* (2008)
NFPA 99 *Standard for Health Care Facilities* (2005)
NFPA 600 *Standard on Industrial Fire Brigades* (2005)
NFPA 921 *Guide for Fire and Explosion Investigations* (2008)
NFPA 1500 *Standard on Fire Department Occupational Safety and Health Program* (2007)
NFPA 1620 *Recommended Practice for Pre-incident Planning* (2003)
NFPA 1901 *Standard for Automotive Fire Apparatus* (2009)

Fire and Emergency Services Higher Education (FESHE) Model Curriculum

National Fire Science Degree Programs Committee (NFSDPC): Associate's Curriculum

Fire Protection Hydraulics and Water Supply

1. Apply the application of mathematics and physics to movement of water in fire suppression activities. [p. 59–102]
2. Identify the design principles of fire service pumping apparatus.
3. Analyze community fire flow demand criteria.
4. Demonstrate, through problem solving, a thorough understanding of the principles of forces that affect water at rest and, in motion. [p. 59–102]
5. List and describe the various types of water distribution systems.
6. Discuss the various types of fire pumps.

Knowledge Objectives

After studying this chapter, you will be able to:
- List the elements needed to calculate pump discharge pressure.
- Describe the concepts underlying theoretical hydraulic calculations.
- Describe the concepts underlying fire-ground hydraulic calculations.
- Describe how to utilize fire-ground hydraulic calculations during an incident.

Skill Objectives

After studying this chapter, you will be able to perform the following skills:

- Calculate the smooth-bore nozzle flow.
- Calculate the friction loss in single hose lines.
- Calculate the friction loss in multiple hose lines.
- Calculate the elevation pressure loss and gain.
- Calculate the friction loss in an appliance.
- Use a Pitot gauge to test the friction loss in a portable master stream appliance through 3″ (76-mm) hose.
- Use in-line gauges to test the friction loss in a specific hose.
- Determine the pump discharge pressure in a wye scenario with equal lines.
- Determine the pump discharge pressure in a wye scenario with unequal lines.
- Determine the pump pressure for a Siamese connection line by the split flow method.
- Calculate the friction loss in Siamese connection lines by coefficient.
- Calculate the friction loss in Siamese connection lines by percentage.
- Calculate the pump discharge pressure for a pre-piped elevated master stream device.
- Calculate the pump discharge pressure for an elevated master stream.
- Calculate the pump discharge pressure for a standpipe during preplanning.
- Calculate the pump discharge pressure for a standpipe.
- Calculate the flow rate (GPM) for a given hose size and nozzle pressure using a slide rule calculator.
- Calculate the friction loss for a given hose size and nozzle pressure using a slide rule calculator.
- Perform calculations by hand.
- Perform the hand method of calculation for 2½″ (51-mm) hose.
- Perform the hand method of calculation for 1¾″ (44-mm) hose.
- Perform the subtract 10 method of calculation.
- Perform the condensed Q method of calculation.

You Are the Driver/Operator

You have been driving the fire apparatus for some time, learning how it maneuvers and becoming accustomed to its size. You have memorized the items in each compartment, you know what each item of equipment is used for, and you know how to use all the tools and equipment. You have started to learn the territory and to memorize alternate routes. In your classes, you were taught basic fire apparatus placement and strategy and tactics, and you have leaned on the company officer and other fire fighters to help you over the hurdles you encountered. You know that soon you will have to perform these tasks alone.

You go over hose and nozzle sizes and flows in your mind. There's a lot of math to learn! You remember your instructor repeating over and over, "Flow versus heat." If you are lucky, your fire officer will take you out and let you practice these calculations in a training scenario somewhere. If you are lucky, your first time actually pumping in the field will be a small incident—for example, a dumpster or trash fire. It's good to get through any nervousness on something small before going into a large working fire.

Then it happens! The tones sound. The other members of your crew begin suiting up. The company officer confirms that you know how to get to the scene. All too soon, you arrive at the incident. "Catch the hydrant!" a voice yells. The fire fighter in back steps out and wraps the hydrant. The company officer gives you a quick order on fire apparatus placement and hose line selection. You pull slowly away and move into position. Because your apparatus is the first-due engine, no one else has arrived on the scene. Anxiety briefly rises within you as you see flames coming from a window and hear people screaming. You set the brake, put the apparatus into pump mode, and dismount the vehicle. You throw the chock under the tire, make the supply connection to the intake, and signal to send the water. The officer and fire fighter pull an attack line and are waiting at the door for you to charge the line.

Now you realize what you have trained for. You are all alone; there is no one to help or remind you how to set the pump. You have to reach inside, look past the adrenalin, and focus on your job. At this point, you have to figure the correct pump discharge pressure (PDP). As other companies arrive, they will pull hose lines and advance into the burning structure. Their lives and safety will be affected by what you do now.

1. How do you determine that you have an adequate water supply?
2. What are the nozzle sizes and flows for the attack lines?
3. What are the nozzle sizes and flows for the master stream devices?

Introduction

Fire service hydraulic calculations are used to determine the required **pump discharge pressure (PDP)** for fire-ground operations. As the driver/operator, you have a great deal of responsibility at each incident. You must maneuver the fire apparatus through traffic and weather and arrive safely at the correct address. You must secure an adequate water supply and position the fire apparatus to strategic advantage. With the strategy and tactics selected, you must apply the appropriate hose lines or combination of hose lines and/or master stream devices to provide enough water to overcome the heat produced by the fire. The hose lines must be charged quickly and at the correct pump discharge pressure (PDP).

Clearly, you as the driver/operator are critical to the success of the firefighting attack. Your actions are also critical to ensuring the safety of the attack team, any occupants within the fire-involved structure, the personnel on the scene, and any exposures. The position of driver/operator is a very serious role whose decisions are often a matter of life and death.

The fundamentals of theoretical hydraulic calculations are a necessary building block in the professional fire fighter's education. Although many concepts, formulas, and constants are used in hydraulic calculations, hydraulics can be described as an art rather than a science because of the large number of variables that influence the ultimate decisions made. Calculations are influenced by specific factors such as the apparatus manufacturer's design specifications; the characteristics of the fire apparatus, hose, nozzles, appliances, adapters, and couplings; and the relevant departmental policies. The way in which these influences interact is specific to each fire department; as a consequence, no standard set of rules will apply. You must know the requirements of the equipment on your fire apparatus to make the appropriate calculations correctly. This chapter will cover the most common calculations that you will need to know as a driver/operator.

Fire service **hydraulics** is the study of the characteristics and movement of water as they pertain to calculations for fire streams and fire-ground operations. These calculations are generally categorized as theoretical hydraulics and fire-ground hydraulics. The scientific or more exact calculations are commonly referred to as **theoretical hydraulics**. The concepts underlying this field are fundamental to the understanding of fire streams and enable you to correctly calculate pump discharge pressures for fire streams and increase your ability to troubleshoot problems and perform professionally on the fire ground. Fire-ground operations are very dynamic, however, and situations can change very quickly. **Fire-ground hydraulics** is the term for the less exact, but certainly more user-friendly and forgiving, calculation methods used on the fire ground.

While much of hydraulics is scientific, much is inexact. The balance is an art. The major variable affecting hydraulics is the many numerical values influencing calculations.

Manufacturers offer a wide variety of nozzles, appliances, and hose, each with individual specifications that may vary from those identified on traditional charts. Different brands of hose will flow slightly more or less water than other brands, and different nozzles will vary in their performance. To be as accurate as possible when performing hydraulics calculations, you should use the manufacturer's recommended nozzle pressure specific to the actual nozzle used on your hose or appliance, with your fire apparatus and under the guidelines of your fire department. Flow tests with your equipment will confirm the most precise pressures to use.

As the driver/operator, you apply your knowledge of the specific equipment on your fire apparatus to the methodology underlying hydraulic theory. There is an art to figuring out this relationship, however, and you must be prepared to adjust for the variables in the scenario at hand. Practice and experience with hydraulics calculations are imperative so that you can perform as a skilled driver/operator.

Driver/Operator Tip

The driver/operator works alone most of the time; however, sometimes teamwork may be used. A new driver/operator may arrive first on the fire ground. If the second-arriving engine has a more experienced driver/operator who does not have a more pressing assignment, it is often a good idea for the experienced driver/operator to assist the less experienced operator. This can be an unwritten policy or determined by the company officer.

Driver/Operator Tip

When two or more fire pumpers are in use, a water supply officer may be assigned to verify adequate supply and support operations.

Pump Discharge Pressure

Fighting fire with water is a matter of water flow (gallons per minute [GPM] or liters per minute [L/min]) versus heat generation (**British thermal units [Btu]** or kilocalories [kcal]). There must be sufficient flow applied to overcome the heat generated by the fire—a rate referred to as the **critical rate of flow**. For this reason, one of the most important attack decisions after securing the water supply is the proper hose and nozzle selection to produce adequate flow to extinguish the fire. The hose line or lines then must be placed strategically, using the correct tactics, so as to protect lives and confine and extinguish the fire.

You supply the hose lines with the optimal PDP. Too great a pressure will cause the stream to break up and lessen its effectiveness. It will also create a greater nozzle reaction, which will hinder the attack team's ability to advance the hose line and cause the team members to become fatigued more quickly. Inadequate pressure will produce insufficient flow to overcome the fire, possibly endangering the safety of the attack team. It may also create more kinks in the hose line, further restricting the water flow. Each hose and nozzle combination has an optimal delivery pressure. As the driver/operator, you must make the necessary calculations to determine the correct PDP for the scenario to be supplied.

The PDP is the total pressure needed to overcome all friction, appliance, and elevation loss while maintaining adequate nozzle pressure to deliver effective fire streams. Several factors must be addressed to achieve this goal, including nozzle pressure, friction loss in hose lines, elevation gain or loss, and friction loss in appliances. Calculations lengthen relative to the complexity of the fire attack operation, but all must start with the basic PDP formula.

The basic PDP formula is

$$PDP = NP + FL$$

Where: PDP = pump discharge pressure
NP = nozzle pressure
FL = friction loss

Safety Tip

Always place wheel chocks to prevent the fire apparatus from rolling when pumping. There have been cases when the fire apparatus slipped out of pump and into drive, causing it to lurch forward.

Nozzle Pressures

To determine the PDP, use the formula **PDP = NP + FL**. Next you must establish the nozzle pressure (NP) from the scenario to inject into the formula. Nozzle pressure is the pressure required at the nozzle to deliver the fire stream and flow rate for which the nozzle was designed. This pressure is defined by the manufacturer of the nozzle and is determined through testing to deliver an efficient amount of water while maintaining an NP at or below safe thresholds. Combination nozzles (fog nozzles) traditionally have a NP of 100 pounds per square inch (psi) (700 kilopascal [kPa]), but come with different features that have varied NP levels. Low-pressure fog nozzles have a NP that varies from 50 to 75 psi (350 to 525 kPa) Figure 4-1. Smooth-bore nozzles, also referred to as solid-tip nozzles, have a NP of 50 psi (350 kPa) for handlines and 80 psi (560 kPa) for master streams Figure 4-2. Smooth-bore nozzles on a handline are normally flowed at a NP of 50 psi (350 kPa). Some master stream smooth-bore nozzles are rated at less than 80 psi (560 kPa), however Figure 4-3. The actual maximum flow from a master stream nozzle may decrease when the nozzle is detached from the fire apparatus and used from the portable base Figure 4-4.

The manufacturer's specifications for maximum water flow from an aerial device are listed along with the nozzle specifications. If the nozzle is rated at a higher flow than the aerial can handle, then fire fighters should limit the nozzle to the capability of the aerial; usually, this scenario arises only with older aerial devices. Unless otherwise directed, limit the nozzle pressure for smooth-bore nozzles on elevated master streams to a maximum of 80 psi (560 kPa).

Broken-stream nozzles are available in both fog and smooth-bore varieties. Fog distributors are generally rated at 100 psi (700 kPa) and smooth-bore nozzles at 50 psi (350 kPa) Figure 4-5. Distributor nozzles constantly spin and present centrifugal forces that are symmetrical. Because these forces occur perpendicular

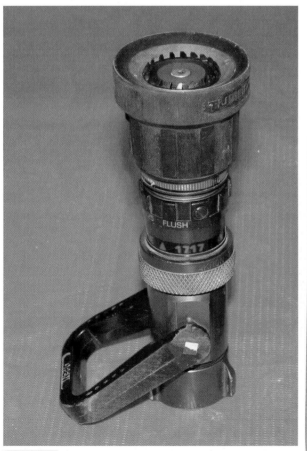

Figure 4-1 Low-pressure fog nozzles are typically rated at a nozzle pressure of 50 to 75 psi (350 to 525 kPa).

Figure 4-2 Smooth-bore handline nozzles are commonly rated at a nozzle pressure of 50 psi (350 kPa).

Figure 4-3 Smooth-bore nozzles on prepiped elevated master streams are commonly rated at a nozzle pressure of 80 psi (560 kPa).

Figure 4-4 Master stream devices that are attached to a deck gun are commonly rated at a higher flow (GPM) than devices that are attached to the portable base because the truck provides added stability to counteract nozzle reaction.

Figure 4-5 Fog distributor nozzles are commonly rated at a nozzle pressure of 100 psi (700 kPa).

to the hose, they do not produce a corresponding nozzle reaction. For this reason, smooth-bore distributors normally rated at 50 psi (350 kPa) should not present a safety hazard to flow water at increased pressures up to 100 psi (700 kPa).

It would be difficult to attempt instruction using every type of product on the market. Therefore this text will work from the most common specifications to encourage you to learn the fundamental concepts of hydraulic calculations. Once you understand the fundamentals, you will be able to master the variables you encounter on the actual fire ground.

Three standard nozzle pressures (SNPs), all of which are dictated by the type and use of the nozzle, are generally sufficient for most fire-ground operations. For the purpose of this text, we will use these three SNPs:

- 100 psi (700 kPa) NP for all fog nozzles
- 50 psi (350 kPa) NP for smooth-bore handline nozzles
- 80 psi (560 kPa) NP for smooth-bore master stream nozzles

Determining Nozzle Flow

<u>Flow rate</u> is the volume of water moving through the nozzle; it is measured in units of either GPM or L/min. Fog nozzles are designed with predetermined flow rates based on a set NP. For example, a fog nozzle might have the specification that it will flow 175 GPM (660 L/min) at a NP of 100 psi (700 kPa). The flow and pressure will differ by design and purchaser's selection (i.e., it will be more or less flow at a greater or lesser pressure). A label describing the performance ratio of flow to pressure can be found on the face of the nozzle.

To determine the flow rate of a smooth-bore nozzle, you need to employ your mathematical skills. Common smooth-bore nozzle flows are listed in Table 4-1. To calculate the flow (GPM or L/min) from smooth-bore nozzles, use the formula presented in Skill Drill 4-1:

1. Determine the diameter of the nozzle tip (it is usually stamped into the side of the nozzle). Convert the diameter fraction to a decimal and square it. For example:

$$(1\tfrac{1}{4}")^2 = (1.25)^2$$

Standard Table 4-1 **Smooth-Bore Nozzle Flows**

Common Smooth-Bore Nozzle Pressures by Application and Their Square Root

Application	Nozzle Pressure	Square Root
Handline	50	7.07
Master stream	80	8.94

Smooth-Bore Nozzle Flows

Tip Size	At 50 psi	At 65 psi	At 80 psi	Round to
¾"	118	135		
⅞"	162	183		
15⁄16"	185	210		
1"	210	239		
1⅛"	266	303		
1¼"	328		414	400
1⅜"			502	500
1½"			597	600
1¾"			813	800
2"			1062	1000

Smooth-bore nozzle tips for a 2½" handline may be rounded off as follows:
1" = 209 GPM; round to 200 GPM
1⅛" = 266 GPM; round to 250 GPM
1¼" = 326 GPM; round to 300 GPM

Metric Table 4-1 **Smooth-Bore Nozzle Flows**

Common Smooth-Bore Nozzle Pressures by Application and Their Square Root

Application	Nozzle Pressure (kPa)	Square Root
Handline	350	18.7
Master stream	560	23.7

Smooth-Bore Nozzle Flows

Tip Size (mm)	At 350 kPa	At 450 kPa	At 560 kPa	Round to
19	446	505		
22	598	678		
24	711	806		
25	772	875		
29	1038	1177		
32	1264		1599	1600
35			1913	1900
38			2255	2250
45			3163	3150
50			3905	3900

Smooth-bore nozzle tips for a 65-mm handline may be rounded off as follows:
25 mm = 772 L/min; round to 750 L/min
29 mm = 1038 L/min; round to 1050 L/min
32 mm = 1264 mm; round to 1250 L/min

2. Find the square root of the desired nozzle pressure ($\sqrt{NP}$).
3. Multiply the diameter squared (d^2) times the square root of the nozzle pressure ($\sqrt{NP}$), and then multiply the result by the constant 29.7 to find the GPM flow of the nozzle.

To calculate the flow from smooth-bore nozzles using metric measurements, follow these steps **Figure 4-6**:

1. Determine the diameter of the nozzle tip (it is usually stamped into the side of the nozzle) and square it. For example:

$$(32)^2$$

2. Find the square root of the desired nozzle pressure ($\sqrt{NP}$). Note that $\sqrt{350}$ kPa = 18.7 and $\sqrt{560}$ kPa = 23.7 for future reference.
3. Multiply the diameter squared (d^2) times the square root of the nozzle pressure ($\sqrt{NP}$), and then multiply the result by the constant 0.066 to find the L/min flow of the nozzle.

Driver/Operator Tip

Rule of thumb: the nozzle should never be larger than half the diameter of the hose.

Driver/Operator Tip

A secured water supply is essential, but it is wise to think about an escalated situation that may require a secondary or supplemental water supply. Make a habit of thinking ahead.

Friction Loss

After determining the correct NP, you must next determine the friction loss. Friction loss (FL) is the pressure lost from turbulence as water passes through pipes, hoses, fittings, adapters, and appliances. Friction loss is measured in units of either psi or kPa.

It is important to know that FL calculations know no bounds. You can continue calculating with no end in sight;

Figure 4-6 Smooth-bore distributor nozzles are commonly rated at a nozzle pressure of 50 psi (350 kPa).

Skill Drill 4-1

Standard (5.2.1)

Calculating the Flow of a 1¼″ Smooth-Bore Nozzle on a 2½″ Handline

Example 1: Calculate the flow rate for a 1¼″ tip used on a 2½″ handline.

GPM = 29.7 × d^2 × $\sqrt{NP}$
GPM = 29.7 × (1 1/4)² × $\sqrt{50}$
GPM = 29.7 × (1.25)² × $\sqrt{50}$
GPM = 29.7 × 1.56 × 7.07
GPM = 327.567 (round to 300 GPM)

Example 2: Calculate the flow rate for a 1½″ smooth-bore master stream nozzle.

GPM = 29.7 × d^2 × $\sqrt{NP}$
GPM = 29.7 × (1.5)² × $\sqrt{80}$
GPM = 29.7 × 2.25 × 8.94
GPM = 597.415 (round to 600 GPM)

Skill Drill 4-1

Metric (5.2.1)

Calculating the Flow of a 32-mm Smooth-Bore Nozzle on a 64-mm Handline

Example 1: Calculate the flow rate for a 32-mm tip used on a 64-mm handline.

L/min = 0.066 × d^2 × $\sqrt{NP}$
L/min = 0.066 × (32)² × $\sqrt{350}$
L/min = 0.066 × 1024 × 18.7
L/min = 1263.821 (round to 1250 L/min)

Example 2: Calculate the flow rate for a 38-mm smooth-bore master stream nozzle.

L/min = 0.066 × d^2 × $\sqrt{NP}$
L/min = 0.066 × (38)² × $\sqrt{560}$
L/min = 0.066 × 1444 × 23.7
L/min = 2258.7 (round to 2250 L/min)

however, reality says there are limitations to FL. For example, you might calculate a FL for 500 GPM (1900 L/min) through 1¾″ (44-mm) hose; you will get an answer, but it is evident that this is not a realistic expectation because 1¾″ (44-mm) hose cannot actually flow that amount of water. If you exceed 50 psi (350 kPa) of FL per 100′ (30 m) of hose, then the flow may begin to decrease either from the increasing turbulence in the hose or from a reduction in the pump capacity owing to the excessive PDP.

Over the years, many equations and other means of calculating FL have been suggested, most of which were designed for specific hoses and equipment. As fire service equipment improved, hydraulic calculations reflected the changes and evolved into the modern FL equation: **FL = C Q² L**. This equation first appeared in the *NFPA Fire Protection Handbook*, fifteenth edition, in 1981; it is now widely accepted as *the* fire service standard. The success of this equation is due to its ability to adapt to changes in hose diameter, water flow, and length of hose.

Standard Table 4-2 Friction Loss Coefficients

Friction Loss Coefficients	
Hose Diameter	Coefficient
¾"	1100
1"	150
1¼"	80
1½"	24
1¾"	15.5
2"	8
2½"	2
3" with 2½" couplings	0.8
3" with 3" couplings	0.67
3½"	0.34
4"	0.2
4½"	0.1
5"	0.08
6"	0.05
Siamese 2½" Hose	
Two	0.5
Three	0.22
Four	0.12
Siamese 3" Hose with 2½" Couplings	
Two	0.2
Three	0.09

Metric Table 4-2 Friction Loss Coefficients

Metric Friction Loss Coefficients	
Hose Diameter (mm)	Coefficient
19	1736
25	237
32	126
38	38
44	24.5
51	12.6
64	3.16
76 mm with 65-mm couplings	1.26
76 mm with 77-mm couplings	1.07
89	0.54
102	0.315
115	0.158
127	0.126
152	0.079
Siamese 64-mm Hose	
Two	0.789
Three	0.347
Four	0.189
Siamese 76-mm Hose with 65-mm Couplings	
Two	0.316
Three	0.142

The coefficients in Table 4-2 were derived from the standard measures using the following formula:

$$C_m = C_s \times (6.894757)/[(3.785412)^2 \times 0.3048]$$
$$= C_s \times 1.57862$$

Where: C_m = coefficient (metric)
C_s = coefficient (standard)
Given that: 1 psi = 6.894757 kPa
1 GPM = 3.785412 L/min
1' = 0.3048 m

Calculating Friction Loss

The friction loss formula in its simplest form is expressed as

$$FL = C Q^2 L$$

Where: FL = friction loss
C = the **coefficient**, a numerical measure that is a constant for each specific hose diameter **Table 4-2**
Q = the *quantity* of water flowing (GPM or L/min) divided by 100
L = the *length* of hose in feet divided by 100, or in meters divided by 100

The FL formula may be expressed as $FL = C \times (Q/100)^2 \times L/100$. Given that driver/operators use the formula so frequently, they typically divide Q by 100 and L by 100 mentally to simplify and shorten the written form of the formula to **FL = C Q² L**.

Table 4-2 lists the FL coefficients used in the FL formula. Although this list of coefficients is both typical and reliable, you must be aware of the improvements in hose technology and use the manufacturer's recommendations for these values when purchasing new hose and equipment.

To determine the FL in a hose lay, follow the steps in **Skill Drill 4-2**:

1. Write the FL formula: $FL = C Q^2 L$.
2. Select the coefficient (C) from Table 4-2 that matches the hose in the given scenario.
3. The coefficient for 2½" hose is 2. Therefore C = 2.
4. Determine the flow rate (quantity) of water through the hose line and divide the amount by 100 to find Q. Next, square Q.
5. The flow is determined by knowing the flow of each nozzle. We know that a 1¼" handline nozzle tip will flow 328 GPM. On paper you may use 328, but on the fire ground you may round this value off to 300 GPM. If you are flowing 300 GPM, then 300/100 = 3.
6. Because Q = 3, $Q^2 = 3^2 = 9$.
7. Determine the length of the hose you are calculating and then divide by 100 to find L. How long is the hose line? If it is 200' long, then 200/100 = 2. Therefore L = 2.
8. Multiply the results from steps 1, 2, and 3 to determine FL:

$$FL = C \times Q^2 \times L$$
$$FL = 2 \times 9 \times 2$$
$$FL = 36$$

Skill Drill 4-2

Calculating Friction Loss in Single Hose Lines

Example 1: What is the PDP for 300′ of 2½″ hose with a 1⅛″ smooth-bore nozzle?

In this problem, we see a smooth-bore nozzle on a handline, so we know to assign it a NP of 50 psi:

PDP = NP + FL
PDP = 50 + FL

Now we must calculate FL:
- For a 2½″ hose, the coefficient is 2 (C = 2).
- The nozzle has a 1⅛″ tip, which we know will flow 250 GPM (rounded).
- We know the nozzle flow from Table 4-1 or by the formula, GPM = 29.7 d^2 $\sqrt{NP}$.
- The flow is 250 GPM, so 250/100 = 2.5 and Q = 2.5.

Insert these values into the formulas:

FL = C Q^2 L
FL = 2 × $(2.5)^2$ × 300/100
FL = 2 × 6.25 × 3
FL = 37.5 psi

Now we return to our PDP formula and insert the FL:

PDP = NP + FL
PDP = 50 + 37.5
PDP = 87.5 psi

Example 2: Calculate the FL in 200′ of 1¾″ hose with a ⅞″ smooth-bore nozzle.

A smooth-bore handline has a NP of 50 psi, so NP = 50. A ⅞″ tip flows 161 GPM, so Q = 1.6.

PDP = NP + FL   FL = C Q^2 L
PDP = 50 + FL   FL = 15.5 × $(1.6)^2$ × 2
PDP = 50 + 80   FL = 15.5 × 2.56 × 2
PDP = 130 psi   FL = 79.36 psi (round to 80 psi)

Calculating PDP requires the operator to determine FL so as to arrive at the PDP.

Multiple Hose Lines of Different Sizes and Lengths

On the fire ground, you will undoubtedly encounter a scenario where multiple water pressures are needed from one fire pumper on the fire scene. This may not sound like a problem until you understand that only one pressure—the highest pressure—can be created by the fire pumper. Knowing this fact, you need to know how to control the water pressures on the discharges requiring less pressure. When flowing two identical lines that have the same nozzle pressure and flow and are of equal size, length, and elevation, both lines require the same pressure. Operating lines of unequal size and length with different nozzles and flows can be challenging.

Under normal conditions, you always fully open a valve to prevent excess turbulence and FL through the valve; however, situations may occur where partially closing a valve is necessary to prevent excessive pressure from being delivered. For those discharges requiring a lesser pressure, you must open the valve just enough to deliver the desired pressure.

For example, suppose you have three discharges flowing. The first requires 160 psi (1100 kPa), the second requires 130 psi (900 kPa), and the third requires 200 psi (1400 kPa). First, you open discharge number 3 to establish a pressure of 200 psi (1400 kPa)—the highest pressure—on the master discharge gauge. This will adequately supply all three discharge valves but will over-pressurize the first and second discharge valves. Next, you

Chapter 4 Mathematics for the Driver/Operator

Metric (5.2.1)

Skill Drill 4-2

Calculating Friction Loss in Single Hose Lines

Example 1: What is the PDP for 90 m of 64-mm hose with a 29-mm smooth-bore nozzle?

In this problem, we see a smooth-bore nozzle on a handline, so we know to assign it a NP of 350 kPa:

PDP = NP + FL
PDP = 350 + FL

Now we must calculate FL:
- For a 64-mm hose, the coefficient is 3.16 (C = 3.16).
- The nozzle has a 29-mm tip, which we know will flow 1050 L/min (rounded).
- We know the nozzle flow from Table 4-1 or by the formula, L/min = 0.066 d^2 $\sqrt{NP}$.
- The flow is 1050 L/min, so 1050/100 = 10.5 and Q = 10.5.

Insert these values into the formulas:

FL = C Q^2 L
FL = 3.16 × (10.5)^2 × 90/100
FL = 3.16 × 110 × 0.9
FL = 313 (round to 300 kPa)

Now we return to our PDP formula and insert the FL:

PDP = NP + FL
PDP = 350 + 300
PDP = 650 kPa

Example 2: Calculate the FL in 60 m of 44-mm hose with a 22-mm smooth-bore nozzle.

A smooth-bore handline has a NP of 350 kPa; therefore NP = 350. A 22-mm tip flows 598 L/min; therefore Q = 5.98 (round to Q = 6).

PDP = NP + FL   FL = C Q^2 L
PDP = 350 + FL   FL = 24.5 × (6)^2 × 0.6
PDP = 350 + 530   FL = 24.5 × 36 × 0.6
PDP = 880 kPa   FL = 529.2 kPa
                     (round to 530 kPa)

---

slowly open discharge valves 1 and 2 until the desired pressure is reached. After you open one discharge and start flowing, any additional discharge you flow will lower the pressure of lines already open, so be sure to increase the throttle as additional lines are opened to maintain the original pressure.

To calculate the FL for multiple hose lines of different size and length, follow the steps in **Skill Drill 4-3**:

1. Determine the correct PDP for each line in the scenario.
2. Use the pump discharge formula: PDP = NP + FL (nozzle pressure + friction loss). Also use the FL formula: FL = C Q^2 L, where C = coefficient for the line diameter, Q^2 = quantity of flow (GPM or L/min) divided by 100 and squared, and L = length divided by 100.
3. Determine which line requires the highest pressure. This pressure is the correct answer for a written problem.
4. On the fire ground, pump to the highest pressure and gate back any lines requiring lower discharge pressures.

To calculate FL in multiple hose lines of different sizes and lengths using the metric system, follow these steps:

1. Determine the correct PDP for each line in the scenario.
2. Use the pump discharge formula: PDP = NP + FL (nozzle pressure + friction loss). Also use the FL formula: FL = C Q^2 L, where C = coefficient for the line diameter, Q^2 = quantity of flow (L/min) divided by 100 and squared, and L = length in meters divided by 100.

Skill Drill 4-3

Calculating Friction Loss in Multiple Hose Lines of Different Sizes and Lengths

Example 1: A pumper is supplying two attack lines. The first is 150′ of 1¾″ hose flowing 150 GPM with a fog nozzle. The second line is 200′ of 2½″ hose flowing 250 GPM with a smooth-bore nozzle. What is the PDP?

With this scenario, the operator will calculate the pressure for each line, then set the pump to the highest pressure and gate the other discharge back to its required pressure. The answer to the written problem is the highest pressure.

The PDP for the first line is

$FL = C Q^2 L$
$FL = 15.5 (1.5)^2 1.5$
$FL = 15.5 \times 2.25 \times 1.5 = 52.3$

Insert the FL into the PDP formula:

$PDP = 100\ NP + 52.3\ FL = 152.3\ psi$

The PDP for the second line is

$FL = C Q^2 L = 2 (2.5)^2 2$
$FL = 2 \times 6.25 \times 2 = 25$

Insert the FL into the PDP formula:

$PDP = 50\ NP + 25\ FL = 75\ psi$

The answer for this scenario is the highest pressure, which is 152.3 psi (round to 152 psi).

Now that you are acquainted with the PDP formula, begin writing it with just the values for NP and FL, such as PDP = 50 + 25, so PDP = 75 psi.

Example 2: A pumper is supplying three hose lines. The first attack line is 300′ of 2½″ hose flowing 300 GPM with a smooth-bore nozzle. The second line is 200′ of 1¾″ hose flowing 150 GPM with a fog nozzle. The third line is 200′ of 1¾″ hose flowing 161 GPM with a ⅞″ smooth-bore tip. What is the PDP?

With this scenario, the operator will calculate the pressure for each line, then set the pump to the highest pressure and gate the other discharges back to their required pressures. The answer to the written problem is the highest pressure.

The PDP for the first line is

$FL = C Q^2 L$
$FL = 2 (3)^2 3 = 54$

Insert the FL into the PDP formula:

$PDP = 50 + 54 = 104\ psi$

The PDP for the second line is

$FL = C Q^2 L$
$FL = 15.5 (1.5)^2 2$
$FL = 15.5 \times 2.25 \times 2 = 69.75$

Insert the FL into the PDP formula:

$PDP = 100 + 69.5 = 169.75\ psi$

The PDP for the third line is

$FL = C Q^2 L$
$FL = 15.5 (1.61)^2 2$
$FL = 15.5 \times 2.59 \times 2 = 80$

Insert the FL into the PDP formula:

$PDP = 50 + 80 = 130\ psi$

The answer for this scenario is the highest pressure, which is 169.75 psi (round to 170 psi).

Chapter 4 Mathematics for the Driver/Operator

Skill Drill 4-3

Calculating Friction Loss in Multiple Hose Lines of Different Sizes and Lengths

Metric (5.2.1)

Example 1: A pumper is supplying two attack lines. The first is 45 m of 44-mm hose flowing 550 L/min with a fog nozzle. The second line is 60 m of 64-mm hose flowing 950 L/min with a smooth-bore nozzle. What is the PDP?

With this scenario, the operator will calculate the pressure for each line, then set the pump to the highest pressure and gate the other discharge back to its required pressure. The answer to the written problem is the highest pressure.

The PDP for the first line is

$FL = C Q^2 L$ $FL = 24.5 \times (5.5)^2 \times 0.45$
$FL = 24.5 \times 30.25 \times 0.45 = 333.5$ (round to 334)

Insert the FL into the PDP formula:

$PDP = 700 \text{ NP} + 334 \text{ FL} = 1034 \text{ kPa}$

The PDP for the second line is

$FL = C Q^2 L = 3.16 \times (9.5)^2 \times 0.6$ $FL = 3.16 \times 90.25 \times 0.6 = 171$

Insert the FL into the PDP formula:

$PDP = 350 \text{ NP} + 171 \text{ FL} = 521 \text{ kPa}$

The answer for this scenario is the highest pressure, which is 1034 kPa (round to 1050 psi).

Now that you are acquainted with the PDP formula, begin writing it with just the values for NP and FL, such as PDP = 350 + 171, so PDP = 521 kPa.

Example 2: A pumper is supplying three hose lines. The first attack line is 90 m of 64-mm hose flowing 1100 L/min with a smooth-bore nozzle. The second line is 60 m of 44-mm hose flowing 550 L/min with a fog nozzle. The third line is 60 m of 44-mm hose flowing 600 L/min with a 22-mm smooth-bore tip. What is the PDP?

With this scenario, the operator will calculate the pressure for each line, then set the pump to the highest pressure and gate the other discharges back to their required pressures. The answer to the written problem is the highest pressure.

The PDP for the first line is

$FL = C Q^2 L$
$FL = 3.16 \times (11)^2 \times 0.9 = 344$

Insert the FL into the PDP formula:

$PDP = 350 + 344 = 694 \text{ kPa}$

The PDP for the second line is

$FL = C Q^2 L$
$FL = 24.5 \times (5.5)^2 \times 0.6$
$FL = 24.5 \times 30.25 \times 0.6 = 445$

Insert the FL into the PDP formula:

$PDP = 700 + 445 = 1145 \text{ kPa}$

The PDP for the third line is

$FL = C Q^2 L$
$FL = 24.5 \times (6)^2 \times 0.6$
$FL = 254.5 \times 36 \times 0.6 = 529$

Insert the FL into the PDP formula:

$PDP = 350 + 529 = 879 \text{ kPa}$

The answer for this scenario is the highest pressure, which is 1145 kPa (round to 1150 kPa).

3. Determine which line requires the highest pressure. This pressure is the correct answer for a written problem.
4. On the fire ground, pump to the highest pressure and gate back any lines requiring lower discharge pressures.

> **Safety Tip**
>
> Use the intake or discharge outlets on the pump panel as a last resort. It is safer if you are not standing next to these pressurized lines in case one bursts.

Elevation Pressure

Calculations must be adjusted for the distance the nozzle is above or below the pump, which is referred to as **elevation pressure (EP)**. The pressure lost when the nozzle is above the pump is known as **elevation loss**; it requires pressure to be added to the discharge pressure to compensate for the loss. The pressure gained when the nozzle is below the pump is known as **elevation gain**; it requires pressure to be subtracted from the discharge pressure. After calculating the EP, it must be added to or subtracted from the PDP.

Elevation is relative to grade; altitude is relative to sea level. You will regularly encounter situations where water will have to be discharged at an elevation higher or lower than the position of the fire pumper—for example, down a hill to supply a fire apparatus or up to the upper floor of a home or office building for fire suppression. The change in elevation affects the PDP because water has weight and the weight must be compensated for. Thus EP may be either a gain or a loss.

Figure 4-8 Floors in high-rise structures are typically spaced 12' to 14' (3.5 to 4.25 m) apart.

Figure 4-7 Floors in residential structures are typically spaced 10' (3 m) apart.

Water exerts a pressure of 0.434 psi per 1' (9.817 kPa/m) of water column. This can be a positive or a negative factor in calculating PDP. If water is discharging below the center line of the pump, then subtract 0.434 psi per 1' (9.817 kPa/m) of decline; conversely, add 0.434 psi per 1' (9.817 kPa per meter) of incline if water is discharging above the pump. To make multiplication simpler, multiply by 0.5 as opposed to 0.434 (10 as opposed to 9.817 in the metric system).

To speed the calculation process further, simply determine the elevation change in 10' (3 m) increments and multiply your findings by 5 psi (5 psi per 10') [30 kPa (10 kPa per 3 m)]. A common application for this rule would be 5 psi per 30 kPa of gain or loss for each floor of elevation change in a residential structure where floor spacing is commonly 10' (3 m) **Figure 4-7**. However, not all buildings have floors spaced every 10' (3 m). High-rise buildings are spaced in 12' to 14' (3.5- to 4.25-m) increments **Figure 4-8**. Just remember that ±5 psi (±30 kPa) per floor may lead you to underestimate the EP due to alternate floor spacing in different types of structures.

For the purposes of this text, use a variance of ±5 psi (±30 kPa) per floor when calculating the elevation gain/loss. For multistory buildings, use the following formula: EP = 5 psi × (num-

Skill Drill 4-4

Calculating the Elevation Pressure (Loss and Gain)

Example 1: Determine EP when the nozzle is on the eleventh floor of a building.

EP = (11 − 1) × 5 psi
EP = 10 × 5 psi
EP = 50 psi

You can also determine the EP for a grade using the formula EP = 0.5 H, where 0.5 is a constant and H = height:

1. Estimate the elevation gain or loss. When the nozzle is higher than the pump, the EP is a positive number (compensate for the loss by increasing pressure). When the nozzle is below the pump, the elevation is a negative number (compensate for the gain by decreasing pressure).
2. Multiply 0.5 × height, where the height will be a positive or negative number.

Example 2: Determine the EP when the nozzle is 30′ below the pump.

EP = −30 × 0.5 EP = −15

Example 3: Determine the EP when the nozzle is 50′ above the pump.

EP = 50 × 0.5 EP = 25

---

ber of stories − 1) [or 30 kPa × (number of stories − 1)]. Do not count the first story of a multistory building. To calculate EP for a grade, use the following formula: EP = 0.5 H, where 0.5 is a constant and H = height (or EP = 10 H, where 10 is a constant and H = height in meters).

To calculate the EP loss or gain, follow the steps in **Skill Drill 4-4**:

1. Determine the EP when operating in a multistory building.
2. Use the formula: EP = 5 psi × (number of stories − 1).
3. Determine the number of stories to the nozzle.
4. Multiply 5 psi times the number of stories minus 1.

To calculate the EP using the metric system, follow these steps:

1. Determine the EP when operating in a multistory building.
2. Use the formula: EP = 30 kPa × (number of stories − 1)
3. Determine the number of stories to the nozzle.
4. Multiply 30 kPa times the number of stories minus 1.

Driver/Operator Tip

Mark the preconnect pressures. Preconnects are the most commonly used attack lines, so you should be able to quickly establish the needed pressures with minimal calculations. These pressures should be established before a fire occurs. This is a good time to do the math, especially if you are an inexperienced driver/operator.

The best way to determine correct pressures is to flow your hose lines with the nozzles on your fire apparatus. This is the only way to be sure of the exact performance of your equipment. Take your fire apparatus out onto the parking lot or drill ground. Pull the preconnect lines and charge them. Establish the pressure for each single hose line and mark it on the individual pressure gauge with a red pen stripe on automotive tape. Now it will be effortless to pump a preconnect hand line. This is a great practice tool while you are mastering the calculations needed for pumping.

Skill Drill 4-4

Calculating the Elevation Pressure (Loss and Gain)

Metric (5.2.1)

Example 1: Determine the EP when the nozzle is on the eleventh floor of a building.

EP = (11 − 1) × 30 kPa
EP = 10 × 30 kPa
EP = 300 kPa

You can also determine the EP for a grade using the formula EP = 10 H, where 10 is a constant and H = height in meters:

1. Estimate the elevation gain or loss. When the nozzle is higher than the pump, the EP is a positive number (compensate for the loss by increasing pressure). When the nozzle is below the pump, the elevation is a negative number (compensate for the gain by decreasing pressure).
2. Multiply 10 × height, where the height will be a positive or negative number.

Example 2: Determine the EP when the nozzle is 9 m below the pump.

EP = −9 × 10   EP = −90 kPa

Example 3: Determine the EP when the nozzle is 15 m above the pump.

EP = 15 × 10   EP = 150 kPa

Appliance Loss

Appliances are devices that are used to connect and adapt hoses, and direct and control water flow in various hose layouts. Appliances consist of, but are not limited to, adapters, reducers, gated wyes, Siamese connections, water thieves, monitors, manifolds, and elevated master stream devices **Figure 4-9**. Like any other water supply device, appliances may add to FL. Also, as in fire hose, the FL in an appliance is directly proportional to the volume (GPM) of water flowing through the system. Generally, the appliance FL is considered insignificant when water flows are less than 350 GPM (1300 L/min). However, it is recommended that you test your appliances for FL at reasonable flows to determine how you will account for the FL associated with them.

For the purposes of this text, we will follow these guidelines:

- Allow 10 psi (70 kPa) FL in appliances when the flow is 350 GPM (1300 L/min) or greater.
- Allow 25 psi (175 kPa) FL for all master stream appliances.
- Friction loss in appliances will not be calculated when flows are less than 350 GPM (1300 L/min).
- The 25 psi (175 kPa) allows for FL in the intake, the internal piping, and nozzle.
- Remember that the loss in internal piping is specific to the apparatus and may actually be quite large. One discharge may be a straight pipe with almost no FL, whereas the piping of another discharge on the same pump may create 10 to 30 psi (70 to 210 kPa) of FL depending on the number of turns the pipe makes.
- Written tests or quizzes for this text cannot account for specific piping FL, so they will use these guidelines.

It is important to remember that fire streams of less than 350 GPM (1300 L/min) are considered handlines and that fire streams of 350 GPM (1300 L/min) or greater are considered to

A.

B.

C.

D.

Figure 4-9 Appliances include adapters, reducers, gated wyes, Siamese connections, water thieves, monitors, manifolds, and elevated master stream devices, among other items. **A.** A typical gated wye device. **B.** A hydrant valve. **C.** A typical Siamese connection device. **D.** A typical monitor.

be <u>master streams</u>. Assign 25 psi (175 kPa) in appliance loss (AL) for master streams, and 10 psi (70 kPa) in AL for handlines with a flow of 350 GPM (1300 L/min) or greater.

Table 4-3 lists ALs for some commonly encountered devices. It is highly recommended that you conduct FL testing of your appliances and determine how they will be used on the fire ground.

Determining Friction Loss in Appliances

It is recommended that you check the appliances in your department for FL at the water flows that are most likely to be encountered on the fire ground. The process to check for FL is very simple. You will need the following equipment:

- The hose or appliance to be tested.
- Two in-line pressure gauges for testing hose. If you are

Standard Table 4-3 Appliance Losses for High-Flow Devices

Device	Appliance Loss (psi)
Wye, Siamese connection, FDC, manifold (all if the flow is greater than 350 GPM)	10
Elevated master stream and monitor	25
Standpipe riser if no preplanning information is available	25
Prepiped elevated master stream	As determined by testing

Metric Table 4-3 Appliance Losses for High Flow Devices

Device	Appliance Loss (kPa)
Wye, Siamese connection, FDC, manifold (all if the flow is greater than 1300 L/min)	70
Elevated master stream and monitor	175
Standpipe riser if no preplanning information is available	175
Prepiped elevated master stream	As determined by testing

testing a master stream device, you may use one or two in-line gauges, or they may be replaced by a manual or threaded Pitot gauge **Figure 4-10**. A flow meter may also be used.

- Nozzle of an appropriate size for the hose to control the water discharged.
- Hose necessary to make all connections.
- Adapters and fittings necessary to make all connections.

Choose which method you will use and document the results. The FL in appliances is minimal until the pump is flowing 350 GPM (1300 L/min) or greater. It is recommended that tests be conducted on level ground and that you complete the process at various flows to see how the FL changes.

To determine the FL in appliances, follow the steps in **Skill Drill 4-5**:

1. Attach 100′ (30 m) of 2½″ (64-mm) or larger hose to the selected discharge. (**Step 1**)
2. Attach one in-line pressure gauge to the hose in Step 1. (**Step 2**)
3. Attach the appliance to be tested to the in-line gauge. (**Step 3**)
4. Attach the second in-line gauge to the discharge side of the appliance or Pitot gauge if testing a smooth-bore nozzle on a master stream device.
5. If not testing a nozzle, connect additional hose and a nozzle to the second in-line gauge, enough to direct the discharged water to the desired location.
6. Engage the pump, flow water, and increase the throttle until the desired amount of pressure is achieved on the second in-line gauge or the Pitot gauge.
7. Record the pressure reading from both gauges.
8. Subtract the pressure on the second gauge from the pressure on the first gauge; this is your appliance FL. (**Step 4**)

To use a Pitot gauge to test the FL in a portable master stream appliance through 3″ (76-mm) hose, follow the steps in **Skill Drill 4-6**:

1. Connect 100′ (30 m) of 3″ (76-mm) hose to the pump. (**Step 1**)
2. Place and safely secure a smooth-bore 1¼″ (32-mm) tip on the end of the hose. (**Step 2**)
3. Charge the line. Using a Pitot gauge, increase the pressure until it reads 80 psi (560 kPa). Record the PDP. (**Step 3**)
4. Remove the nozzle and place the portable master stream appliance on the end of the hose; attach the same tip to the discharge. (**Step 4**)
5. Use a Pitot gauge and increase the pressure until the gauge reads 80 psi (560 kPa). Record the PDP.
6. Compare the PDP before the appliance was added to the PDP after the appliance was added. This pressure difference represents the FL within the appliance. With a 1¼″ tip, the PDP before the appliance was added was 93 psi (640 kPa) and the PDP after adding the appliance was 113 psi (780 kPa). This represents 20 psi (140 kPa) FL in the appliance. (**Step 5**)

A.

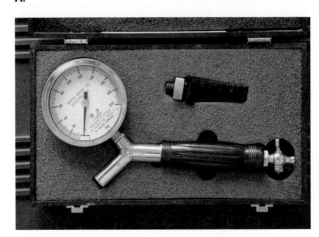

B.

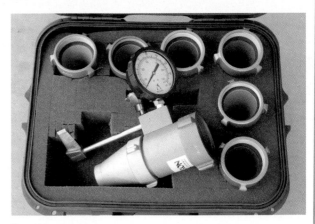

C.

Figure 4-10 Two in-line pressure gauges are needed for testing hose. If testing a master stream device, you may use one or two in-line gauges, or they may be replaced by a manual or threaded Pitot gauge. **A.** In-line pressure gauge. **B.** Manual Pitot gauge. **C.** Threaded Pitot gauge.

To use in-line gauges to test the FL in a specific hose, follow the steps in **Skill Drill 4-7**:

1. Connect 50′ (15 m) of hose to the pump discharge with an in-line gauge on the end. (**Step 1**)
2. Connect 200′ (60 m) of hose to the in-line gauge. (**Step 2**)

Skill Drill 4-5

Determining the Friction Loss in Appliances

1. Attach 100′ (30 m) of 2½″ (64-mm) or larger hose to the selected discharge.

2. Attach one in-line pressure gauge to the hose.

3. Attach the appliance to be tested to the in-line gauge.

4. Attach the second in-line gauge to the discharge side of the appliance or Pitot gauge if testing a smooth-bore nozzle on a master stream device. If not testing a nozzle, connect additional hose and a smooth-bore nozzle to the second in-line gauge, enough to direct the discharged water to the desired location. Engage the pump, flow water, and increase the throttle until the desired amount of pressure is achieved on the second in-line gauge or the Pitot gauge. Record the pressure reading from both gauges. Subtract the pressure on the second gauge from the pressure on the first gauge; this is your appliance FL.

3. Attach a second in-line gauge to the end of the 200′ (60-m) hose. (**Step 3**)
4. Add 50′ (15 m) to 100′ (30 m) of hose with a smooth-bore nozzle on the end. (**Step 4**)
5. Compare gauges.
6. Using the gauge reading, compare the pressure from the first gauge to the pressure from the second gauge. This pressure loss will illustrate the FL in the amount of hose between the gauges for this specific hose. (**Step 5**)

Driver/Operator Tip

When calculations are more complex and extend beyond a simple single line, it is often helpful to break the problem into sections. In the classroom, it is helpful to sketch the scenario, put parentheses around each section of the problem, and work from one end to the other. This way you are less likely to miss a calculation. Many driver/operators find it helpful and efficient to do this type of "chunking" mentally on the fire ground as well.

Skill Drill 4-6

Using a Pitot Gauge

1 Connect 100′ (30 m) of 3″ (76-mm) hose to the pump.

2 Place and safely secure a smooth-bore 1¼″ (32-mm) tip on the end of the hose.

3 Charge the line. Using a Pitot gauge, increase the pressure until it reads 80 psi (560 kPa). Record the pump discharge pressure.

4 Remove the nozzle and place the portable master stream appliance on the end of the hose; attach the same tip to the discharge.

5 Use a Pitot gauge and increase the pressure until it reads 80 psi (560 kPa). Record the PDP. Compare the PDP before the appliance was added to the PDP after the appliance was added. This pressure difference represents the FL within the appliance. With a 1¼″ tip, the PDP before the appliance was added was 93 psi (640 kPa) and the PDP after adding the appliance was 113 psi (780 kPa). This represents 20 psi (140 kPa) FL in the appliance.

Total Pressure Loss

Example 1 (Standard): Determine the PDP of 250′ of 2½″ hose flowing 300 GPM where the hand-held smooth-bore nozzle is 40′ below the pump.

$$PDP = NP + FL \qquad FL = C Q^2 L + EP$$
$$PDP = 50 + FL \qquad FL = [2 \times (3)^2 \times 2.5] - 20$$
$$PDP = 50 + 25 \qquad FL = (45) - 20$$
$$PDP = 75 \text{ psi} \qquad FL = 25$$

Example 1 (Metric): Determine the PDP of 75 m of 64-mm hose flowing 1100 L/min where the hand-held smooth-bore nozzle is 12 m below the pump.

$$PDP = NP + FL \qquad FL = C Q^2 L + EP$$
$$PDP = 350 + FL \qquad FL = [3.16 \times (11)^2 \times 0.75] - 120$$
$$PDP = 350 + 167 \qquad FL = (287) - 120$$
$$PDP = 517 \text{ kPa} \qquad FL = 167$$

Total Pressure Loss

In *any* pumping scenario, if there are appliances used or elevation gain or loss, you *must* insert AL and EP into the FL formula. First start with the equation PDP = NP + FL and determine the NP. Next determine the **total pressure loss (TPL)**. The FL formula in expanded form becomes **FL = C × Q² × L + AL + EP**. This expanded formula reflects the TPL, so it may be expressed as follows: **TPL = C × Q2 × L + AL + EP**. The TPL equation does not include a value for the NP.

Skill Drill 4-7

Using In-Line Gauges to Test Friction Loss

1. Connect 50' (15 m) of hose to the pump discharge with an in-line gauge on the end.
2. Connect 200' (60 m) of hose to the in-line gauge.
3. Attach a second in-line gauge to the end of the 200' (60-m) hose.
4. Add 50' (15 m) to 100' (30 m) of hose with a smooth-bore nozzle on the end.
5. Compare the gauges.

After calculating the TPL, apply it to the PDP formula, which can now be expressed as **PDP = NP + TPL**. You may use either form of the formula as long as you have inserted all contributing factors to the calculation.

Wyed Hose Lines

A wye is used to split a single line into two lines. This requires a series of calculations to find the final FL. For a wyed hose lay where all discharge lines are of equal size and length, follow the steps in **Skill Drill 4-8** to determine the PDP:

1. Add the flow rate for all discharges from the wye together to give you the quantity (Q) for the supply line to the wye device.
2. Calculate the FL for the supply line to the wye device using the Q value determined in Step 1.
3. Calculate the FL for one of the discharge lines from the wye device and add the nozzle pressure.
4. If the scenario is flowing 350 GPM (1300 L/min) or greater, add 10 psi (70 kPa) friction loss for the wye device.
5. Add the results of Steps 2 through 4 to determine the PDP.

For a wyed hose lay where each discharge line from the device is different (unequal), follow the steps in **Skill Drill 4-9** to determine the PDP:

1. Add the flow rate for all discharges from the wye together to give you the quantity (Q) for the supply line to the wye device.
2. Calculate the FL for the supply line to the wye device using the Q value determined in Step 1.
3. Calculate the FL and NP for each of the discharge lines from the wye device. Add the NP to the FL for the hose line requiring the greatest pressure.
4. If the scenario is flowing 350 GPM (1300 L/min) or greater, add 10 psi (70 kPa) FL for the wye device.
5. Add the results of Steps 2 through 4 to determine the PDP.

Skill Drill 4-8

Determining the Pump Discharge Pressure in a Wye Scenario with Equal Lines

Example 1: A pumper is supplying 100′ of 2½″ hose to a wye that has two 1¾″ lines 200′ long flowing 150 GPM each through fog nozzles. What is the PDP?

PDP = NP + FL

FL = C Q² L

FL for the 2½″ = 2 (3)² 1 = 18 psi

FL for a 1¾″ line:

FL = C Q² L = 15.5 (1.5)² 2

FL = 15.5 × 2.25 × 2 = 69.75 (round to 70 psi)

It takes 70 psi to flow the first 1¾″ line; the second 1¾″ line is identical. Thus both lines experience the same FL and have the same NP. Insert the values in the formula:

PDP = NP + FL

PDP = 100 (NP) + 18 (for the 2½″ line) + 70 (for the 1¾″ lines)

PDP = 100 + 18 + 70 = 188 psi

As the flow did not exceed 350 GPM, do not add 10 psi AL for the wye.

Example 2: A pumper is supplying 100′ of 3″ (with 2½″ couplings) to a wye with two equal lines of 2½″ hose, each 200′ flowing 250 GPM through smooth-bore nozzles. What is the PDP?

PDP = NP + FL

FL for the 3″ line:

FL = C Q² L = 0.8 (5)² 1

FL = 0.8 × 25 = 20

FL for the 2½″ lines:

FL = C Q² L = 2 (2.5)² 2

FL = 2 × 6.25 × 2 = 25

PDP = 50 NP + (20 for the 3″ line) + 25 (for the 2½″ lines) + 10 (AL)

PDP = 50 + 20 + 25 + 10 = 105 psi

Siamese Hose Lines

A Siamese connection is a device that allows multiple hose lines to converge into one hose line. Such a device is often used on the intake side of the pump, thereby allowing multiple lines to supply the fire pumper. A Siamese connection is used by fire departments that do not have large-diameter hose (LDH) and need to supply large amounts of water at a reasonable amount of pump pressure. On the discharge side of the pump, this device may be used to bring two or three lines into one attack line, thereby reducing the FL in a long reach. **Portable master stream devices** (such as a removable deck gun) and the inlet of aerial waterways on squirts, quints, and ladder trucks may be supplied by a single LDH supply or a Siamese connection. The **fire department connection (FDC)** on buildings with standpipe sprinkler systems will have a Siamese connection consisting of two 2½″ (64-mm) connections or one LDH connection.

The three methods used to calculate the FL in the lines to the Siamese connection when the lines are of equal size and length are the split flow method, the coefficient method, and the percentage method:

Skill Drill 4-8

Metric (5.2.4)

Determining the Pump Discharge Pressure in a Wye Scenario with Equal Lines

Example 1: A pumper is supplying 30 m of 64-mm hose to a wye that has two 44-mm lines 60 m long flowing 550 L/min each through fog nozzles. What is the PDP?

PDP = NP + FL

FL = C Q² L

FL for the 64-mm line = 3.16 (11)² 0.3 = 115 kPa

FL for a 44-mm line = C Q² L = 24.5 (5.5)² 0.6

FL = 24.5 × 30.25 × 0.6 = 444.6
    (round to 445 kPa)

It takes 445 kPa to flow the first 44-mm line; the second 44-mm line is identical. Thus both lines experience the same FL and have the same NP. Insert the values in the formula:

PDP = NP + FL

PDP = 700 (NP) + 115 (for the 64-mm line) + 445
    (for the 44-mm lines)

PDP = 700 + 115 + 445 = 1260 kPa

As the flow did not exceed 1300 L/min, do not add 70 kPa AL for the wye.

Example 2: A pumper is supplying 30 m of 76-mm hose (with 64-mm couplings) to a wye with two equal lines of 64-mm hose, with each 60 m flowing 950 L/min through smooth-bore nozzles. What is the PDP?

PDP = NP + FL

FL for the 76-mm line:

FL = C Q² L = 1.26 (19)² 0.3

FL = 1.26 × 361 × 0.3 = 136.4 (round to 136)

FL for the 64-mm lines:

FL = C Q² L = 3.16 (9.5)² 0.6

FL = 3.16 × 90.25 × 0.6 = 171.1 (round to 171)

PDP = 350 NP + (136 for the 76-mm line) + 171
    (for the 64-mm lines) + 70 (AL)

PDP = 350 + 136 + 171 + 70 = 727 kPa

If the lines are the same diameter and length, have nozzles with the same flow and NP, and are at the same elevation, they require the same PDP. The pressure from the pump will be the same in both lines.

- In the split flow method, you divide the flow among the lines equally and calculate the pressure to supply that one line as that pressure will support all the supply lines.
- The coefficient method utilizes the FL equation using the appropriate coefficient for the lines entering the Siamese connection.
- In the percentage method, you calculate the total flow through one line. The FL in two hose lines will be approximately 25 percent of the total flow through one hose line, and the FL in three hose lines will be approximately 10 percent of the total flow through one hose line.

To use the split flow method, follow the steps in **Skill Drill 4-10 ▶**:

Skill Drill 4-9

Determining the Pump Discharge Pressure in a Wye Scenario with Unequal Lines

Example 1: A pumper is supplying 100′ of 3″ hose (with 2½″ couplings) to a wye. The first attack line from the wye is 200′ of 2½″ hose with a 1¼″ smooth-bore nozzle flowing 300 GPM. The second attack line is 150′ of 2½″ hose with a 1″ smooth-bore nozzle flowing 200 GPM. What is the PDP? (Remember that when the flow is 350 GPM or greater in a wye, you must add 10 psi AL.)

FL for 3″ hose = C Q² L
= 0.8 (5)² 1
= 0.8 × 25 × 1
= 20 psi

FL for first 2½″ hose = C Q² L = 2 (3)² 2 = 36 psi

PDP = 50 (NP) + 20 (FL in 3″ hose) + 36 (FL in first 2½″ hose) + 10 (AL) = 116 psi

FL for second 2½″ hose = C Q² L = 2 (2)² 1.5
= 12 psi

PDP = 50 (NP) + 20 (FL in 3″ hose) + 12 (FL in second 2½″ hose) + 10 (AL) = 92 psi

The second attack line requires less pressure; gate this line back to 92 psi. As the first attack line has a greater FL, pump to its pressure of 116 psi:

PDP = 116 psi

Example 2: A pumper is supplying 100′ of 3″ hose to a wye. From the wye, the first attack line is 200′ of 1¾″ hose flowing 150 GPM through a fog nozzle. The second attack line is 200′ of 2½″ hose flowing 300 GPM through a fog nozzle. What is the PDP?

FL for the 3″ hose = C Q² L
= 0.67 (4.5)² 1
= 0.67 × 20 × 1
= 13.5 (round to 14)

FL for first 1¾″ attack line: FL = C Q² L
= 15.5 (1.5)² 2
= 15.5 × 2.25 × 2
= 69.75 (round to 70 psi)

PDP = 100 (NP) + 14 (FL in 3″ hose) + 70 (FL in 1¾″ hose) + 10 (AL) = 194 psi

FL for second attack line (2½″ hose): FL = C Q² L
= 2 (3)² 2
= 2 × 9 × 2
= 36 psi

PDP = 100 (NP) + 14 (FL in 3″ hose) + 36 (FL in 2½″ hose) + 10 (AL) = 160 psi

The PDP for the 1¾″ line is greater, so pump at 194 psi and gate the 2½″ hose down to 160 psi:

PDP = 194 psi

Chapter 4 Mathematics for the Driver/Operator

Skill Drill 4-9

Determining the Pump Discharge Pressure in a Wye Scenario with Unequal Lines

Example 1: A pumper is supplying 30 m of 76-mm hose (with 64-mm couplings) to a wye. The first attack line from the wye is 60 m of 64-mm hose with a 32-mm smooth-bore nozzle flowing 1150 L/min. The second attack line is 45 m of 64-mm hose with a 25-mm smooth-bore nozzle flowing 750 L/min. What is the PDP? (Remember that when the flow is 1300 L/min or greater in a wye, you must add 70 kPa AL.)

FL for 76-mm hose = C Q² L
= 1.26 (19)² 0.3
= 1.26 × 361 × 0.3
= 136.4 kPa (round to 136 kPa)

FL for first 64-mm hose = C Q² L = 3.16 (11.5)² 0.6
= 250.7 kPa (round to 251 kPa)

PDP = 350 (NP) + 136 (FL in 76-mm hose) + 251 (FL in first 64-mm hose) + 70 (AL)
= 807 kPa

FL for second 64-mm hose = C Q² L
= 3.16 (7.5)² 0.45 = 79.9 kPa (round to 80 kPa)

PDP = 350 (NP) + 136 (FL in 76-mm hose)
+ 80 (FL in second 64-mm hose) + 70 (AL)
= 636 kPa

The second attack line requires less pressure; gate this line back to approximately 636 kPa. As the first attack line has a greater FL, pump to its pressure of 807 kPa:

PDP = 807 kPa

Example 2: A pumper is supplying 30 m of 76-mm hose to a wye. From the wye, the first attack line is 60 m of 44-mm hose flowing 570 L/min through a fog nozzle. The second attack line is 60 m of 64-mm hose flowing 1150 L/min through a fog nozzle. What is the PDP?

FL for the 76-mm hose = C Q² L = 10.7 (17.2)² 0.3
= 10.7 × 296 × 0.3 = 95 kPa

FL for first 44-mm attack line: FL = C Q² L
= 24.5 (5.7)² 0.6
= 24.5 × 32.5 × 0.6
= 477.75 (round to 478 kPa)

PDP = 700 (NP) + 95 (FL in 76-mm hose)
+ 478 (FL in 44-mm hose) + 70 (AL)
= 1343 kPa

FL for second attack line (64-mm hose): FL = C Q² L
= 3.16 (11.5)² 0.6
= 3.16 × 132.25 × 0.6
= 250.7 (round to 251 kPa)

PDP = 700 (NP) + 95 (FL in 76-mm hose)
+ 251 (FL in 64-mm hose) + 70 (AL)
= 1116 kPa

The PDP for the 44-mm line is greater, so pump at 1343 psi and gate the 64-mm line down to 1116 kPa:

PDP = 1343 kPa

Skill Drill 4-10

Determining the Pump Pressure for a Siamese Line by the Split Flow Method

Example: A pumper is supplying 1200 GPM through three 2½" lines that are 300' long to a Siamese connection. What is the FL to the Siamese connection as calculated using the split flow method?

1200 GPM/3 hose lines = 400 GPM each

$FL = C Q^2 L$

$C = 2$

$Q = 400/100 = 4$

$L = 300'/100 = 3$

$FL = 2 \times 4^2 \times 3$

$FL = 96$ psi

Skill Drill 4-10

Determining the Pump Pressure for a Siamese Line by the Split Flow Method

Example: A pumper is supplying 4500 L/min through three 64-mm lines that are 90 m long to a Siamese connection. What is the FL to the Siamese connection as calculated using the split flow method?

4500 L/min / 3 hose lines = 1500 L/min each

$FL = C Q^2 L$

$C = 3.16$

$Q = 1500/100 = 15$

$L = 90$ m$/100 = 0.9$

$FL = 3.16 \times 15^2 \times 0.9$

$FL = 640$ kPa

---

1. Divide the total amount of flow desired by the number of lines in the Siamese connection.
2. Use the resulting flow amount as the quantity (Q).
3. Complete the modern FL equation with the calculated Q value.

To calculate the FL in Siamese lines by use of the coefficient method, follow the steps in **Skill Drill 4-11**:

1. Determine the diameter and number of hose lines going to the Siamese connection.
2. Find the corresponding arrangement on the Siamese coefficient table to assign the correct Siamese coefficient.
3. Insert the coefficient into the FL formula ($C Q^2 L$) to determine the FL in the Siamese supply part of the calculation.

Skill Drill 4-11

Calculating Friction Loss in Siamese Lines by the Coefficient Method

Standard (5.2.1)

Example: A fire pumper is supplying 1200 GPM through three 2½" lines that are 300' long to a Siamese line. What is the FL to the Siamese connection as calculated using the Siamese coefficient?

$FL = C Q^2 L$
$C = 0.22$
$Q = 1200/100 = 12$

$L = 300'/100 = 3$
$FL = 0.22 \times 12^2 \times 3$
$FL = 95$ psi

Skill Drill 4-11

Calculating Friction Loss in Siamese Lines by the Coefficient Method

Metric (5.2.1)

Example: A fire pumper is supplying 4500 L/min through three 64-mm lines that are 90 m long to a Siamese line. What is the FL to the Siamese connection as calculated using the Siamese coefficient?

$FL = C Q^2 L$
$C = 0.347$
$Q = 4500/100 = 45$

$L = 90 \text{ m}/100 = 0.9$
$FL = 0.347 \times 45^2 \times 0.9$
$FL = 632$ kPa

4. If there is no Siamese equation for the configuration, use one of the other methods to calculate the FL.

To calculate the FL in Siamese lines by use of the percentage method, follow the steps in **Skill Drill 4-12**:

1. Determine the total flow of the lines supplying the Siamese connection.
2. Figure out the quantity (Q) from the total flow and calculate the FL as if the total flow were going through one line.
3. If two lines supplied the flow, divide the FL based on total flow through one line by 25 percent to find the actual FL.
4. If three lines supplied the flow, divide the FL based on total flow through one line by 10 percent to find the actual FL.

Calculating Elevated Master Streams

Prepiped Elevated Master Stream

The term **prepiped elevated master stream** refers to an aerial fire apparatus (ladder truck) with a fixed waterway attached to

Skill Drill 4-12

Calculating Friction Loss in Siamese Lines by the Percentage Method

Example 1: A fire pumper is supplying 1200 GPM through three 2½" lines that are 300' long to a Siamese connection. What is the FL to the Siamese connection as calculated using the percentage method?

FL = C Q² L

C = 2

L = 300'/100 = 3

Q² = (1200/100)² = 12² = 144

FL = 2 × 144 × 3 = 864: 10% of 864 = 86.4

FL = 86.4

Working the same problem with each method produced the following results:
- Split flow method: FL = 96 psi
- Coefficient method: FL = 95
- Percentage method: FL = 86.4

Example 2: A pumper is supplying 1000 GPM through two 2½" lines that are 200' long to a Siamese connection. What is the FL to the Siamese connection?

Percentage method:

FL = C Q² L

C = 2

L = 200'/100 = 2

Q = 1000/100 = 10² = 100

FL = 2 × 100 × 2 = 400: 25% of 400 = 100

FL = 100 psi

Split flow method:

FL = 2 × 5² × 2

FL = 100 psi

Skill Drill 4-12

Calculating Friction Loss in Siamese Lines by the Percentage Method

Example 1: A fire pumper is supplying 4500 L/min through three 64-mm lines that are 90 m long to a Siamese connection. What is the FL to the Siamese connection as calculated using the percentage method?

$FL = C Q^2 L$

$C = 3.16$

$L = 90 \text{ m}/100 = 0.9$

$Q^2 = (4500/100)^2 = 45^2 = 2025$

$FL = 3.16 \times 2025 \times 0.9 = 5759$: 10% of 5759
$= 575.9$ (round to 576 kPa)

$FL = 576$ kPa

Working the same problem with each method produced the following results:
- Split flow method: FL = 640 kPa
- Coefficient method: FL = 632 kPa
- Percentage method: FL = 576 kPa

Example 2: A pumper is supplying 3800 L/min through two 64-mm lines that are 60 m long to a Siamese connection. What is the FL to the Siamese connection?

Percentage method:

$FL = C Q^2 L$

$C = 3.16$

$L = 60 \text{ m}/100 = 0.6$

$Q = 3800/100 = 38^2 = 1444$

$FL = 3.16 \times 1444 \times 0.6 = 2738$: 25% of 2738
$= 684$

$FL = 684$ kPa

Split flow method:

$FL = 3.16 \times 19^2 \times 0.6$

$FL = 684$ kPa

Voices of Experience

During my fire service career, I responded to and performed as a driver/operator on many of the same common fires as every other fire department. Often, especially during the initial stages of fire attack, you do not have time to think about, much less calculate complex hydraulic calculations, so I depended heavily on the preplanned pump pressures that I had determined for basic and complex fire attack scenarios.

I committed some of these simpler preplanned pressures to memory. The more complex calculations were outlined on pump charts that I could immediately consult on the fireground to save time. Without the knowledge of friction loss and the steps to accurately calculate pump discharge pressures, I would not have been able to develop those dependable and complete pump pressures for either basic or complex fire attack scenarios.

Thorough hydraulic calculations can be very challenging when you are first learning to become a driver/operator; however, it is well worth both your time and effort. Your knowledge of hydraulics calculations will not only make you a better pump operator, but it will ensure the safety of your fire attack crew. As a driver/operator, you must supply an adequate amount of pressure and flow to extinguish the fire and protect their lives. Hydraulic calculations will ensure the integrity of your water supply and will ensure their safety.

Daryl Songer
Roanoke Fire-EMS
Roanoke, Virginia

> *"Hydraulic calculations will ensure the integrity of your water supply."*

Figure 4-11 A typical prepiped elevated master stream device.

Standard Table 4-4 Coefficients for Multiple Lines

Siamese 2½" Hose	
Two lines	0.5
Three lines	0.22
Four lines	0.12
Siamese 3" Hose with 2½" Couplings	
Two lines	0.2
Three lines	.09

Metric Table 4-4 Coefficients for Multiple Lines

Siamese 64-m Hose	
Two lines	0.789
Three lines	0.347
Four lines	0.189
Siamese 76-mm Hose with 64-mm Couplings	
Two lines	0.316
Three lines	0.142

the underside of the ladder with a water inlet at the base supplying a master stream device at the end. The master stream device has a locking mechanism that places the nozzle on the bed section for rescue or on the fly section for water tower operations **Figure 4-11**. There is no single FL amount that can adequately encompass all prepiped elevated master stream devices; therefore it is imperative that each aerial device be tested for its unique amount of FL.

When water comes into the intake, often it is split (in the form of a "T"), turns with an "L", then goes toward the front of the truck, turns again with an "L", and finally with another "L" turns into the pump. All of these turns increase FL. Unless the loss in the intake, piping, and nozzle has been established for the specific apparatus, use a minimum of 25 psi (175 kPa) FL for all master stream appliances.

To calculate the PDP for a prepiped elevated master stream device, follow the steps in **Skill Drill 4-13**:

1. Determine the NP by observing which nozzle is on the tip. If it is a smooth-bore nozzle, the NP will be 80 psi (560 kPa) unless the manufacturer's specifications differ. If it is a fog nozzle, is it an automatic that delivers the standard pressure of 100 psi (700 kPa) or does it meet some specification?
2. Determine the FL in the supply line(s) from the pumper to the aerial device. Use the split flow, Siamese coefficient **Table 4-4**, or percentage method to find the FL in the supply line.
3. Determine the FL of the prepiped master stream. Add 25 psi (175 kPa) FL for any master stream device.
4. Calculate the EP.
5. Add the results of Steps 1 through 4 to obtain the PDP.

Calculate PDP for a Non-prepiped Elevated Master Stream

Before the days of prepiped waterways, detachable ladder pipes were clamped onto the end of a ladder, with hose being laid from the appliance down the ladder bed and to a Siamese connection on the ground. The only difference in these calculations is that you must add the FL for the hose running up the ladder to the tip and allow for both appliances.

To calculate the PDP for an elevated master stream, follow the steps in **Skill Drill 4-14**:

1. Determine the NP.
2. Calculate the FL in the hose from the pumper to the master stream device.
3. Determine the FL in the master stream device.
4. Adjust for elevation changes between the nozzle and the pumper.
5. Add the results of Steps 1 through 4 to determine the PDP.

■ Standpipe Systems

<u>Standpipe</u> operations pose a huge firefighting and logistical challenge. Many different strategic and tactical decisions for fire attack may be made depending on the various types of standpipe-equipped buildings encountered. To make good decisions and to increase the chance of a successful operation, fire fighters must know the buildings, their occupancy, the fire load, special hazards, and hydrant locations. They should be familiar with the FDC and standpipe locations, know whether there are pressure-regulating valves (and their types, if they are present), and know whether a fire pump is available in the building. In addition, they should have established a preincident plan for these high hazard targets.

For these types of buildings, the water supply is critical. When the FDC has two 2½" (64-mm) connections, it is a good idea to advance a 2½" (64-mm) hose line to the FDC and charge it before attaching the second hose line. Get water into the <u>standpipe system</u> as soon as possible. When possible, a

Skill Drill 4-13

Calculating the Pump Discharge Pressure for a Prepiped Elevated Master Stream Device

Example 1: A pumper supplying an elevated master stream is on the hydrant, supplying one 5″ line that is 200′ long to the inlet of the ladder truck, which has the ladder extended 80′ vertically; it is equipped with a 1½″ smooth-bore master stream tip (600 GPM). There is 30 psi of FL in the fixed waterway of this particular truck. Assign 25 psi FL for the master stream appliance. What is the PDP?

$PDP = NP + FL$

$PDP = 80 + FL$

$FL = C \times Q^2 \times L + AL + EP$

$FL = 0.08 \times (6)^2 \times 2 + 25 + 40 = 71$

$FL = 71 + 30$ specific to the waterway

$FL = 101$ psi

$PDP = 80 + 101$

$PDP = 181$ psi

Example 2: Determine the PDP for supplying a prepiped elevated master stream through 200′ of 4″ hose, where the aerial is raised to 80′ and has a 2″ tip rated at 80 psi, and the appliance has 25 psi FL.

$PDP = NP + FL$

$PDP = 80 + FL$

$FL = C \times Q^2 \times L + AL + EP$ ($FL = 0.2 \times 1062/100^2 \times 200'/100 + AL + EP$)

$FL = 0.2 \times 10.62^2 \times 2 + AL + EP$

$FL = 0.2 \times 113 \times 2 + AL + EP$

$FL = 45$ psi $+ AL + EP$

$FL = 45 + 25 + EP$

$FL = 45 + 25 + 80' \times 0.5$

$FL = 45 + 25 + 40$

$PDP = 80 + (45 + 25 + 40)$

$PDP = 190$ psi

---

different pumper should supply each FDC on a building with multiple FDCs. Buildings that pose a high risk should have one fire pumper on the hydrant supplying another on the standpipe connection. Whenever possible, a water supply independent of the supply of the building should be available as well.

Incidents requiring use of a standpipe system can easily be one of the most demanding pumping scenarios that you as the driver/operator will encounter on the fire ground. Standpipe systems not only require the greatest number of calculations to be made, but also require very demanding pump pressures. Standpipe pump pressures may be calculated, identified from markings on or at the FDC, or be determined by departmental standard operating procedures (SOP). It is extremely beneficial to predetermine pump discharge pressures for high-rise structures to limit the amount of calculations needed on the fire ground during the excitement of an actual incident. As a general rule, do not exceed a PDP of 200 psi (1400 kPa) unless allowed by the system rating.

As the driver/operator at a high-rise incident, you can control only one pressure to the standpipe, even though multiple discharge hose lines may be in use in the same structure on different floors. The one pressure that you control from the pump panel is the highest pressure needed at any one discharge (regardless of the floor it is on). As explained in the discussion of supplying lines of unequal size or length, the operator must pump to the highest pressure required. This same rule also applies to standpipes in multistory buildings: Calculate the PDP for each line being supplied, and then pump

Skill Drill 4-13

Metric (5.2.4)

Calculating the Pump Discharge Pressure for a Prepiped Elevated Master Stream Device

Example 1: A pumper supplying an elevated master stream is on the hydrant, supplying one 127-mm line that is 60 m long to the inlet of the ladder truck, which has the ladder extended 24 m vertically; it is equipped with a 38-mm smooth-bore master stream tip (2250 L/min). There is 210 kPa of FL in the fixed waterway of this particular truck. Assign 175 kPa FL for the master stream appliance. What is the PDP?

PDP = NP + FL

PDP = 560 + FL

FL = C × Q² × L + AL + EP

FL = 0.126 × (22.5)² × 0.6 + 175 + 240 = 453

FL = 453 + 210 specific to the waterway

FL = 663 kPa

PDP = 560 + 663

PDP = 1223 kPa

Example 2: Determine PDP for supplying a prepiped elevated master stream through 60 m of 102-mm hose, where the aerial is raised to 24 m with a 50-mm tip rated at 560 kPa, and the appliance has 175 kPa FL.

PDP = NP + FL

PDP = 560 + FL

FL = C × Q² × L + AL + EP (FL = 0.315
 × (3900/100)² × 60 m/100 +AL +EP)

FL = 0.315 × 39² × 0.6 + AL + EP

FL = 0.315 × 1521 × 0.6 + AL + EP

FL = 288 kPa + AL + EP

AL = 175

EP = 24 m × 10

EP = 240 kPa

PDP = NP + (FL + AL + EP)

PDP = 560 + 288 + 175 + 240

PDP = 1263 kPa

---

to the highest pressure. This will not necessarily be the highest in elevation.

For example, suppose a fire pumper is supplying two attack lines on the top floor of a five-story building, each with 200′ (60 m) of 2½″ (64-mm) hose and 1⅛″ (29-mm) smooth-bore nozzles requiring a PDP of 141 psi (987 kPa). The same fire pumper is also supplying a 1¾″ (44-mm) hose with a fog nozzle performing overhaul on the third floor, which has a PDP of 181 psi (1267 kPa). The fire pumper can send only one pressure into the standpipe; it cannot pump two different pressures. Therefore the driver/operator must pump to the highest pressure, which in this case is 181 psi (1267 kPa).

The actual numbers would be rounded, but note that the hose requiring the most pressure is not on the highest floor in this example. Also note that NP is important, as a fog nozzle requires a nozzle pressure of 100 psi (700 kPa) NP, whereas a smooth-bore nozzle requires only 50 psi (350 kPa).

In commercial buildings and multistory structures, the attack hose should be 2½″ lines (64 mm) with smooth-bore nozzles. In multistory buildings, the attack lines should be supplied from the floor below with an in-line gauge on the standpipe outlet. Any lower pressures needed at subsequent discharges will have to be controlled by a fire fighter staged at the discharge valve of the **standpipe riser** used Figure 4-12 ▶. A standpipe riser is the vertical portion of the system piping within a building that delivers the water supply for fire hose connections, and sprinklers on combined systems, vertically from floor to floor.

Skill Drill 4-14

Calculating the Pump Discharge Pressure for an Elevated Master Stream

Example: Determine the PDP for supplying a non-prepiped elevated master stream that is raised to 60′ and is flowing 600 GPM to a 1½″ tip rated at 80 psi through 300′ of 3″ hose with 2½″ couplings. The aerial and master streams have 25 psi of FL each.

PDP = NP + FL
PDP = 80 + FL
FL = 0.8 × (600/100)² × 300′/100 (+ AL + EP)
FL = 0.8 × (6)² × 3 = 0.8 × 36 × 3
FL = 86 psi
AL = 25 psi

EP = 60′ × 0.5
EP = 30 psi
PDP = NP + FL + AL + EP
PDP = 80 + 86 + 25 + 30
PDP = 221 psi

The fire fighter needs to know the required operating pressure of the attack hose lines, so he or she can adequately control the pressure from that outlet. The attack hose line *must* be flowing water if the crew member is to obtain an accurate reading from the gauge. Increase the pressure until the in-line pressure gauge at the riser meets the predetermined pressure for the attack hose/nozzle combination. The use of the in-line gauge will compensate for any discrepancies if calculations are off or circumstances have changed.

Use of an in-line gauge is a preferred advantage for standpipe operations. If in-line gauges are not available, you are well advised to test the flow from your high-rise hose line and nozzle combination in a realistic simulation to set performance expectations. To calculate the discharge pressure, follow the steps in **Skill Drill 4-15**:

1. Determine how you will connect to the standpipe and the amount of hose needed.
2. Use the appropriate FL calculation given the size and amount of hose needed. An LDH fire department connection may require only a single line, whereas the more common 2½″ (64-mm) FDC will require multiple hoses to be used. When using two 2½″ lines, split the flow to determine the quantity value (Q) for the FL equation. For multiple 2½″ hose lines, use the coefficient for Siamese lines.
3. Calculate the FL in the riser using the applicable coefficient for standpipe risers from **Table 4-5**. The height of the highest discharge used will be the value for L in the equation. Add 10 psi (70 kPa) FL for the FDC when flowing 350 GPM (1300 L/min) or greater. If the amount of water flowing is less than 350 GPM (1300 L/min), then do not add the 10 psi (70 kPa).
4. Calculate the amount of pressure lost due to the increase in elevation as described in the discussion of EP.
5. Calculate the friction loss in the attack hose lines, and then add the associated NP.
6. Add the results from Steps 1 through 5 to determine the PDP.

Figure 4-12 Lower pressures needed at subsequent discharges should be controlled by a fire fighter staged at the discharge valve of the standpipe riser used.

Skill Drill 4-14

Calculating the Pump Discharge Pressure for an Elevated Master Stream

Metric (5.2.4)

Example: Determine the PDP for supplying a non-prepiped elevated master stream that is raised to 18 m and is flowing 2250 L/min to a 38-mm tip rated at 560 kPa through 90 m of 76-mm hose with 64-mm couplings. The aerial Siamese connection has 70 kPa of friction loss and the master stream has 175 kPa of FL.

PDP = NP + FL

PDP = 560 + FL

FL = 1.26 × (2250/100)² × 90 m/100 (+ AL + EP)

FL = 1.26 × (22.5)² × 0.9 = 1.26 × 506.25 × 0.9

FL = 574 kPa

AL = 70 + 175 = 245 kPa

EP = 18 m × 10

EP = 180 kPa

PDP = NP + FL + AL + EP

PDP = 560 + 574 + 245 + 180

PDP = 1559 kPa

Standard Table 4-5 Coefficients for Standpipe Risers

Friction Loss Coefficients for Standpipe Risers	
Riser Diameter	Coefficient
4"	0.374
5"	0.126
6"	0.052

Metric Table 4-5 Coefficients for Standpipe Risers

Friction Loss Coefficients for Standpipe Risers	
Riser Diameter	Coefficient
100 mm	0.590
125 mm	0.199
150 mm	0.082

The PDP for standpipe operations requires that calculations be made for the supply lines to the standpipe, FL within the standpipe system, FL for the attack handlines, and pressure loss due to an increase in elevation.

If there is no preincident plan and the size of the riser is unknown, then it is too late to try to measure the riser. Instead, simply add 25 psi (175 kPa) to account for the standpipe riser. When personnel are in place, the in-line gauge is on the riser outlet, and water is flowing, then the person on the gauge will radio you to increase or decrease the pressure as needed.

To calculate the PDP for standpipe operations, follow the steps in **Skill Drill 4-16**:

1. Determine how you will connect to the standpipe and the amount of hose needed.
2. Use the appropriate FL calculation given the size and amount of hose needed. An LDH fire department connection may require only a single line, whereas the more common 2½" (64-mm) FDC will require multiple hoses to be used. When using two 2½" (64-mm) lines, split the flow to determine the quantity value (Q) for the FL. For multiple 2½" (64-mm) hose lines, use the Siamese coefficient.
3. Calculate the FL in the riser using the applicable coefficient from Table 4-5. The height of the highest discharge used will be the value of L in the equation. Add 10 psi (70 kPa) FL for the FDC if flowing 350 GPM (1300 L/min) or greater. If the amount of water flowing is less than 350 GPM (1300 L/min), do not add the 10 psi (70 kPa). If the pump pressure has not been predetermined and the riser size and length are not known, add 25 psi (175 kPa) FL for the FDC.
4. Calculate the amount of pressure lost due to the increase in elevation as described in the discussion of elevation loss/gain.
5. Calculate the FL in the attack hose lines, and then add the associated NP.
6. Add the results of Steps 1 through 5 to determine the PDP.

Skill Drill 4-15

Calculating the Discharge Pressure for a Standpipe During Preplanning

Example 1: You have been assigned to develop a preincident plan for the use of a standpipe system for fire attack on the top floor of a 17-story building. The attack hose will consist of two lines, each of which comprises 200′ of 2½″ hose with a smooth-bore 1⅛″ tip nozzle. A third 2½″ line with a 1¼″ tip is the backup line. The total flow is 858 GPM. The FDC is supplied with two 100′ lengths of 2½″ hose. The building is equipped with a 5″ riser. (The coefficient is 0.126.) What is the PDP to operate these three lines on the seventeenth floor?

Supply FL = C Q^2 L = 2 × $(4.29)^2$ 1
   = 36.8 (round to 37 psi)

FDC = 10 psi AL when flowing 350 GPM or greater and calculating for standpipe FL

Riser FL = C Q^2 L = 0.126 $(8.58)^2$ 1.6
   = 9.27 × 1.6 = 14.84 (round to 15 psi)

Two attack lines with 200′ of 2½″ hose with 1⅛″ smooth-bore nozzles:

FL = C Q^2 L = 2 $(2.66)^2$ 2 = 28 psi
PDP = 50 + 28 = 78 psi for these two lines
   (round to 80 psi)

One attack line with 200′ of 2½″ hose with a 1¼″ smooth-bore nozzle:

FL = C Q^2 L = 2 $(3.26)^2$ 2 = 42.5 psi
   (round to 43 psi)
PDP = 50 + 43 = 93 psi

Pump to the line requiring the highest pressure (93 psi) and add it all together:

   93 highest-pressure attack line
   15 for riser (14.84 actual)
   10 FDC
   + 37 supply to the FDC
   155 psi total PDP

Pressure-Regulating Valves

<u>Pressure-regulating valves (PRV)</u> are installed on standpipe risers where static pressures exceed 175 psi (1225 kPa) per NFPA 13. If pressures while flowing exceed 100 psi (700 kPa), then NFPA 14 requires the installation of a device at the outlet to restrict or reduce the flow pressure to a maximum of 100 psi (700 kPa). The height of a column of water above the discharge is known as the head; the pressure in that column of water is referred to as head pressure. To determine the head pressure, divide the number of feet (height) by 2.304; for calculations in the metric system, divide the number of meters (height) by 9.812.

In this case, the column of water is in the standpipe riser. As the height of the riser increases, the head pressure becomes so great that it could easily exceed the burst pressure of fire hose. Without the control established by restricting or regulating valves, excessive elevation pressures within the building standpipe system could burst hose lines, halting the suppression attack and possibly injuring fire fighters.

Standpipe test documentation should be on file in the building's maintenance office. The design pressure of the building's standpipe system should meet the minimum PRV pressure. You may boost the pressure to the maximum design pressure; however, if calculations require a pressure lower than the system minimum, default to the designed pressure so that the PRV can operate properly. These valves have huge implications for fire fighters, whether installed properly or not, and can severely hinder firefighting operations. One of the most notable examples of this problem came during the One Meridian Plaza Fire in Philadelphia in 1991, in which three fire fighters lost their lives. Fire fighters in that incident struggled due to the lack of pressure within the standpipe system.

Driver/Operator Tip

In multistory buildings, the PDP is set to support the line requiring the highest pressure regardless of its location.

Skill Drill 4-15

Calculating the Discharge Pressure for a Standpipe During Preplanning

Example 1: You have been assigned to develop a preincident plan for the use of a standpipe system for fire attack on the top floor of a 17-story building. The attack hose will consist of two lines, each of which comprises 60 m of 64-mm hose with a smooth-bore 29-mm tip nozzle. A third 64-mm line with a 32-mm tip is the back-up line. The total flow is 3350 L/min. The FDC is supplied with two 60-m lengths of 64-mm hose. The building is equipped with a 125-mm riser. (The coefficient is 0.199.) What is the PDP to operate these three lines on the seventeenth floor?

Supply FL = C Q^2 L = 3.16 × $(16.75)^2$ × 0.6
  = 532 kPa

FDC = 70 kPa AL when flowing 1300 L/min or greater and calculating for standpipe FL

Riser FL = C Q^2 L = 0.199 $(33.5)^2$ (16 × 3 m/100)
  = 0.199 × 1022.25 × 0.48 = 97.6
  (round to 98 kPa)

Two attack lines with 60 m of 64-mm hose with 29-mm smooth-bore nozzles:

FL = C Q^2 L = 3.16 $(10.5)^2$ 0.6 = 209.1 kPa
  (round to 209 kPa)
PDP = 350 + 209 = 559 kPa for these two lines

One attack line with 60 m of 64-mm hose with a 32-mm smooth-bore nozzle:

FL = C Q^2 L = 3.16 $(13)^2$ 0.6 = 320.4 kPa
  (round to 320 kPa)
PDP = 350 + 320 = 670 kPa

Pump to the line requiring the highest pressure (670 kPa) and add it all together:

  670 highest-pressure attack line
  98 for riser
  70 FDC
  + 532 supply to the FDC
  1370 kPa

Net Pump Discharge Pressure

A water supply can be either static or dynamic. Static sources such as lakes and streams require pulling a draft. Dynamic sources such as hydrants or another fire apparatus are positive-pressure sources. **Net pump discharge pressure (NPDP)** from a positive-pressure source (pps) is referred to as $NPDP_{pps}$. Net pump discharge pressure from a static source is referred to as $NPDP_{draft}$. Incoming pressure means that the pump does not have to work as hard to achieve proper PDPs; it merely has to add to the pressure that is received from the pressurized source.

Thus $NPDP_{pps}$ is the amount of pressure created by the pump after receiving pressure from a hydrant or another pump. For example, assume you are pumping a PDP of 170 psi (1190 kPa) and you have 50 psi (350 kPa) of pressure coming in from a hydrant. The $NPDP_{pps}$ is the PDP minus the incoming pressure. In this case, the calculation is 170 psi (1190 kPa) minus 50 psi (350 kPa), giving a $NPDP_{pps}$ of 120 psi (840 kPa).

Safety Tip

There is an old fire service saying: "As goes the first line, goes the fire!" It is much better to be "over-gunned" than "under-gunned." When selecting a handline, always use a 2½" (64-mm) line for fires in commercial buildings or well-involved fires. Remember—small fire = small water and big fire = big water!

Fire Service Hydraulic Calculations

Fire service hydraulic calculations may generally be categorized as either theoretical hydraulics or fire-ground hydraulics. Theoretical hydraulic calculations present more exact calculations, generally require more mathematical skills, and take more time to compute. Away from the fire ground, they may be computed on paper, calculator, or computer.

Skill Drill 4-16

Calculating Pump Discharge Pressure for a Standpipe

Example: You are supplying three 200′ attack lines of 2½″ hose flowing 1⅛″ tips rated at 50 psi. The highest attack line is on the fifteenth floor. The other two lines are on the fourteenth floor. Use the split flow method for the two 2½″ supply lines that are 200′ long to the FDC. What is the PDP?

Friction Loss: Supply

Total flow = 266 GPM × 3
  = 798 GPM (round to 800)
800 GPM/2 supply lines = 400 GPM each
FL (supply) = 2 × (4)² × 2 = 64 psi

Friction Loss: Standpipe

The standpipe is made of 6″ pipe, and the connection on the fifteenth floor is 180′ in elevation (L).

FL = C Q² L
FL (riser) = 0.052 × (798/100)² × 180′/100
  (round 798 to 800)
FL (riser) = 0.052 × (8)² × 1.8 (round to 2)
FL (riser) = 6 psi
FDC = 10 psi
AL (FDC and standpipe) = 16 psi

Elevation Pressure

EP = (15 − 1) × 5 = 14 × 5 psi
EP = 70 psi

Friction Loss: Attack Line

FL (attack line) = 2 × (266/100)² × 200′/100
FL (attack line) = 2 × (2.66)² × 2
FL (attack line) = 2 × 7.08 × 2
FL (attack line) = 28 psi

Nozzle Pressure

NP = 50 psi

Pump Discharge Pressure

PDP = NP + FL (supply) + AL (FDC and standpipe)
  + EP + FL (attack line)
PDP = 50 + 64 + 16 + 70 + 28
PDP = 228 psi

Theoretical hydraulic calculations may be used on the fire ground or for preincident planning purposes, such as determining the appropriate pump pressures for high-risk or high-hazard targets as well as buildings requiring complex hose layouts and standpipe systems. On the fire ground, theoretical calculations are performed mentally. As a driver/operator, you should practice these mental calculations extensively until you are competent and comfortable using the formulas.

Theoretical hydraulics calculations are primarily used on the fire ground by more experienced driver/operators. You should be able to effortlessly calculate FL and PDP. More-complex problems may require a few moments of thought and often a notepad to note calculations. These problems should be performed during the preincident planning stage.

Fire-ground hydraulic calculations employ methods devised to estimate calculations more quickly due to the urgency of the incident. These methods may also be used as a backup to mental theoretical calculations, as a system to check results, and as a means to ensure reasonable accuracy. They also reinforce the learning process as you develop the ability to compute theoretical hydraulics calculations in the field. These methods yield working approximations and are not as accurate as theoretical computation.

Skill Drill 4-16

Calculating Pump Discharge Pressure for a Standpipe

Example: You are supplying three 60-m attack lines of 64-mm hose flowing 29-mm tips rated at 350 kPa. The highest attack line is on the fifteenth floor. The other two lines are on the fourteenth floor. Use the split flow method for the two 64-mm supply lines that are 60 m long to the FDC. What is the PDP?

Friction Loss: Supply

Total flow = 1050 L/min × 3 = 3150 L/min

3150 L/min / 2 supply lines = 1675 L/min each

FL (supply) = 3.16 × $(16.75)^2$ × 0.6 = 532 kPa

Friction Loss: Standpipe

The standpipe is made of 152-mm pipe, and the connection on the fifteenth floor is 42 m in elevation (L).

FL = C Q^2 L

FL (riser) = 0.082 × $(3150/100)^2$ × 42/100

FL (riser) = 0.082 × $(31.5)^2$ × 0.42

FL (riser) = 34 kPa

FDC = 70 kPa

AL (FDC and standpipe) = 104 kPa

Elevation Pressure

EP = (15 − 1) × 3 m × 10 kPa/m = 42 × 10 kPa

EP = 420 kPa

Friction Loss: Attack Line

FL (attack line) = 3.16 × $(1050/100)^2$ × 60/100

FL (attack line) = 3.16 × $(10.5)^2$ × 0.6

FL (attack line) = 3.16 × 110.25 × 0.6

FL (attack line) = 209 kPa

Nozzle Pressure

NP = 350 kPa

Pump Discharge Pressure

PDP = NP + FL (supply) + AL (FDC and standpipe) + EP + FL (attack line)

PDP = 350 + 532 + 104 + 420 + 209

PDP = 1618 kPa

Driver/Operator Tip

When performing fire-ground hydraulic calculations to determine PDP, use a second method to confirm the results of the first. This cross-checking decreases the chance of error.

The two fundamental formulas for determining PDP are as follows:

- PDP = NP + FL
- FL = C Q^2 L + AL + EP

Most fire-ground hydraulic methods account for FL only. The acquired FL number is then added to the other components of the formula to determine the correct PDP. All of the essential components of hydraulics—those determining factors influencing PDP—must be considered when using the fire-ground hydraulic methods. Each provides a way to reduce the math required, although the NP, flow (GPM or L/min), size and length of hose, and FL including AL and elevation gain or loss must be accounted for. Common fire-ground hydraulic calculation methods include the following techniques:

- Charts
- Hydraulic calculators
- Hand method
- "Subtract 10 method," also known as the "GPM flowing method"
- Condensed Q method
- Flow meters
- Preincident plan data

Near Miss REPORT

Report Number: 08-0000085
Report Date: 02/12/2008 11:00

Synopsis: LDH adapter comes apart and cause fire fighter injury.

Event Description: We responded to an 800-square foot house fire. Upon arrival, there was heavy smoke coming from the C-side of the building. We pulled a pre-connected hose line and then a 5-inch LDH hose about 100 feet to a hydrant and made a connection for the water supply. We then waited for the driver to give the signal to charge the LDH. When it was OK to charge the line, we opened the hydrant. When the water charged the line, the 4.5-inch to LDH adapter came apart at the threaded connection. This caused a sight injury to an officer and could have resulted in loss of water supply during the fire attack. Due to quick thinking, we were able to restore the water supply in just a few moments. We have since told our personnel to check all LDH adapters for this problem.

Lessons Learned: Check all LDH adapters for this problem and I would recommend that these have a set screw and that the manufacturer look at using left handed threads on these connection. This would help when charging the line to reduce the amount of force that is applied to the adapter by the counter clock wise rotation of the hose line.

Demographics

Department type: Paid Municipal
Job or rank: Training Officer
Department shift: 24 hours on—48 hours off
Age: 43–51
Years of fire service experience: 17–20
Region: FEMA Region IV
Service Area: Suburban

Event Information

Event type: Fire emergency event: structure fire, vehicle fire, wildland fire, etc.
Event date and time: 02/11/2008 14:00
Event participation: Involved in the event
Weather at time of event: Clear and Dry

What were the contributing factors?
- Equipment

What do you believe is the loss potential?
- Property damage
- Life-threatening injury
- Lost time injury
- Minor injury

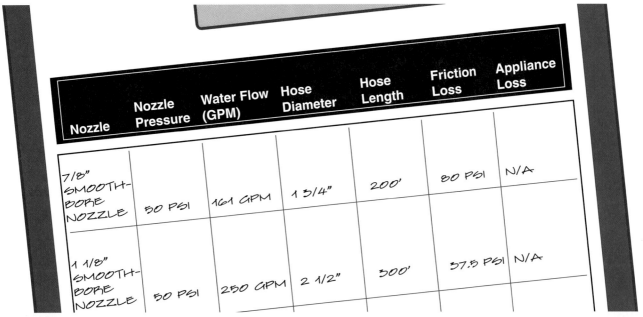

Figure 4-13 Fire fighters often create a chart of typical handline and master stream calculations for the hoses, nozzles, and devices specific to the fire apparatus within their fire department.

Charts

Fire fighters often create a **chart** of typical handline and master stream calculations for the hoses, nozzles, and devices specific to the fire apparatus within their department. These charts list the most commonly used or reasonably expected hose lays. The charts are designed so that you may easily find the needed discharge pressure without having to perform calculations on the fly on the fire ground **Figure 4-13**.

The chart lists the common pressures for the attack hose and nozzle combinations at different lengths as well as the pressures required for the use of master stream appliances. These represent different scenarios with different tip sizes and flows, elevations, and corresponding FL. Such charts, whether homemade or purchased from an outside source, may be pocket-size or larger. Often the charts consist of a laminated sheet of paper that is taped inside a compartment door, within sight of the pump panel. When creating your own pump chart, list the nozzle, NP, flow (GPM or L/min), hose diameter, hose length, FL, and AL when applicable. All elements of hydraulic calculations must be considered when compiling this kind of chart. Most importantly, the calculations must be accurate for the chart to be reliable. When using these charts, do not allow pressures to exceed hose test pressures and be sure to stay within the pressures mandated by your fire department's SOP.

Charts may also be prepared for complex fire problems such as multistory buildings or other high-hazard target locations. Preincident planning affords you the opportunity to establish baseline calculations before an emergency response becomes necessary. These pressures may then be placed on a chart, entered into the preincident plan book or computer, and/or labeled on the FDC.

Charts for high-rise buildings, such as the one by Battalion Chief David McGrail of the Denver Fire Department, suggest a pressure for a range of floors **Table 4-6**. As the driver/operator, you will pump to the pressure on the chart for a given floor, subsequently increasing or decreasing the pressure as instructed by personnel on that floor when they confirm the location and begin suppression operations. The members of the attack team

Table 4-6 Friction Loss Pump Chart for High-Rise Standpipe Operations

Floors	Pump Pressure
1 to 5 stories	125 psi
6 to 10 stories	150 psi
11 to 15 stories	175 psi
15 to 20 stories	200 psi
21 to 25 stories	225 psi
26 to 30 stories	250 psi
31 to 35 stories	275 psi
36 to 40 stories	300 psi
41 to 45 stories	325 psi
45 to 50 stories	350 psi
51 to 55 stories	375 psi
56 to 60 stories	400 psi
61 to 65 stories	425 psi

Note: Friction Loss for 150′ of 2 ½″ Hose Line and a 1⅛″ Smooth Bore Nozzle.
Source: David McGrail of the Denver, Colorado Fire Department.

will first flush the standpipe riser and then connect the in-line pressure gauge and gated wye. They will then connect the attack lines and advance toward the fire. As soon as water is flowing, the fire fighter assigned to the riser must read the in-line pressure gauge at the standpipe riser.

The attack team should know in advance what pressure is needed from a riser for the particular attack line and nozzle combination in their high-rise/standpipe bag. For example, 200′ (60 m) of 2½″ (64-mm) hose with a 1¼″ (32-mm) solid tip may require a pressure of 90 psi (630 kPa) to achieve the desired flow. If you increase the pressure until the in-line pressure gauge at the riser reads 90 psi (630 kPa), then the nozzle should deliver the desired flow. The fire fighter reading the gauge will advise you by radio to increase or decrease the pressure until the target pressure is met. Be aware that this pressure is flow pressure, not static pressure. Water must be flowing to make this determination. Use of this fire-ground method simplifies a situation requiring complicated theoretical hydraulics calculations.

Hydraulic Calculators

Hydraulic calculators may be either manual (mechanical) or electronic. The manual hydraulic calculator may consist of a sliding card or a slide rule. These items can handle calculations involving a variety of nozzle pressures, flow rates (in GPM or L/min), and hose diameters, and indicate FL per 100′ (30 m) of hose. Slide rule hydraulic calculators offer a fire stream calculator (GPM or L/min) on one side and a FL calculator on the other side.

To find the flow rate with a fire stream calculator, follow the steps in **Skill Drill 4-17**:

1. Slide the interior until the arrow at the top points to the NP selected.
2. Choose the NP: 50-psi (350 kPa) smooth-bore nozzle, 100-psi (700-kPa) fog nozzle, or 75-psi (525-kPa) fog nozzle.
3. For a smooth-bore nozzle, find the tip size; the flow will be indicated below it.
4. For a 100-psi (700-kPa) fog nozzle, find the rating at 100 psi (700 kPa); the flow will be indicated below it.
5. For a 75-psi (525-kPa) fog nozzle, find the rating at 75 psi (525 kPa); the flow will be indicated below it.

To calculate the flow for a given hose size and nozzle pressure using a slide rule calculator and measurements in the metric system, follow these steps:

1. Slide the interior until the arrow at the top points to the NP selected.
2. Choose the nozzle: smooth-bore nozzle, 700-kPa fog nozzle, or 525-kPa fog nozzle.
3. For a smooth-bore nozzle, find the tip size; the flow will be indicated below it.
4. For a 700-kPa fog nozzle, find the rating at 700 kPa; the flow will be indicated below it.
5. For a 525-kPa fog nozzle, find the rating at 525 kPa; the flow will be indicated below it.

To use a manual friction loss calculator, follow the steps in **Skill Drill 4-18**:

1. Slide the interior to set the flow to the arrow at the top.
2. Look for the hose size (whether single or connected to a Siamese connection) and read the FL indicated.

Electronic calculators may be either mounted or hand-held devices. The mounted calculators are seldom encountered today, whereas the hand-held type is quite common. The hand-held pocket-size electronic calculator allows you to calculate engine pressure, FL, application rate, flow rate, and reaction force (RF, also known as NR). All in all, the pocket calculator is a very useful tool.

Hand Method

The purpose of the hand method (the counting fingers method) is to quickly determine the amount of FL per 100′ (30 m) of hose. The FL must then be multiplied by the number of 100′ (30-m) lengths of hose. Next, add any AL, the EP, and the NP to provide the correct PDP.

To use the hand method to calculate friction, follow the steps in **Skill Drill 4-19**, and then complete Skill Drill 4-20:

1. Multiply the first digit of the flow by the number of sections used.
2. Add the other variables of AL and elevation gain or loss if applicable. This will determine the PDP for that hose line.
3. With the FL established, add the other variables of AL and elevation gain or loss, if applicable, to the NP to determine the PDP for that hose line.

Hand calculation methods have been established for almost every size of hose. The primary attack lines in the fire service today are 1¾″ and 2½″ hose, so naturally the methods for these two sizes are most common. To perform the hand method calculation for the 2½″ hose, follow the steps in **Skill Drill 4-20**:

1. Start with the left hand open and the thumb on the left.
2. Assign the flow in hundreds of GPM to the base of each finger, from left to right, in 100-GPM increments, designating the thumb as 100 GPM, then 200 GPM, 300 GPM, 400 GPM, and finally 500 GPM on the little finger.
3. Assign the spaces between fingers as the half-hundred figures (i.e., 150, 250, 350, 450). Thus the flow designations have been assigned to the hand for 2½″ hose.
4. Assign to the fingertips the multiplier numbers of 2 for the thumb, and 4, 6, 8, and 10 across the hand to the fingertips.
5. Assign 3, 5, 7, and 9 to the spaces between the fingers.
6. The FL for 100′ of 2½″ hose is obtained by selecting the desired flow and multiplying the number at the tip of the finger by the first digit of the number at the base of the finger. Half-hundred flows such as 250 GPM may be translated as 2.5.

The hand method for 2½″ hose works so nicely because the coefficient for 2½″ hose is 2 when using imperial units. In metric units, use the quick method for 64-mm hose. The coefficient for 64-mm hose (same hose, different name) is 1 if you use 30-m lengths, which makes things even easier. The FL for each 30-m length of 64-mm hose can be determined by dividing the flow in L/min by 100 and squaring that number to obtain

Skill Drill 4-19

Performing the Hand Method of Calculation

Example: Calculate the PDP for 200′ of 2½″ hose flowing 250 GPM through a fog nozzle up a hill with an elevation of 40′.

The hand method calculates the FL per 100′ of hose. When flowing 250 GPM through a 2½″ line, we find the number on the hand that corresponds to the flow of 250 GPM. The number 5 corresponds to the flow, so we multiply 2.5 (from the flow of 250) times 5: 2.5 × 5 = 12.5 per 100′.

From the scenario we know we have 200′ of hose, so FL = 2 × 12.5 = 25 psi.

Insert the values into the formula:

PDP = 100 NP + 25 FL + 20 EP = 145 psi

If there were AL, it would have been added into the calculation.

Skill Drill 4-20

Performing the Hand Method of Calculation for 2½″ Hose

Example 1: When flowing 250 GPM in 100′ of 2½″ hose, multiply the first digit of the flow, 250 GPM = 2.5 (found at the space between the index and middle finger), by the corresponding multiplier of 5 = 12.5 psi FL for 100′ of 2½″ hose.

Example 2: If flowing 300 GPM through 100′ of 2½″ hose, multiply the first digit of the flow, 300 GPM = 3 (found at the base of the middle finger), by the corresponding multiplier of 6 = 18 psi FL per 100′ of 2½″ hose.

---

the FL in kPa. This is the same as setting both C and L equal to 1 in the FL equations.

To perform the hand method calculation for 1¾″ hose, follow the steps in **Skill Drill 4-21**. Note that there is no simple metric equivalent to this method.

1. Start with the left hand open and the thumb on the left.
2. Assign the flow in GPM to the base of each finger, from left to right, in 25 GPM increments, designating the thumb as 100 GPM, then 125 GPM, 150 GPM, 175 GPM, and finally 200 GPM on the little finger.
3. Assign numbers to the tips of the fingers beginning with the thumb as 1, then 2, 3, 4, and 5 on the little finger.
4. Assign the multiplier number of 12 to the base of the thumb, and 9 to the base of each of the other fingers.
5. To calculate the FL for a given flow through a 1¾″ hose, find the GPM on the hand, and multiply the number at the tip times the number at the bottom of the finger.
6. The hand method calculates the *FL per 100′ of hose only*. Multiply this value times the length and add any AL loss, elevation gain/loss, and NP to find the correct PDP.

Skill Drill 4-21

Using the Hand Method Calculation for 1¾" Hose

Example: When flowing 150 GPM, note that 150 GPM is the middle finger and has the number 3 at the tip and the number 9 at the base. When multiplied, 3 × 9 = 27 psi FL in 100' of 1¾" hose.

Subtract 10 Method (GPM Flowing Method)

The subtract 10 method determines FL in 2½" hose only and for flows of 160 GPM or greater. It is useful for either fog or smooth-bore nozzles. The simplicity of this method is its strength; using this technique, you can determine FL in 2½" hose very quickly. To use this method, follow the steps in **Skill Drill 4-22**. Note that there is no simple metric equivalent to this method.

1. Assume a flow of 200 GPM in a 2½" line.
2. Subtract 10 from the first two digits of the flow to obtain the FL per 100' of hose: 200 GPM = 20 − 10 = 10 psi FL per 100'.
3. If flowing 250 GPM, subtract 10 from 25 = 15 psi FL per 100'.
4. If flowing 300 GPM, subtract 10 from 30 = 20 psi FL per 100'.
5. After obtaining the FL, multiply it times the number of 100' sections and add AL, EP (if any), and NP to arrive at the PDP.

Condensed Q Method

The **condensed Q method** is a quick method for calculating FL per 100' in 3" to 5" hose line only. It is especially useful when the apparatus is part of a relay and is supplying another pumper. Note that there is no simple metric equivalent to this method. The condensed Q formulas are as follows:

Skill Drill 4-22

Using the Subtract 10 Method (GPM Flowing Method)

Example 1: Calculate the PDP for 300' of 2½" hose with a 1" smooth-bore tip.

The 1" tip flows 200 GPM and has a NP of 50 psi. Subtract 10 from the first two digits of the flow:

For 200 GPM: 20 − 10 = 10 psi FL per 100'
300' of hose with 10 psi loss per 100' = 30 psi FL
PDP = NP + FL = 50 + 30 = 80 psi

Example 2: Do the same problem using the Q formula.

FL = C Q² L or 2 × (2)² × 3 = 24 psi FL
PDP = 50 + 24 = 74 psi

$$3'' \text{ hose FL} = Q^2$$
$$3\tfrac{1}{2}'' \text{ hose FL} = Q^2/3$$
$$4'' \text{ hose FL} = Q^2/5$$
$$5'' \text{ hose FL} = Q^2/15$$

To use the condensed Q method, follow the steps in **Skill Drill 4-23**:

1. Use the appropriate formula for the size hose.
2. To calculate FL in 3″ hose, use the formula FL = Q^2.
3. When flowing 300 GPM, $Q^2 = 3^2 = 9$ psi FL per 100′ of hose.
4. To calculate FL in 3½″ hose, use the formula FL = $Q^2/3$. For example, when flowing 300 GPM through 3½″ hose, $Q^2/3 = 3^2/3$ or $9/3 = 3$ psi per 100′ of 3½″ hose.
5. To calculate FL in 4″ hose, use the formula FL = $Q^2/5$. For example, when flowing 500 GPM, $Q^2/5 = 5^2/5 = 25/5 = 5$ psi per 100′ of 4″ hose.
6. To calculate FL in 5″ hose, use the formula $Q^2/15$. For example, when flowing 700 GPM, $Q^2/15 = 7^2/15 = 49/15 = 3.2$ or 3 psi per 100′ of 5″ hose.

Driver/Operator Tip

As an alternative calculation on the fire ground for 5″ hose, some fire fighters use a 5″ condensed Q formula of FL = $Q^2/10$, simply because the lengths and GPM are easily divided by 10. For example, if the flow is 700 GPM, $Q^2/10 = 7^2/10$ or $49/10 = 5$ psi per 100′ of 5″ hose. A calculation using the relationship $Q^2/15$ would result in 3.26 psi per 100′ of hose. Approximations allow you to set up the scene quickly.

The Condensed Q Method in a Relay Operation

When two pumpers arrive at a fire scene at the same time or within seconds of each other, the first pumper often positions itself for attack while the other lays a line from the attack pumper to the hydrant to establish the water supply. In a rural setting it is often essential to establish a relay owing to the lack of hydrants or the long distance to a hydrant.

Whenever one pumper is supplying another apparatus, the operator of the supply pumper must quickly calculate the PDP for the supply line to the attack pumper. Because most fire departments use a supply line with a diameter in the range of 3″ to 5″, the condensed Q method is a quick and easy way to determine the PDP.

For example, suppose Engine 1 is on a hydrant supplying 600′ of 5″ hose to Engine 2, which is pumping 1200 GPM. What is the PDP for Engine 1? Engine 1 needs to know only the size and length of the hose and the rate at which Engine 2 is flowing water. It doesn't matter which hose configuration Engine 2 is using; instead, Engine 1 needs to know the total flow so Q can be determined. The driver operator for Engine 2 tells the driver/operator of Engine 1 on the radio that he is flowing 1200 GPM. The driver/operator for Engine 1 will calculate the FL for the 5″ hose with the knowledge that Q = 12. As there is no NP, allow 20 psi for the intake into Engine 2 and insert this value in place of NP.

Calculating PDP Using the Q Formula

$$PDP = NP + FL$$
$$PDP = 20 + FL$$
$$FL = C Q^2 L$$
$$FL = 0.08 \times (12)^2 \times 6$$
$$FL = 0.08 \times 144 \times 6 = 69.12 \text{ or } 70 \text{ psi}$$
$$PDP = 20 + 70 = \mathbf{90} \text{ psi}$$

Alternative: Calculating PDP Using the Condensed Q Formula

With 5″ FL = $Q^2/15$:

$$FL = 12^2/15 = 144/15 = 9.6 \text{ per } 100'$$
$$FL = 9.6 \times 6 = 57.6$$
$$PDP = 20 + 58 = 78 \text{ psi or } \mathbf{80 \text{ psi}}$$

With 5″ FL = $Q^2/10$:

$$FL = 12^2/10 = 144/10 = 14 \text{ per } 100'$$
$$FL = 14 \times 6 = 84 \text{ psi}$$
$$PDP = 20 + 84 = \mathbf{104 \text{ psi}}$$

Although the numbers vary, from this example you can see that the condensed Q method will quickly present a close workable calculation. If the pressure supplied to the attack pumper is too low or too high, then the driver/operator of the attack pumper will radio the driver/operator of the supply pumper and simply say, "Give me 10 more pounds" or "Drop it down 10 pounds," instead of offering new numbers to recalculate the needed pressure.

Flow Meters and Electronic Pump Controllers

The **flow meter** is basically a straight section of pipe, threaded on each end, that is placed between two sections of hose. A sensing device on top of the meter is designed to measure the flow (GPM) through the hose. In the fire service, these sensors are generally of spring probe or paddlewheel design. The paddlewheel version has the paddlewheel on top of the pipe where sediment cannot settle on the paddlewheel; the speed of the wheel indicates the flow. The spring probe flow meter injects a stainless steel spring probe into the stream to measure resistance and read the amount of flow (in either GPM or L/min), with measurements being read in increments of 10 GPM (40 L/min) or less.

Flow meters give an indication of the actual volume of water being discharged through each line. This flow is sent as a signal to the gauge. The indicator gauge may be either digital or analog.

The use of a flow meter can lessen the amount of calculations required to determine PDP. To do so, the driver/operator communicates with the attack team and, by reading the flow meter, increases or decreases the pressure to achieve the desired flow as directed.

Many types of flow meters are available, from a variety of manufacturers. NFPA 1901 allows flow meters on discharges from 1½" (38 mm) to 3" (76 mm) in diameter. Larger-diameter hose may use a flow meter but must have a pressure gauge as well. Combination models offer both flow and pressure gauges. Some indicate the flow of only one discharge, whereas others provide a total flow button that displays the total amount flowing from the pump. Flow meters generally boast accuracy within 1 to 3 percent, but require periodic calibration to maintain this accuracy.

One important characteristic of the flow meter is that it can deliver the desired flow without the operator knowing the length of the hose lines (L), the FL, or the EP. As a consequence, use of flow meters is associated with several advantages:

- Flow meters prove to be very useful in multistory buildings, as hose length, friction loss, and elevation vary in incidents involving these structures and may be undetermined.
- The flow from master stream appliances with automatic nozzles may look good but may be actually flowing less than required. Flow meters eliminate this problem.
- Flow meters eliminate the math needed for a relay operation because the flow meter is automatically set to the desired flow.

Electronic Pump Controllers and Pressure Governors

In recent years, technology has reduced the amount of calculations that the driver/operator must perform. In addition, technology is constantly changing the electronic systems on fire trucks and has influenced the way the pump panel operates. Multiplexing is a term used in electrical engineering to refer to the process of combining information from several different sources. In the fire service, this means different sensors and controls on a pumper are brought to one control board or indicating panel. The trend in modern fire apparatus is to use electronic control and monitoring equipment. All new pumping apparatus, for example, include electronic throttle and pressure controls.

Electronic components are often designed by an independent contractor and integrated into the manufacturer's apparatus. Two examples are the Electronic Fire Commander (EFC) offered by Class 1 and the Incontrol mechanism developed by Fire Research Group. These companies provide a small computer panel (keypad and display) for controlling the pump. Placed on the pump panel, these panels contain both a pressure governor and an engine speed governor, and include a preset function that eliminates the need for a throttle or pressure relief valve. They also feature an LED information center to display engine data such as temperature, oil pressure, speed (rpm), and voltage. A one-touch button advances the pressure to the preprogrammed pressure; another returns the engine speed to the idle level. If the preprogrammed pressure is not needed, the pressure may be increased or decreased by pushing an increase pressure button or decrease pressure button, respectively.

While this technological innovation assists the driver/operator, the concern that arises with its use is the possibility of an electrical failure disabling the fire apparatus. The threat of an electrical malfunction is increased with such technology because of the frequent absence of manual overrides on fire apparatus with these features. This fact further highlights the necessity of the driver/operator having a thorough understanding of the fire apparatus and pumping procedures should problems occur.

Preincident Plan

When developing a suppression strategy, there is no more useful tool than knowledge of the target structure and a well-developed **preincident plan**. When conducted with permission from the owner and at a convenient time, a preincident survey yields valuable information to the fire department's knowledge base. The resulting preincident plan data may be stored on paper, in a book, on the station computer, or in the onboard computer on the fire apparatus. Gathering building blueprints or diagrams and as much relevant data as possible during the survey process will enrich the plan.

The FDC on the structure, for example, may indicate whether it serves the standpipe system or sprinkler system or both. It may indicate the pump pressures required and the maximum pressure the system will allow. It may also indicate which wing or part of a building the FDC serves.

Information collected as part of preincident planning and the knowledge gained by the company while conducting the preplanning process are priceless assets. This information and experience will better prepare the company and department for an emergency response.

Wrap-Up

Chief Concepts

- Fire service hydraulic calculations are used to determine the required pump discharge pressure (PDP) for fire-ground operations.
- Fighting fire with water is a matter of water flow versus heat generation.
- Sufficient flow must be applied to overcome the heat generated by the fire.
- Friction loss is the pressure lost from turbulence as water passes through pipes, hoses, fittings, adapters, and appliances. It is measured in pounds per square inch (psi) or kilopascals (kPa).
- The pump discharge pressure formula is PDP = NP (nozzle pressure) + FL (friction loss).
- Fire service hydraulic calculations may generally be categorized as either theoretical hydraulics or fire-ground hydraulics.
- Theoretical hydraulic calculations entail more exact calculations, generally use more mathematical skills, and take more time to compute.
- Away from the fire ground, theoretical hydraulic calculations may be computed on paper, calculator, or computer.
- Fire-ground hydraulic calculations involve methods devised to estimate the needed data more quickly due to the urgency of the fire scene.
- Fire-ground hydraulic calculations may also be used as a backup to mental theoretical calculations, as a system to check results, and as a means to ensure reasonable accuracy.
- Fire-ground hydraulic calculations also reinforce the learning process as the driver/operator develops the ability to compute theoretical hydraulics in the field.
- Fire-ground hydraulic calculations yield working approximations and are not as accurate as theoretical hydraulic computations.

Hot Terms

British thermal unit (Btu) The quantity of heat required to raise the temperature of one pound of water by 1°F at the pressure of 1 atmosphere and a temperature of 60°F. A British thermal unit is equal to 1055 joules, 1.055 kilojoules, and 252.15 calories. (NFPA 921)

chart A document summarizing the typical handline and master stream calculations for the hoses, nozzles, and devices specific to the fire apparatus. It lists the most commonly used or reasonably expected hose lays.

coefficient (C) A numerical measure that is a constant for a specified hose diameter.

condensed Q method Fire-ground method used to quickly calculate friction loss in 3″ (76-mm) to 5″ (127-mm) hose lines.

critical rate of flow The essential flow, measured in gallons per minute, that is needed to overcome the heat generated by fire.

elevation gain The pressure gained when the nozzle is below the pump; it requires pressure to be subtracted from the discharge pressure.

elevation loss The pressure lost when the nozzle is above the pump; it requires pressure to be added to the discharge pressure to compensate for the loss.

elevation pressure (EP) In hydraulic calculations, this is the distance the nozzle is above or below the pump.

fire department connection (FDC) A connection through which the fire department can pump supplemental water into the sprinkler system, standpipe, or other system, thereby furnishing water for fire extinguishment to supplement existing water supplies. (NFPA 25)

fire-ground hydraulics Simpler fire-ground methods for performing hydraulic calculations.

flow meter A device for measuring volumetric flow rates of gases and liquids. (NFPA 99)

flow rate The volume of water moving through the nozzle measured in gallons per minute or liters per minute.

hydraulics Study of the characteristics and movement of water as they pertain to calculations for fire streams and fire-ground operations.

master stream A portable or fixed firefighting appliance supplied by either hose lines or fixed piping that has the capability of flowing in excess of 300 GPM (1140 L/min) of water or water-based extinguishing agent. (NFPA 600)

net pump discharge pressure (NPDP) Amount of pressure created by the pump after receiving pressure from a hydrant or another pump.

nozzle pressure (NP) Pressure required at the inlet of a nozzle to produce the desired water discharge characteristics. (NFPA 14)

portable master stream device A master stream device that may be removed from a fire apparatus, typically to be placed in service on the ground.

preincident plan A written document resulting from the gathering of general and detailed data to be used by responding personnel for determining the resources and actions necessary to mitigate anticipated emergencies at a specific facility. (NFPA 1620)

prepiped elevated master stream An aerial ladder with a fixed waterway attached to the underside of the ladder, with a water inlet at the base supplying a master stream device at the tip.

pressure-regulating valve (PRV) The type of valve found in a multistory building, which is designed to limit the pressure at a discharge so as to prevent excessive elevation pressures under both flowing (residual) and nonflowing (static) conditions.

pump discharge pressure (PDP) Pressure measured at the pump discharge needed to overcome friction and elevation loss while maintaining the desired nozzle pressure and delivering an adequate fire stream.

standpipe The vertical portion of the system piping that delivers the water supply for hose connections, and sprinklers on combined systems, vertically from floor to floor in a multistory building. Alternatively, the horizontal portion of the system piping that delivers the water supply for two or more hose connections, and sprinklers on combined systems, on a single level of a building. (NFPA 14)

standpipe riser The vertical portion of the system piping within a building that delivers the water supply for fire hose connections, and sprinklers on combined systems, vertically from floor to floor in a multistory building.

standpipe system An arrangement of piping, valves, hose connections, and allied equipment with the hose connections located in such a manner that water can be discharged in streams or spray patterns through attached hose and nozzles, for the purpose of extinguishing a fire and so protecting designated buildings, structures, or property in addition to providing occupant protection as required. (NFPA 14)

theoretical hydraulics Scientific or more exact fire-ground calculations.

total pressure loss (TPL) Combination of friction loss, elevation loss, and appliance loss.

Driver/Operator in Action

Practice your mathematical skills with these word problems.

1. A pumper is flowing 250 GPM through 200′ of 2½″ hose with a fog nozzle. What is the PDP?

2. A pumper is flowing water through 300′ of 2½″ hose with a 1″ smooth-bore nozzle. What is the PDP?

3. Using the hand method of calculation, what is the FL for 175 GPM through 200′ of 1¾″ hose?

4. A pumper is flowing two lines. Line A is flowing 300 GPM through 200′ of 2½″ hose with a smooth-bore nozzle. Line B is flowing 161 GPM through 300′ of 1¾″ hose with a smooth-bore nozzle. What is the PDP for each line?

5. A pumper has a 1¾″-diameter line that is 200′ long, going up a 30′ hill, with a 185-GPM smooth-bore nozzle on the end. What is the PDP?

6. A pumper has a 2½″-diameter line that is 300′ long, going down a 50′ hill, with a 250-GPM fog nozzle. What is the PDP?

7. A pumper is relaying water from a hydrant through 600′ of 5″ hose to an attack pumper flowing water at 1000 GPM. What is the PDP for the supply pumper?

8. A pumper is supplying two 2½″ lines that are 200′ long to a portable master stream device on the ground that has a 1¾″ tip. What is the PDP?

Practice your mathematical skills with these word problems.

1. A pumper is supplying three 64-mm lines, each 30 m long, to a portable master stream device on the ground that is flowing 3800 L/min through a fog nozzle. What is the PDP?

2. A pumper is supplying three 64-mm lines, each 60 m long, to a Siamese appliance that is flowing water at 3800 GPM. Using the Siamese coefficient method of calculation, what is the TPL?

3. A pumper is supplying three 64-mm lines, each 30 m long, to a portable master stream device with a 45-mm tip. What is the PDP calculated using the percentage method?

4. A pumper is supplying two 100-mm lines, each 90 m long, to an aerial ladder elevated 24 m and flowing water at 2250 L/min through a smooth-bore nozzle. What is the PDP?

5. The supply pumper is supplying an attack pumper in a relay with two 150-m lines of 89-mm hose. The attack pumper is flowing water at 1900 L/min. What is the PDP for the attack pumper?

6. A pumper is supplying an aerial master stream with two 100-mm lines, each 45 m long. The ladder is extended 30 m and flowing water at 5700 L/min through a fog nozzle. What is the PDP?

7. A pumper is supplying a standpipe system in a multistory building with two 64-mm lines, each 30 m long. There is 60 m of 64-mm attack line with a 32-mm tip on the fifteenth floor. There is also 45 m of 64-mm attack line with a 29-mm tip on the fourteenth floor. What is the PDP?

8. You are conducting a preplanning survey of a standpipe system. The pumper has two 64-mm lines, each 60 m long, to the FDC. There are two 64-mm attack lines, each 60 m long, with 29-mm tips on the seventh floor. The riser is 150 mm in diameter. What is the PDP?

Pumper Apparatus Overview

CHAPTER 5

NFPA 1002 Standard

There are no specific NFPA objectives for this chapter.

Additional NFPA Standards

NFPA 20 *Standard for the Installation of Stationary Pumps for Fire Protection* (2007)

Knowledge Objectives

After studying this chapter you will be able to:

- Explain the importance of understanding the fire pump and its systems.
- Describe the exterior and interior features of a pumper.
- Define the term *pump*.
- Explain the basic operations of positive-displacement pumps and centrifugal pumps.
- Explain the different types of positive-displacement pumps.
- Explain the different types of centrifugal pumps.
- Describe a single-stage pump and a two-stage pump.

Skill Objectives

There are no skill objectives for this chapter.

You Are the Driver/Operator

As the driver/operator of the fire apparatus, you are assigned to give a short presentation to some new fire fighters on how the fire apparatus pump works. Your officer has asked that you give this presentation to all of the station's members in the next 30 minutes. This means that you had better be ready for any questions that the other members of the firehouse might throw at you and that you have some questions to ask them in return. You decide to look at some of your old class notes on the different types of pumps that are frequently used in the fire service. While reviewing the material, you come up with some good questions to ask the other members of the company.

1. What does a positive-displacement pump do?
2. How does a centrifugal pump work?
3. Does this engine have a single-stage or multistage pump?

Introduction

This chapter discusses the basic principles underlying the use of pumps. Pumps are vital to the operations of firefighting. Without them, fire fighters would be unable to discharge water under pressure to extinguish fire Figure 5-1 . As the driver/operator of the pump apparatus, you must understand how pumps operate so that you are able to fix any problems that you may encounter on the fire ground.

The fire service has an old saying, "Don't just be a knob puller—know what you are doing." Do not just memorize a specific sequence of tasks to operate the fire pump; make sure that you have a thorough understanding of how the pump operates. For example, if you understand that a **centrifugal pump** is capable of pumping only fluids (and not air), than you will know that you have to **prime the pump** before it can operate.

Exterior of the Pumper

The pumper used by an engine company is a very basic fire apparatus. Essentially, it is a large fire pump with hose and tools to extinguish fires. The pumper is the most common fire apparatus, being found in almost every fire department's fleet. The cab of the fire apparatus, where the fire fighters ride to the emergency, sits on a steel frame. Attached to the frame are the storage compartments and the fire pump. The compartments on the fire apparatus store all of the various tools and equipment needed for firefighting operations. Ground ladders are either mounted on the side of the fire apparatus or stored inside a special compartment at the rear of the fire apparatus.

Supply hose and attack lines are stored in the hose bed, which may be either left uncovered or covered with a tarp to protect it from the elements. Some fire apparatus may provide a metal cover to further protect the hose. Most pumpers carry preconnected attack lines for quick deployment of attack lines; these hose lines are usually located above or on the side of the pump panel.

The pump panel is the most notable device on the fire apparatus Figure 5-2 . It is usually covered in stainless steel and has multiple levers and gauges mounted on it. Although this panel may seem intimidating to the new driver/operator, once you have a thorough understanding of the pump behind it and how it functions, you will be capable of confidently operating the fire pump.

Figure 5-1 Pumps are essential to a successful fire attack.

Figure 5-2 The pump panel may seem very confusing. Once you learn the basics of pump operations you will become more comfortable with its components. **A.** A 1998 KME Quint 1500 GPM single stage pump with no pressure governor. **B.** 1500 GPM waterous two stage pump.

Interior of the Pumper: The Cab

Inside the cab of the pumper, you will find all of the necessary controls to operate the fire apparatus **Figure 5-3**. When you are sitting in the driver's seat, many of the controls and features will seem similar to those found in any other large vehicle. There are switches to operate the seats, mirrors, headlights, and windshield wipers, just as in any other vehicle. Some of the controls, however, are very different. For example, the controls to engage the fire pump or operate the emergency lights will not be found in other types of vehicles. It is important for the new driver/operator to understand that the fire apparatus is not like any other vehicle—and, therefore, should not be operated like other vehicles.

Pumps

Fires are extinguished when the proper amount of water (or gallons per minute [GPM] rate) is applied to the fire. This volume of water is directed through a nozzle at the required pressure to give the fire stream enough reach so that it can penetrate to the seat of the fire. A pump pressurizes the water used to attack the fire. According to the 2007 Edition of NFPA 20, *Standard for the*

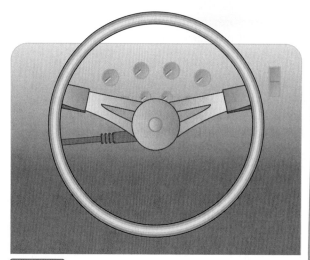

Figure 5-3 The interior vehicle controls on an engine apparatus are very similar to those found on many other large vehicles with a few exceptions.

Installation of Stationary Pumps for Fire Protection, a **fire pump** is defined as a provider of liquid flow and pressure dedicated to fire protection. It is a mechanical device used to move fluids. This concept is very important for fire fighters because they want the pump to move the water from the source, such as the onboard water tank, to the fire through the attack lines. To do so successfully, you must understand how each different type of pump works.

One of the first things to understand is that pumps alone do not create pressure. Pumps displace the fluid, which causes this fluid to move or flow; resistance to this flow, in turn, creates pressure. With higher pressures, you will have less volume or flow; with higher flows, you will have less pressure. Pumps cannot provide both high pressure and high volume of water at the same time.

The two main sides of all pumps are the intake and discharge sides. The **intake side** of the pump is where the water enters the pump. It is also referred to as the "supply side" of the pump because it is where water is supplied to the pump. On most pumps, the pipes and connections for the intake are set lower than the discharge piping and connections. The location where the water exits the pump is called the **discharge side**.

Types of Fire Pumps

The fire service primarily uses two types of pumps: positive-placement pumps and centrifugal pumps. Each type relies on very different operating principles and offers a very different set of features.

Positive-Displacement Pumps

According to NFPA 20, a **positive-displacement pump** produces a flow by capturing a specific volume of fluid per pump revolution and reducing the fluid void by a mechanical means to displace the pumping fluid. It traps a fixed amount of fluid and forces that amount into the discharge stream during every revolution of the pumping element. The pump displaces liquid by creating a space between the pumping elements and trapping liquid in the space. As the pumping element moves, it reduces the size of this space, which in turn forces the fluid out of the pump.

Positive-displacement pumps rely on tightly fitting parts to function properly. Although these models are self-priming, they still require proper conditions, such as no excessive air leaks, to function properly. They can move air or water during every revolution because the parts fit together so closely. This makes the positive-displacement pump an ideal choice for use as a priming pump for centrifugal pumps.

As a priming pump, the positive-displacement pump is connected to the top of the centrifugal pump. When the positive-displacement pump is operated, it draws air and water from the top of the centrifugal pump until a constant flow of water is achieved, thereby ensuring that only water is inside the centrifugal pump casing.

Positive-displacement pumps may also be used as high-pressure auxiliary pumps or portable pumps. Because their efficiency depends on their close-fitting moving parts, the performance of these pumps will begin to deteriorate as they wear down over time and with excessive use. Sand and other debris can cause the moving parts to wear out prematurely, further diminishing their performance.

The two broad classifications of positive-displacement pumps for firefighting purposes are the piston pump and the rotary pump. Rotary pumps exhibit a circular motion, whereas piston pumps have an up-and-down action.

Piston Pumps

Piston pumps use a cylinder to contain the fluid; a piston then moves back and forth inside the unit to move the liquid out of the cylinder. As the piston moves, water is drawn in from the intake side of the pump and expelled out the discharge side of the pump. This output of fluid is directly related to the revolutions per minute (rpm) of the piston: As the piston moves faster, the liquid will be discharged faster.

Piston pumps have three moving parts: the piston, the intake valve, and the discharge valve. Inside the cylinder is the piston, which creates a seal in the cylinder to separate the intake side and the discharge side of the pump. The intake and discharge valves create or block their respective openings so as to direct the water toward the target.

Two types of piston pumps are available: single-acting piston pumps and double-acting piston pumps.

Single-Acting Piston Pump

The operation of a **single-acting piston pump** is very simple. This type of pump has one intake side and one discharge side, which are located at the same ends of the cylinder. As the piston moves upward, away from the intake side, a vacuum is created in the cylinder. This vacuum causes the intake valve to open and fill the cylinder with water. Water continues to fill the cylinder until the piston stops. At the same time, at the discharge side of the pump, the discharge valve remains in the closed position. Once the piston moves in the opposite direction, the intake valve will close and the discharge valve will open to expel the water.

The single-acting piston pump is similar to a squirt gun. As you depress the trigger, water inside a cylinder is compressed and discharged out the tip **Figure 5-4**. When you release the trigger, the cylinder refills with water and waits to be discharged again **Figure 5-5**.

> **Driver/Operator Tip**
>
> The first hand pump used in the first fire brigades consisted of a single-acting piston pump.

With every pump on the lever, a definite amount of fluid is discharged from the single-acting piston pump. Water is discharged only on the downward movement of the piston, which creates a pulsating effect because water flow is not constant. As a consequence, single-acting piston pumps are not suitable for use on an attack line. The fire stream from this type of pump would be inconsistent and have moments without any water discharged. An additional piston can be added to increase the

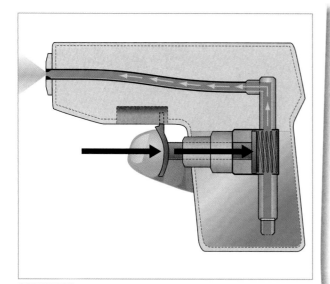

Figure 5-4 As you depress the trigger, water inside a cylinder is compressed and discharged out the tip.
Adapted from Hale Products, Inc.

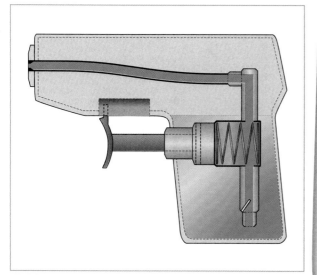

Figure 5-5 When the trigger is released, the cylinder refills with water and waits to be discharged again.
Adapted from Hale Products, Inc.

water flow and provide a more constant flow. With this setup, the two pistons would operate in opposite directions to maintain the water flow. While one piston is filling with water, the other would be discharging the water.

Double-Acting Piston Pump

A <u>double-acting piston pump</u> allows water to flow more continuously than a single-acting piston pump while using only one piston. This double-acting piston pump still has periods of limited flow—specifically, when the pistons are at the top of their stroke before changing direction. Instead of having only one intake valve and discharge valve at one end of the cylinder, however, this pump has both valves at each end. This allows the

Driver/Operator Tip

Steam-powered pumps used the same pump design as piston pumps to move water. Some of the first motorized fire apparatus used piston pumps that were able to deliver a flow volume of as much as 500 GPM (1892.7 liters per minute [L/min]). The Aherns-Fox fire apparatus was a popular piston pump fire apparatus for many years **Figure 5-6**.

Figure 5-6 The Aherns-Fox fire apparatus was a popular piston pump fire apparatus for years.

cylinder to fill with water on both the up and down strokes of the piston. When the piston is pulled upward, the water flows into the cylinder through the intake valve; at the same time, the discharge valve on the other side of the piston flows water through it.

Rotary Pumps

<u>Rotary pumps</u> are typically used as the priming pump for a centrifugal pump. These pumps operate in a circular motion and discharge a constant flow of water with each revolution. To accomplish this task, some <u>pumping element</u>, such as a gear or vane encased in the pump casing, rotates, expanding the volume inside to allow water to enter the pump. As the pumping elements rotate, the area in which the fluid is contained is reduced, which in turn forces the water out of the pump. This results in a smooth continuous flow of water discharged from the pump. The internal components fit together tightly and allow for very little water to slip back from the discharge side to the intake side of the pump during operation.

Rotary Gear Pumps

The type of positive-displacement pump most commonly encountered in the fire service is the <u>rotary gear pump</u>, which is typically used as a priming pump **Figure 5-7**. The rotary gear pump uses two gears, most often driven by a 12-volt electric motor. Inside the pump casing, two gears rotate in opposite directions. These gears are closely positioned to each other and to the inside of the pump casing, forming a watertight seal. As the two gears mesh together, they trap water and move it to the

Figure 5-7 Rotary gear pump.

discharge side of the pump. Usually, a motor powers one gear, which then powers the second gear. Oil may be fed into the pump to maintain as watertight a seal as possible. The oil will fill any open spaces to make the pump work more efficiently.

Rotary Vane Pumps

A **rotary vane pump** uses small moveable elements called **vanes**, which freely move in and out of the slots of the **rotor** to maintain a tight seal against the pump casing **Figure 5-8**. The vanes are positioned off-center inside the pump casing and automatically

Driver/Operator Tip

Operating the priming valve on the pump panel opens a valve from the priming pump to the centrifugal pump and engages the electric motor of a rotary pump. With every revolution of the rotary pump, the pump moves air or water. As a priming pump, it is designed to pump the air from the centrifugal pump, thereby ensuring that only water is available inside the centrifugal pump.

Driver/Operator Tip

Other priming devices that do not use a pump may also be used to prime the main pump. For example, an intake manifold priming device uses the vacuum of the truck's engine on the intake manifold to prime the pump. The two chambers of the device prevent water from entering the engine—an arrangement commonly known as an **exhaust primer**. The engine's exhaust is directed through a device that causes a **Venturi effect**, which creates a vacuum. This vacuum is connected to the main fire pump, drawing air out of it. While some fire apparatus use this type of device, it is most often found on portable pumps.

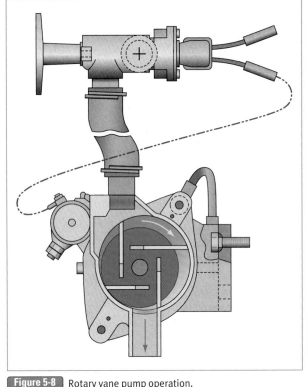

Figure 5-8 Rotary vane pump operation.
Adapted from Hale Products, Inc.

maneuver in and out to compensate for changes in the pump casing. As the vanes approach the intake side of the pump, the void space increases and the vanes slide farther out of their slots. Water then flows in between these vanes, becoming trapped there. When the vanes approach the discharge side of the pump, the void space decreases and the vanes slide farther back into their slots. Water that is trapped in between the vanes is then discharged out the pump. Because the vanes in the rotor move independently, as the surface of the vane itself wears down over time, it will continue to maintain a watertight fit: It can automatically adjust itself to compensate for the wear and tear in the unit. Centrifugal force keeps the vane tightly pressed against the pump casing, thereby ensuring a tight seal. This makes the pump operate very efficiently.

■ Centrifugal Pump

The most common fire pump in use today is the centrifugal pump **Figure 5-9**. This type of pump has largely replaced positive-displacement pumps on modern fire apparatus. Centrifugal pumps do not flow a definite amount of water with each revolution; instead, the flow or amount of water discharged is based on the pressures at the discharge side of the pump. At higher flow rates (rpm), the pump will flow less volume but will create higher pressures. Consequently, as the pump spins more slowly, it will create less pressure on the discharge side but will flow a greater volume of water. This flexibility makes the centrifugal pump a very versatile piece of equipment on the fire ground. As the driver/operator, you need to understand how this pump works to maximize its potential as part of firefighting operations.

Voices of Experience

One of the most valuable skills I have learned is being calm while under extreme duress. A driver/operator who is traumatized and panics has a detrimental affect on his or her performance, endangers the lives of team members, and can endanger everyone on the incident. Driver/operators must be able to function within the team, which is partially why the camaraderie is so strong among us. Literally, you must be willing to trust your life with your fellow fire fighters. You must train extensively to develop a sense of calm so that you can function effectively when everyone and everything around you appears to be in a chaotic state. Driver/operators routinely operate in environments that endanger life and property. The old adage that, "fire fighters run in, when everyone else is running away," still holds true.

The best driver/operators are always seeking knowledge to enhance their ability to perform. This knowledge should be obtained from a variety of sources:

- College based instruction and formal education
- Local, Regional, State, National Fire schools and academies
- Institutional knowledge such as thorough knowledge of the department and local mutual aid resources available
- Experiential learning by hands-on practical application of techniques
- Community knowledge: pertinent information about community demographics, water mains, water sources, utilities, building design and construction, potential community hazards, traffic routes, evacuation routes, and location of fire protection systems

> "The best driver/operators are always seeking knowledge to enhance their ability to perform."

The importance of each driver/operator's contribution to the overall mission of public safety and the department's role must be extolled. Fire fighters are called when people do not know what to do; it may involve a seemingly insignificant operation such as water in the basement, or a devastating fire with multiple fatalities. The only way a driver/operator can effectively respond to this task is through being clam and constantly learning.

Charles Garrity
Berkshire Community College
Pittsfield, Massachusetts

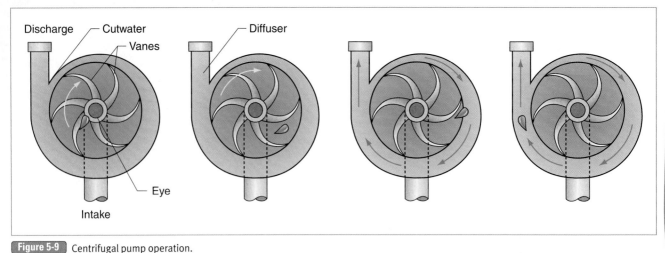

Figure 5-9 Centrifugal pump operation.
Adapted from Hale Products, Inc.

This pump operates on the basic principle of **centrifugal force**—that is, the outward force from the center of rotation. The centrifugal pump receives water into the center or eye of an **impeller** (Figure 5-10) that is mounted (as an offset) inside the pump casing. The pump impeller forms the heart of the pumping device: This metal rotating component transfers energy from the vehicle's motor to discharge the incoming water. Inside the impeller are **impeller vanes**, which divide the impeller. As the impeller spins, it accelerates the movement of water between the vanes and discharges the fluid radially outward into a collection area called a **volute**. The volute gradually decreases in area, causing the water pressure to increase; it then directs the water into the **discharge header**, where piping is attached and valves are arranged to deliver water to the intended hose lines or devices.

A centrifugal pump can pump only water or other liquids. This kind of pump is completely devoid of any valves from the intake to the discharge side. Water is free to move between either side of the impeller when it is not spinning. When the impeller is spinning, the water is taken from the intake side and flows out the discharge side. If the discharge valves are all closed, the water inside the pump will simply churn around inside. The centrifugal pump relies on the movement of water from the intake side to the discharge side to operate effectively. Unlike a positive-displacement pump, it is not self-priming; therefore the centrifugal cannot pump air—only liquids (typically water).

Also unlike positive-displacement pumps, centrifugal pumps are able to take advantage of any incoming pressure on the intake side of the pump. This additional intake pressure will simply increase the discharge pressure of the pump. For example, if the pump is receiving water at a pressure of 60 pounds per square inch (psi) (413 kilopascals [kPa]) from a hydrant at the intake side of the pump, and you need to supply the attack line at 100 psi (700 kPa), the pump will have to create only an additional 40 psi (275 kPa) to achieve the desired pressure.

Single-Stage and Multistage Pumps
Centrifugal pumps can be single-stage models (one impeller) or multistage units (two or more impellers within one pump housing turning on the same shaft).

Single-Stage Pump
A **single-stage pump** has one impeller, which both takes the water in and discharges it out of the pump. This single impeller is responsible for supplying 100 percent of the total amount of wa-

> **Driver/Operator Tip**
>
> The eye of the impeller is where the water enters the centrifugal pump. Some pumps have water enter from only one side of the impeller. Other models, called double-intake impellers, have the water enter from each side.

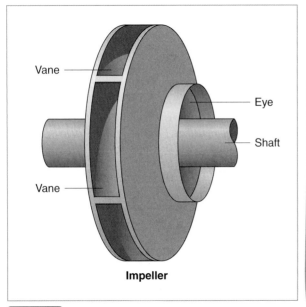

Figure 5-10 A centrifugal pump impeller.

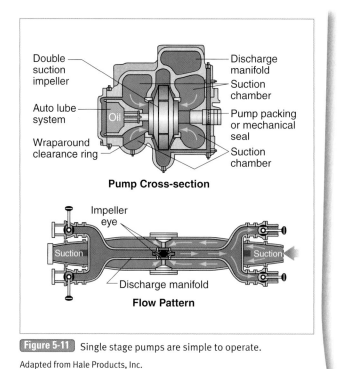

Figure 5-11 Single stage pumps are simple to operate.
Adapted from Hale Products, Inc.

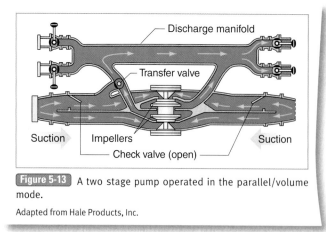

Figure 5-13 A two stage pump operated in the parallel/volume mode.
Adapted from Hale Products, Inc.

ter for the pump. This design makes the pump easier to operate because the pump can operate in only one mode **Figure 5-11**.

Two-Stage Pump

The most common type of multistage pump is the two-stage pump. A two-stage pump has two impellers, which are enclosed in their own pump casings but are part of a common housing. Each impeller is identical in size and capacity, and each is mounted on the same drive shaft and, therefore, spins at the same speed. At the discharge side of the first impeller, a **transfer valve** directs the water to either the pump's discharge header or the intake side of the second impeller **Figure 5-12**. A transfer valve determines whether the pump will be operated in series/pressure mode or parallel/volume mode. Both positions have their own strengths and weaknesses.

Parallel/Volume Mode

When water is directed in the **parallel/volume mode**, it enters each impeller from a common intake side and is discharged into a common discharge header **Figure 5-13**. This results in the pump's maximum volume of water being discharged. In this mode, each of the two impellers pumps 50 percent of the pump's total capacity. For example, if the pump is rated as a 2000 GPM pump, then each impeller would discharge a flow of 1000 GPM. When more than 50 percent of the pump's rated capacity is needed on the fire ground, the pump should be operated in parallel/volume mode to ensure that the desired flow can be achieved.

Series/Pressure Mode

When water is directed in the **series/pressure mode**, it travels through one impeller at a time or in series **Figure 5-14**. Water will enter the first-stage pump impeller's intake side, gain pressure, and be discharged to the second-stage impeller's intake side. The second impeller will then pump the water out the discharge header. The water will have more pressure as it enters the second-stage impeller than when it entered the first impeller, and it will gain additional pressure after being discharged from the second impeller. Series/pressure mode is the position in which two-stage pumps are most commonly operated. Most of the fire responses do not require the pump to deliver more than 50 percent of the pump's rated capacity. Therefore, the pump is left in series/pressure mode so that the desired pressure is achieved by operating at a lower rate (rpm).

Figure 5-12 Multi-stage pumps are more efficient than single stage pumps when operating at lower flows.

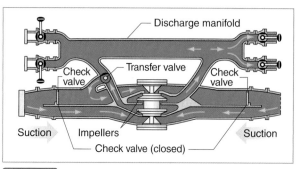

Figure 5-14 A two stage pump operated in the series/pressure mode.
Adapted from Hale Products, Inc.

Near Miss REPORT

Report Number: 05-0000270
Report Date: 05/27/2005 14:10

Synopsis: At the time of the call, our local interstate was under major construction with multiple lanes being added.

Event Description: At the time of the call, our local interstate was under major construction with multiple lanes being added. During that time of construction, which lasted for about 4 years, the lanes that occupied traffic were lined with jersey walls to protect construction crews. During construction it was predicted that there would be multiple deaths related to the narrowing of the traffic corridor due to the jersey walls and the length of interstate involved in this project; about 8 continuous miles.

Our station was dispatched to a multiple vehicle accident on the interstate on January 25 at 2:30 AM. Enroute, we had our normal discussion of duties and our assignments once on the scene. Upon arrival, we noticed the accident was just past the interstate on-ramp.

We noticed that there was a tractor trailer stopped short of the on-ramp, parked in the slow outside lane. At this point that lane was shut down. We parked our apparatus at the top of the on-ramp, just short of the interstate, and asked the police department to block traffic. There were very few vehicles driving at the time, which was very unusual for the traffic to be that light and that quiet.

Our crew dismounted our apparatus and proceeded to the accident scene with the proper turnout gear and equipment to perform the task of accessing the injured and assessing for any possible hazards that may be present.

Just before the crew approached the vehicles, out of nowhere, two vehicles entered our emergency scene at a high rate of speed. They sped right into the accident vehicles. Luckily there were no patients in the original vehicles and none of the crew or bystanders were injured. If our crew had dismounted the apparatus sooner or if we had arrived a little faster, we would have possibly had three fire personnel fatalities.

Lessons Learned: The lesson learned here was to secure the emergency scene before committing yourself and your crew. All lanes were not secured before emergency crews committed themselves to this incident scene.

We suspect that due to the position of the parked tractor trailer and the position of the fire apparatus, the fast lane traffic could not readily see the emergency lights of our vehicle. We, as an emergency unit, did not make our presence known to traffic or warn on-coming traffic of the hazards ahead.

Every individual in your crew must take responsibility for situational awareness and ensure the safety of others as well as themselves. Since this incident, we ensure that we have control of the interstate when an incident occurs. Make your presence known in a big way. Emergency personnel should wear their turnout gear to be highly visible, especially at night.

We now have multiple lanes to deal with, five to six lanes in both directions of the interstate. If we have an incident on the interstate, we shut down up to three lanes utilizing the apparatus along with traffic cones and the police department. When we do this, traffic is bottlenecked down to two to three lanes of travel. This slows traffic down to a manageable level and makes the emergency scene much safer for crews to operate in. If traffic begins to speed up to a level that is unacceptable, either the police department or fire department will stop traffic momentarily and then release it again.

Take control of the situation and make it as safe as it can be. If necessary, shut all traffic down. You may have the police department question this action, but it's better to stop traffic than to have an additional motor vehicle accident.

We don't know if placing the apparatus across both lanes would have prevented the second collision or not. This is due to the fact that the two vehicles that entered our emergency scene were racing each other and may not have had time to stop anyway.

With this in mind, apparatus placement is crucial. Park the apparatus diagonally to the lanes of travel and far enough back from the incident scene to give yourself a buffer zone in case it is hit. When weather is a factor such as snow or ice, increase this distance. Use common sense and think safety. Know that there are no routine incidents and that we, as emergency workers, cannot afford to take our safety for granted.

Near Miss REPORT (Continued)

Report Number: 05-0000270
Report Date: 05/27/2005 14:10

Demographics

Department type: Paid Municipal

Job or rank: Driver/Engineer

Department shift: 24 hours on—48 hours off

Age: 34–42

Years of fire service experience: 21–23

Region: FEMA Region IV

Service Area: Urban

Event Information

Event type: Non-fire emergency event: auto extrication, technical rescue, emergency medical call, service calls, etc.

Event date and time: 01/25/2003 02:30

Hours into the shift: 17–20

Event participation: Involved in the event

Do you think this will happen again? Uncertain

What were the contributing factors?
- Procedure
- Situational Awareness
- Protocol

What do you believe is the loss potential?
- Lost time injury
- Life-threatening injury
- Property damage

A fire pump is rated by and tested to Underwriters Laboratories (UL) specifications. To satisfy the UL standards, it must produce 100 percent of its rated capability at 150 psi (1034 kPa) for 20 minutes, 70 percent of its rated capability at 200 psi (1400 kPa) for 10 minutes, and 50 percent of its rated capability at 250 psi (1723 kPa) for 10 minutes. No additional rating is provided for pressures greater than the 250 psi (1723 kPa) rating; such pressures can be generated by using one pump impeller or multiple pump impellers inside the pump. Alternatively, the pump can be configured to provide pressures well above 250 psi (1723 kPa) by directing the water from the first impeller directly to the second impeller and then to the discharge, thereby doubling the pressure. With this configuration, 50 percent of the flow rate of the pump is lost because the first impeller does not send its water to the discharge side, but rather sends it to the second impeller (second stage).

As a driver/operator, you will find a control valve on the pump panel of a multistage pump indicating whether the impellers are being operated in parallel/volume mode or series/pressure mode. If the pump operation requires a large volume of water, then the pump should be operated in parallel/volume mode. In most pumping operations, you will supply one or two handlines with a total flow of 300–400 GPM (1135–1514 L/min) of water. A 1000-gal (3785-L) two-stage pump can easily supply this amount while operating in series/pressure mode. The advantage of this configuration is not attributable to any need for excessive pressure; rather, this setting allows the pumper engine and power supply to produce the water at lower engine speed (rpm), which is more efficient.

Think of a two-stage pump as similar to working from a hydrant water supply and pumper. When you start operating from the onboard water tank and pump water at a pressure of 100 psi (700 kPa), the total pressure available on the discharge is 100 psi (700 kPa). When a hydrant line is opened, it provides 100 psi (700 kPa) to the intake side of the pump, which is combined with the 100 psi (700 kPa) flow that the pump is

Driver/Operator Tip

When pumping requirements exceed 50 percent of the rated capacity of a two-stage pump, then you should operate the pump in parallel/volume mode. This will allow for the rated capacity of the pump to be used based on the availability of water on the intake side of the pump. A 1500 GPM (5678 L/min) two-stage pump would provide this capacity at 150 psi (1034 kPa) when operating from draft. When operating from a residual source, such as a hydrant, it will produce additional flow (GPM) and pressure. There is a point at which only so much water can be pushed through an opening, however—so do not expect the pump to deliver heroic amounts of water or water at an unusually high pressure simply because you are connected to a hydrant with a good flow rate and pressure.

already supplying. Therefore the pressure will be increased at the discharge to 200 psi (1380 kPa). The same concept applies when water is directed from the first impeller to the second impeller instead of being allowed to flow out the discharge. The pressure is doubled in the pump, but the total flow capability depends on the capability of the first impeller because the second impeller now receives its water from the first impeller.

Special multistage pumps have been built to produce extraordinarily high pressures for special pumping requirements. In addition, some fire departments want to preserve the ability to use high-pressure booster lines while still having volume pumping capabilities of 500 GPM (1892 L/min) or more. One method of achieving this goal is to place both a piston pump and a volume pump on the same fire apparatus. Another approach is to use a three-stage pump; the first two stages operate as a two-stage pump as previously described, while a small amount of water is directed to a third stage that produces higher pressures, typically supplying one or two booster hose lines.

Pump capacity over the years has increased significantly as larger pumps with larger impellers have been developed. In the 1950s, pumps typically delivered flows at 500 or 750 GPM (1892 or 2839 L/min). By comparison, today's fire pumps can flow water at 2000 GPM (7570 L/min) or larger. The other element driving the increases in pump capacity is the increase in diesel engines' horsepower, which has doubled from 250 rpm to more than 500 rpm since the 1950s. To turn a pump fast enough to produce a flow of 2000 GPM (7570 L/min) at 150 psi (1034 kPa) from draft requires a significant power supply, which is readily achieved with modern fire apparatus.

■ Power Supplies for Pumps

As the driver/operator, you must be familiar with how the various types of pumps receive their power. Given that much of this equipment is out-of-sight, you should study the fire apparatus schematics and manuals and examine the underside of the fire apparatus to observe how the power train is designed.

The simplest form of power supply is available with the **portable pump** Figure 5-15 ▶. This pump is typically carried by two or more fire fighters to a water source and used to pump water from that source. It has a small engine attached directly to the pump and provides power on a one-to-one basis: For every revolution of the engine, there is one revolution of the pump. With this type of unit, there is typically no gear box, clutch,

Figure 5-15 A portable pump is designed for use in remote areas.

shifting level, or other devices to operate. You simply start the engine and the pump starts. This same concept is often used for aircraft crash rescue apparatus so they can have a pump-and-roll capability—that is, the ability to discharge water as the apparatus is moving.

Some fire apparatus may have a pump mounted on the front bumper, known as a **front mount** Figure 5-16 ▼. The power for this pump is taken from the crankshaft on the front of the engine. It is transferred via a drive shaft to the pump transmis-

Figure 5-16 Front mount pumps are mounted on the front bumper of the fire apparatus.

Driver/Operator Safety

With both the front-mount pump and the pump that uses a power take-off supply, it is absolutely critical that the fire apparatus be secured when it is parked and the pump is being operated. The brakes need to be set and locked, the wheels chocked, and the gear shift locked into neutral, especially with automatic transmissions. Imagine what might happen if a front-mount pump is operating at 2000 rpm when someone accidentally knocks the shift lever into drive—the fire apparatus could become a lethal weapon.

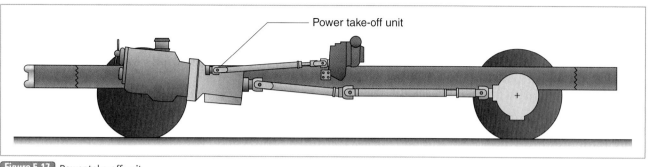

Figure 5-17 Power take-off unit.
Adapted from Waterous Co.

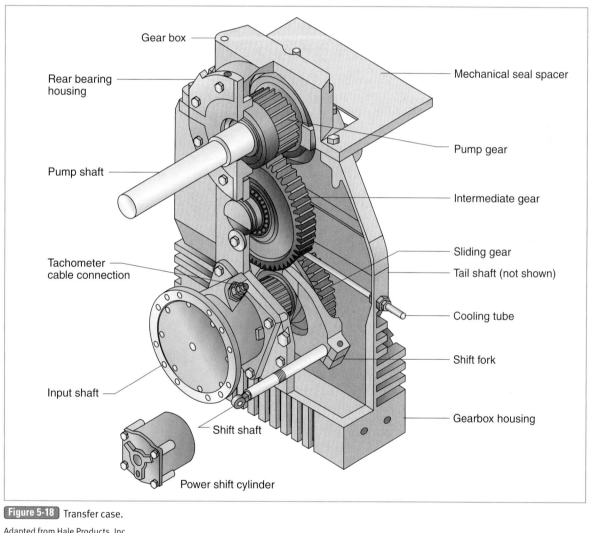

Figure 5-18 Transfer case.
Adapted from Hale Products, Inc.

sion that operates the pump. This same concept is often used to power large snowplow trucks, cement mixers, and other heavy equipment.

Power take-off units Figure 5-17 are commonly used for small pumps such as those found on tankers or tenders. A power take-off unit is mounted to the side of the transmission and through a shaft directed to the gear case on the pump. It provides a less expensive method of developing power for the pump, especially when the pump has only limited capacity. This approach also provides a pump-and-roll capability for apparatus such as wildland firefighting trucks.

The most common power system found in pumps is the **transfer case** Figure 5-18. This gearbox is mounted to the fire apparatus frame between the transmission and the rear axle. A drive shaft is connected from the fire apparatus transmission to the transfer case, so that the transfer case has the ability to direct the power to the rear axle or to the pump. When you place the pump into pump gear in the cab, you are transferring the power from the rear axle to the pump. The gear ratio is typically 1:1. The purpose of this configuration is simply to divert the power from one usage to another.

The speed of the pump is directly related to the speed of the transmission. A transmission placed in first gear will propel the pump at a slow pace; by comparison, a transmission placed in fifth gear will turn it at a much faster rate. The fire apparatus manufacturer determines the most effective gear to use for the engine and the optimal transmission arrangement for each fire apparatus. The manufacturer must also consider how much power is required for the pump application.

Automatic transmissions are specifically made for pumping operations that will lock into the gear intended for pump operations once the fire apparatus is placed in pumping mode. This setup prevents the transmission from shifting gears and ensures that the pump operates only in the selected gear (typically the highest gear).

Wrap-Up

Chief Concepts

- As the driver/operator of the pump apparatus, you must understand how pumps operate so that you are able to fix any problems that you may encounter on the fire ground.
- The pump panel is the most notable device on the fire apparatus. It is usually covered in stainless steel and has multiple levers and gauges mounted on it. Once you have a thorough understanding of the pump behind it and how it functions, you will be capable of confidently operating the fire pump.
- It is important for the new driver/operator to understand that the fire apparatus is not like any other vehicle—and, therefore, should not be operated like other vehicles.
- According to the NFPA, a pump is defined as a provider of liquid flow and pressure dedicated to fire protection. It is a mechanical device used to move fluids.
- This concept is very important for fire fighters because they want the pump to move the water from the source, such as the onboard water tank, to the fire through the attack lines. To do so successfully, you must understand how each different type of pump works.
- The fire service primarily uses two types of pumps: positive-displacement pumps and centrifugal pumps. Each type relies on very different operating principles and offers a very different set of features.
- Positive-placement pumps produce a flow by capturing a specific volume of fluid per pump revolution and reducing the fluid void by mechanical means to displace the pumping fluid.
- There are two broad classifications of positive-displacement pumps: the piston pump and the rotary pump.
- The most common fire pump in use today is the centrifugal pump.
- Centrifugal pumps do not flow a definite amount of water with each revolution; instead, the flow or amount of water discharged is based on the pressures at the discharge side of the pump.
  - At higher flow rates (rpm), the pump will flow less volume but will create higher pressures.
  - Consequently, as the pump spins more slowly, it will create less pressure on the discharge side but will flow a greater volume of water.
  - This flexibility makes the centrifugal pump a very versatile piece of equipment on the fire ground.

Hot Terms

centrifugal force The outward force that is exerted away from the center of rotation or the tendency for objects to be pulled outward when rotating around a center.

centrifugal pump A pump in which the pressure is developed principally by the action of centrifugal force. (NFPA 20)

discharge header The piping and valves on the discharge side of the pump.

discharge side The side of a pump where water is discharged from the pump.

double-acting piston pump A positive-displacement pump that discharges water on both the upward and downward strokes of the piston.

exhaust primer A means of priming a centrifugal pump by using the fire apparatus's exhaust to create a vacuum and draw air and water from the pump.

fire pump (fire apparatus) A device that provides for liquid flow and pressure and that is dedicated to fire protection. (NFPA 20)

front mount An apparatus pump that is permanently mounted to the front bumper of the apparatus and is directly connected to the motor.

impeller A metal rotating component that transfers energy from the fire apparatus's motor to discharge the incoming water from a pump.

impeller vanes Sections that divide the impeller.

intake side The side of a pump where water enters the pump.

parallel/volume mode Positioning of a two-stage pump in which each impeller takes water in from the same intake area and discharges it to the same discharge area. This mode is used when pumping water at the rated capacity of the pump.

piston pump A positive-displacement pump that operates in an up-and-down action.

portable pump A type of pump that is typically carried by hand by two or more fire fighters to a water source and used to pump water from that source.

positive-displacement pump A pump that is characterized by a method of producing flow by capturing a specific volume of fluid per pump revolution and reducing the fluid void by a mechanical means to displace the pumping fluid. (NFPA 20)

power take-off unit A direct means of powering a pump with the fire apparatus's transmission and through a shaft directed to the gear case on the pump.

prime the pump To expel all air from a pump.

pumping element A rotating device such as a gear or vane encased in the pump casing of a rotary pump.

rotary gear pump A positive-displacement pump that uses two gears encased inside a pump casing to move water under pressure.

rotary pump A positive-displacement pump that operates in a circular motion.

rotary vane pump A positive-displacement pump characterized by the use of a single rotor with vanes that move with pump rotation to create a void and displace liquid. (NFPA 20)

rotor A metal device that houses the vanes in a rotary vane pump.

series/pressure mode Positioning of a two-stage pump in which the first impeller sends water into the second impeller's intake side and water is then discharged out the pump's discharge header, thereby creating more pressure with less flow.

single-acting piston pump A positive-displacement pump that discharges water only on the downward stroke of the piston.

single-stage pump A pump that has one impeller that takes the water in and discharges it out using only one impeller.

transfer case A gear box that transfers power, thereby enabling a fire apparatus's motor to operate a pump.

transfer valve An internal valve in a multistage pump that enables the user to change the mode of operation to either series/pressure or parallel/volume.

vane A small, movable, self-adjusting element inside a rotary vane pump.

Venturi effect The creation of a low-pressure area in a chamber so as to allow air and water to be drawn in.

volute The part of the pump casing that gradually decreases in area, thereby creating pressure on the discharge side of a pump.

Driver/Operator in Action

While presenting a lecture on how the fire pumper operates, several members of the firehouse join you. Each fire fighter and fire officer gives his or her own unique perspective on the issues at hand. A senior member of the crew describes how the older pumps used to operate, while the lieutenant describes a brand-new pumper that another engine company has just put into service. The entire training session sparks a renewed interest in the pumper and how it operates.

1. Which type of pump traps a fixed amount of fluid and forces the fluid into a discharge during every revolution of the pumping element?
   A. Positive displacement
   B. Centrifugal
   C. Two stage
   D. Single stage

2. Which of the following is *not* one of the three moving parts of a piston pump?
   A. Piston
   B. Volute
   C. Intake valve
   D. Discharge valve

3. In a rotary vane pump, which type of force keeps the vanes pressed tightly against the pump casing to ensure a tight seal?
   A. Circular
   B. Residual
   C. Centrifugal
   D. Static

4. Which mode is a two-stage pump operating in when water enters each impeller from a common intake side and is discharged into a common discharge header?
   A. Series
   B. Volume
   C. Pressure
   D. Perpendicular

5. A fire pump produces 100 percent of its capacity while operating at _____ psi.
   A. 200
   B. 250
   C. 100
   D. 150

Performing Apparatus Check-Out and Maintenance

CHAPTER 6

NFPA 1002 Standard

4.2 Preventive Maintenance. [p. 128–148]

4.2.1* Perform routine tests, inspections, and servicing functions on the systems and components specified in the following list, given a fire department vehicle, its manufacturer's specifications, and policies and procedures of the jurisdiction, so that the operational status of the vehicle is verified:

(1) Battery(ies)
(2) Braking system
(3) Coolant system
(4) Electrical system
(5) Fuel
(6) Hydraulic fluids
(7) Oil
(8) Tires
(9) Steering system
(10) Belts
(11) Tools, appliances, and equipment [p. 130–148]

(A) **Requisite Knowledge.** Manufacturer specifications and requirements, policies, and procedures of the jurisdiction. [p. 129–130]

(B) **Requisite Skills.** The ability to use hand tools, recognize system problems, and correct any deficiency noted according to policies and procedures. [p. 130–148]

4.2.2 Document the routine tests, inspections, and servicing functions, given maintenance and inspection forms, so that all items are checked for operation and deficiencies are reported. [p. 130–148]

(A) **Requisite Knowledge.** Departmental requirements for documenting maintenance performed and the importance of keeping accurate records. [p. 130–148]

(B) **Requisite Skills.** The ability to use tools and equipment and complete all related departmental forms. [p. 130–148]

5.1.1 Perform the routine tests, inspections, and servicing functions specified in the following list in addition to those in 4.2.1, given a fire department pumper, its manufacturer's specifications, and policies and procedures of the jurisdiction, so that the operational status of the pumper is verified:

(1) Water tank and other extinguishing agent levels (if applicable)
(2) Pumping systems
(3) Foam systems [p. 130–148]

(A) **Requisite Knowledge.** Manufacturer's specifications and requirements, and policies and procedures of the jurisdiction. [p. 130–148]

(B) **Requisite Skills.** The ability to use hand tools, recognize system problems, and correct any deficiency noted according to policies and procedures. [p. 130–148]

6.1.1 Perform the routine tests, inspections, and servicing functions specified in the following list in addition to those specified in 4.2.1, given a fire department aerial apparatus, and policies and procedures of the jurisdiction, so that the operational readiness of the aerial apparatus is verified:

(1) Cable systems (if applicable)
(2) Aerial device hydraulic systems
(3) Slides and rollers
(4) Stabilizing systems
(5) Aerial device safety systems
(6) Breathing air systems
(7) Communication systems [p. 130–148]

(A) **Requisite Knowledge.** Manufacturer's specifications and requirements, and policies and procedures of the jurisdiction. [p. 130–148]

(B) **Requisite Skills.** The ability to use hand tools, recognize system problems, and correct any deficiency noted according to policies and procedures. [p. 130–148]

Additional NFPA Standards

NFPA 1720 *Standard for the Organization and Deployment of Fire Suppression Operations, Emergency Medical Operations, and Special Operations to the Public by Volunteer Fire Departments* (2004)

NFPA 1901 *Standard for Automotive Fire Apparatus* (2009)

NFPA 1911 *Standard for the Inspection, Maintenance, Testing, and Retirement of In-Service Automotive Fire Apparatus* (2007)

Knowledge Objectives

After studying this chapter, you will be able to:

- Describe the inspection and maintenance procedures required by your fire department.
- Describe the inspection and maintenance procedures recommended by the manufacturer on each of the fire apparatus that you will be required to inspect, test, or maintain.
- Describe the items on the written inspection and maintenance forms required to be completed by your fire department.
- Describe the procedures to be followed when an inspection reveals maintenance problems beyond the scope of the driver/operator's abilities.
- Describe the type of problems found during the inspection and routine maintenance of fire apparatus that warrant taking the fire apparatus or equipment out of service.
- Describe the equipment carried on fire apparatus that requires inspection and maintenance.

- Describe the routine maintenance procedures or adjustments to be completed by the driver/operator.
- Describe the maintenance procedures and items that will be performed by specially trained personnel other than the driver/operator.
- Describe the process to initiate required maintenance procedures.
- Describe the schedule for routine inspection and maintenance procedures for all fire apparatus and equipment that the driver/operator will be responsible for inspecting, maintaining, or testing.

Skill Objectives

After studying this chapter, you will be able to:

- Perform the daily inspection of fire apparatus and equipment in a safe and effective manner.

You Are the Driver/Operator

You are the driver/operator of a very busy engine company. Your fire apparatus has been scheduled for maintenance at the local repair facility and will be out of service for several weeks. To prepare for this event, all of the tools and equipment from your fire apparatus must be moved to a reserve fire apparatus provided by the fleet maintenance division of the fire department. The fire department uses this reserve fire apparatus while the regular fire apparatus receives any necessary repairs—a practice that ensures the engine company can remain in service.

The reserve fire apparatus to which you have been assigned while your rig is being repaired is much older than your regular fire apparatus. Although the same company built both fire apparatus, the two vehicles differ in terms of their features. The reserve fire apparatus does not have as many electrical devices as the newer fire apparatus; it also has some older features that you have not encountered since you were training to become a driver/operator. Your fire officer has asked that you thoroughly inspect this reserve fire apparatus before the rest of the crew starts to move any equipment onto it.

1. As you walk around the reserve fire apparatus, which features should you inspect?
2. When you check the engine oil, transmission fluid, and power steering fluid levels, what are some problems that may indicate that this fire apparatus may or may not be serviceable?
3. How are the emergency lights and sirens operated on the reserve fire apparatus?

Introduction

Being assigned as the driver/operator of a fire apparatus is a great responsibility. Duties assigned to this position include safely driving the fire apparatus and operating the equipment on the fire apparatus such as the pump or aerial device. In addition, the driver/operator is often given the responsibility to inspect and maintain the fire apparatus in as perfect condition as possible. Each fire apparatus must always be ready to respond and perform on the emergency scene in the manner it was designed to do so. If the fire apparatus is equipped with a **fire pump** **Figure 6-1**, it must be capable of flowing water at the required pressures. If the fire apparatus has an **aerial device**, the driver/operator is responsible for making sure that equipment is capable of operating as required **Figure 6-2**. Ensuring serviceability of all components of the apparatus is a key aspect of the **preventive maintenance program** that each fire department establishes for its fire apparatus. A quality preventive maintenance program ensures that the fire apparatus in the fire department's fleet are adequately maintained by qualified and trained personnel, the vehicles are inspected on a regular basis by the members who use the fire apparatus, and all documentation is accurate and complete.

Many fire departments—and especially career-oriented departments—may perform these inspections at least daily. In contrast, in a **volunteer fire department** where no specific

Figure 6-1 If the fire apparatus is equipped with a fire pump, it must be capable of flowing water at the required pressures.

Figure 6-2 A fire apparatus equipped with an aerial device.

member is assigned as the fire apparatus driver/operator, such inspections may be done on a less frequent basis. The bottom line is that the fire apparatus must be maintained in a state of readiness to respond to emergencies on a moment's notice Figure 6-3 ▼. To ensure their readiness, the fire apparatus and equipment must be inspected, tested, and maintained according to the manufacturer's recommendations. Safety is the most important and obvious reason for inspecting the fire apparatus regularly.

Inspection

The <u>fire apparatus inspection</u> is an evaluation of the fire apparatus and its equipment to ensure its safe operation. This inspection should be planned, methodical, and performed in an organized manner. Driver/operators typically conduct such inspections at the start of the shift, when the fire apparatus is being put back into service after repairs were made, and after a large incident during which the fire apparatus was used extensively at the scene of an emergency. This process identifies deficiencies with the fire apparatus or the equipment that might limit or incapacitate the fire apparatus from performing as required. Although most of the inspection can be performed by a single individual, thoroughly inspecting some of the features of the fire apparatus requires two crew members. For example, when inspecting the brake lights on the fire apparatus, one crew member will need to be behind the fire apparatus while the other member operates the brake pedal inside the cab Figure 6-4 ▼.

To perform the inspection, you must have some basic knowledge and skills related to vehicle maintenance. That is not to say that you should be capable of *performing* the actual maintenance of the fire apparatus; rather, you should be capable of *identifying* any potential problems before they become critical safety issues. Some fire departments prohibit their members from making any repairs on the fire apparatus, whereas other fire departments encourage fire station crews to make simple repairs to these apparatus. If the fire apparatus is under factory warranty, be aware that you may void the warranty by completing any repairs on the fire apparatus. Always refer to your fire department's inspection procedures before attempting to make any repairs to the fire apparatus.

Conducting an inspection may require using basic vehicle maintenance equipment such as tire pressure gauges, screwdrivers, wrenches, flashlights, and other small tools Figure 6-5 ▶. Every fire station should have a basic set of tools to aid the driver in performing the fire apparatus inspection. You should also have access to replacement fluids if you are required to maintain the fluid levels of the fire apparatus.

It is critical that the driver/operator performing the inspections, tests, or maintenance be familiar with the operating procedures of the fire department as well as the recommendations of the fire apparatus manufacturer. NFPA 1901, *Standard for Automotive Fire Apparatus*, requires each fire apparatus manu-

Figure 6-3 Fire apparatus should always be prepared to respond at a moment's notice.

Figure 6-4 Some parts of the fire apparatus inspection may involve more than one crew member. Teamwork is essential to completing a thorough inspection.

Figure 6-5 The inspection of a fire apparatus involves many of the same tools that you may use to maintain your own vehicle.

ample, if the fluid level is low in a radiator cooling system, should you add water or antifreeze coolant? If the oil level is found to be low following a <u>dip stick</u> test, which type of oil should be added? Specific fluids are required to ensure proper functioning of specific systems. For instance, oil for a hydraulic system may need to be of a different type or viscosity than oil for the engine. Likewise, the transmission will likely require a different type of oil than either the engine or the hydraulic system. You must adhere to the manufacturer's specification when adding fluids to the fire apparatus; otherwise, you risk damaging the equipment. Each department is responsible for educating members on how to properly maintain the fire apparatus.

Before you are assigned to perform the task of inspecting the fire apparatus, it is critical that the tasks are made clear and are well understood. It is also important that you consider your safety. Loosening a <u>radiator cap</u> on a hot engine, for example, may lead to a sudden release of hot liquid and steam. Working near a running engine may cause a shirt sleeve or body part to become entangled in a belt. Getting battery acid on your skin will cause burns and could damage your eyesight if it gets into your eyes. Each department is responsible for training its members on how to inspect a fire apparatus both safely and thoroughly. Always wear the appropriate personal protective equipment (PPE) while performing the fire apparatus inspection, including safety glasses, work gloves, and hearing protection **Figure 6-6**.

facturer to provide documentation of the following items for the entire fire apparatus and each major operating system of the fire apparatus:

- The manufacturer's name and address, for contact purposes
- Country of manufacture
- Source for service and technical information regarding the fire apparatus
- Parts replacement information
- Descriptions, specifications, and ratings of the chassis, pump (if applicable) and aerial device (if applicable)
- Wiring diagrams for low-voltage and line-voltage systems
- Lubrication charts
- Operating instructions for the chassis, any major components such as a pump or aerial device, and any auxiliary systems
- Precautions related to multiple configurations of aerial devices, if applicable
- Instructions regarding the frequency and procedure for recommended maintenance
- Overall fire apparatus operating instructions
- Safety considerations
- Limitations of use
- Inspection procedures
- Recommended service procedures
- Troubleshooting guide
- Fire apparatus body, chassis, and other component manufacturer's warranties
- Copies of required manufacturer test data or reports, manufacturer certifications, and independent third-party certifications of test results
- A material safety data sheet (MSDS) for any fluid that is specified for use on the fire apparatus

As a driver/operator, you should use the fire department's procedures and the manufacturer's recommendations as a reference to help you properly maintain the fire apparatus. For ex-

Figure 6-6 Personal safety is always the first priority when conducting an inspection of a fire apparatus.

DAILY ENGINE INSPECTION SHEET

Week of _____   Unit I.D. _____
                                          Shop # _____

Date:	Mon.	Tues.	Wed.	Thurs.	Fri.	Sat.	Sun.
Name:							
Knox box serial number							
Fuel level							
Motor oil							
Radiator							
Wipers							
Gauges							
Brakes							
Starter							
Lights/siren							
Generator							
Mirrors							
Body condition							
Water level							
Pump controls/gauges							
Press control device							
Hydrant tools							
Hose/nozzles							
Appliances							
Tools/ladders							
SCBA—PPE							
Radios							
Box-lights							
Map-books/computer							
Keys							
Accountability							
Clipboard							
Tire pressure							
Batteries							
Transmission fluid							
Bleed air tanks							
Primer fluid							
Drain valves							
Tool box							
Power tools							

Comments:
_____
_____
_____
_____
_____
_____

Figure 6-7 The fire apparatus inspection form is essential for maintaining an accurate record of the condition of the fire apparatus.

Safety Tip

When performing any maintenance functions on the fire apparatus or equipment, appropriate safety precautions must always be taken. Appropriate eye protection, hand protection, hearing protection, and inspection procedures must be utilized.

Driver/Operator Tip

If the inspection form is not organized in a logical format and process, then the inspection will not be performed properly. Make sure you take the time to organize the process you will use to complete the inspection process. As an example, similar items that will be examined in one process may be color-coded on the inspection form. Groupings such as electrical system components, drive-train components, pump-related components, and so on can all be logical methods of organizing your inspection. Being organized can help you manage your time effectively and ensure that items are not overlooked.

Inspection Process

The inspection process should begin with a review of the **apparatus inspection form** that was completed after the previous inspection **Figure 6-7**. This document identifies who performed the inspection and when it was performed. It also identifies any equipment that is damaged or has been repaired and points out any other preventive maintenance procedures performed on the fire apparatus. For example, the last member who inspected the fire apparatus may have documented items such as "one quart oil added to the engine," "right rear outside dual tire low and reinflated to the correct psi," "right rear warning light located on top rail not working and bulb replaced by mechanic." If during your inspection process you note that the engine is low on oil again, this finding may be an indication of some type of ongoing mechanical problem or leak. If the rear right dual tire is low again, it would indicate the tire is leaking and in need of repairs.

In many fire departments, the fire apparatus inspection form is simply attached to a clipboard and stored in the station, usually in the main office. Other departments may have an electronic version of the fire apparatus inspection form and use a computer program to track the inspections of their fire apparatus. Either way the driver/operator must review this material before beginning the inspection. Without reviewing the previous inspection report, some information identified in the current inspection may not seem relevant. Driver/operators from different shifts should also communicate to one another about the fire apparatus and any problems they encounter. Sometimes talking to the other members of the fire department who inspect the same fire apparatus can help to locate a potential problem. Taken in conjunction with the information from the previous report, your inspection process could reveal a need for an appropriate mechanic or qualified person to inspect the vehicle for defects.

Next, you must perform the actual inspection of the fire apparatus. This investigation may take some time depending on the size and complexity of the fire apparatus and its components. Fire apparatus inspections should be performed in a systematic manner. In other words, you should perform the inspection the same way each time to reduce the likelihood that something will be missed. While you are conducting your survey, the fire apparatus should be located in a safe area—either inside the fire apparatus bay, on the fire station driveway, or on an open lot devoid of traffic. The fire apparatus should be parked on a flat, level surface if possible. Check the area around the fire apparatus to determine if it is safe to operate the various components of the fire apparatus, such as lifting the cab of the fire apparatus or lowering the ladder rack on the side of the apparatus **Figure 6-8**.

During the inspection process, you should thoroughly document your findings on the fire apparatus inspection report. This step will help ensure that all documentation is as accurate as possible and no items are overlooked. Most of the items are inspected visually. In doing so, you are looking for any signs of damage, excessive wear, or defects. Some items must be operated during the inspection to ensure they function properly. For example, the emergency lights can be inspected visually by simply turning them on and walking around the fire apparatus looking for any inoperable lights.

Remember, it does not always take a mechanic to recognize a problem. By visually inspecting and operating the equipment during every shift, you will become familiar with the fire apparatus and its normal condition. You will then be able to quickly recognize when components break down or are in need of maintenance and can recommend the fire apparatus undergo repairs when necessary. Always be guided by your fire department's policies regarding fire apparatus inspections, and do not be afraid to ask for a second opinion if you are unsure about something you find in the inspection.

After the inspection is complete, review the report and make sure that no items were missed. Many fire departments require their members to complete the inspection of the fire apparatus

Figure 6-8 Position the fire apparatus in a safe location prior to starting the inspection process.

by a certain time each day or else face disciplinary actions. The inspection of fire apparatus should be taken very seriously—because it is a serious matter. Failure to complete a thorough inspection may result in an unsafe fire apparatus operating on the road and the emergency scene.

> **Driver/Operator Tip**
>
> Each fire department's fire apparatus inspection forms should reflect the policies of the fire department and identify which items need to be checked. Some fire departments have set up inspection procedures that include items to be inspected daily, items to be inspected weekly, items to be inspected monthly, and items to be inspected quarterly.
>
> While you may be responsible for the daily and weekly items, fire officers may be responsible for carrying out the more comprehensive inspections, tests, or maintenance procedures to be performed. For example, the fire officer may be accountable for scheduling the annual fire hose and ground ladder tests for the fire apparatus, though it is your responsibility as the driver/operator to ensure that the equipment is prepared for this inspection. You must be familiar with all of your fire department's required inspection and maintenance procedures.

> **Driver/Operator Tip**
>
> A good inspection of a fire apparatus will get you dirty. If you do not walk into the station after checking out the fire apparatus with dirt and grease from your elbows to your fingertips, then you probably did not do an adequate inspection.

Fire Apparatus Sections

Often the inspection process is broken down into sections. Dividing the inspection in this way allows you to focus on a single aspect of the fire apparatus and discourages you from jumping from one element of the inspection to another without a plan. Jumping around randomly leads to the possibility that critical elements may be missed. Each driver/operator should use whatever system or sequence is recommended by his or her fire department.

The following is a suggested fire apparatus inspection procedure that has been broken down into several sections. The first five sections apply to all fire apparatus. The last two apply only to those fire apparatus that meet the criteria for that section. Not all of the items in each section will apply to every fire apparatus; it is up to the driver/operator to determine which items are applicable to his or her fire apparatus. These sections should be inspected in the following order unless otherwise stated by your fire department:

- Exterior inspection
- Engine compartment
- Cab interior
- Brake inspection
- General tools/equipment inspection
- Pump inspection
- Aerial device inspection

Exterior Inspection

The first section to be completed is the inspection of the fire apparatus exterior. Physically walk around and look at the fire apparatus's general condition. Is the fire apparatus clean and well maintained, or is it worn and in need of several minor repairs? Determine if the fire apparatus is leaning to one side; this may indicate a broken suspension system or tires not inflated to the correct pressure. Visually inspect under the fire apparatus for any fresh oil, coolant, or other fluid leaks. Check for any damage to the body such as dents, scratches, and paint chips. Look for signs of stress or cracks on the body. Inspect the compartment doors, hinges, and latches for proper operation.

> **Driver/Operator Tip**
>
> As part of the exterior inspection, you may need to tilt the cab or crawl under the fire apparatus to properly inspect the items.

> **Safety Tip**
>
> If the fire apparatus will be moved outside to the apron for engine and pump operation, first make sure it is safe to do so. Confirm that all compartment doors are closed, all personal protective equipment is secure, and all cab doors are closed. Whenever the vehicle is operated inside the station, make sure that it is connected to an extractor exhaust system—a system designed to draw the fire apparatus exhaust to the outside so that it does not fill the fire apparatus bay with harmful gases. When the inspection is complete, before backing the vehicle into the station, verify that no one is behind the fire apparatus and have a spotter maintain a visual observation behind the fire apparatus while it is operating in reverse.

Tires are critical to many aspects of the safe operation of the fire apparatus—namely, proper stability of the fire apparatus, stopping capability, and ability to carry loads. When inspecting the tires, use a flashlight to get a better look at their overall condition. It is very dangerous to drive with tires in bad condition. Look for problems such as cuts, cracks, or fabric showing through the tread or sidewall.

The valve stems on all tires should be accessible and devoid of cracks and cuts, with the valve caps being securely fastened. The size and make of all tires should match those recommended by the manufacturer. Dual tires should not be in contact with each other or with other parts of the fire apparatus. Determine if the tread on the tire is wearing unevenly, as this may indicate a possible problem with the suspension system, an issue related to the steering system, or inflation of the tire to an incorrect pressure. When inspecting the wear, the tire should have at least 1/8″ of tread depth in every major groove on all tires. U.S. coins can be substituted for a tire tread depth gauge as tires wear to the critical final few thirty-seconds of an inch of their remaining tread depth. To use this technique for measuring tread depth, place a quarter into several tread grooves across the tire. If part of George Washington's head is always covered by the tread, the tire has more than 1/8″ of tread depth remaining.

Refer to the manufacturer's recommendations to determine the appropriate tire pressures for each fire apparatus. Using a pressure gauge, check the tire pressure by removing the valve cap and applying the pressure gauge. If the tire pressure is adequate per the manufacturer's or department's specifications, then return the valve cap. If the tire pressure is low, then use an air hose to inflate the tire to the correct pressure. Check the tire again to obtain the pressure level. If necessary, add more air until the desired pressure level is obtained. Replace the valve cap and note that air was added to the tire on the inspection form.

A damaged wheel or rim can cause a tire to lose pressure or even slip off. This event can cause an accident if it occurs while the apparatus is operating on the roadway. Look for any sign of damage, including dents or large scratches along the edge that meets the tire. If rust is found around the wheel nuts, it may be an indication that the nuts are loose and need to be retightened. The wheel should not be missing any clamps, spacers, studs, lugs, or protective covers.

Today's fire apparatus are equipped with a **power steering system**. This system reduces the effort required to steer the vehicle by using an external power source to assist in turning the apparatus's wheels. While inspecting the steering system, look for any bent, loose, or broken parts, such as the steering column or tie rods. With the engine compartment exposed, examine the power steering pump, hoses, and fittings for leaks. While in the cab of the fire apparatus, inspect the amount of free play in the steering wheel. If the steering has more than 10 degrees of free play, a mechanic should service the fire apparatus. Ten degrees of free play is equivalent to 2″ of movement at the rim of a 20″ steering wheel **Figure 6-9**.

The suspension system keeps the vehicle's axles in place and holds up the fire apparatus and its load. A defect in this system may cause problems with the fire apparatus's braking or power steering system. Inspect the frame assembly to ensure that no parts are cracked, loose, broken, or missing. Look for any spring hangers or other axle positioning parts that are broken and might allow the axle to move out of position. Also, look for any broken sections or sections that have shifted out of place in the leaf springs. Identify whether the shock absorbers are leaking fluids or are bent out of shape. Torque rods and torsion bars should be free of damage.

Visually inspect the fire apparatus exhaust system to check for any loose, broken, or missing mounting brackets or parts. The exhaust piping should not rub against the tires or other moving parts of the fire apparatus, and no leaks should be found. A broken exhaust system may allow poisonous fumes to enter the cab, harming the crew members aboard the fire apparatus.

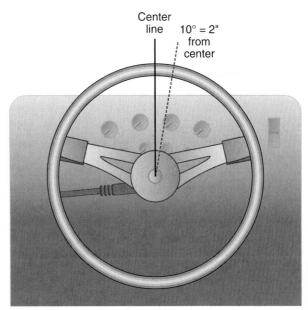

Figure 6-9 Ten degrees of play is equivalent to 2″ of movement at the rim of a 20″ steering wheel.

Driver/Operator Tip

Cleanliness is a very important part of proper fire apparatus maintenance. A clean fire apparatus is not only a source of pride in the station and its crew, but is also safer than a dirty fire apparatus. Dirt and grime build-up will damage moving parts and cover defects. By keeping the fire apparatus clean, you are gaining intimate knowledge of the equipment, thereby ensuring that any defects will be identified sooner.

To clean the fire apparatus, you must first rinse it with clean water to remove any loose dirt. This action also reduces the chance of scratching the paint during the remainder of the clean-up procedure. Wash the fire apparatus with an automotive soap, as recommended by the manufacturer. The entire fire apparatus should be thoroughly washed, including the top of the cab, wheel wells, and diamond plate surfaces, among other components. Rinse the vehicle with clean water, and then dry the fire apparatus with an approved chamois or towel. All trash should be removed from the cab's interior; this compartment should then be dusted, swept, or vacuumed and dressed with the appropriate surface treatment.

Glass should be cleaned and all painted surfaces waxed as necessary after the fire apparatus is completely dry. Metal surfaces should be polished to prevent tarnish and dull surfaces. Compartments should be cleaned out and all equipment maintained as necessary. If the engine compartment is clean, it makes the inspection process easier. Never use gasoline or other unapproved solvents to clean painted surfaces, as they may cause discoloration and damage.

The fuel cap should be securely fastened to prevent any spillage or fumes leaking from the tank. This cap should also be labeled with the appropriate fuel. Although larger fire apparatus use diesel fuel, some vehicles may require a specific grade or bio-diesel. Always consult the operator's manual provided by the manufacturer to determine which type of fuel to add to the fire apparatus. The fuel tank should be checked to make sure that no leaks are present and the mounting brackets are properly secured.

Engine Compartment

This section focuses on the process of inspecting the fluid levels, battery charge, cooling system, motor components such as belts, charging system, and drive train elements. NFPA 1901, *Standard for Automotive Fire Apparatus*, requires that all fire apparatus be designed so that the manufacturer's recommended routine maintenance checks of lubricants and fluid levels can be performed through a limited-access port without lifting the cab of the fire apparatus or without the need of special tools for routine maintenance checks of lubricants and fluid levels Figure 6-10. On most fire apparatus, you will still need to raise the cab to inspect most portions of the engine, including belts, hoses, and fan blades. Older fire apparatus may not have an access door through which to check the fluids; as a consequence, the cab must be tilted to determine the fluid levels in these vehicles. If this is the case, the cab should be secured with a locking device so that the cab does not fall on anyone operating underneath it Figure 6-11. While the engine is off, you can inspect the engine compartment. Examine this area for any fluid leaks; broken, cracked, or damaged hoses; and electrical wiring that shows signs of wear, chaffing, or damage from heat. Also, confirm that the cooling fan is free of any obstructions or defects. The air intake filter should be replaced as necessary and its housing should not have any cracks, loose fasteners, or broken supports.

Depending on your fire department's policy, the driver/operator may be required to maintain the appropriate fluid levels in the fire apparatus. Remember, when adding any fluids, you

Figure 6-11 Always use a locking device to secure the cab when operating underneath it.

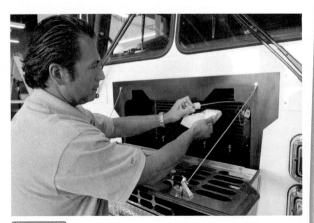

Figure 6-10 Access doors on newer fire apparatus allow you to inspect the fluids in the engine compartment without having to lift the cab.

must record the amount on the fire apparatus inspection form. You should not rely solely on sensors and computer systems to give an accurate reading of fluid level—always physically check the fluid levels.

The engine oil level is checked with a dip stick, usually after the engine has been turned off for at least 15 minutes; this delay allows the oil to settle back down and gives an accurate reading. The dip stick is pulled, wiped clean, and then replaced. It is then pulled a second time and the oil level compared to the marking on the dip stick. Generally a range is provided between "low" and "full." If the engine was operated just before the oil is checked, the dip stick level may appear low, as oil is still in the engine components and not totally drained to the crankcase. If the oil level is truly low, then the recommended amount of oil should be added via the correct port. After waiting a few minutes, check the level again to ensure the proper level has been achieved and record the amount of oil added on the inspection form.

The coolant level should be measured, observed, or checked in the manner recommended by the manufacturer.

Most systems in use today do not require the removal of the radiator cap and provide an exterior coolant reservoir that is marked with the appropriate level. Some manufacturers provide a sight glass on the radiator to determine the coolant level

Chapter 6 Performing Apparatus Check-Out and Maintenance

Figure 6-12 A sight glass can be very useful to check the coolant levels.

Figure 6-12 . If the coolant system is low on coolant, then you should consult your fire department's procedures to determine which coolant should be added to bring the reservoir to the proper level. Adding water will diminish the protection from freezing. Additionally, antifreeze fluid contains other materials designed to prevent rusting and act as a lubricant; it may be the only appropriate liquid to add to the cooling system. If the fire apparatus has been run recently, be aware that the coolant itself may be hot.

Always use caution when removing the radiator cap, as the coolant may be under pressure and be hot; if it boils over, it could cause an injury. For this reason, it is not recommended that you remove the cap when the fire apparatus is running or when the fire apparatus is hot.

Power steering system fluid is checked in the same way as the engine oil. A small dip stick is inserted to determine the fluid level. If the level is low, add power steering fluid as required.

The transmission fluid is the only fluid that may need to be inspected while the engine is running, although some manufacturers also recommend checking power steering fluid at operating temperature. Many manufacturers recommend that the transmission be operated at the normal operating temperature (usually 170°F) after the apparatus has run through all of its gears and that the vehicle be parked in neutral when the fluid is checked. To check the transmission fluid, use a dip stick in a similar manner as when performing an engine oil check.

Other fluid levels, such as the rear differential fluid (axle), hydraulic oil, and pump gear box oil levels, are often checked by a fire department mechanic on a periodic basis. Always refer to the manufacturer's recommendations when determining the correct levels of these fluids.

Belts that drive engine components such as alternators, the power steering pump, the air compressor, and other equipment may become loose due to wear. To check the tension on a belt, push against the belt in an area where there is no pulley. Depending on the manufacturer's recommendations, the belt may be able to be pushed to some extent, but should not have any excess slack. In some fire departments, a mechanic or specially trained inspector will check items such as belts on a frequent basis; thus the driver/operator may not be required to check them. Nevertheless, you should ensure that the belts do not show any signs of excessive wear or fraying. Always refer to the manufacturer's recommendations regarding belt inspections.

The fire apparatus's (one or more) batteries should be examined for signs of corrosion on the terminals where the wires connect to the battery post. It is important that you protect yourself from corrosion or the liquid inside the battery. Appropriate eye, hand, and/or body protection should be provided and worn during this part of the fire apparatus inspection. Corrosion may be removed by scraping the terminal with a wire brush; always wear eye protection when performing this activity.

The physical process of removing the electrical wire connection from the terminal is normally performed by a mechanic or other person specially trained to perform this task. Given that most of today's fire apparatus have computerized systems on board, severe damage could occur if removal of the battery cables is done improperly Figure 6-13 . For example, the computer system that operates the fire apparatus may have to be reset by a qualified technician if it loses power for a significant amount of time. Although older vehicle batteries may have fluids cells that can be refilled as often as needed, many newer batteries are sealed, meaning there is no way to check the fluid cells. You should refer to the manufacturer's recommendations when inspecting batteries on any fire apparatus.

Voltage levels may be checked by observing the **voltage meter** on the dashboard if the vehicle is so equipped. A volt meter registers the voltage of the battery system. For example, the voltage of a 12-volt battery will typically be recorded as a number such as 14 volts. Batteries that are equipped with removable caps on the cells of the battery should be checked for appropriate liquid levels. Liquid should cover the cells, albeit not to overflowing. If the liquid level is low, follow the fire department's or manufacturer's recommendations regarding which liquid to use to refill the battery.

Cab Interior

When inspecting the cab interior, first check that all cab-mounted equipment is present and accounted for, including the following items:

Figure 6-13 Do not disconnect the batteries on newer-model fire apparatus, as this step may cause damage to the computer system.

Figure 6-14 NFPA 1901 requires that seat belts be red or orange so that fire fighters can tell them apart from the waist belts on SCBA units.

- Portable radios
- Self-contained breathing apparatus (SCBA)
- Maps
- Traffic vests
- Hearing protection
- Medical gloves
- Box lights

Also, check for worn or torn seats, cushions, dashboards, and headliners. Ensure that all seat belts are functioning properly and are free of cuts and frays. NFPA 1901 requires that all seated positions in the modern-day fire apparatus cab be equipped with bright orange or red seat belts so they are not confused with seat-mounted SCBA belts and straps Figure 6-14 ▸.

The interior of the cab is where most of the controls for the fire apparatus are located. For this section of the inspection, set the parking brake and start the fire apparatus by engaging its normal starting sequence. This provides the opportunity to observe the gauges such as those measuring oil pressure, electrical system, engine temperature, and air pressure. Once you are sitting inside the cab, adjust the seat and the mirrors and familiarize yourself with the functional controls. As a driver/operator, you must be familiar with all of the controls in the cab so that you do not need to take your eyes off the road to make any adjustments while driving.

Make sure that all of the gauges indicate performance within the normal operating ranges. Each fire apparatus may have different ranges for normal operations; refer to the manufacturer's recommendations. Determine whether any indicator lights are activated and need attention. For example, if the oil light is activated and you have recently added more oil to the engine, either the sensor is faulty or another problem may exist. Check the operation of the functional control switches—that is, the controls that operate interior functions as well as those that operate exterior functions.

Interior Functional Control Switches

Interior functional control switches include the controls for items that are located inside the cab itself—for example, the heater, air conditioner, defroster, map lights, dash lights, MDT, radio, and other devices. All of these items should be inspected to ensure that they are operating correctly. Some fire departments require members to check the MDT and radio to ensure that they are transmitting information appropriately. Always refer to your department's policies when conducting this part of the fire apparatus inspection.

Exterior Functional Control Switches

Exterior functional control switches include the controls for items that are located outside of the cab but are operated from controls

Figure 6-15 You will have to walk around the fire apparatus and visually inspect the operation of the lights on the fire apparatus.

Driver/Operator Tip

The vehicle's fuel level can easily be checked by observing the fuel gauge when the electrical system is energized. Be familiar with your fire department's policies stating when the fuel tank must be replenished or topped off. Many fire departments have a policy that states the fuel level for any fire apparatus should never go below half a tank.

in the cab—for example, the emergency lights, headlights, directional lights, brake lights, side marker lights, spotlights, and taillights. All of the lights should be clean and operating correctly. To inspect the lights of the fire apparatus, you must activate the lights and walk around the fire apparatus visually inspecting their operation Figure 6-15 . Another fire fighter should assist you in checking the brake and reverse lights.

Check the mirrors, windows, and windshield. All of the windows of the fire apparatus should be clean and free of cracks or chips. If any defects are found during the inspection, document them on the report. All of the windows should roll down properly, whether they are manually or electronically controlled. Windshield wipers and fluid level should be checked as well. The wiper blades should be soft, flexible, and clean. If the blades are too hard or cracked, then they should be replaced. The wiper fluid level should be topped off with the appropriate liquid as recommended by the manufacturer.

Driver/Operator Tip

The emergency warning equipment must be checked to confirm that all components are functioning appropriately. Driving lights (head, tail, marker) and warning lights should be checked for appropriate function, and audible warning devices should be tested to ensure they operate correctly. Likewise, all other electrical equipment—such as scene lighting devices and traffic directional arrows—should be tested to make sure they are functioning correctly. If the emergency lights do not operate properly, for example, the fire apparatus is not capable of operating safely during an emergency response.

Brake Inspection

The brakes of the fire apparatus should be given special attention when performing the inspection, as this equipment is clearly vital to the safe and efficient operation of the fire apparatus. First, you should inspect the brakes for the following conditions:

cracked drums or rotors; shoes or pads contaminated with oil, grease, or brake fluid; and shoes or pads that are worn dangerously thin, missing, or broken. Next, you should test the brakes of the apparatus. Your department's policies and procedures will dictate if and when the following tests are to be completed.

Parking Brake Test

Make sure that the fire apparatus is in a safe position and has plenty of room to perform this simple brake test. With the fire apparatus turned on, allow it to move forward at a speed less than 5 mi/h (8 km/h). Apply the parking brake. If the fire apparatus does not stop, bring it to a halt using the service brakes and have the fire apparatus inspected by a qualified mechanic.

Brake Pedal Test

Make sure the fire apparatus is in a safe position and has plenty of room to perform this test. Accelerate to 5 mi/h (8 km/h). Push the brake pedal firmly. If the fire apparatus demonstrates excessive pulling to one side or the other, if it exhibits a delayed stopping action, or if the "feel" of the pedal is off, this may indicate a potential problem. In such a case, have the fire apparatus inspected by a qualified mechanic.

The following tests are applicable only to fire apparatus equipped with air brake systems. Your department's policies and procedures will dictate if and when the following tests are to be completed. Those vehicles without air brake systems may not be required to complete any additional brake tests. For the following tests, you should chock both sides of the front left wheel with the fire apparatus wheel chocks Figure 6-16 .

Figure 6-16 Chock the wheels of the fire apparatus when conducting a dual air brake system warning and buzzer test.

Dual Air Brake System Warning Light and Buzzer Test

Many fire apparatus are operated with a dual air brake system. With this system, a mechanically operated parking brake is activated in the event of a service brake failure. A dual air brake system actually consists of two separate air brake systems that use a single set of brake controls—a setup that provides more air capacity and, therefore, a safer system. Each system has its own air tank and hoses; they are split so that one system operates the front of the fire apparatus and the other operates the rear. On the dash of the fire apparatus, the gauges for each system are labeled as "front" and "rear," respectively.

To complete the test, follow these steps. First, turn the fire apparatus on and allow time for the air compressor to build up to a minimum of 110 pounds per square inch (psi) in both the front and rear systems. Next, shut the engine off. Leave the battery in the "on" position, and step on and off the brake pedal to reduce air tank pressure. An audible alarm should signal before the pressure drops to less than 60 psi in the air tank with the lowest air pressure. Have the fire apparatus inspected by a qualified mechanic if the audible alarm does not work properly, as such a malfunction could cause the system to lose air pressure without your knowledge. Without the proper air pressure, the brakes will become less effective, thereby increasing the stopping distance of the fire apparatus and leading to unsafe operation.

Spring Brake Test

This test ensures that the parking brake operates as it was designed. The parking brakes should engage whenever brake pressure drops below 40 psi in the rear brake system. The spring brake test is performed by allowing enough air pressure to build up in the braking system to release the parking brake. Depress the parking brake knob to release it. Next, step on and off the brake pedal to reduce the pressure in the system. The parking brake knob should activate when the air pressure drops below 40 psi. The spring brakes will then activate and help to prevent the vehicle from moving.

As a result of normal condensation of moisture in the air and moisture created during the compression phase, some water may enter into the air supply of the braking system. The danger from the presence of moisture in the air supply is that in cold weather it may collect and freeze, thereby preventing the braking system from operating properly. Fire apparatus equipped with air brakes may be equipped with automatic moisture exhaustion valves to overcome this problem. In addition, some fire apparatus is equipped with a manual water drain valve that must be opened to drain moisture from the air system **Figure 6-17**. Refer to the operator's manual to determine which procedures are recommended to maintain the system. Automatic moisture reduction systems will make a spitting noise when they are removing moisture from the system; this is normal and not a cause for alarm. As a driver/operator, you must become familiar with the air brakes and the moisture removal system with which your fire apparatus is equipped.

■ General Tools/Equipment Inspection

All other equipment carried on the fire apparatus—such as breathing apparatus, cascade systems (including compressors), generators, fans, hydraulic rescue tools, hand tools, power tools,

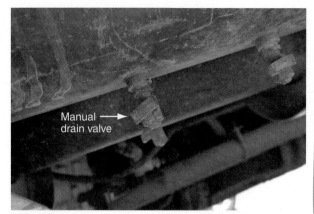

Figure 6-17 Manual valves to drain the moisture in the brake system are found underneath the fire apparatus.

hose, nozzles, ground ladders, and so on—must also be inspected to make sure that it is operational **Figure 6-18**. The SCBA should be checked in accordance with the fire department's respiratory protection program, which may require completion of a separate inspection form. A record of this inspection process

Figure 6-18 Equipment found in the compartments of the fire apparatus must be inspected and the results of the investigation documented on the inspection report.

must be kept by the member in charge of the respiratory protection equipment in accordance with OSHA 29 CFR 1910.134.

Power tools such as saws, fans, and hydraulic rescue tools should be checked for oil level (if appropriate) and fuel supply. All equipment should be started and operated. Equipment such as power saws may have two-cycle engines, in which the lubricating oil is mixed with the gasoline or another fuel; that is, they may not have a crankcase with a separate oil supply serving as the lubricant. Devices such as portable pumps will typically have four-cycle engines and an oil reservoir that must be checked. When refueling power tools, it is critical to make sure the appropriate fuel is used.

Many fire apparatus carry equipment on board that needs to be recharged as the fire apparatus sits in the station bay. This equipment may include portable radios, thermal imaging cameras, or batteries for electrical equipment Figure 6-19 ▶. If the fire apparatus is not operated for an extended duration, its batteries may be drained of power by the ongoing recharging of equipment stored on board the fire apparatus. Therefore, it is critical to keep the vehicle's batteries properly charged at all times. To do so, many fire apparatus are equipped with a charging system that connects to an electrical outlet in the fire station. This electrical line recharges the battery(s) while the fire apparatus is not running, thereby ensuring that the fire apparatus has enough electrical power in its batteries to recharge any equipment and start the fire apparatus as required.

Figure 6-19 Small electrical equipment may be charged while mounted on the fire apparatus.

Safety Tip

Extreme caution should be used when adding fuel to power tools. Never fuel equipment inside the station, and never fuel equipment while it is running. If the engine is hot, allow it to cool before refueling the equipment. Use the appropriate fuel dispensing device(s).

Pump Inspection

The pump inspection process includes the following items:
- Water supply tank
- Foam supply tank
- Intakes and discharges
- Primer pump
- Centrifugal fire pump

A visual inspection of the water supply tank should be conducted, even if it has a water tank level gauge. This step will help verify that the gauge is correct. Tank gauges may not be accurate because of fluctuations caused by the materials in "hard water" and electrical malfunctions. Always visually confirm that the water tank is full.

In fire apparatus that is equipped with foam systems, the foam tank levels should also be checked. This inspection is typically completed at the same time as the water tank check. As with the water tank, always visually confirm the fluid level in the foam tank.

Before opening any valve and allowing water to drain, check the floor area under the fire apparatus for the presence of water.

A puddle of water or dripping water may indicate a loose pump seal or some other leak in the system Figure 6-20 ▼. A loose pump seal may not be considered a significant problem if the pump is always operated from a pressurized water supply, such as a hydrant. It is a problem, however, if the pump needs to be operated in draft mode. It may not be possible to prime the pump in such a case owing to the leaking seal. Repairing this problem requires a mechanic who is specially trained to adjust the seals. In addition, leaking water from the pump seals during operations in cold weather will cause additional ice to build up in and around the fire apparatus.

In areas subject to cold weather, pumps may be kept in a dry state during winter months. If this is the case, do not open the tank-to-pump valve, as this maneuver will cause the pump to fill with water. Always refer to your department's policy regarding the use of wet versus dry fire pumps.

While the pump is not engaged, open and close each discharge valve several times to ensure that it operates properly. Remove all of the caps on the discharges and open the drain valve for each discharge to ensure it functions properly. Confirm that all caps are easily operated and free of corrosion.

Figure 6-20 Water or other liquids dripping from underneath the fire apparatus should be inspected to determine whether the leak is serious.

The intakes—that is, the piping that allows water to enter the fire pump—are inspected next. To do so, remove the plugs, caps, or **piston intake valve (PIV)** and visually inspect the piping. The PIV is a large appliance that connects directly to the pump's intake and controls the amount of water that flows from a pressurized water source into the pump. Hard water and corrosion can cause the plugs or caps from the intakes or at the end of the PIV to stick and make it difficult to open them. In such a case, it is better to fix the problem at the fire station during the fire apparatus inspection than to wait until it happens at the next fire scene.

Intake strainers are located at the front of all intakes directly on the pump. These small screens prevent debris such as rocks from entering the pump and causing damage. These screens should be checked at least weekly or as directed by your department. Be careful when completing this portion of the inspection, because water will be released as the pump partially drains. Check the grids of the strainers to make sure they are clear and no pieces are bent or missing Figure 6-21. Repeat this process for each intake. After the intake strainers are checked and the caps or PIV replaced, it is critical that the fire pump be refilled with water as required.

Because the fire apparatus is equipped with a centrifugal pump, it will have a priming pump; the oil reservoir of this priming pump most likely must be checked daily Figure 6-22. The priming pump is used to draw air out of the centrifugal pump for proper operation. Some of these pumps allow a small amount of oil to enter the centrifugal pump during priming operations to help seal and lubricate the pump.

If the priming pump is wet, then a stream of water should be observed within a few seconds. If the pump is being carried dry, then no water should be seen. If the pump drains are open, then no negative pressure will be recorded on the compound gauge.

If the priming pump is being carried in a wet condition and after the pump valves and components have been operated, the fire apparatus engine should be started and the pump engaged in appropriate gear. Before starting the engine, make sure that the parking brake is set and that the wheels are chocked. Once the engine is started and the pump engaged, observe the fire apparatus's speedometer to see whether a speed is being recorded, indicating the pump is in gear. Some fire apparatus may use a power take-off drive system for the centrifugal pump and, therefore, will not record any speed on the speedometer because the truck is not in gear. On the pump panel, observe the tachometer to observe that the pump is activated. Ensure that all of the gauges and instruments are operating properly.

Figure 6-22 A priming pump is used to draw air out the centrifugal pump.

To set the pressure relief valve, open the tank-to-pump valve and the tank fill valve. Next, increase the engine speed to bring the pump pressure up to an operating pressure, such as 125 psi. Then set the pressure relief valve and confirm that it is functioning properly. To do so, increase the engine speed and the pressure to cause the relief valve to operate or open. If the relief valve is functioning properly, the pressure will not increase more than the set pressure. For engines equipped with a pressure governor both the tank-to-pump and tank fill valves should be open and the pressure governor should be turned to the "on" position. Increase the pressure to about 125 psi, which will in turn set the pressure governor at this pressure. Once water flow is established and the governor set, close the tank fill valve halfway. The governor should adjust the engine speed to maintain the desired pressure. Reopen the tank fill valve all the way.

Be sure not to operate the pump for more than a few minutes without circulating water back to the tank or the other discharge line. Without the cooling provided by the circulating water, friction will cause the water to heat and boil, which may damage the pump.

Figure 6-21 The strainer should be in good condition and should not have any missing or damaged pieces.

Voices of Experience

Routine maintenance is just that: routine. But Daily Apparatus Checks are not an area where short cuts can be taken. That expectation needs to be made clear on the first day a new driver/operator reports to the company. In addition, senior driver/operators must stay diligent in taking these checks seriously every single day of their careers.

The criteria and sequence for fire apparatus maintenance are established by detailed fire department check-off lists. However, the steps for developing a positive safety attitude towards fire apparatus maintenance within the company are not written down—that is up to the driver/operator.

Remember, the fire apparatus is both your office and your toolbox. Daily Apparatus Checks will ensure its safety. And these tips will help ensure that your Daily Apparatus Check is both thorough and effective:

- The vibration that takes place while the fire apparatus is on the road tends to loosen certain pieces of equipment. All attack lines, whether pre-connected or not, have a nozzle attached and have to be physically checked. Ensure that the nozzle is firmly attached to the hose and make sure the tip of the nozzle is screwed down tightly at the ball-valve shut-off.

- Many departments carry their monitor tips "stacked" at the end of the deck gun. Ensure that all tips are firmly screwed on to the appliance.

- Whenever sections of hose are changed out, make sure there is a gasket inside the female swivel of the coupling.

- Make sure that your fire extinguishers are properly charged and that the gauge is "in the green."

- Check the chain on the chainsaw. There have been incidents when a fire fighter has unknowingly put the chain on backwards on the bar.

- The apparatus floors should be swept and mopped right after the Daily Apparatus Checks. Be alert to any puddles forming throughout the shift from water, oil, fuel or other fluids. This may be your only indication that there is a slow leak in the tank, a loose pump seal, an oil, fuel or hydraulic system leak. Some companies have members wipe down the apparatus before shift change with a chamois. Some even go so far as to wipe down the undercarriage. Though this may seem excessive, it gets the members under the rigs with creepers and forces them to take notice of the inside tires, belts, bolts, and mounting brackets, etc.

> "Remember, the fire apparatus is both your office and your toolbox."

Daily Apparatus Checks are more than checking a diesel engine for fuel and oil. Take these checks seriously. The lives of your crewmembers and their families are depending on you to prepare that apparatus for a safe response. Taking short cuts can make your rig unsafe on the road and unreliable on the fireground.

Raul A. Angulo
Seattle Fire Department
Seattle, Washington

Fire Service Pump Operator: Principles and Practice

Near Miss REPORT

Report Number: 09-0000695
Report Date: 07/16/2009 10:16

Synopsis: Faulty brakes almost cause collision.

Event Description: During an emergency response to a reported structure fire, a near-miss incident occurred between an engine company and a ladder company. As the two companies approached a controlled intersection from opposite directions, the engine company driver was unable to slow the apparatus down and narrowly avoided running head-on into the ladder company. Fortunately, the ladder driver was able to quickly react to the situation and the collision was averted. If we would have collided, multiple fire fighters could have been injured and/or killed. Brake failure of the engine contributed to the cause of the incident.

Lessons Learned: Keep the apparatus maintained by performing daily checks.

Demographics

Department type: Paid Municipal
Job or rank: Lieutenant
Department shift: 24 hours on—48 hours off
Age: 34–42
Years of fire service experience: 11–13
Region: FEMA Region V
Service Area: Urban

Event Information

Event type: Vehicle event: responding to, returning from, routine driving, etc.
Event date and time: 07/12/1999 17:00
Event participation: Involved in the event
Weather at time of event: Clear and Dry

What were the contributing factors?
- Human Error
- Decision-Making
- Command
- Situational Awareness
- Equipment

What do you believe is the loss potential?
- Property damage
- Life-threatening injury
- Lost time injury

For fire apparatus with multistage pumps, the <u>transfer valve</u> should be set to change from pressure to volume operation or from series to parallel operation. The pressure should change appropriately with about half the pressure in parallel (volume) from the series (pressure) setting. This change should be made two or three times to exercise the valve.

Aerial Device Inspection

The aerial device should be inspected according to the manufacturer's recommendations. NFPA 1911, *Standard for the Inspection, Maintenance, Testing, and Retirement of In-Service Automotive Fire Apparatus*, requires that aerial devices be tested annually. As part of their inspection, you should record the amount of time it takes to perform each of the three recommended tests: full lift, extension, and 90-degree rotation. An increase in the amount of time it takes to complete a test may indicate problems with the hydraulic system. All tests are meant to check for the proper operation and adjustment of components if necessary. Inspection that ensures the availability of a properly maintained and adjusted aerial device may prevent an accident or a catastrophic failure of the device.

Many departments require the driver/operator to not only inspect the aerial device, but also to clean and lubricate it. Follow the manufacturer's recommended maintenance schedule for the replacement of hydraulic filters and hydraulic fluid replacement.

Always follow your department's policies for this inspection process. Many different aerial devices exist, and each has different features. Before operating the aerial device, always make sure that it is in a safe position and that no overhead obstructions are present in the immediate area.

Some of the components most commonly found on an aerial device include the following items:
- Aerial device hydraulic system
- Stabilizing system
- Cable systems
- Slides and rollers
- Aerial device safety system
- Breathing air system
- Communications system

The main component of the aerial device is the hydraulic system that powers it. Using the hydraulic fluid and large cylinders that constitute this system, the aerial device can be maneuvered into almost any position. The hydraulic system is made up of a reservoir of hydraulic fluid, a pump, pressurized lines, and hydraulic cylinders that power the stabilizers and the aerial device.

To inspect this system, with the aerial device and stabilizers in the stored position, you should first check the fluid stored in the reservoir. Depending on the manufacturer, a simple dip stick or a sight glass may be used to determine the fluid level. The hydraulic lines should also be visually inspected for any leaks or signs of chafing. In addition, engage the hydraulic system and verify that all of the functions and alarms are operating correctly.

Next, place the wheel chocks on the fire apparatus and prepare to deploy the stabilizers. The stabilizers should be checked

Figure 6-23 Clean stabilizers are easier to inspect for defects.

for their full range of motion; they should also operate smoothly and evenly. Check the stabilizer arms for any leaks, cracks, broken welds, or loose parts. The stabilizers should be clean and free of rust, and all working parts should be lubricated as required **Figure 6-23**. All of the controls should be properly labeled.

Once the stabilizers are set, put the aerial device through its full range of operation using the main controls. The controls' response should be smooth and even with no unusual noise or vibration.

Inspect the aerial device for any cracks, loose parts, damage, or signs of heat stress. Check the turntable gears for any missing teeth, broken welds, leaking hydraulic lines/cylinders, or damage to the lifting cylinders. Inspect the cables for looseness, frays, broken strands, or other signs of damage. Next, inspect the slides and rollers of the aerial, which allow the different sections of the aerial device to move in and out with out rubbing against each other **Figure 6-24**. Ensure that there is no metal-to-metal contact and that the slides and rollers are properly lubricated and functioning properly.

Many aerial devices are equipped with safety systems that will not allow the device to perform specific functions if the apparatus is not in a safe position. For example, if the stabilizers are not fully extended on one side of the fire apparatus, the aerial device will not be allowed to operate on that side. A sensor will

A.

B.

Figure 6-24 The slides and rollers prevent the sections of the aerial device from contacting each other. **A.** The roller. **B.** The aerial.

Figure 6-25 A communication system allows the driver/operator at the turntable to remain in contact with the crew members who are working at the tip of the aerial device.

stop the aerial device from operating on that side so as to prevent the fire apparatus from tipping over. Each manufacturer provides different safety systems and overrides for these systems. You must be very comfortable operating the aerial device and know how and why to perform an override of any system. Remember, however, that these safety systems are designed for the protection of the fire fighter.

Many times fire fighters will operate at the tip of the aerial device during firefighting operations. During this time the fire fighters may use the air supply from the fire apparatus rather than the SCBA that they carry on their backs—a strategy that allows them to work for longer periods of time than would be possible with SCBA. To check the functioning of the fire apparatus's air supply system, first make sure the air tanks are full. Many fire departments have the driver/operator document this information on the fire apparatus inspection form. Look for any cracked hoses or loose parts.

A communication system at the turntable enables the driver/operator of the fire apparatus to speak with a fire fighter working at the tip of the aerial device. This system should be checked for proper operation as part of the fire apparatus inspection **Figure 6-25**.

Follow the steps in **Skill Drill 6-1** to perform an inspection of a fire apparatus:

1. Review the previous apparatus inspection reports for information regarding the fire apparatus. **(Step 1)**
2. Inspect the exterior of the apparatus in accordance with the department's policies and procedures. **(Step 2)**
3. Inspect the engine compartment of the fire apparatus in accordance with the department's policies and procedures. **(Step 3)**
4. Inspect the cab interior of the fire apparatus in accordance with the department's policies and procedures. **(Step 4)**
5. Complete a brake inspection of the fire apparatus in accordance with the department's policies and procedures. **(Step 5)**
6. Inspect the tools and equipment of the fire apparatus in accordance with the department's policies and procedures. **(Step 6)**
7. Inspect the pump of the fire apparatus and all of the features associated with its function in accordance with the department's policies and procedures (if applicable). **(Step 7)**
8. Inspect the aerial device and all of the features associated with its function in accordance with the department's policies and procedures (if applicable). **(Step 8)**

Chapter 6 Performing Apparatus Check-Out and Maintenance

Skill Drill 6-1

Performing an Apparatus Inspection

1. Review the previous fire apparatus inspection report.

2. Inspect the exterior of the fire apparatus.

3. Inspect the engine compartment.

4. Inspect the cab interior.

5. Inspect the fire apparatus's brakes.

6. Inspect the tools and equipment carried on the fire apparatus.

7. If applicable, inspect the pump of the fire apparatus and all the features associated with its function.

8. If applicable, inspect the aerial device and all the features associated with its function.

Safety

Not only is it critical that you perform the fire apparatus inspection in a safe manner, but you must also ensure that the fire apparatus is prepared for a safe response. This includes making sure that the fire apparatus is in proper working condition, the emergency warning equipment is operating correctly, tools and equipment are functional, and the vehicle is ready to support sustained operations.

The final portion of the safety evaluation focuses on making sure that the fire apparatus is safe to ride on and operate. As part of that safety check, all tools and equipment should be secured, breathing apparatus secured, equipment properly placed and secured on compartment shelves, and equipment carried on the outside of the apparatus properly secured. Hose lines should be loaded and ready for deployment, ground ladders securely nested, portable tanks secured on water tankers/tenders, and other tools or equipment properly secured to the fire apparatus Figure 6-26 . In case of a sudden stop or use of evasive driving techniques, all equipment in the passenger compartment must stay firmly secured. A sudden stop can cause a hammer, for example, to become a dangerous projectile.

Weekly, Monthly, or Other Periodic Inspection Items

In addition to undergoing its daily inspection, each fire apparatus will typically be subject to other inspections that are performed on a less frequent basis. For example, some elements of the inspection process—such as checking hydraulic oil levels, checking pump bearing oil reservoir levels, and so on—may be done on a weekly, monthly, or even quarterly basis.

Consider tire inspections. In addition to the daily visual inspection and evaluation of air pressure, more comprehensive inspection of tires may include items such as tread depth measurement, age from wear, and examination for cuts or abrasions. This examination may be done on a periodic basis and may be performed by a person other than the driver/operator. Vehicles with hydraulic braking systems may require the hydraulic fluid level to be checked in both the main and backup reservoir. This assessment may be performed by a mechanic.

Completing Forms

The forms recording the inspection and maintenance process are filled out as the inspection process takes place. At the conclusion

Figure 6-26 Exterior-mounted equipment should be properly stored before the fire apparatus moves.

of the fire apparatus and equipment inspection, the forms should be completed and filed in accordance with the fire department's procedures. Any abnormalities should be reported to the officer in charge so that he or she can make a determination on the best method of corrective action. In some instances, a situation found during your inspection may require that the fire apparatus be immediately taken out of service. In other situations, appropriate maintenance and repairs may be scheduled to be performed at a later date.

Wrap-Up

Chief Concepts

- The fire apparatus inspection is an evaluation of the fire apparatus and its equipment that is intended to ensure its safe operation.
- The inspection should be planned, methodical, and performed in an organized manner.
- The inspection process should begin with a review of the apparatus inspection form that was completed after the previous inspection.
- Dividing the inspection into sections allows you to focus on a single aspect of the fire apparatus and discourages you from jumping from one element of the inspection to another without a plan; jumping during an inspection leads to the possibility that critical elements may be missed.
- Not only is it critical that you perform the inspection in a safe manner, but you must also ensure that the fire apparatus is prepared for a safe response. This includes making sure that the fire apparatus is in proper working condition, the emergency warning equipment is operating, tools and equipment are functional, and the vehicle is ready for sustained operations.

Hot Terms

<u>aerial device</u> An aerial ladder, elevating platform, aerial ladder platform, or water tower that is designed to position personnel, handle materials, provide continuous egress, or discharge water. (NFPA 1901)

<u>apparatus inspection form</u> A document that identifies who performed the inspection and when the fire apparatus inspection was performed, identifies any equipment that is damaged and/or repaired, and details other preventive maintenance procedures performed on the apparatus.

<u>dip stick</u> A graduated instrument for measuring the depth or amount of fluid in a container, such as the level of oil in a crankcase.

<u>extractor exhaust system</u> A system used inside the fire apparatus bay that connects to the fire apparatus tailpipe and draws its exhaust outside the building.

<u>fire apparatus inspection</u> An evaluation of the fire apparatus and its equipment that is intended to ensure its safe operation.

<u>fire pump</u> A water pump with a rated capacity of 250 gpm (1000 L/min) or greater at 150 psi (10 bar) net pump pressure that is mounted on an fire apparatus and used for firefighting. (NFPA 1901)

<u>piston intake valve (PIV)</u> A large appliance that connects directly to the pump's intake and controls the amount of water that flows from a pressurized water source into the pump.

<u>power steering system</u> A system for reducing the steering effort on vehicles in which an external power source assists in turning the vehicle's wheels.

<u>preventive maintenance program</u> A program designed to ensure that apparatus are capable of functioning as required and are maintained in working order.

<u>radiator cap</u> The pressure cap that is screwed onto the top of the radiator, and through which coolant is typically added.

<u>transfer valve</u> An internal valve in a multistage pump that enables the user to change the mode of operation to either series/pressure or parallel/volume.

<u>voltage meter</u> A device that registers the voltage of a battery system.

<u>volunteer fire department</u> A fire department in which volunteer emergency service personnel account for 85 percent or more of its department membership. (NFPA 1720)

Driver/Operator in Action

While you are inspecting the reserve fire apparatus for any deficiencies, a fire fighter asks if he can assist you. His help would be very beneficial to you because several items on the fire apparatus are difficult—if not impossible—to inspect by yourself, such as the brake lights. In the past, you have found that it is very useful to have another set of eyes looking for any problems with the fire apparatus or the equipment on the fire apparatus. An extra set of eyes helps to ensure that nothing critical is missed and makes for a better inspection.

1. What is the single most important reason for inspecting the fire apparatus regularly?
   A. Department policy
   B. Locate new items
   C. Maintenance of the fire apparatus
   D. Safety

2. What is the NFPA's Standard for Automotive Fire Apparatus?
   A. NFPA 1931
   B. NFPA 1002
   C. NFPA 1901
   D. NFPA 1500

3. The inspection process should begin with a review of which document?
   A. Fire apparatus inspection form
   B. Fire apparatus owner's manual
   C. Fire department's standard operating procedure
   D. Fire apparatus identification form

4. How long should you wait before checking the engine's oil?
   A. 5 minutes
   B. 15 minutes
   C. 45 minutes
   D. More than 1 hour

5. Many fire departments have a policy that requires that their fire apparatus fuel tanks do not go below what level?
   A. ¼ of a tank
   B. ½ of a tank
   C. ¾ of a tank
   D. None of the answers are correct

Pumper Operations

CHAPTER 7

NFPA 1002 Standard

4.3 Driving/Operating. [p. 155–185]

4.3.1 Operate a fire department vehicle, given a vehicle and a predetermined route on a public way that incorporates the maneuvers and features, specified in the following list, that the driver/operator is expected to encounter during normal operations, so that the vehicle is operated in compliance with all applicable state and local laws, departmental rules and regulations, and the requirements of NFPA 1500, Section 4.2:

(1) Four left turns and four right turns
(2) A straight section of urban business street or a two-lane rural road at least 1.6 km (1 mile) in length
(3) One through-intersection and two intersections where a stop has to be made
(4) One railroad crossing
(5) One curve, either left or right
(6) A section of limited-access highway that includes a conventional ramp entrance and exit and a section of road long enough to allow two lane changes
(7) A downgrade steep enough and long enough to require down-shifting and braking
(8) An upgrade steep enough and long enough to require gear changing to maintain speed
(9) One underpass or a low clearance or bridge

(A) Requisite Knowledge. The effects on vehicle control of liquid surge, braking reaction time, and load factors; effects of high center of gravity on roll-over potential, general steering reactions, speed, and centrifugal force; applicable laws and regulations; principles of skid avoidance, night driving, shifting, and gear patterns; negotiating intersections, railroad crossings, and bridges; weight and height limitations for both roads and bridges; identification and operation of automotive gauges; and operational limits.

(B) Requisite Skills. The ability to operate passenger restraint devices; maintain safe following distances; maintain control of the vehicle while accelerating, decelerating, and turning, given road, weather, and traffic conditions; operate under adverse environmental or driving surface conditions; and use automotive gauges and controls. [p. 156–158, 166–169]

4.3.2* Back a vehicle from a roadway into restricted spaces on both the right and left sides of the vehicle, given a fire department vehicle, a spotter, and restricted spaces 3.7 m (12 ft) in width, requiring 90-degree right-hand and left-hand turns from the roadway, so that the vehicle is parked within the restricted areas without having to stop and pull forward and without striking obstructions. [p. 170–180]

(A) Requisite Knowledge. Vehicle dimensions, turning characteristics, spotter signaling, and principles of safe vehicle operation. [p. 170–185]

(B) Requisite Skills. The ability to use mirrors and judge vehicle clearance. [p. 170–185]

4.3.3* Maneuver a vehicle around obstructions on a roadway while moving forward and in reverse, given a fire department vehicle, a spotter for backing, and a roadway with obstructions, so that the vehicle is maneuvered through the obstructions without stopping to change the direction of travel and without striking the obstructions. [p. 170–180]

(A) Requisite Knowledge. Vehicle dimensions, turning characteristics, the effects of liquid surge, spotter signaling, and principles of safe vehicle operation. [p. 170–185]

(B) Requisite Skills. The ability to use mirrors and judge vehicle clearance. [p. 166–185]

4.3.4* Turn a fire department vehicle 180 degrees within a confined space, given a fire department vehicle, a spotter for backing up, and an area in which the vehicle cannot perform a U-turn without stopping and backing up, so that the vehicle is turned 180 degrees without striking obstructions within the given space. [p. 175, 177, 178]

(A) Requisite Knowledge. Vehicle dimensions, turning characteristics, the effects of liquid surge, spotter signaling, and principles of safe vehicle operation. [p. 166–185]

(B) Requisite Skills. The ability to use mirrors and judge vehicle clearance. [p. 166–185]

4.3.5* Maneuver a fire department vehicle in areas with restricted horizontal and vertical clearances, given a fire department vehicle and a course that requires the operator to move through areas of restricted horizontal and vertical clearances, so that the operator accurately judges the ability of the vehicle to pass through the openings and so that no obstructions are struck. [p. 177, 179]

(A) Requisite Knowledge. Vehicle dimensions, turning characteristics, the effects of liquid surge, spotter signaling, and principles of safe vehicle operation. [p. 166–185]

(B) Requisite Skills. The ability to use mirrors and judge vehicle clearance. [p. 166–185]

Additional NFPA Standards

NFPA 1221 *Standard for the Installation, Maintenance, and Use of Emergency Services Communications Systems* (2007)
NFPA 1901 *Standard for Automotive Fire Apparatus* (2009)
NFPA 1500 *Standard on Fire Department Occupational Safety and Health Program* (2007)
NFPA 1936 *Standard on Powered Rescue Tools* (2005)

Knowledge Objectives

After studying this chapter, you will be able to:
- Describe the role that the driver/operator plays in promoting safety, educating crew members, and promoting team building.
- Describe the driver/operator's responsibility maintaining a safe work environment.
- Describe the functions and limitations of the fire apparatus and its equipment.
- Describe the driver/operator's role when responding to an incident.
- Describe the driver/operator's role when returning from an incident.

Skill Objectives

After studying this chapter, you will be able to:
- Identify dispatch information.
- Perform a 360-degree inspection.
- Start the fire apparatus.
- Perform the serpentine exercise.
- Perform a confined-space turnaround.
- Perform a diminishing clearance exercise.
- Back a fire apparatus into a fire station bay.
- Shut down and secure a fire apparatus.

You Are the Driver/Operator

It's two o'clock in the morning. You are the driver/operator for Engine 60, the first unit on the scene of a fire at a one-story ranch-style home. You can hear the captain calling for the attack lines to be deployed. The supply line is stretched from the hydrant to your fire pumper. You can hear a fire fighter attempting to start the ventilation saw, but it does not cooperate. The interior crew calls for water; at the same time, the water supply snakes its way up the street to your engine. Lighting is needed on side B of the structure; however, it appears that the generator has taken its lead from the saw and will not cooperate.

1. Do you know the equipment and fire apparatus systems well enough to keep the operation going smoothly?
2. Is your knowledge and skill base sufficient that you have enough substance and experience to have a backup plan in place?
3. Has your training prepared you to handle these daunting fire-ground problems?

Introduction

As a driver/operator, you have many obligations to yourself, to your crew, and to the community that you serve. Internal and external expectations hinge on your ability to perform and complete your initial response assignments, as well as to fulfill any extra duties that arise at the scene. You know the ones—those little problems that occur at almost every call. You hear about them time after time: the generator that cannot get running or the power unit that will not start. You are charged with the responsibility to prevent, manage, or rectify those little requests that can make or break the ongoing operations at the scene.

To be successful in your new role, you must first understand and accept the responsibilities that come with the job. You may have moved from the attack crew to the driver/operator position; however, this does not remove you from the frontline responsibilities. As a driver/operator, you now play a greater role in the efficiency and success of your crew. Preventive maintenance, regular inspections of apparatus and equipment, and adherence to jurisdictional operating guidelines will support your success and offer some relief from the pressures of your new challenges.

New driver/operators may not always realize the responsibilities that this position entails. Fire fighters are injured or killed every year as a result of accidents that occur while responding and returning from emergency and nonemergency incidents. You must first acknowledge the hazards associated with this position, and then you must accept the responsibility to prevent these hazards. You have a responsibility to act the role through the "lead by example" process.

When you complete regularly scheduled fire apparatus and equipment inspections, you bring confidence to the other crew members that the fire apparatus and equipment will function when needed. Suppression, rescue, hazardous materials, and emergency medical crews have enough to deal with during the normal course of an emergency. Having confidence in your ability and the reliability of the fire apparatus and equipment creates a situation that is much easier to manage.

Despite your best inspection and preventive maintenance efforts, there may be situations when the equipment fails. If you have a solid foundation of knowledge, experience, and skill, however, a variety of backup plans and alternative methods for accomplishing specific assignments will be readily available to you. Simply stated, more knowledge offers more options, and more options lead to better outcomes.

The Many Roles of the Driver/Operator

A driver/operator is a teacher, a mentor, a vital crew member, and a safety advocate. You are expected to fix what is broken, offer alternative methods, maintain a state of constant readiness, and support every function that your apparatus can provide. In addition, you have a duty to educate other crew members on their roles and responsibilities to support you, much in the same way that you support their efforts on an emergency scene. For example, if you drive a ladder apparatus, you should ensure that all of the fire fighters in your company understand how to assist in the setup and deployment of the aerial device.

Crew members may not always understand the problems created from their small and seemingly insignificant actions during the response and return phases of an assignment. Consider what happens when fire fighters remove a tool from the fire apparatus and do not inform you. If you do not know the tool has been taken away, you may not account for it when the fire apparatus leaves the scene—and then you have a missing tool. It is your job to bring these problems to light so that success is shared by all.

To build crew confidence and efficiency, you must demonstrate your commitment to the department, the crew, your officer, and the community. To accomplish this goal, you must follow operating guidelines and applicable laws and regulations; create and maintain a safe work environment; and follow sound risk management principles. Team synergy begins with confidence and trust. These attributes initially begin with your words, and are later demonstrated through your actions.

Promoting Safety

There are several key roles that an effective driver/operator must play. As with all of the positions within the fire service, safety is your first priority. Your safety and the safety of your crew and the community is an important motivating factor for the manner in which you conduct yourself.

Getting to the incident is important to the operation; however, you should consider the events that occurred prior to your response. Were the required preventive maintenance actions taken prior to your response? Is the fire apparatus in a proper state of readiness? These are all good questions—but what are the real answers?

Driver/operators play an important role in the safety and efficiency of fire service operations. As such, there are many steps you can take to support a safer work environment. You begin by recognizing the associated hazards and then taking measures to reduce or eliminate these hazards.

Many fire departments utilize standard operating procedures (SOPs) to maintain work-safe environments. In most cases, SOPs were developed to prevent injuries, establish uniformity, and serve as a basic foundation for effective operations at an incident. You have an obligation to acknowledge the importance of these procedures and guidelines. Additionally, acceptance and demonstration of these procedures are critical to the overall safety of the crew and the community.

Seat belts are one safety device that is all too often underutilized by many crew members. You can demonstrate the importance of wearing a seat belt by being the first one to buckle up and by insisting that your crew members follow suit. A variety of excuses may be cited for not using these safety devices; however, if you educate crew members on the consequences of ignor-

Figure 7-1 Seat belts have saved many fire fighter lives; you can promote safety by being the first crew member to buckle up.

ing this safety measure, compliance usually prevails and safety is maintained **Figure 7-1**. Safety is an attitude, and changing attitudes may not come easy. Your job is to demonstrate the importance of utilizing vehicle safety systems by being the first to comply.

Safety Tip

Commercial fire and rescue chassis do not always account for the presence of custom seats that accommodate self-contained breathing apparatus (SCBA) units when designing their restraint systems. You should visually inspect the effectiveness and operation of the seat restraint systems with a fire fighter in full turnout gear in place. SCBA-mounting seats may require the use of longer-than-normal seat belts to safely accommodate a properly donned fire fighter.

Driver/Operator Tip

You and your fire officer are a team. By working together and being on the same page, you set an example of teamwork and leadership for the entire department.

Another area of concern is the equipment carried on the fire apparatus to support its function. This equipment, including heavy tools, is often stored in areas that present a significant risk to fire fighters. A dangerous situation can exist when fire fighters are retrieving or restoring this equipment from an elevated area during preventive maintenance activities, training activities,

Chapter 7 Pumper Operations

Figure 7-2 Equipment and heavy tools are often stored in areas that present a significant risk to fire fighters.

and emergency response activities **Figure 7-2**. You should demonstrate to crew members the safest technique to accomplish this task with proper use of the fire apparatus's mounted steps, ladders, and rails.

Heavy equipment, regardless of the location in which it is stored, can present a significant safety problem for fire fighters. To minimize their risk of injury, you can educate crew members on the proper removal and lifting techniques for specific pieces of equipment. Mounting devices such as racks, brackets, or trays are not standardized; each device may have a specific process for tool and equipment removal. It is your role to explain and demonstrate the proper techniques for removing and restoring equipment to its proper location.

As a safety advocate, you can play an important role in influencing future fire apparatus and equipment purchases, as well as modifications to existing equipment. Explaining and demonstrating the potential hazards to management is a proactive approach toward maintaining a safe work environment. For example, you might demonstrate what can be accomplished by installing the proper rails and steps on the fire apparatus so that fire fighters can safely retrieve equipment. By considering fire apparatus design needs, your department will save in the long term on injury and lost-time compensation costs.

Educating Crew Members

You must educate your crew members on the many potential hazards associated with the fire apparatus and its equipment. Fire fighters have been injured while working around fire apparatus-mounted equipment, such as pike poles and ground ladders that stick out on the fire apparatus.

Here is one simple philosophy to follow: Knowing the "why" makes the "how" come easy. In other words, knowing *why* hazards or dangerous conditions exist makes knowing *how* to reduce or prevent these conditions easier. There are many safety tips that you can offer to the crew members; knowing which information to present is the key to ensuring an effective and safe response. For example, when you are positioning the fire apparatus at an emergency scene, you are observing the traffic conditions while approaching the scene. However, the rest of the crew members do not have the luxury of viewing mirrors and facing forward all of the time. The crew may be unaware of hazards as they exit the fire apparatus—and it is up to you to make sure they are properly informed of any dangers before they leave the cab.

You must address rider positioning to the crew members. Crew members may inadvertently create a hazard during the response or return phase simply by blocking your visibility. For example, if a crew member attempts to don SCBA in the front cab area while en route to a call, his or her actions may obstruct your view of the side view mirror. By placing an arm out the open window while sitting in the front passenger position, a fire fighter may obstruct your view of a spot mirror during critical apparatus positioning maneuvers. Explaining or demonstrating the hazards created makes other crew members aware of such dangerous conditions so that these situations are not repeated in the future **Figure 7-3**.

Riding assignments are used by many fire departments, where each seat or riding position represents a specific task or function. For example, most fire departments designate the front passenger seat to the fire apparatus fire officer in charge (OIC). Other assigned positions may include fire fighter on the nozzle, backup line, ventilation, or forced entry (irons). Assigned riding positions are a proactive means to ensure that critical functions are assigned prior to arrival. You can extend the responsibilities

Figure 7-3 Crew members can create a hazard during the response or return phase by blocking your visibility.

of these riding positions by explaining to the crew members the response and return hazards associated with driving the fire apparatus. Fire fighters positioned in the jump seats can communicate when they see vehicles attempting to pass the fire apparatus or vehicles hanging back in one of the fire apparatus's many **blind spots**. Taking advantage of the existing riding positions is an effective safety measure to make the response safer.

Sometimes communicating on the fire ground or en route to the scene can be difficult. You can educate crew members on the various distractions that hinder effective communications, such as sirens, engine noise, equipment noise, and other radio traffic. Collectively, this barrage of noise can make effective communications almost impossible. Having prior knowledge of these possible conditions allows the crew members to develop a backup plan to reduce possible delays in critical communications during emergency situations. Some crews may use a **vehicle intercom system** to communicate en route to an emergency scene. A vehicle intercom system provides a headset and microphone for each rider on the fire apparatus. It enables the fire fighters on the fire apparatus to communicate clearly, without interference from outside noise. This intercom system may also be connected to the fire department's communication system, allowing these personnel to listen to and transmit information over a radio channel.

> **Safety Tip**
>
> All casual chat in the cab must cease while responding so that you can listen to fire-ground reports.

You must be familiar with SOPs regarding emergency communications such as mayday, emergency traffic, urgent, or emergency evacuation signals. In many cases, fire departments rely on air horns from the fire apparatus outside the fire to signal an emergency evacuation of the structure. If you are not familiar with this process or if you are not fully focused on the events of the incident, the outcome could be catastrophic. Paying strict attention to the **tactical benchmarks** of the incident will keep you ready to initiate life-preserving actions.

Trust and Team Building

The more the crew knows about the driver/operator's roles and responsibilities, the stronger the crew synergy becomes. The best way to impart this knowledge and ensure cohesiveness is to educate the other members of the crew on what you do. Show them how to operate the aerial device, the fire pump, and other features of the fire apparatus. The crew members may not be responsible for the actual operation of the equipment at a fire scene, but they will have more of an appreciation of what you and the rest of the team are trying to accomplish.

Staying focused on the tasks at hand makes for a more effective effort. Team building begins with confidence and finishes with trust. Trusting someone means that you believe in their abilities. In the fire service, crew members must believe in your ability as a driver/operator. They must believe that you have prepared the fire apparatus, the equipment, and yourself to the best of your abilities. They must believe that your decisions are calculated and that your actions are precise. This all sounds good, but how do you make it a reality? Trust is built through your communications and your actions. Consistency is the foundation needed to gain someone's trust. The more consistent you become with your decisions and your actions, the more trust you will gain.

Building a solid team takes time and patience. It is important to have support when you need it most. A working team makes successful outcomes look easy, even though that is not always the case. Consider this scenario: Do you and your crew perceive packing hose as a meaningless task, or do you perceive it as an action of readiness to meet the demands of the next alarm? If you have ever pulled a "spaghetti mess" hose load, you can relate to the importance of team building—effective teams do not demonstrate such carelessness.

How does this issue relate to the roles and responsibilities of a driver/operator? All members of the crew must be able to perform their assignments without reservation or hesitation. As the nozzle person at a fire scene, have you ever stood at the front door of a burning building waiting for the driver to charge the hose line? When you are in this position and ready to make an attack on the fire, it is frustrating to not have the line immediately charged with water. Crews want a driver they can trust to get them a sustained water supply quickly and efficiently. As the driver/operator, you are obligated to your crew members to provide them with the best possible on-scene support that you can muster up. Members of sports teams rely on one another to win championships; fire fighters rely on one another to survive.

Maintaining a Safe Work Environment

Maintaining a safe work environment is a tough job. Keeping fire apparatus and equipment service-ready can be a challenge to the newly appointed driver/operator. To make the process flow seamlessly, experienced driver/operators should start training future candidates for this position early in their careers.

Educating crew members is a major responsibility for any driver/operator. You must teach them how to recognize operational errors or equipment malfunctions that occur at emergency and nonemergency events. In addition, you should help them understand the **stressors** that challenge the driver/operator on a regular basis and emphasize that everyone plays a role in maintaining a safe work environment. Buy-in is what you wish to achieve; have the entire crew onboard with your work-safe philosophy. In time, you will create a work-safe ethic that will accommodate future changes in emergency and nonemergency operations.

Safety is embedded in all aspects of firefighting. As a driver/operator, you are charged with performing apparatus and equipment inspections and undertaking basic preventive maintenance operations. When the hydraulic rescue system will not start, things can become very stressful, very quickly. Regularly scheduled inspections, operational service checks, and scheduled preventive maintenance actions make major contributions to ensuring the reliability of the fire apparatus and equipment carried onboard. Routine inspections must be taken seriously.

Fire apparatus and equipment operational checks must not only be performed, but also documented by the driver/operator—after all, it's your job.

What should you be doing in the course of a day? First, you should actively promote the work-safe philosophy by completing daily inspections of equipment and fire apparatus systems such as hydraulic rescue systems; electrical power plants and cables; pneumatic equipment and supply hoses; portable and mobile radios; fire pump operation; gas detection meters; ladder main or boom operation; thermal imaging cameras; SCBA; hand lights; apparatus braking systems; tires and wheels; and audible and visual warning devices, just to name a few. The trick to completing a thorough inspection is to use a **job aid** such as a checklist or inspection sheet Figure 7-4 ▶. If supported by an inspection process, these inspection sheets or checklists will quickly become routine parts of the job. Their ongoing use will provide the consistency that you desire for fire apparatus and equipment reliability. Cutting corners has no place in the fire service. An upfront commitment to completing your assigned responsibilities is paramount.

Other informational sources that could be used during routine inspection and preventive maintenance processes include operating and general service manuals. These manuals are generally provided by the manufacturer with the purchase of its fire apparatus or equipment. Operating manuals are key tools for any preventive maintenance and general service program. Selling fire apparatus and equipment is a competitive market for manufacturers; the continuing stream of apparatus system upgrades, equipment functionality revisions, product design changes, and newly introduced safety features makes it imperative for you to stay on top of the ongoing process of upgrading and redesigning equipment and fire apparatus. Unfortunately, not all fire pumps, rescue and suppression equipment, and apparatus systems are created equal. Different manufacturers require specific procedures, service and maintenance schedules, and lubricants to comply with their warranty obligations and life-cycle expectations. This task becomes much easier when you remember to refer to the manufacturer's recommendations. In other words, always check the manual!

Your day is not done yet. Training and education continues at the fire station. There, you should educate crew members on the importance of using the safety equipment. Demonstrate how to pre-set, start, and operate specialized equipment. This task can be accomplished during the daily inspection and maintenance process. Get everyone involved—after all, their involvement supports their own personal safety. Teach company members how to troubleshoot problems in the field and how to correct problems as they occur through the use of a process or routine.

Consider the gasoline-powered equipment carried on the apparatus. "I started it this morning but it wouldn't start at the rescue scene"—does that complaint sound familiar? Give crew members something to work with, something to remember. In basic training, you were taught the PASS mnemonic as a tool to help you execute the procedure for operating a fire extinguisher: Pull the pin, aim the nozzle, squeeze the handle, and sweep the base of the fire. Even to this day, you remember that memory aid. If this technique works for fire extinguishers, it should also work for starting power units. Try something like FCSP ("fire fighters can save people") to help crew members remember the right sequence: fuel, choke, switch (start/engine cutoff switch to "on"), and pull. This acronym may help fire fighters remember to first check the fuel level and fuel shut-off valve, set the choke, turn the start/engine cutoff switch to "start" or "run," and pull the cord or engage the start/ignition switch. To reinforce this sequence, you might write the acronym on the top of the engine cover.

Tools such as phrases and acronyms are used often in the fire service to help fire fighters remember important information. Commercial driving instructors use ABCDE to emphasize key points to be covered in fire apparatus tire inspections: abrasions, bulging, cuts, dry rot, and even-wear inspection. Remember the "everything under pressure is worse" philosophy? That statement holds true for fire fighters whether they are attempting to do ten things at once while people are screaming for help or whether they are performing their routine preventive maintenance procedures. Your role is to relieve some of the stress on the job by educating the crew on the basic mechanics of how equipment operates and functions.

With experience, you may discover other roles and responsibilities that are charged to the driver/operator position. For example, while operating at the scene of a motor vehicle accident, you may not only position the fire apparatus so as to protect the scene, but also act as a lookout for the crew in heavy traffic conditions. If you anticipate and prepare for these additions and changes in your role, you will maintain your consistency and reliability by completing your assignments. Know your fire apparatus, your equipment, and your community; how you perform will reflect your knowledge of these three areas. In time, you will gain confidence and experience. Your reliability will be acknowledged and your consistency will be steadfast.

■ Lead by Example

As a driver/operator, you must lead by example. Follow the rules yourself. SOPs are for everyone—so lead by adhering to them. Follow and promote compliance with all SOPs. For example, by completing and documenting your fire apparatus inspections as part of your normal routine, you demonstrate to the crew your commitment to everyone's success. This will give you credibility with the other crew members and keep you safe at the same time. SOPs cover many fire department service areas, such as responding to and departing from incidents, required personal protective equipment for highway incidents, fire apparatus positioning, and general company assignments. Examples of such procedures include having the first-due engine lay a supply line and connect to the fire department connection (FDC) and having the first-due ladder go to the address side of the structure.

There are many critical life-preserving guidelines that you must know. Evacuation signals, mayday or urgent communications, and water supply signals are a few examples. Remember, each fire department operates in a manner intended to best meet the needs of its community and the situation at hand. You must fully understand the strategies and tactics used by your department's fire officers. For example, some fire departments may have their driver/operators perform horizontal ventilation of a structure fire by breaking a window while the attack team enters the building.

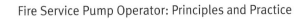

DAILY ENGINE INSPECTION SHEET

Week of _____   Unit I.D. _____
                                    Shop # _____

Date:	Mon.	Tues.	Wed.	Thurs.	Fri.	Sat.	Sun.
Name:							
Knox box serial number							
Fuel level							
Motor oil							
Radiator							
Wipers							
Gauges							
Brakes							
Starter							
Lights/siren							
Generator							
Mirrors							
Body condition							
Water level							
Pump controls/gauges							
Press control device							
Hydrant tools							
Hose/nozzles							
Appliances							
Tools/ladders							
SCBA—PPE							
Radios							
Box-lights							
Map-books/computer							
Keys							
Accountability							
Clipboard							
Tire pressure							
Batteries							
Transmission fluid							
Bleed air tanks							
Primer fluid							
Drain valves							
Tool box							
Power tools							

Comments:
_____
_____
_____
_____
_____

Figure 7-4 To complete thorough equipment and apparatus inspections, a driver/operator can use a job aid such as a checklist or inspection sheet. This sheet also serves as a legal document and maintains accountability.

> **Safety Tip**
>
> Always follow the rules of your fire department to the best of your ability. Whether you are driving, pumping, or operating a specialized piece of equipment, your actions invariably fall under a law, regulation, or SOP. Do the right thing; exercise due regard on and off the street. You should always advocate safety and compliance to SOPs to others.

> **Driver/Operator Tip**
>
> Personal protective equipment requirements for driver/operators always seem to be a point of discussion; however, this debate can be easily resolved by following your department's SOPs. Get involved in the process of developing these SOPs. Demonstrate safety concerns as they relate to the process of evaluating the effectiveness of each SOP.

If you put a concerted effort into learning the fire apparatus and equipment systems and acknowledge and comply with SOPs, you will be recognized as a leader and a valued crew member. Your colleagues will respond to your successes and actions by following your example. In the long haul, equipment may begin to run a little better or the tactical operations may be executed a little more smoothly.

> **Driver/Operator Tip**
>
> You can make a difference by educating your crew, maintaining your knowledge and skill levels, and following the rules that are in place for your safety.

Fire Apparatus and Equipment: Functions and Limitations

Successful driver/operators understand the primary function of each fire apparatus as well as recognize the limitations of the apparatus and the specialized equipment that it carries. For example, the primary mission for a pumper is to supply water to attack lines or other units. Water tenders haul large volumes of water, whereas brush units are primarily used to suppress vegetation fires. Regardless of the fire apparatus's function, you must know the expectations of the crew and the fire officers. Missing an assignment or not completing an assignment properly is not an option in the firefighting business. You must communicate immediately if a problem arises that cannot be quickly managed in the field. The only way to accomplish this is to fully understand the functions and limitations of the fire apparatus.

Consider the different water supplies that may be available at incident: onboard supplies, municipal water supplies, and natural or human-made water supplies. How much fire can a specific fire apparatus handle? How many hose lines can be deployed until your demand for water surpasses your available water supply? Will your process for supplying water change with different types of water supplies? The fire ground is not the place to seek out the answers to these questions. Although the answers are in reach, it takes the proper training to identify them. Before these questions arise on the fire ground, take the fire apparatus out and flow some water with the equipment that you have on it. Try out different scenarios that you and your crew may face, such as a broken hydrant or a burst hose line. Understand that the complexity of knowledge and skills needed to become proficient in the driver/operator position is high. Remember, great driver/operators can troubleshoot the problem and find a way.

Have you ever seen a fire fighter use a tool for something other than the purpose for which it was intended? For example, a fiberglass handle pike pole would be a poor choice of equipment for trying to pry open a door or window. This application would be using the wrong tool for the job and would present a safety hazard. Because safety is critical to a successful operation, you must know the functions and limitations of the equipment and the onboard systems. What can each rescue tool cut and what can it not cut? What are the weight limitations for the stabilizers? What are the critical angles for the ladder truck? You need to know these answers or, at the very least, know where to find them quickly. Operating manuals are the best sources for retrieving the answers to specific operational questions. Educate your crew members on the limitations of the equipment established by the manufacturers. Over the long term, this kind of education and training will save your fire department money by keeping the equipment in proper working condition. It will also reduce the possibility that personnel will be exposed to avoidable hazards because they have used the equipment for unintended assignments.

Not all restrictions are associated with the fire apparatus's equipment and systems, however: The fire apparatus itself presents some challenges. What happens when you cross a bridge with a 6-ton (5.4 metric ton) weight limit while driving a 15-ton (13.6 metric ton) truck? Knowing the height and weight restrictions for each fire apparatus is always helpful. This is easily accomplished by displaying the height and weight in the cab so that it is visible to the driver/operator and the fire officer in charge of the unit. **Vehicle dynamics** are also critical to the safe operation of the fire apparatus. Weight added or subtracted will change the handling characteristics of the fire apparatus as well as the vehicle dynamics. Simply stated, a fire apparatus may not handle in the same way with a full tank of water as it does when it carries a quarter tank of water. Prior knowledge of these conditions will provide for a much safer return.

> **Driver/Operator Tip**
>
> You should know your response area as well as you know the back of your hand so that there are no vertical clearance surprises. Also be aware of hills or dips where the fire apparatus could potentially "bottom out" from a low chassis clearance.

This sounds like a tall order; however, it makes perfect sense. You should know the ins and outs of the fire apparatus that you drive and operate, as well as the equipment that it carries and the systems that support operations. The driver/operator is the crew's "go-to person." When you fill this position, you always need to be on your game—to have the correct answers or the perfect backup plan. Education, training, and repetition will support your ability and enable you to develop alternative options during critical situations.

Fire Apparatus and Equipment Inspections

Routine inspections include all onboard systems and all carried equipment. If equipment is not routinely used, it may be forgotten. As a consequence, it may not always be in service-ready condition.

The best place to start is with a complete inventory of the fire apparatus. This exercise will provide you with a working document; at the same time, it will ensure that you and your crew learn where the equipment is kept and how to perform operational checks for each device. Although this inventory process sounds basic, it really is not. You may be surprised by the knowledge and skill levels of your crew members when it comes to the equipment carried on the rig; it is not always at the level you might expect.

Cleaning equipment is another method of inspection. Teach crew members to inspect the equipment as they clean it Figure 7-5. This "double duty" saves time and offers another level of protection from personal injury.

As a driver/operator, you are responsible for routine inspections and basic preventive maintenance procedures of the fire apparatus. To assist in this effort, you should identify resources from which to retrieve accurate information, proper specifications, and timelines for periodic inspections or maintenance schedules. For instance, when should you bring the fire apparatus in to have the transmission fluid changed? When does the pump oil need to be changed? At what point does tire wear become a major concern? Other areas of fire apparatus safety include coolant systems, steering components, batteries and charging systems, engine lubrication, and fuel systems. This information can be easily obtained from emergency vehicle technicians. General-service truck mechanics may be another good source of information, although their knowledge and experience may be limited to the main chassis and driveline components of the fire apparatus. For specialty equipment such as fire pumps, piping, valves, and generators, you will need to seek out advice from a certified emergency vehicle technician Figure 7-6.

Figure 7-6 Specific information on maintenance schedules and indications of component wear can be obtained from a certified emergency vehicle technician.

Safety Across the Board

Many different types of fire apparatus are used in the fire service. Some fire apparatus, such as the vehicles used by engine companies, are equipped with the water and hose necessary to extinguish a fire. Ladder companies' apparatus are equipped to help crew members gain access to structures and effect support functions at a fire scene. A heavy technical rescue company will be equipped for confined-space rescue and special operations.

No matter what the fire apparatus is used for or which type of emergency scene it may respond to, all fire apparatus are operated in two basic ways:

- **Emergency response.** This process starts with the initial dispatch and ends when the fire apparatus is back inside the fire station. It may include both emergency and non-emergency responses.
- **On-scene operations.** These processes include the operation of any of the fire apparatus-mounted equipment at the incident.

Both operations are critical components of fire services and will dramatically affect the outcome of the incident. As the driver/operator, you will set the tone for the rest of the team members with your response to the incident. If you drive the fire apparatus in a calm, defensive, and safe manner, your behavior will have an effect on the other members of your crew. The fire officer will be able to focus on the communications and scene size-up instead of worrying about your driving abilities. The fire fighters in the back of the apparatus will be less likely to get too excited and lose focus on the upcoming tasks. Conversely, if you drive in an erratic, out-of-control manner and allow your emotions to get the best of you, the entire crew may be in jeopardy. If the

Figure 7-5 Encourage fire fighters to inspect equipment as they clean it. This "double duty" saves time and offers another level of protection from personal injury.

crew does not get to the call in one piece, they won't be able to help anyone. When responding to an emergency, remember this mantra: *It's not my emergency*. Always respond in a safe manner no matter what the situation is.

Once at the scene, you will have to effectively operate the equipment of the fire apparatus. You must be capable of operating alone. While the other members of the crew may be tasked with various duties at the scene, your job is to support the crew. If you are part of an engine company, you may need to establish a water supply and charge the attack lines for the crew. Members of a ladder company may rely on you to set up the aerial device so that the ladder company can perform vertical ventilation. Whatever the circumstance, your job is not done once you reach the scene. You must be well trained and use your equipment effectively to support the crew.

Emergency Vehicle Response

Dispatch

The communications center will give the information for the emergency to you in the form of a dispatch. Dispatch is the process of sending out emergency response resources promptly to an address or incident location for a specific purpose. This step is usually performed by a telecommunicator at the communications center. Fire departments use a variety of dispatch systems, ranging from telephone lines to radio systems. The communications center must have at least two separate ways of notifying each fire station. The primary method may consist of a hard-wired circuit, a telephone line, a data link, a microwave link, or a radio system.

The majority of fire departments use a verbal message to those units that are responding to the incident. This dispatch may be announced from speakers located in the station or via an apparatus-mounted radio. In some fire departments, response vehicles may be equipped with mobile data terminals (MDT)—that is, computers located on the fire apparatus Figure 7-7 ▼ . With this approach, the dispatch can be transmitted to the fire apparatus through both the radio speakers and the MDT. Sensitive information, however, may be transmitted to the MDT without announcing it over a radio frequency, as the radio communications may be monitored by the media.

When dispatched for an emergency, whether the message is delivered at the fire station or in the fire apparatus, you should always pay attention to the following information:

- **Type of emergency.** This may be a structure fire, EMS call, or nonemergency call for service. The information given will determine how you will respond—that is, emergency mode or nonemergency mode. Each fire department should have a predetermined response mode for all types of incidents to which its personnel will respond. For example, when dispatched to a structure fire, the unit may respond to the scene with lights and sirens. If the fire apparatus is dispatched to a call that is not life threatening, it may respond without lights and sirens. Always follow your local SOPs in determining the mode of response.
- **Location of the emergency.** Each dispatch includes the physical address of the emergency. Usually, the dispatch also contains information such as cross streets, the geographical location of the response district, and possibly grid numbers to locate the incident on a map. These data are used to pinpoint the location of the incident.
- **Description of the incident.** The dispatcher should clearly describe what is happening at the incident. This information may include the victim's condition during an EMS incident or the location of a fire in a multistory occupancy.
- **Other responding units.** You should be familiar with your fire department's normal response to emergencies. If your fire apparatus is dispatched to a structure fire in your response district and the units that normally respond with your company are not dispatched as well, this can be a problem. Perhaps your district's normal backup units are on another call or out of service for training. Although other units may respond in their place, there could be a delay in response time. At the scene of a fire, this factor can drastically change the tactics of the initial units.
- **The assigned tactical radio frequency.** NFPA 1221, *Standard for the Installation, Maintenance, and Use of Emergency Services Communications Systems*, states that each fire department should be capable of assigning units to a separate radio channel for on-scene tactical communications. During the initial dispatch to a structure fire or as part of other multi-company responses, most fire departments assign a tactical channel to the responding units. You need to ensure that you are on the correct channel while responding.
- **Additional information from the dispatcher.** Most dispatchers provide some additional information to a responding fire apparatus while the unit is going to the call. This information can be used to determine the appropriate action to take while responding to the incident. For example, if the dispatcher advises that there is a *violent scene,* you may adjust your response. In some fire departments, responding units are required to stage several blocks from the scene until it is deemed safe by

Figure 7-7   Mobile data terminal.

the local police department. During this type of response, the driver/operator would turn off the emergency lights and siren so that the unit is not identified until the scene is safe to enter.

You should carry a writing utensil and a notepad at all times. When a call goes out, you can then write down the information from the initial dispatch, especially the location of the emergency. Some calls and their locations may become well known to fire fighters. This is no excuse to become complacent, however. Even veteran driver/operators can be stumped with an unusual address or location in their response area. Being prepared can help avoid confusion and ensures that valuable time is not wasted by asking for the communications center to retransmit the information. Write it down the first time!

Maps

Because it can take years to learn a response area, you must be familiar with your maps. Some fire departments may not have maps for their response area and instead rely solely on the fire fighter's knowledge of the response area. While this approach may be effective for some fire departments, most metropolitan fire departments have a map that driver/operators can use to locate incidents.

Fire department maps may be accessed several different ways. The most common type of map is found at the fire station. Usually it comprises a large paper map, which may be placed in various areas around the station. Most of these maps show details of the fire station's primary response area. Some are color-coded to identify the response districts of other fire stations **Figure 7-8**. The maps may have a grid system to divide the entire response area into more specific areas. These areas may be sectioned off according to the fire station's response area. For example, in the Albuquerque Fire Department, the entire city and surrounding county is divided into fire boxes. These areas may be several square blocks or several square miles depending on the density of the area. Each fire station has any number of fire boxes within its response area. When a call is dispatched, the incident's location is identified with an address and a fire box. This practice allows Albuquerque fire fighters to look on the map, locate the station's response area, the fire box, and then the actual street address of the call. Before you leave the station, you should reference the station-mounted map. Knowing where you are going saves valuable time during the response.

Driver/Operator Tip

Never hesitate to reference your station's map!

A smaller version of the station-mounted maps may also be found on the fire apparatus. These maps are sometimes grouped with all of the neighboring districts' response areas. When responding to out-of-station calls, these maps can prove very useful. While these maps may not be as detailed as those found at the fire station, they are still effective.

You should never try to read a map while responding to an emergency. Firefighting is a *team concept*. Another member of the crew should reference the map and guide you.

Another map that some fire departments are now using is found on the MDT. When used in tandem with a **global positioning system (GPS)**, the MDT may be capable of pinpointing the exact location of the emergency in relation to your location. A GPS is a device that uses satellite technology to locate the fire apparatus anywhere in a specified area. Some fire departments use this technology to dispatch the closest unit to an incident. When a call is dispatched, the information is transmitted to the MDT, so the location of the emergency appears on the screen and aids you in locating the scene. Of all of the maps, this one is usually the easiest to update and may be the most current.

Each of these maps used by fire fighters may have similar features. Sometimes the maps identify the location of hydrants, enabling crews to locate a water supply while en route to the fire scene. The maps may also identify parks, schools, and other important features of the response area.

Whichever type of map your fire department uses, it is important to select a safe and efficient route for each call. The fire apparatus must arrive to the correct location in a timely fashion. This responsibility will always fall upon you.

To identify the critical information received from the dispatch, locate the emergency scene, and determine the proper route on the map and the response mode, follow the steps in **Skill Drill 7-1**:

1. Start with the dispatch for an emergency. Document on a notepad the type of emergency, the location of the emergency incident, and the assigned tactical radio frequency (if applicable).
2. Locate the emergency scene on a map using the information given during the dispatch.
3. From the fire station, determine the most efficient response route using the map.
4. Determine the fire apparatus response mode for this emergency, using the information from the dispatch.

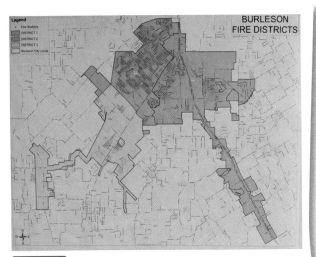

Figure 7-8 Some firefighting-specific maps are color-coded.

Voices of Experience

I began my career in the fire service and was trained as a driver/operator in an urban area with wide streets and plenty of lanes. Later in my career, my wife and I decided to relocate to the southern part of our state which happens to be more rural. When I joined my current department, I felt confident in my abilities as a driver/operator and informed the chief that I was ready to drive as soon as he was ready to let me. Back then, our certification process back was non-existent.

During training one day, the chief decided that he was ready to let me drive and prove my abilities. I confidently assumed the driver's seat and began the work of proving to the chief that I was everything that I claimed to be behind the driver's wheel. Just as I had predicted, everything went fine and there were no problems in my driving proficiency. However, during this evaluation time, we spent all of our time on wider "in-town" roads and kept off the more narrow country roads.

Finally it was time for me to drive beyond the confines of training. My sergeant told me where we were going. I was familiar with the road and knew that it was one of the most narrow in our district, but I felt confident that I could handle it. We turned onto the narrow road and at first there were no issues; however, there was no oncoming traffic yet either.

Then it happened, I met my first oncoming vehicle. Unfortunately, the vehicle did not feel like yielding to our much larger fire engine. So I moved over to the side of the road a little bit more than I should have and accidentally dropped the rear tire into a shallow ditch. I thought that my heart was going to jump out of my chest and that my sergeant was going to pass out.

Luckily, we were able to get the rear tire back up onto the road without any incident or damage, but I learned that day that there is a huge difference in driving a large fire engine in the city then in the country. I also learned to never yield too much road. So now if a motorist decides not to give me enough room, I stop and allow the car to pass before I proceed. It's better to get there a few seconds later than not at all.

Jesse Vacra
New Mexico Firefighters Training Academy
Socorro, New Mexico

> "It's better to get there a few seconds later than not at all."

360-Degree Inspection

Once you know where you are going, you must prepare for the response. Before entering the cab, you must complete a preliminary inspection of the fire apparatus, also known as a 360-degree inspection. A 360-degree inspection is a quick check of the fire apparatus and its surroundings to ensure that the apparatus is prepared for a response, either to an emergency or a nonemergency. Failure to complete this inspection may result in damage to life and/or property. For example, an open compartment door may be sheared off by the fire station walls as the fire apparatus leaves the bay. Unsecured equipment stored on the outside of the fire apparatus may fly off and strike a civilian. The preliminary inspection must be performed every time that you move the fire apparatus, regardless of the emergency. Remember—complacency kills.

If the fire apparatus is responding from inside the fire station, the first step in a preliminary inspection is to open the fire apparatus bay door. While the door is opening, continue with the inspection: Do not waste valuable time waiting for the door to open later. Walk around the fire apparatus and physically check that the cab doors are completely shut. All compartments should be checked to ensure that they are secure for travel. Ground ladders and equipment mounted to the exterior must be properly secured. Many fire apparatus have an interior warning light or buzzer to notify you that a compartment is open or not completely latched. Inspect the hose for any signs that it may come loose during a response. Remove any cups, equipment, or debris that may have been improperly placed on the running boards, front bumper, or tailboard. Visually verify that the area underneath the fire apparatus is free of debris. All of this should take only a matter of seconds. By making this inspection a habit, you will ensure that the fire apparatus is safe to respond.

To perform a 360-degree inspection, follow the steps in **Skill Drill 7-2**:

1. Open the fire apparatus bay door completely. **(Step 1)**
2. Walk completely around the fire apparatus. **(Step 2)**
3. Check that all doors are secured. **(Step 3)**
4. Check that all compartment doors are secured. **(Step 4)**
5. Check that all exterior-mounted equipment is properly secured. **(Step 5)**
6. Remove any debris from the running boards, front bumper, and tailboard. **(Step 6)**
7. Check that the area underneath the fire apparatus is clear of debris. **(Step 7)**
8. Check that the area in front of the fire apparatus is clear and free of debris. **(Step 8)**

Starting the Fire Apparatus

After the 360-degree inspection is complete, you will enter the cab and initiate the sequence to start the fire apparatus. Modern fire apparatus are usually powered by a diesel engine. This engine requires a significant amount of current during the starting process. Before starting the fire apparatus, always ensure that unnecessary electrical loads are shut off (i.e., headlights, heater, and air conditioning). Verify that the **parking brake** is set. This brake is required by NFPA 1901, *Standard for Automotive Fire Apparatus*, to hold the fire apparatus on at least a 20 percent grade. If the fire apparatus has an automatic transmission (most modern fire apparatus do), ensure it is in the neutral position.

For most fire apparatus, the battery selector switch is turned on next. The **battery selector switch** is used to disconnect all electrical power to the fire apparatus to prevent discharge while it is not in use. Once the power has been transferred to the electrical system, in some fire apparatus it will initiate the prove-out sequence. A **prove-out sequence** is a series of checks that an electrical system completes to ensure that all systems are functioning properly before the fire apparatus is started. This check usually takes only a few seconds to complete. If the fire apparatus is started without allowing the prove-out sequence to complete, it may cause intermittent alarms to occur. For any nonemergency response, always allow the prove-out sequence to continue until the cycle is complete. During an emergency situation, this time delay may not be practical, however.

Once the prove-out sequence is complete, the fire apparatus is ready to start. Some newer models are equipped with an ignition switch and one or two starter switches. The **ignition switch** delivers operational power to the chassis; engage this switch. The **starter switch** engages the starter motor for cranking. If two starter switches are available, they are provided for redundancy. Engage either or both of these switches to operate the starter motor. When the engine starts, release the starter switch. If the engine does not start within 15 seconds, release the starter switch and allow the starter motor to cool off for 60 seconds before attempting to start it again.

The seats and mirrors should all be adjusted during the daily fire apparatus inspection. If they are not adjusted correctly, readjust them now while the fire apparatus is not in motion. It is very dangerous to attempt to make any changes to the seating or mirror configurations while operating a moving vehicle. Many driver/operators prefer to adjust the mirrors so that the side of the fire apparatus is barely in view of the mirror; this way they can see the side of the apparatus if necessary and can observe any objects to the side of the vehicle **Figure 7-9**.

Before you leave for the incident, you must look over the instrument panel. The following items on the instrument panel should always be referenced:

- **Fuel gauge**. This gauge indicates the amount of fuel in the apparatus's tank. All fire apparatus should be in a constant state of readiness. No fire apparatus fuel tank should ever drop below the halfway level. Always monitor the fuel gauge during operations.
- **Air pressure gauges**. These gauges identify the air pressure stored in the tanks or reservoirs of apparatus equipped with a pneumatic braking system. This air pressure is used to slow down and stop the apparatus while it is responding on the roadway. NFPA 1901, *Standard for Automotive Fire Apparatus*, requires that the fire apparatus have a quick build-up section in the air reservoir system arranged so that if the apparatus has a completely discharged air system, it is able to move within 60 seconds of start-up. On a chassis that cannot be equipped with a quick build-up air brake system, an onboard automatic electric compressor or a fire station compressed air shore-

Skill Drill 7-2

Performing a 360-Degree Inspection

1. Open the fire apparatus bay door completely.

2. Walk completely around the fire apparatus.

3. Check that all doors are secured.

4. Check that all compartment doors are secured.

5. Check that all exterior-mounted equipment is properly secured.

6. Remove any debris from the running boards, front bumper, and tailboard.

7. Check that the area underneath the fire apparatus is clear of debris.

8. Check that the area in front of the fire apparatus is clear and free of debris.

line hookup is permitted so as to maintain full operating air pressure while the vehicle is not running. The fire apparatus is also required to have a warning alarm to indicate a low level of air pressure in the system. This alarm activates below 60 pounds per square inch (psi) (413 kPa) and remains active until adequate pressure has built up to release the parking brake. For most fire apparatus equipped with pneumatic braking systems, a pressure between 100 to 120 psi (690 to 827 kPa) is adequate. Always consult the manufacturer's recommendations to determine the appropriate pressure.

- **Voltmeter**. The voltmeter measures the voltage across the battery terminals of the apparatus and gives an indication of the electrical condition of the battery **Figure 7-10**.

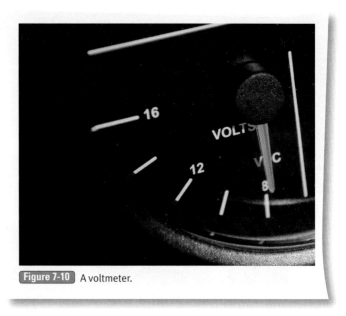

Figure 7-10 A voltmeter.

Figure 7-9 Adjust your mirrors for safety.

Operating voltage while the alternator is charging may vary among vehicles depending on the regulator setting. You should use this gauge for direct observation of the system voltage. If this kind of monitoring is provided, an alarm will sound if the system voltage drops below 11.8 V for 12-V nominal systems or below 23.6 V for 24-V nominal systems for more than 120 seconds.

- **Oil pressure gauge**. This gauge identifies the pressure of the lubricating oil in the engine. When the fire apparatus is started, it should provide a reading within a few seconds. If it does not, stop the engine and have a trained technician check the oil pressure.

Seat Belt Safety

Before the fire apparatus moves, you must visually check that all members are wearing a seat belt. You should never move an emergency vehicle with members unsecured. NFPA 1901, *Standard for Automotive Fire Apparatus*, requires that all seats of fire apparatus be enclosed and provided with an approved seat belt.

NFPA 1500, *Standard on Fire Department Occupational Safety and Health Program*, requires the driver/operator not to move the fire apparatus until all persons on the vehicle are seated and secured with seat belts in approved riding positions. While the vehicle is in motion, fire fighters shall not release or loosen their seatbelts for any purpose, including the donning of personal protective equipment (PPE) or SCBA. NFPA 1500 does recognize that there may be certain instances when fire fighters may need to be unsecured while a vehicle is in motion. Strict guidelines must be followed while operating under these conditions, however. While the vehicle is in motion, fire fighters may be in positions other than an approved riding position in three circumstances:

- When providing medical care
- When loading hose
- During tiller training

Safety Tip

Newer fire apparatus have alarms to notify you if a fire fighter is not wearing a seat belt. Some fire apparatus will not move unless a weight sensor indicates that all fire fighters are properly wearing their seat belts.

Providing Necessary Emergency Medical Care

Providing medical care to a victim while the vehicle is in motion can be difficult while the fire fighter is restrained in a seat belt. The preferred method would be to have a harness system to restrain the fire fighters while they provide care for the patient. This approach may not be practical in some situations, including chest compressions during cardiopulmonary resuscitation (CPR).

Loading Hose

Loading hose while the fire apparatus is in motion is dangerous and usually unnecessary. Fire departments that condone the practice must have written a SOP in place that covers certain safety aspects of the operation. All members involved in this hose loading process should be trained in the following procedures.

Figure 7-11 Before leaving the bay, the driver/operator verified that all exterior compartment doors, ladder racks, telescoping scene lights, and any other fire apparatus mounted equipment was secure.

During this operation, one member will be assigned as a safety observer. He or she is not permitted to physically load the hose. Instead, this crew member must have an unobstructed view of the hose loading operation and be in visual and voice contact with the driver/operator at all times.

During this procedure, no non-fire department vehicular traffic should be permitted in the area; if such traffic is allowed, it should be controlled by a trained traffic control person. The fire apparatus is allowed to drive only in a forward direction at a speed of no more than 5 mi/h. No crew members should be allowed to stand on the tailboard, sidesteps, running boards, or any other location on the fire apparatus while the apparatus is in motion. Members may be permitted in the hose bed, but must refrain from standing while the vehicle is in motion.

Prior to beginning any hose loading operation where the fire apparatus is in motion, the overall situation must be evaluated. The safety of the fire fighters and the public should always be paramount concerns. If a safer way is available, then it should be considered before hand.

Tiller Training

Operating a tiller apparatus is a unique skill that must be taught while the vehicle is in motion. This type of operation cannot be learned in a classroom setting, but rather must be experienced and mastered in a safe environment (i.e., a parking lot or training ground). It is helpful for the instructor to aid the trainee while the vehicle is in motion. The fire apparatus should be equipped with seating positions for both the tiller instructor and the tiller trainee. Both seating positions should be equipped with seat belts for each individual. The instructor's seat may be detachable, but will need to be structurally sufficient to support and secure the instructor. This seat should be attached and used only during training purposes. If the apparatus is called to an actual emergency during the training session, the training session will be terminated and all members of the crew must be seated and belted in approved riding positions.

Safety Tip

NFPA standards are not exclusive to just emergency vehicle responses. During funeral processions, parades, or public relations/education events, standing or riding on the tailboard, sidesteps, running board, or any other exposed position should be specifically prohibited. This action is unprofessional and needlessly jeopardizes the safety of fire fighters and the public they serve. Whenever the fire apparatus moves, everyone should be seated and belted—no exceptions!

Getting Underway

Once all members are secure in the cab of the fire apparatus, verify once again that all exterior compartment doors, ladder racks, telescoping scene lights, and any other fire apparatus-mounted equipment is secure Figure 7-11. Some fire apparatus are equipped with compartment-door indicator lights. These lights are activated only if the fire apparatus compartment doors are open and the parking brake is in the *off* position. Other fire apparatus may also have a digital display that shows any open compartment doors or other equipment that may be damaged if the vehicle moves Figure 7-12. A 360-degree inspection is always required to ensure safe operation; use these apparatus-mounted systems only as a secondary resource. Double-check the sides of the fire apparatus by using the mirrors.

Interior compartment doors also need to be secured in the closed position while the apparatus is responding to an incident. Otherwise, the items that are stored in these compartments may fly out if the fire apparatus is involved in an accident. No fire fighter should be injured because someone failed to secure equipment that is stored inside the cab. In particular, equipment such as axes and Halligan tools should never be stored in the cab. If a compartment or mounting system is provided for equipment, then use it!

When applicable, the engine should be allowed to warm up before the response begins. Of course, this may not always

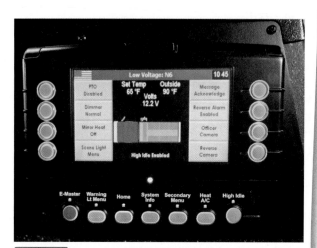

Figure 7-12 The fire apparatus may have a digital display that shows open compartment doors or other equipment that may be damaged if the vehicle moves.

be possible depending on the nature of the emergency. Be aware that operating a cold engine and transmission under very hard conditions (emergency response) may have a damaging effect to the engine. Whenever possible, allow the engine to warm up for 3 to 5 minutes. This will help to extend the life of the fire apparatus and may prevent future breakdowns.

The headlights should be turned on whenever the fire apparatus is moving, not just during the night. Research has shown that a vehicle is easier to identify during any conditions if it has its headlights on.

Once the overhead garage door is fully open, the fire apparatus may be driven out of the station. Be aware that other fire apparatus at the station may be leaving at the same time or that the exhaust extractor system and electrical cords may not eject properly. For these reasons, do not exceed 5 mi/h (8 km/h) while pulling out of the fire station.

Look to the sides of the fire apparatus and ensure that the cords and extractor are clear. Always remember to close the doors after clearing the building. If responding in an emergency mode, activate the emergency lights and audible warning devices at this time.

To start a fire apparatus and ensure that it is ready for a response, follow the steps in **Skill Drill 7-3**:

1. Enter the fire apparatus cab. (**Step 1**)
2. Verify that the parking brake is set in the "on" position. (**Step 2**)
3. Check that the transmission is in the neutral position. (**Step 3**)
4. Verify that any unnecessary electrical loads (including the air conditioner/heater and headlights) are turned off before activating the battery. (**Step 4**)
5. Turn on the battery selector switch. (**Step 5**)
6. Allow the fire apparatus to complete the prove-out sequence.
7. Engage the ignition switch. (**Step 6**)
8. Engage the starter switch. (**Step 7**)
9. Adjust the seat and mirrors, if necessary. (**Step 8**)
10. Check the gauges: fuel, air pressure, voltmeter, and oil pressure. (**Step 9**)
11. Ensure that each crew member is wearing a seat belt before moving the fire apparatus. (**Step 10**)
12. Check the compartment-door indicator light, if applicable, to ensure that no compartments are open. (**Step 11**)
13. Turn on the headlights. (**Step 12**)
14. Allow the engine to warm up before moving, if applicable.
15. Place the transmission in "drive." (**Step 13**)
16. Release the parking brake. (**Step 14**)
17. Activate the emergency lights and audible warning devices. (**Step 15**)
18. Drive the fire apparatus out of the fire station at a speed less than 5 mi/h (8 K/h).
19. Check the sidewalk to make certain that all pedestrian traffic has stopped. (**Step 16**)
20. Activate any signal control system to stop traffic in front of the fire station. (**Step 17**)
21. Close the fire station doors with the remote control to ensure security. (**Step 18**)

Safety Tip

According to a Texas State Fire Marshal's Office Fire Fighter Fatality Investigation Report, a fire fighter fell from a moving fire apparatus while responding to a structure fire. On April 23, 2005, Fire fighter Brian Hunton of the Amarillo Fire Department was riding in the left-rear seat of a Ladder 1 apparatus when the door opened and he fell out. Hunton struck his head on the street and sustained severe head injuries; he died two days later. During the response, he had been donning his SCBA and not wearing his seat belt.

All fire fighters are responsible for their own safety. During an emergency response, all members should wear their seat belts. This type of accident is preventable with the proper use of seat belts. Each fire department should have a SOP in place that requires the use of seat belts, and this policy must be enforced by all members. Requiring fire fighters to wear seat belts is a simple rule that could prevent 10 to 15 fatalities every year.

Driver/Operator Tip

Driving a fire apparatus while wearing turnout pants and boots can be very cumbersome for some fire fighters. For this reason, some fire departments do not require their driver/operators to dress in turnout gear while driving to the incident. Evaluate your ability to safely operate the fire apparatus while wearing turnout pants and boots. If you do not feel safe operating under these conditions, notify your supervisor and determine the best course of action. Remember—safety always comes first.

Driving Exercises

Riding in a fire apparatus as it is responding to an emergency scene can be very exciting. Driving the fire apparatus to a scene is even more exciting. You may develop a rush of adrenaline while operating the fire apparatus on the roadway. This excitement, however, should not make you lose sight of the task at hand—transporting the fire apparatus and its members to the emergency scene in a safe and efficient manner. The fire apparatus should not be driven faster than existing conditions permit or at a speed greater than can be maintained with safety. At all times, you must be able to maintain control of the fire apparatus. Do not allow the situation or other members of the crew to push you into driving the fire apparatus beyond your abilities. Instead, always use common sense and good judgment. A speedy response is achieved through a safe and efficient means of operation—not by taking unnecessary risks. Never endanger life or property, under any circumstances.

During the emergency response, you may have to maneuver the fire apparatus around objects at the scene or parked vehicles that are blocking access to a preferred location. This type of ma-

Skill Drill 7-3

Starting a Fire Apparatus

1. Enter the fire apparatus cab.

2. Verify that the parking brake is set in the "on" position.

3. Check that the transmission is in the neutral position.

4. Verify that any unnecessary electrical loads (including the air conditioner/heater and headlights) are turned off before activating the battery.

5. Turn on the battery selector switch.

6. Allow the apparatus to complete the prove-out sequence. Engage the ignition switch.

(continues)

Skill Drill 7-3 Continued

Starting a Fire Apparatus

7. Engage the starter switch.

8. Adjust the seat and mirrors, if necessary.

9. Check the gauges: fuel, air pressure, voltmeter, and oil pressure.

10. Ensure that each crew member is wearing a seat belt before moving the fire apparatus.

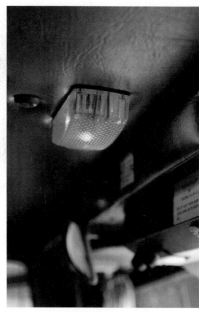
11. Check the compartment-door indicator light, if applicable, to ensure that no compartments are open.

12. Turn on the headlights.

Skill Drill 7-3 Continued

Starting a Fire Apparatus

14. Release the parking brake.

13. Allow the engine to warm up before moving, if applicable. Place the transmission in "drive."

15. Activate the emergency lights and audible warning devices.

18. Close the fire station doors with the remote control to ensure security.

16. Drive the fire apparatus out of the fire station at a speed less than 5 mi/h (8 K/h). Check the sidewalk to make certain that all pedestrian traffic has stopped.

17. Activate any signal control system to stop traffic in front of the fire station.

Near Miss REPORT

Report Number: 06-0000572
Report Date: 11/19/2006 12:14

Synopsis: Fire stream delays—result is compromised safety.

Event Description: During "routine" single family frame structure fire (with known victims), the pumper took 4 attempts to engage causing a delay in water to the hose for approximately 5–6 minutes. During this delay the interior suffered a flashover (4 minutes after arrival) forcing all interior crew members back downstairs. (The attack line was manned by one fire fighter and one officer. The remaining fire fighter from the engine was attached to the ladder crew for rescue due to four known victims.) One member of the ladder crew was unable to make it back to stairs and was forced to self rescue from second floor window to ground. He was operating in room next to fire room approximately 10'–15' from stairs. He had just vented the only window in the room and was starting to search when the flashover occurred. He went back to the window and waited for us to start knocking the fire down. He decided to bail when conditions weren't getting any better. He suffered 10% 2nd degree burns to his back, right arm, and face (hood was pulled from around his mask exposing a ring of skin). A radio report claiming a partial collapse of second floor ceiling with members still in area initiated a personnel accountability report (PAR) and attempt to re-enter area to find missing member. A radio report from incident commander (IC) verified missing members self rescue. A second line was now in place, (first line had burnt through due to being hung up on railing upstairs), and entry was attempted again. This was delayed due to first line (burnt through) still flowing and causing a reduction in pressure. Radio communications where intermittent and the pumper/operator wasn't getting the message to shut down the first line. This was accomplished by sending a runner. Once proper pressure was established extinguishment was accomplished and the 4 victims were found (none survived).

Initial crews on scene: 1 Ladder company w/one officer and 3 fire fighters (3 entered structure while operator placed ground ladders to porch); 1 Engine company w/one officer and 3 fire fighters (3 entered with officer and fire fighter on line and remaining member w/Ladder company to initiate rescue; and 1 ALS unit w/2 fire fighter medics (we are a fire-based EMS system).

Conditions on arrival: Smoke showing from 2 windows (fire room and victims location) and no fire visible. Occupants on sidewalk claiming victims on second floor. Moderate to high heat encountered at top of stairs. Smoke down to approx. 8"–10" from floor.

Operator error ruled out and no problem could be found with pumper. Initial thoughts were the twist the chassis experienced turning corner at fire scene (recent storm drain work had left approx. 18" drop from pavement. Unable to replicate. Shop changed alternator, batteries, and cables thinking a low voltage issue was responsible.

Problem experienced: Delay in water; flashover; burnt hose line; intermittent radio communications; low pressure due to open line.

It is my firm belief that the injured fire fighter's training, experience, and level headedness prevented this from becoming a line of duty death (LODD).

Lessons Learned: The importance of self rescue training can't be overstated. The rapid intervention team (RIT) companies had just arrived on scene and were unable to respond without some delay. Two-in-two-out wasn't an option here due to KNOWN multiple victims.

Be the eyes and ears of the IC when you are inside. The call of a partial collapse (it didn't register that we had just had a flash-over initially) and resulting PAR call reduced the discovery time of a missing member.

If there's any delay in getting water to the line let all companies know. This will let interior crews back out of the hazard area and will alert incoming companies that a shift in positioning may be warranted. Our SOPs have the 2nd arriving engine stop at the nearest hydrant and await further instructions. If they know coming in that the on-scene engine is having difficulties they can respond directly to the scene and take over pump operations with a minimum of delay.

This was a "routine" fire on arrival but resulted in a cascade of problems that, I believe, were a direct result of the delay in water. If we had gotten water on the first attempt this situation would probably still have resulted in the civilian casualties but not the injuries suffered by a member of my station.

Near Miss REPORT (Continued)

Report Number: 06-0000572
Report Date: 11/19/2006 12:14

Demographics

Department type: Paid Municipal
Job or rank: Lieutenant
Department shift: 24 hours on—48 hours off
Age: 34–42
Years of fire service experience: 17–20
Region: FEMA Region V
Service Area: Urban

Event Information

Event type: Fire emergency event: structure fire, vehicle fire, wildland fire, etc.
Event date and time: 10/27/2006 05:38
Hours into the shift: 21–24
Event participation: Involved in the event
Do you think this will happen again? Uncertain

neuver is done at a reduced speed with an emphasis on the safety of pedestrians, other objects, and the fire apparatus. NFPA 1002, *Standard for Fire Apparatus Driver/Operator Professional Qualifications*, requires that all driver/operators complete a serpentine exercise that simulates these conditions. The exercise measures the driver/operator's ability to maneuver in close quarters without stopping the fire apparatus.

Performing a Serpentine Maneuver

As part of the serpentine maneuver exercise, a minimum of three marker cones are spaced 30 to 38 feet (9 to 12 m) apart in a line. The spacing of the marker cones should be equal to the fire apparatus's wheel base. The space on the sides of the marker cones must provide adequate space for the fire apparatus to travel freely. To perform the serpentine maneuver, drive the fire apparatus along the left side of the marker cones in a straight line and stop with the fire apparatus just past the final marker cone. You are now in position to begin the exercise. Back the fire apparatus to the left of marker cone 1, to the right of marker cone 2, and to the left of marker 3. Once the front of the fire apparatus is past marker cone 3, drive the fire apparatus forward between the marker cones by passing to the right of marker cone 3, to the left of marker cone 2, and to the right of marker cone 1.

During the entire exercise, the marker cones should not be struck and the fire apparatus should move in a continuous motion, except when required to change direction of travel. A spotter is necessary for this exercise. To perform the serpentine exercise, follow the steps in **Skill Drill 7-4** ▶:

1. Position the fire apparatus past marker cone 1.
2. Activate the emergency lights on the fire apparatus.
3. Ensure that a spotter is in position at the rear of the fire apparatus.
4. Place the fire apparatus in reverse.
5. Proceed in reverse through the exercise area, to the left of marker cone 1.
6. Proceed in reverse through the exercise area, to the right of marker cone 2.
7. Proceed in reverse through the exercise area, to the left of marker cone 3.
8. Position the front of the fire apparatus past marker cone 3.
9. Proceed forward through the exercise area, to the right of marker cone 3.
10. Proceed forward through the exercise area, to the left of marker cone 2.
11. Proceed forward through the exercise area, to the right of marker cone 1.
12. Once the rear of the fire apparatus passes marker cone 1, the exercise is complete.

Performing a Confined-Space Turnaround

When responding to an incident, you may inadvertently pass a street on which you should have turned. When this situation occurs, the best course is to drive the fire apparatus around the block rather than try to turn the fire apparatus around in the confines of a roadway. Other motorists may be confused by your actions as you attempt to move a large fire apparatus around 180 degrees only to proceed back in the direction from which you just came. However, even seasoned driver/operators can go down a wrong road or find themselves in a position where they need to turn around and go the other way.

NFPA 1002 requires that all driver/operators complete an exercise that simulates these circumstances, during which they turn a fire apparatus 180 degrees within a predetermined area where the apparatus cannot complete a U-turn. The exercise,

Skill Drill 7-4

Performing the Serpentine Exercise

1. Position the fire apparatus past marker cone 1.
2. Activate the emergency lights on the fire apparatus.
3. Ensure a spotter is in position at the rear of the fire apparatus.
4. Place the fire apparatus in reverse.
5. Proceed in reverse through the exercise area, to the left of marker cone 1.
6. Proceed in reverse through the exercise area, to the right of marker cone 2.
7. Proceed in reverse through the exercise area, to the left of marker cone 3.
8. Position the front of the fire apparatus past marker cone 3.
9. Proceed forward through the exercise area, to the right of marker cone 3.
10. Proceed forward through the exercise area, to the left of marker cone 2.
11. Proceed forward through the exercise area, to the right of marker cone 1.
12. Once the rear of the fire apparatus passes marker cone 1, the exercise is complete.

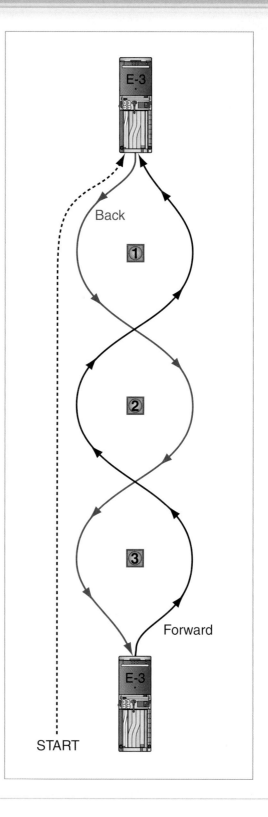

which is called a confined-space turnaround, measures your ability to turn the fire apparatus around in a confined space without going outside a set boundary. This exercise is completed in a 50′ × 100′ (15.25 m × 30.5 m) area, where marker cones may be used to identify the set boundary. The fire apparatus enters the area through an opening, no more than 12′ (3.6 m) wide, in the center of one of the 50′ (15.25 m) sides. The fire apparatus proceeds forward, turns around 180 degrees, and returns through the same opening. You may maneuver the fire apparatus forward and reverse as many times as needed to accomplish the task, but the fire apparatus must remain inside the set boundary lines. During the entire exercise, the fire apparatus must remain in the marked boundary and move in a continuous motion, except when required to change direction of travel. A spotter is necessary for this exercise.

To perform the confined-space turnaround exercise, follow the steps in **Skill Drill 7-5 ▶**:

1. Position the fire apparatus outside the boundary opening.
2. Proceed forward through the opening.
3. Maneuver the fire apparatus so that it is turned 180 degrees. This will require driving forward, swinging the tail of the apparatus almost 90 degrees, and then maneuvering forward. "Forward, swing, forward."
4. Proceed forward through the opening and out of the set boundary.

Performing a Diminishing Clearance Exercise

Once at the scene, you may have to operate the fire apparatus in tight quarters. Sometimes other fire apparatus, trees, storefront signs, or buildings can obstruct your path. However, the most common obstruction on an emergency scene is parked cars. Any one of these obstructions may restrict the horizontal or vertical clearance of the fire apparatus and make your task as driver/operator difficult at best. During these situations, you must judge the distance you have available to clear through openings and not cause any damage.

NFPA 1002 requires that all driver/operators complete an exercise that simulates a restricted horizontal and vertical clearance for a fire apparatus. This "diminishing clearance" exercise measures your ability to maneuver the fire apparatus in a straight line and judge the distance of the fire apparatus to an object in front of it. The fire apparatus should proceed at a speed that requires your quick judgment. The fire apparatus moves forward down two rows of marker cones to form a lane 75′ (23 m) long. The lane's width diminishes from a starting point of 9′ 6″ (2.9 m) to a width of 8′ 2″ (2.5 m). Marker cones identify the lanes boundary. To perform the exercise successfully, you must proceed forward down the lane without striking the marker cones. Bring the apparatus to a complete stop 50′ (15.25 m) past the last marker cone at a predetermined finish line. No portion of the fire apparatus can protrude past this set line. The fire apparatus then proceeds in reverse past the point where it entered the diminishing clearance exercise.

During the course of the fire apparatus moving forward and reversing, a vertical crossbar prop is positioned to determine your ability to judge the height of the fire apparatus. The crossbar may be positioned at several heights, including one that is lower than the fire apparatus. You must be capable of judging the vertical and horizontal clearances of the fire apparatus.

During the entire exercise, the fire apparatus must remain within the marker cones and move in a continuous motion, except when required to change direction of travel. A spotter is necessary for this exercise.

To perform the diminishing clearance exercise, follow the steps in **Skill Drill 7-6 ▶**:

1. Position the fire apparatus at the opening of the diminishing clearance.
2. Proceed forward through the lane.
3. Stop the fire apparatus at the finish line. (**Step 1**)
4. Proceed in reverse through the lane.
5. Identify any vertical heights on the crossbar that may strike the fire apparatus. (**Step 2**)

Driver/Operator Tip

When operating the fire apparatus, you must always drive with "due regard for the safety of others." This statement or something similar is found in most state laws that pertain to emergency vehicle operations. An emergency vehicle operator may have the right to disregard certain traffic laws but does not have the right to jeopardize the safety of other motorists or pedestrians. For example, some fire departments and local laws allow emergency vehicles to exceed the speed limit on roads and highways. Usually, certain conditions must be met to do this—namely, light traffic and good weather conditions. If the emergency vehicle operator exceeds the speed limit under adverse conditions, he or she is not "driving with due regard for the safety of others." During the entire response, you must be constantly aware of other motorists and pedestrians as well as your own safety.

Driver/Operator Tip

The National Fallen Firefighters Foundation (NFFF) summarizes the issues pertaining to vehicle safety:
- It's not a race.
- Safe is more important than fast.
- Stop at red lights and stop signs! There are no excuses.
- If they do not get out of your way, do not run them over! Think and react carefully.

Safety Tip

The use of sirens during gridlock merely confuses other drivers when they have nowhere to move their cars. Use the public address system and give the drivers directions.

Skill Drill 7-5

Performing a Confined-Space Turnaround

1. Position the fire apparatus outside the boundary opening.

2. Proceed forward through the opening.

3. Maneuver the fire apparatus so that it is turned 180 degrees. This will require driving forward, swinging the tail of the apparatus almost 90 degrees, and then maneuvering forward. "Forward, swing, forward."

4. Proceed forward through the opening and out of the set boundary.

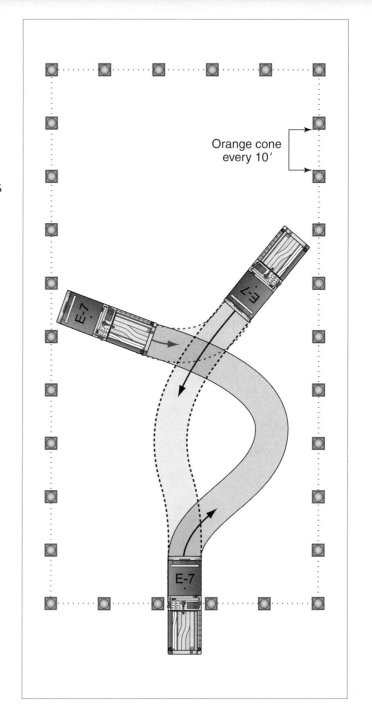

Orange cone every 10'

Skill Drill 7-6

Performing a Diminishing Clearance Exercise

4.3.5

1. Position the fire apparatus at the opening of the diminishing clearance. Proceed forward through the lane. Stop the fire apparatus at the finish line.

2. Proceed in reverse through the lane. Identify any vertical heights on the crossbar that may strike the fire apparatus.

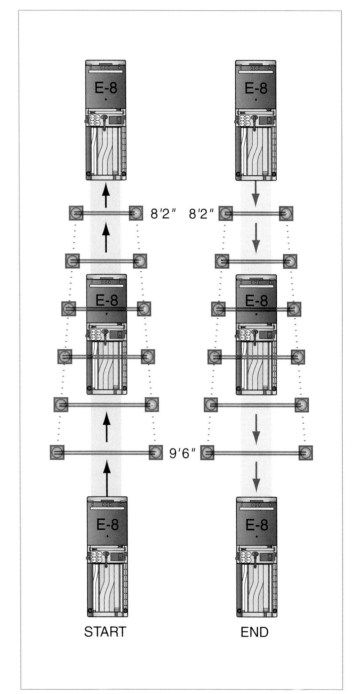

Safety Tip

You should understand that a prompt, safe response is obtained by adhering to the following guidelines:

- Leave the station in a standard manner:
  - Perform a 360-degree walk around the fire apparatus. Check for open compartments.
  - Quickly mount the fire apparatus.
  - Make sure that all personnel are on board, seated with seat belts on.
  - Make sure the station doors are fully open.
  - Close the fire station doors with a remote control upon leaving.
  - Drive defensively and professionally at reasonable speeds.
  - Know where you are going.
  - Use warning devices to move around traffic and to request the right-of-way in a safe and predictable manner.

Do not

- Leave quarters before the crew has mounted safely and before the fire apparatus doors are fully open.
- Drive too fast for the current conditions.
- Drive recklessly or without regard for safety.
- Take unnecessary chances with negative right-of-way intersections.
- Intimidate or scare other drivers.

Returning to the Station

After completing the overhaul and when returning to the fire station, you cannot become complacent. Operating an emergency vehicle on the roadway can be dangerous even when driving back from a call. You may be fatigued and unable to react appropriately. Civilian drivers may think that fire apparatus are responding to an emergency and unexpectedly stop in front of them. Some civilian drivers, while trying to be courteous, may wave or invite fire apparatus into traffic ahead of them. While their intentions are good, you must always obey the rules of the road and be cautious of other drivers. Do not allow anyone to force you into traffic.

When the fire apparatus is a few blocks from the fire station, allow the engine to cool down as much as possible. Do not force the engine to work harder than it has to while driving back to the firehouse. This helps extend the life of the engine and transmission. If the fire apparatus is driven hard everywhere it goes and does not have adequate time to cool down before being shut off, damage to the power train may occur.

Reversing the Fire Apparatus

For most fire departments, the largest number of fire apparatus accidents are related to backing up fire apparatus. Very few fire departments respond with only one member on the fire apparatus. If two or more members are on the fire apparatus, this type of accident is entirely preventable. Every fire department should have a SOP that covers fire apparatus backing procedures. Each member of the fire department needs to be trained and held accountable for these procedures.

Most modern fire stations are now built with drive-through bays so that fire apparatus do not have to back into the station. While this may reduce the number of backing-up accidents at the station, it may also contribute to the lack of backing-up skills.

Whenever an emergency vehicle is backing up, its emergency lights should be turned on and the driver/operator should use a spotter to assist with the procedure. A **spotter** is a person who guides the driver/operator into the appropriate position while the apparatus is operating in a confined space or in reverse mode. The spotter should always be in the driver/operator's full view Figure 7-13 . No vehicle should be backed into an intersection, around a corner, into the fire station, or in traffic unless emergency lighting is used and a spotter precedes it to safely direct such movement. During poor visibility conditions, rear apparatus spotlights may be used as necessary. Never hesitate to use more than one spotter to assist with backing-up maneuvers if the situation warrants.

Driver/Operator Tip

According to a National Incident for Occupational Safety and Health (NIOSH) report:

On August 14, 2004, a 25-year-old female career fire fighter (the victim) died when she apparently fell from the tailboard and was backed over by an engine. The victim and her crew had been released from the scene of a residential fire. The road was blocked by other apparatus, so the victim's crew began backing to an intersection approximately 300 feet away in order to proceed forward. The victim took her position on the tailboard as the "Tailboard Safety Member" and signaled the driver to begin backing. A Captain acting as the "Traffic Control Officer" guided the backing operation from the road on the driver's side, behind the apparatus, by using hand signals. When the Captain turned and walked into the intersection to stop cross-traffic, the victim apparently fell from the tailboard and was run over by the engine. Members on the scene provided advanced life support and the victim was transported to a local hospital, where she was pronounced dead.

You should always be aware of the hazards involved with moving fire apparatus. Even when the situation is not deemed urgent, remain alert to your surroundings. An injury can occur at any time—not just during emergency situations.

Safety Tip

At night, spotters should use flashlights. You should stop immediately if the spotter disappears from the rear-view mirror.

Figure 7-13 The spotter should always be in your full view.

- When the spotter is directing the fire apparatus backward or forward in a straight line, the spotter should have both arms extended forward and slightly wider than the body, parallel to the ground. The palms should face the direction of desired travel. The spotter will bend both arms repeatedly toward the head and chest to direct the fire apparatus **Figure 7-14**.
- When the spotter is directing the fire apparatus either to the right or to the left while it is moving, the spotter should have the directional arm extended from the side of the body, parallel to the ground, indicating the direction in which to travel. The motioning arm should be extended in the opposite direction (palm upward) and repeatedly bent toward the head, indicating the desired direction of travel **Figure 7-15**.
- To provide you with a visual reference for the distance to the stopping point, the spotter should have both arms extended sideways with elbows bent upward at 90 degrees. The spotter's palms should face forward, hands above the head, and the spotter should bring the elbows forward as the distance narrows. As the elbows reach the straight-forward position, the hands continue coming together above the head to indicate that the stopping point is being reached. Upon reaching the stopping point, the spotter should give a loud "Stop!" signal **Figure 7-16**.
- The spotter may signal to you to stop the fire apparatus when it has reached the desired objective or if the safety of the operation is compromised. To do so, the spotter will cross the arms at the wrist (forearms) above the head, and then maintain this position until the fire apparatus comes to a complete stop. While motioning with the arms, the spotter should also shout as loudly as

It is your responsibility to ensure that all spotters know in advance what is expected, where you intend to back up the fire apparatus, and which signals will be used. Before backing up the fire apparatus, roll down the windows and turn off any audio equipment to ensure any orders to stop the fire apparatus given by the spotter are clearly heard and understood. The spotter may use a portable radio to communicate with you.

The following are some suggested procedures for spotters:
- The rear spotter must stay behind and to the left of the fire apparatus. This spotter should position his or her body as a landmark for you to line up on. The spotter should always be able to see you in the side mirror. If the spotter cannot see you, then you cannot see the spotter. Stop immediately!
- Any spotter can give a loud "Stop!" order directly to you when it is deemed necessary.
- During high noise conditions, one spotter must be equipped with a radio to inform you to stop. Select an appropriate radio channel for this communication.
- All verbal signals must be loud enough and clear enough to be heard by you.
- All hand signals must be in "large" movements, and as simple as possible. Signals given in front of the body can be difficult to see.
- While the fire apparatus is backing up at night or in limited-visibility conditions, the spotters should use flashlights to illuminate the surrounding area and hand signals.

Figure 7-14 Arm placement when the spotter is directing the fire apparatus in a straight line.

Figure 7-15 Arm placement when the spotter is directing the fire apparatus right or left.

Figure 7-17 The signal to halt.

Figure 7-16 This spotter is indicating how much room the driver/operator has to maneuver before he needs to stop.

possible, "Stop!" These actions should catch your attention and cause you to bring the fire apparatus to a halt Figure 7-17 ▸.

The spotter must understand that in most circumstances, you are able to operate the fire apparatus in reverse without direction. In these cases, it is the spotter's responsibility to ensure that you do not hit anything or anyone. Regardless of the number of spotters utilized, you should receive directions only from one spotter in the rear of the vehicle and, if utilized, from one spotter to the front of the fire apparatus. If the fire apparatus is equipped with a rear-mounted camera, it should also be used as well Figure 7-18 ▸. This technology should not take the place of an actual spotter, but is simply meant to augment your ability to see what is behind you.

NFPA 1002 requires that all driver/operators complete an exercise that simulates backing up a fire apparatus into a fire station: the station parking procedure drill. A fire apparatus bay is simulated by allowing for a 20′ (6.1 m) minimum setback from a street that is 30′ (9.14 m) wide; barricades are placed at the end of the setback, spaced 12′ (3.66 m) apart to simulate the garage door. The setback distance should accurately reflect those distances found during normal duties. A marker placed on the ground indicates the proper position of the left-front tire of the fire apparatus once stopped and parked. A reference line may be used to facilitate using the fire apparatus mirrors. The minimum depth of the fire apparatus bay is determined by the length of the fire apparatus. During the entire exercise, the fire apparatus must remain within the marked boundary and the fire apparatus should move in a continuous motion, except when required to change direction of travel. A spotter is necessary for this exercise.

To perform the correct procedure for backing a fire apparatus into a fire station bay, follow the steps in Skill Drill 7-7 ▸ :

1. Position the rear of the fire apparatus past the fire station bay's opening and 90 degrees to the station. Ensure a spotter is correctly positioned behind the fire apparatus. (Step 1)
2. Activate the emergency lights.
3. Turn off any fire apparatus mounted stereo equipment. (Step 2)
4. Disengage the parking brake, if set. (Step 3)
5. Shift the transmission to reverse. (Step 4)
6. Proceed in a reverse mode and turn the fire apparatus to align it with the objective. (Step 5)
7. Continue backing the fire apparatus until it has reached the desired objective or the spotter signals, "Stop." (Step 6)

Although it is not recommended, during an emergency situation a fire apparatus may have to be backed up without a spotter. In this situation, you should do a preliminary inspection; check for obstructions, vertical and horizontal clearances, and power lines; and ensure that all compartments, doors, latches, and gates are closed. Proceed with extreme caution and back the unit just far enough to where it can be turned around and then driven forward.

Driver/Operator Tip

Some fire apparatus may be equipped with a 15′ to 20′ (4.5 to 6.1 m) coiled wire and push button for the spotter to use while the fire apparatus is being operated in reverse. This device plugs into the rear of the fire apparatus and allows the spotter to directly communicate with the driver/operator.

Shutting Down the Fire Apparatus

Once the fire apparatus is inside the fire station, it should be properly shut down. All of the electrical loads should be turned off first. This includes the headlights, air conditioner/heater, emergency lights, and any other electrical load that has an on/off switch. If these switches are left in the "on" position, these components will put an unnecessary load on the system the next time the fire apparatus is started.

Next, shift the transmission into neutral and engage the parking brake. If needed, allow the fire apparatus to cool down before shutting it off. This is best done with the fire apparatus sitting at an idle rate for 3 to 5 minutes. When doing so, make sure that the exhaust is properly vented and is not allowed to fill the fire station bay with harmful fumes. The engine's lubricating oil and coolant fluid are able to carry heat away from the combustion chamber, bearings, shafts, and other engine components when the engine is run at idle speed. This step is particularly important with turbocharged engines, because the delicate bearings and seals inside the turbocharger are subject to the high heat of combustion exhaust gases. This heat is carried away by normal oil circulation while the engine is operating. If the engine is abruptly stopped, however, the turbocharger temperature may increase considerably and perhaps result in seized bearings or loose oil seals. Failure to adequately cool the engine for the proper length of time before shutdown can lead to reduced engine life and engine component failure.

Finally, turn the ignition and battery switches to the *off* position. Reconnect any electrical shore lines, exhaust, and extractor systems.

As the driver/operator, you are responsible for the readiness of the fire apparatus at all times. Before anything else is done at the station, the fire apparatus must be returned to a ready state. Any equipment that was used during the emergency must be replaced, cleaned, or repaired. If the call was EMS related and only a few bandages were used, then replace them. If the fire apparatus returned from a small rubbish fire and the on-board water tank is now low, then fill the water tank Figure 7-19 ▼. If the SCBA and

A.

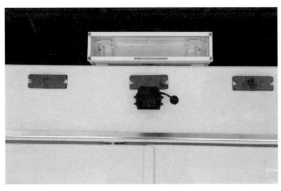

B.

Figure 7-18 A camera is an important piece of safety equipment. **(A)** Side-mounted camera. **(B)** Rear-mounted camera.

A.

B.

Figure 7-19 Fill the on-board tank if it is low. This can be done two ways. A. With a garden hose. B. With a fire department hose line.

Skill Drill 7-7

Backing a Fire Apparatus into a Fire Station Bay

1 Position the rear of the fire apparatus past the fire station bay's opening and 90 degrees to the station. Ensure a spotter is correctly positioned behind the fire apparatus. Activate the emergency lights.

2 Turn off any fire apparatus mounted stereo equipment.

3 Disengage the parking brake, if set.

4 Shift the transmission to reverse.

5 Proceed in a reverse mode and turn the fire apparatus to align it with the objective.

6 Continue backing the fire apparatus until it has reached the desired objective or the spotter signals, "Stop."

Driver/Operator Tip

Some fire apparatus may be equipped with an emergency shutdown feature. If such an apparatus is operated in an atmosphere that is rich in fuel vapors, this switch will provide for a positive shutdown of the engine. A clapper device, activated electrically or pneumatically from inside the cab or at the pump panel, will block air in the intake pipe to the intake manifold. Activation of this device starves the engine of air and shuts it down. Once the valve is set, however, it must be reset manually inside the engine compartment. You must know how to reset this valve if the vehicle is equipped with one. Its use should be limited to emergency use only; excessive use may damage the vehicle's engine.

other equipment need to be cleaned after a structure fire, then clean these items and return the equipment to service.

To perform the procedure for shutting down and securing a fire apparatus, follow the steps in **Skill Drill 7-8**:

1. Turn off the emergency lights, headlights, and any other electrical loads. (**Step 1**)
2. Shift the transmission into neutral. (**Step 2**)
3. Engage the parking brake. (**Step 3**)
4. Ensure that the apparatus has adequately cooled down before shutting it off. (**Step 4**)
5. Turn the ignition switch to the off position. (**Step 5**)
6. Turn the battery to the off position. (**Step 6**)
7. Reconnect any applicable electrical cords and extractor system. (**Step 7**)
8. Replace, clean, and repair any used equipment. (**Step 8**)

Skill Drill 7-8

Shutting Down and Securing a Fire Apparatus

4.3.1

1. Turn off the emergency lights. Turn off the headlights and any other electrical loads.

2. Shift the transmission into neutral.

3. Engage the parking brake.

4. Ensure that the apparatus has adequately cooled down before shutting it off.

5. Turn the ignition switch to the off position.

6. Turn the battery to the off position.

7. Reconnect any applicable electrical cords and extractor systems

8. Replace, clean, and repair any used equipment.

Wrap-Up

Chief Concepts

- To be successful in your role as a driver/operator, you must fully understand and accept the responsibilities that come with the job.
- As a driver/operator, you have many obligations to your crew and to the community. Internal and external expectations hinge on your ability to perform your initial response assignments.
- You have a duty to educate other crew members on their roles and responsibilities to support your function.
- You are a teacher, a mentor, a vital crew member, and a safety advocate. You must lead by example—remember that you are a safety advocate.
- Compliance with standard operating procedures or guidelines is essential. Always follow these procedures, and promote compliance among your crew members.
- All fire apparatus are operated in two basic ways: emergency vehicle response and on-scene operations.
- The communications center gives the dispatch to the responding fire apparatus.
- You must disseminate the dispatch information and identify which information will aid your crew in locating the emergency and responding to it.
- A variety of maps may be used to locate emergency incidents.
- A preliminary inspection must be completed before the fire apparatus is ready to depart for the response.
- The driver/operator should follow the recommended starting sequence for the fire apparatus.
- Before the fire apparatus can get under way, you must ensure all members are seated and belted, unless they meet an exception under NFPA 1500.
- While responding to the scene, the driver/operator may need to complete several maneuvers so that the fire apparatus can access the incident. NFPA 1002 requires the driver/operator to complete several exercises to prove his or her skill in managing these situations.
- Do not become complacent when returning to the station.
- Follow safe procedures when backing up the fire apparatus. Always use spotters to assist with backup operations.
- When operating a fire apparatus in reverse, stop immediately if you lose sight of the spotter.
- Follow the standard sequence for shutting down the fire apparatus.

Hot Terms

<u>air pressure gauges</u> Gauges that identify the air pressure stored in the tanks or reservoirs of an apparatus equipped with a pneumatic braking system.

<u>battery selector switch</u> A switch used to disconnect all electrical power to the vehicle, thereby preventing discharge of the battery while the vehicle is not in use.

<u>blind spots</u> Areas around the fire apparatus that are not visible to the driver/operator.

<u>communications center</u> A building or portion of a building that is specifically configured for the primary purpose of providing emergency communications services or public safety answering point (PSAP) services to one or more public safety agencies under the authority or authorities having jurisdiction. (NFPA 1221)

<u>dispatch</u> To send out emergency response resources promptly to an address or incident location for a specific purpose. (NFPA 450)

<u>due regard</u> The care exercised by a reasonably prudent person under the same circumstances.

<u>fuel gauge</u> A gauge that indicates the amount of fuel in the fire apparatus's fuel tank.

<u>global positioning system (GPS)</u> A satellite-based radio navigation system consisting of three segments: space, control, and user. (NFPA 414)

<u>ignition switch</u> A switch that engages operational power to the chassis of a motor vehicle.

<u>job aid</u> A tool, device, or system used to assist a person with executing specific tasks.

<u>mobile data terminal (MDT)</u> A computer that is located on the fire apparatus.

<u>oil pressure gauge</u> A gauge that identifies the pressure of the lubricating oil in the fire apparatus's engine.

<u>parking brake</u> The main brake that prevents the fire apparatus from moving even when it is turned off and there is no one operating it.

<u>prove-out sequence</u> A series of checks that an electrical system completes to ensure that all of the systems are functioning properly before the fire apparatus is started.

<u>spotter</u> A person who guides the driver/operator into the appropriate position while operating in a confined space or in reverse mode.

<u>starter switch</u> The switch that engages the starter motor for cranking.

<u>stressors</u> Conditions that create excessive physical and mental pressures on a person's body; a stimulus that causes stress.

<u>tactical benchmarks</u> Objectives that are required to be completed during the operational phase of an incident.

<u>vehicle dynamics</u> Vehicle construction and mechanical design characteristics that directly affect the handling, stability, maneuverability, functionality, and safeness of a vehicle.

<u>vehicle intercom system</u> A communication system that is permanently mounted inside the cab of the fire apparatus to allow fire fighters to communicate more effectively.

<u>voltmeter</u> A device that measures the voltage across a battery's terminals and gives an indication of the electrical condition of the battery.

Driver/Operator in Action

Your captain has just announced that the department will be replacing all ventilation and rescue saws over the next eight months. In addition to this action, the department will be posting openings for several driver/operator positions for the proposed Westside station. All interested and qualified members should complete the application process by the end of next month. The applications will be reviewed and the selections will be based on training levels, service time, and a solid understanding of the roles and responsibilities of the driver/operator.

1. One method used to instill confidence in yourself and the other crew members regarding the apparatus and equipment reliability is to
   A. Replace apparatus and equipment every five years.
   B. Complete regularly scheduled apparatus and equipment inspections.
   C. Service apparatus and equipment when breakdowns occur.
   D. Only check apparatus and equipment after each use.

2. One of the most important roles associated with the position of driver/operator is
   A. Chauffeur of the apparatus.
   B. Water supply officer.
   C. Apparatus and equipment mechanic.
   D. Safety advocate.

3. In an effort to maintain a safe work environment, driver/operators should be involved in
   A. Identifying possible hazards associated with the use and installation of new equipment.
   B. Completing major repairs to apparatus and equipment.
   C. The budgeting process for apparatus and equipment.
   D. Scheduling overhaul schedules throughout the life cycle of the apparatus.

4. The driver/operator is charged with performing which of the following tasks?
   A. Performing equipment and apparatus repairs
   B. Performing apparatus and equipment inspections and basic preventive maintenance operations
   C. Identifying possible crew members to complete tactical objectives
   D. Maintaining all apparatus and equipment records

5. Which resources can be used by the driver/operator to ensure consistency and accuracy during the inspection and general maintenance procedures for apparatus and equipment?
   A. Job aids such as checklists and reference sheets
   B. Manufacturer-provided owner's manuals
   C. Certified emergency vehicle technicians
   D. All of the above

Approaching the Fire Ground

CHAPTER 8

NFPA 1002 Standard

4.3.1* Operate a fire department vehicle, given a vehicle and a predetermined route on a public way that incorporates the maneuvers and features, specified in the following list, that the driver/operator is expected to encounter during normal operations, so that the vehicle is operated in compliance with all applicable state and local laws, departmental rules and regulations, and the requirements of NFPA 1500, Section 4.2:

(1) Four left turns and four right turns
(2) A straight section of urban business street or a two-lane rural road at least 1.6 km (1 mile) in length
(3) One through-intersection and two intersections where a stop has to be made
(4) One railroad crossing
(5) One curve, either left or right
(6) A section of limited-access highway that includes a conventional ramp entrance and exit and a section of road long enough to allow two lane changes
(7) A downgrade steep enough and long enough to require down-shifting and braking
(8) An upgrade steep enough and long enough to require gear changing to maintain speed
(9) One underpass or a low clearance or bridge [p. 193–214]

(A) Requisite Knowledge. The effects on vehicle control of liquid surge, braking reaction time, and load factors; effects of high center of gravity on roll-over potential, general steering reactions, speed, and centrifugal force; applicable laws and regulations; principles of skid avoidance, night driving, shifting, and gear patterns; negotiating intersections, railroad crossings, and bridges; weight and height limitations for both roads and bridges; identification and operation of automotive gauges; and operational limits. [p. 193–214]

(B) Requisite Skills. The ability to operate passenger restraint devices; maintain safe following distances; maintain control of the vehicle while accelerating, decelerating, and turning, given road, weather, and traffic conditions; operate under adverse environmental or driving surface conditions; and use automotive gauges and controls. [p. 193–214]

4.3.6* Operate a vehicle using defensive driving techniques under emergency conditions, given a fire department vehicle and emergency conditions, so that control of the vehicle is maintained. [p. 194–199]

(A) Requisite Knowledge. The effects on vehicle control of liquid surge, braking reaction time, and load factors; the effects of high center of gravity on roll-over potential, general steering reactions, speed, and centrifugal force; applicable laws and regulations; principles of skid avoidance, night driving, shifting, and gear patterns; negotiation of intersections, railroad crossings, and bridges; weight and height limitations for both roads and bridges; identification and operation of automotive gauges; and operational limits. [p. 194–199]

(B) Requisite Skills. The ability to operate passenger restraint devices; maintain safe following distances; maintain control of the vehicle while accelerating, decelerating, and turning, given road, weather, and traffic conditions; operate under adverse environmental or driving surface conditions; and use automotive gauges and controls. [p. 194–199]

Additional NFPA Standards

NFPA 1500 *Standard on Fire Department Occupational Safety and Health Program* (2007)

NFPA 1561 *Standard on Emergency Services Incident Management System* (2008)

NFPA 1620 *Recommended Practice for Pre-incident Planning* (2003)

NFPA 1901 *Standard for Automotive Fire Apparatus* (2009)

NFPA 1936 *Standard on Powered Rescue Tools* (2005)

Knowledge Objectives

After studying this chapter, you will be able to:

- Describe the effects of vehicle control during a liquid surge on the roadway.
- Describe the effect of a wet roadway on the braking reaction time.
- Describe the effect of the fire apparatus's load on the control of the vehicle on a wet roadway.
- Describe the risk of a fire apparatus roll-over due to the effects of the apparatus's high center of gravity.
- Describe general steering reactions of a fire apparatus.
- Describe the effect of speed in controlling a fire apparatus on the roadway.
- Describe the effect of centrifugal force in controlling a fire apparatus on the roadway.
- Describe the applicable laws and regulations in operating a fire apparatus.
- Describe the principles of skid avoidance when operating a fire apparatus.
- Describe the principles of safe night driving of a fire apparatus.
- Discuss proper shifting and gear patterns of a fire apparatus.
- Describe how to safely cross intersections, railroad crossings, and bridges.
- Describe the weight and height limitations of a fire apparatus for both roads and bridges.
- Identify and describe the operation of automotive gauges on the fire apparatus.

- Describe the operational limits of the fire apparatus.
- Describe how to control the fire apparatus using defensive driving techniques under emergency conditions.
- Describe your responsibilities as you approach an emergency scene.
- Describe Level I and II staging procedures.
- Identify potential hazards while approaching emergency incidents.
- Describe the driver/operator's role when positioning the fire apparatus to operate at various emergency incidents.
- Describe how to perform four left turns and four right turns with the fire apparatus.
- Describe how to drive a fire apparatus on a straight section of an urban business street or a rural two-lane road for 1 mile (1.6 km).
- Describe how to drive a fire apparatus safely through a one through-intersection.
- Describe how to drive a fire apparatus through two intersections where a stop has to be made.
- Describe how to drive a fire apparatus across a railroad crossing.
- Describe how to drive a fire apparatus on a curve, either left or right.
- Describe how to drive a fire apparatus on a section of limited-access highway with an entrance and exit ramp.
- Describe how to drive a fire apparatus on a highway and perform two lane changes.
- Describe how to operate a fire apparatus on a downgrade requiring a down-shift and braking.
- Describe how to operate a fire apparatus on an upgrade requiring a gear change to maintain speed.
- Describe how to operate a fire apparatus and travel safely under either an underpass or a low-clearance bridge.
- Describe how to operate a fire apparatus and maintain safe following distances.
- Describe how to accelerate in a fire apparatus.
- Describe how to decelerate in a fire apparatus.
- Describe how to turn in a fire apparatus.
- Describe how to operate the fire apparatus safely under adverse driving conditions.

Skills Objectives

After studying this chapter, you will be able to:

- Perform the alley dock exercise with the fire apparatus.

You Are the Driver/Operator

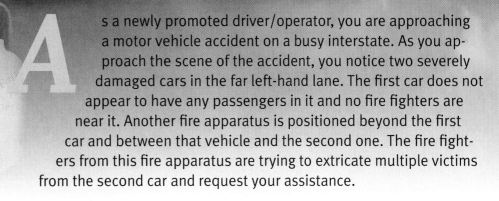

As a newly promoted driver/operator, you are approaching a motor vehicle accident on a busy interstate. As you approach the scene of the accident, you notice two severely damaged cars in the far left-hand lane. The first car does not appear to have any passengers in it and no fire fighters are near it. Another fire apparatus is positioned beyond the first car and between that vehicle and the second one. The fire fighters from this fire apparatus are trying to extricate multiple victims from the second car and request your assistance.

1. Where should you position the fire apparatus?
2. What is your reasoning for positioning the fire apparatus there?
3. Does this position make the scene safer for everyone?

Introduction

The driver/operator must safely get the members of the crew, the fire apparatus, and the equipment that it carries to the scene of the incident. This response should not be taken lightly. Fire apparatus are not designed like many other vehicles on the road. They are usually bigger and heavier, which will change the characteristics of the vehicle—and, therefore, change how they should be driven. Do not assume that the fire apparatus can be driven like your own personal vehicle. Fire apparatus can be very dangerous machines when they are operated by individuals who have not been properly trained in safe and efficient operational techniques. This chapter discusses the laws and regulations that pertain to driving the fire apparatus to an emergency scene. It also discusses some of the safe driving practices for the driver/operator to follow in both good and adverse weather conditions.

Emergency Vehicle Laws

No member of any fire department should be allowed to drive an emergency vehicle or fire apparatus until that person has completed a training course approved by the fire department. Simply allowing any member to drive these vehicles without the proper training is irresponsible. Several Emergency Vehicle Operations courses are available, and each fire department should ensure that its members are trained to operate the fire apparatus that they will be driving. The days of allowing any fire fighter with a driver's license to jump in the front seat and drive the fire apparatus to the emergency scene are behind us. The fire service can and should provide fire fighters with qualified training to ensure that all driver/operators drive safely and act responsibly while operating emergency vehicles.

Each fire department is governed by different sets of state laws and local regulations that apply to emergency vehicles. Some of these laws and regulations are very detailed and descriptive; others are vague and leave much to be interpreted by the members of the fire department. All members of the fire department should be familiar with their department's procedures. A lack of knowledge regarding the applicable emergency vehicle driving laws is not an excuse for disobeying them. You must understand that you will be held responsible for your actions while driving an emergency vehicle. Fire fighters who are found to be at fault for an accident involving a fire apparatus may be prosecuted in both criminal and civil court—which can create many problems for the members both at work and at home. As a driver/operator, you must understand the local laws and regulations with which you must comply.

The use of sirens and warning lights does not automatically give the right-of-way to the fire apparatus. These devices simply request the right-of-way from other drivers, based on their awareness of the emergency vehicle's presence. You must take every step possible to make your presence and intended actions known to other drivers, and you must drive defensively so that you will be prepared for unexpected, inappropriate actions of others. The driver/operator of a fire apparatus is not the only person allowed to use the roadway even under emergency conditions: You must follow many of the same laws as any other driver on the road.

Many states provide certain privileges to emergency vehicle driver/operators while they are responding to an emergency. They also specify the conditions under which these privileges are granted. For example, in New Mexico, the motor vehicle code states that the following four privileges may be granted as long as the emergency vehicle sounds an audible signal and the vehicle is operating with its emergency lights activated; all other laws apply during the emergency vehicle's response:

1. **Park or stand irrespective of the provisions of the motor vehicle code.** This privilege allows for the driver/operator to position the fire apparatus in the roadway to block the scene and provide for the safety of the fire fighters and those involved in the emergency.
2. **Proceed past a red or stop signal or stop sign, but only after slowing down as necessary for safe operation.** The fire apparatus should come to a complete stop before proceeding through the intersection.
3. **Exceed the maximum speed limits so long as the driver/operator does not endanger life or property.** Many fire departments further refine this law by allowing the fire apparatus to exceed the posted speed limit only by 10 miles per hour (mi/h) during an emergency response and only in light traffic and good weather conditions.
4. **Disregard regulations governing direction of movement or turning in specified directions.** This provision allows for the driver/operator to drive or position the vehicle against the flow of traffic. This maneuver should be done only in a safe and controlled manner. Some fire departments refer to this practice as "bucking" traffic and strictly prohibit it. Other fire departments allow moves against the normal flow of traffic to be made within one block of an incident and only for the purpose of positioning the fire apparatus.

These laws do not relieve you from the duty to drive with due regard for the safety of all persons, nor do they protect you from the consequences of reckless disregard for the safety of others. Each driver/operator is accountable for his or her own actions.

Other state laws that may apply to emergency vehicle operators include the following provisions:

- Comply with any lawful order or direction of any police officer invested by law with the authority to direct, control, or regulate traffic.
- Do not pass school buses that are loading or unloading passengers.
- Do not leave the scene of an accident that you are involved in.

Driver/Operator Tip

State laws and many other rules and regulations may be defined in greater detail by local fire departments. Each driver/operator is responsible for knowing the state and local laws governing the operation of fire apparatus.

Driver/Operator Tip

"Due regard" is a legal term. Would a reasonably careful person performing similar duties under the same circumstances react in the same manner? As the driver/operator of a fire apparatus, you should ask yourself this question. If you are performing in an unsafe manner and putting others at risk, you are not driving with due regard for the safety of others. Be aware that you will be held responsible for your actions. Drive safely or do not drive at all.

Safe Driving Practices

When starting the emergency response, you should accelerate slowly and in a controlled manner. Allow the engine to gradually build up speed instead of flooring the accelerator from a stopped position. Both of your hands should remain on the steering wheel at all times unless you are operating air horns, sirens, or other essential equipment. For this reason, you should check and memorize the position of instruments and controls before moving the fire apparatus. It should never be necessary to search for instruments or controls while the fire apparatus is in motion, as this behavior will draw your attention away from the road. Taking your eyes off of the roadway for too long to search for controls may result in a collision.

While driving down hills, you should rely on the engine and the auxiliary braking device to slow down the fire apparatus. If only the fire apparatus brakes are used to control the speed of the fire apparatus, they may heat up to the point where they become ineffective in stopping the fire apparatus. This situation is commonly referred to as **brake fade**. Brake fade occurs when the brake drums become hot and expand away from the brake shoes, so that the stroke of the slack adjusters becomes less effective. To avoid this problem, you should use the brakes only in short, 5- to 10-second bursts rather than as a continuous application. Also, while traveling down a steep slope, you may need to use the auxiliary braking system and down-shift the transmission to a lower gear. To complete this action properly, shift into one gear lower than the gear that would be used when going up the hill. Down-shifting will make it easier to control the fire apparatus and will save wear on the brakes.

When traveling up hills, you should avoid over-throttling the fire apparatus. This action will result in a loss of power while the engine works to catch up to the amount of work that is being demanded. Instead, you should slowly build up speed and use the lower gears of the transmission until adequate horsepower is achieved. To do so, you may have to manually shift the transmission into these lower gears while climbing a hill.

Local jurisdictions determine which level of response is appropriate for the emergency situation. During an emergency response, the fire apparatus should be driven with both the emergency lights and the sirens activated—a practice commonly referred to as a **Code 3 response**. When the fire apparatus is operating with only the emergency lights and no audible siren, it is engaging in a **Code 2 response**. During normal operations or nonemergency responses, the fire apparatus does not have any emergency lights or sirens operating; this situation is considered a **Code 1 response**. A Code 2 response is usually prohibited in many fire departments because it does not properly alert other drivers to the presence of the fire apparatus. Although some fire fighters may argue that the sirens are unnecessary at 3 A.M. when the streets may be devoid of traffic, sirens should still be used to alert others of the fire apparatus's approach and avoid any potential liability should an accident occur.

The fire apparatus should be driven with the intention of passing other vehicles on the left side. On streets with multiple lanes, the fire apparatus should be driven in the far left lane to provide ample room for the other drivers to pull to the right and stop. Weaving in and out of the other vehicles should be avoided.

Safety Tip

Because of their design, diesel engines have very little compression of backpressure to assist in stopping the fire apparatus. Auxiliary braking systems, however, provide braking torque through the driveline to the rear wheels. Use of an auxiliary brake reduces brake wear, reduces brake heat build-up, and can help to minimize the occurrence of brake fade during heavy or frequent braking. Be aware that an auxiliary braking system should not be used during slippery road conditions or inclement weather; doing so may cause the rear wheels to lock up, resulting in a loss of control of the fire apparatus.

Several types of auxiliary braking devices are available for diesel-powered engines. The most popular include the following options:

- **Compression brake.** The compression brake is a mechanical system added to the engine valve train that is electronically actuated. This system alters the operation of the engine's exhaust valves so that the engine works as a power-absorbing air compressor.
- **Transmission retarder.** An efficient means of slowing a vehicle down, the transmission retarder utilizes the transmission fluid to create backpressure to assist in slowing the fire apparatus. Using this type of device will help avoid engine damage. If the transmission fluid is overheated, however, major transmission damage can result.
- **Exhaust brake.** A shutter valve activated in the exhaust system just behind the turbo charger will engage this type of device. The closed valve causes a build-up of pressure in the exhaust system that passes back through the turbo and the valves and into the combustion chamber of the cylinder. The pressure build-up creates braking horsepower, which is then used to slow the vehicle down. With a transmission that offers the same type of direct interface, this system is quite efficient because the same horsepower used to keep the vehicle in motion can help slow it down. The maximum efficiency is reached at the maximum engine RPMs for which the exhaust brake is rated.
- **Electromagnetic retarder.** This is the most efficient, but most expensive, means of slowing a fire apparatus. When engaged, the electromagnet around the driveshaft creates an opposing magnetic field around the driveshaft that causes the driveshaft to resist turning, thereby slowing the fire apparatus. The system may be applied in stages either manually or by combinations of brake and accelerator pedal settings. Any heat that is generated by this system is dissipated by cooling fins on the retarder.

Figure 8-1 When operating the fire apparatus on straight sections of roadway, you must remain in your own lane.

This action confuses other drivers and may cause a collision with other vehicles as they try to avoid the fire apparatus. Always pass other vehicles on the left side.

When operating the fire apparatus on straight sections of roadway, you must remain in your own lane **Figure 8-1**. Some driver/operators have a tendency to occupy parts of two lanes. This practice should be avoided, because it confuses other drivers by not making your intentions clear while traveling down the road. When making lane changes, use your turn signal and try to remain in the far left lane until you need to turn. While performing a two-lane change, make sure that you clear each lane individually before proceeding. Other vehicles may attempt to race or even pass the fire apparatus while you are responding to the scene.

When making left or right turns, ensure that you have enough clearance to completely make the turn without striking curbs, trees, and other objects that may be in the blind spot of the fire apparatus. Fire apparatus may be top heavy from aerial ladders, water, and other equipment. To avoid the potential for roll-over owing to the apparatus's load, you must slow down before making the turn. How fast you take a turn will depend on the road conditions and the condition of the fire apparatus. If you are turning right from the far left lane, you may not be able to see whether vehicles on your right have completely stopped. In this case, you should have the fire officer or other crew member in the front seat look out the window into the blind spot and determine if it is clear to proceed with the turn. When you are alone, you will have to slow down, make your intentions very clear to other motorists, and proceed with extreme caution while making such a turn.

When turning to the left, you should turn wide enough to make the turn but not so wide that you travel into an outside lane and strike another vehicle. During the response, you may have to make a U-turn to position the fire apparatus or maneuver down a one-way street. In such a case, you should make sure that the area is wide enough to complete the turn and that adequate clearance for both the front and rear of the fire apparatus is available. Remember, the back of the fire apparatus will swing out away from the turn. Therefore, if the vehicle is too close to the outside curb, the back end may swing out and strike objects such as poles, trees, and parking meters along the curb. With some fire apparatus, you may not realize you are turning too fast until it is too late. When in doubt, slow down.

Curved roadways are found all over this country—in the mountains, along the coastlines, and in urban areas of major cities. Curved sections can be found on small dirt roads and on highway entrance and exit ramps. No matter where they are, curved roadways can pose a very serious problem for emergency vehicles. Many of these ramps are designed with extreme curves and can be difficult to judge when traveling at normal speeds. As a driver/operator, you must realize that traveling on these curves with a fire apparatus is very different than traveling on the same route with a passenger car.

Drivers of fire apparatus should be familiar with the term **centrifugal force**, including how it relates to driving on curved roads. Centrifugal force is the tendency for objects to be pulled outward when rotating around a center. The fire apparatus is subject to this force when making a turn; the centrifugal force will try to keep the fire apparatus going in a straight line while you are making the turn. If it is to avoid losing control, the fire apparatus must have proper traction with the roadway. To keep the fire apparatus in the turn without allowing centrifugal force to pull the vehicle off the road, you should slow down before the curve and brake gently while making the turn.

Driving along curves will also severely affect the weight transfer of the fire apparatus. As with any vehicle, the weight carried on a fire apparatus is usually evenly distributed between all four tires of the apparatus while it travels down the road. This keeps the vehicle's center of gravity in the middle of the vehicle depending on its height and weight. When the apparatus travels on a curve, however, the center of gravity will shift, causing the weight of the fire apparatus to shift to one side or the other. If this shift of weight is too great, the fire apparatus will roll over. Remember to slow the fire apparatus down well before entering the turn; otherwise, you risk losing control of the fire apparatus.

Every curve in a road has a **critical speed**. If the fire apparatus is traveling faster than the critical speed, it will not be capable of completing the turn and will go off the road or roll over. Either way, you will lose control of the fire apparatus and wreck. To determine the critical speed, you should know how sharp the curve is and how slippery the road conditions are. As the curve gets sharper and the road more slippery, the critical speed goes down; as a consequence, the fire apparatus must travel around the curve at a slower speed. This does not have to be a complicated process. For example, if it is raining and the roads are slippery, you will not be able to drive around the curve at the same speed as if it were dry. Once the fire apparatus is beyond the critical speed, it is too late to try to correct the problem. Prevention is the key—so slow down **Figure 8-2**.

Responding to an emergency during the night is usually more difficult than during the day. The streets and landscape are difficult to see, and the reduced visibility makes the response area look dramatically different. Landmarks and street signs that are easily seen during the day may be missed as you respond in darkness. The lights of other vehicles, street lights, and traffic signals can make it difficult to maneuver the fire apparatus along the roadway. To reduce the risks of responding during the night, follow these guidelines:

Figure 8-2 When approaching a curve, slow down!

- Keep your eyes constantly scanning the roadway. Scan carefully for pedestrians, cyclists, and animals on the road.
- To avoid glare from oncoming lights, glance to the right edge of the road.
- Keep the fire apparatus's front windshield and headlights clean.
- Keep the interior cab lights off and adjust the instrument panel lights to a low setting.
- Many fire apparatus cabs are equipped with red lights inside the cab. These lights should be used during the night because they do not interfere with your vision during the response.
- Stay alert at night. You must be well rested and prepared to operate the fire apparatus.
- Slow down and increase your following distance.

Bridges and overpasses are obstacles that you must be aware of *before* the emergency response begins. Know beforehand which overhead obstructions your fire apparatus can safely pass beneath. Fire apparatus are required by NFPA 1901, *Standard for Automotive Fire Apparatus*, to have labels in the cab that identify the height of the vehicle. You should ensure that the fire apparatus is capable of fitting underneath any obstacle before attempting to pass under it. When in doubt, find another access point or use a spotter to prevent damaging both the obstacle and the fire apparatus.

Another potential problem concerning bridges is their weight limitation. Some bridges are not designed to carry the heavy loads associated with certain fire apparatus. You should identify these potential problems and, if they present obstacles to the fire apparatus, access the scene of the emergency from another route or use smaller fire apparatus to respond to the scene. Under no circumstances should you attempt to operate a fire apparatus over a bridge that is not capable of supporting its weight.

Defensive Driving Practices

While responding to the incident, you should ask the following questions:

- Can I stop this fire apparatus?
- What will I do if someone pulls in front of me?

- Is another vehicle in my blind spot?
- Am I too close to the vehicle in front of me?

All of these questions relate to the need to operate the fire apparatus in a safe manner and to drive defensively. If you drive cautiously and anticipate that the unexpected may happen, then you will be more prepared when it does. Defensive drivers are safer than aggressive drivers.

To drive defensively, you must always be aware of your surroundings. Scan the area in front of the fire apparatus to determine which hazards may lay ahead. Is another vehicle about to pull in front of the fire apparatus? If you are checking the road ahead, then you will see any potential problems and be more prepared to avoid a collision.

While traveling behind other vehicles, consider how close you can safely be to the vehicles ahead. While driving in good weather conditions (during daylight with good, dry roads, and low traffic volume), you should ensure a safe distance from the vehicle ahead of you by following the "three-second rule." This distance will change depending on the speed at which the fire apparatus is traveling and the conditions of the road. To determine the appropriate following distance, first select a fixed object on the road ahead such as a sign, tree, or overpass. When the vehicle that you are following passes the object, slowly count, "One one thousand, two one thousand, three one thousand." If you reach the object before completing the count, then you are following too closely. Making sure that there are at least three seconds between the fire apparatus and the car ahead of it gives you enough time and distance to respond to any problems that occur ahead of you. While responding in poor driving conditions (in inclement weather, in heavy traffic, or at night) you should double the three-second rule to six seconds, for added safety. As the driving conditions worsen, you will have to increase the distance that you travel behind other vehicles.

During the response, consider how fast you are going in relation to the street on which you are traveling. Although many fire departments allow fire apparatus to travel faster than the speed limit, it may not always be safe to do so. For example, in a residential neighborhood, the speed limit is usually 25 mi/h (40.2 km/h) or less. You must consider that small children and other pedestrians might be in such an area and adjust your speed accordingly.

Being able to control the fire apparatus and stop when required should be your goal. The distance that it takes for you to recognize the hazard, process the need to stop the fire apparatus, apply the brakes, and then come to a complete stop is referred to as the **total stopping distance**. The distance that the fire apparatus travels after you recognize the hazard, remove your foot from the accelerator, and apply the brakes is referred to as the **reaction distance**. The distance that the fire apparatus travels from the time that the brakes are activated until the fire apparatus makes a complete stop is known as the **braking distance**. These distances are very important for you to consider while responding in the fire apparatus, and they will differ for each fire apparatus depending on several factors:

- **The size and weight of the fire apparatus.** The more weight the fire apparatus is carrying, the more energy it will take to bring it to a complete stop. Larger vehicles generally will have a greater total stopping distance than smaller vehicles that are traveling at the same speed.
- **The fire apparatus' overall condition, including brakes, tires, and suspension.** Fire apparatus that are maintained in optimal condition are capable of coming to a complete stop more quickly than fire apparatus that are poorly maintained. Tires should always have adequate tread life and be inflated to the proper pressure. The brakes should be capable of stopping the fire apparatus as required. The suspension system of the fire apparatus should not allow excessive bounce; this will prevent the tires from making the proper contact with the road and delay the braking process.
- **The speed at which the fire apparatus is traveling.** Faster speeds will require the fire apparatus to travel a greater distance before it comes to a stop.
- **The surface condition of the road.** Dry, paved roads provide the best surface for both driving and stopping conditions. When roads are wet or covered with ice and snow, they have less friction with the tires of the apparatus. The lower the friction value, the longer it will take for the fire apparatus to come to a complete stop.

Driver/operators must also be aware of the term **liquid surge** and know how it relates to driving, especially for fire apparatus that carry water. Liquid surge is the movement of liquid inside a container as the container is moved. As the fire apparatus accelerates or decelerates, the water carried inside the tank on the fire apparatus will slosh around. This movement can be very hazardous owing to the large amount of water that is carried on many fire apparatus. If the fire apparatus has to brake quickly, then the liquid surge will try to force the fire apparatus forward—putting additional strain on the fire apparatus braking system and increasing the total stopping distance. To reduce the effects of liquid surge, the water tanks on fire apparatus are designed with baffles inside them. These plates inside the tank slow down the movement of the water and displace some of the energy transferred during the movement of the fire apparatus.

As a driver/operator, you should know how to prevent the fire apparatus from skidding. Skids can happen whenever the tires lose grip on the road. This problem can be caused by slippery surfaces, changing speed or direction too suddenly, or lack of maintenance on the tires. A road that is safe under good conditions may, however, be very dangerous when it is wet or has snow or ice on it. Traveling at high speeds even under normal conditions will increase the possibility of a skid if the fire apparatus must complete a turn or stop suddenly. If the fire apparatus is responding in rain or snow conditions, this problem is even more profound. Although adverse weather conditions such as rain and ice contribute to skidding, poor driving skills are the main cause of skidding.

If the fire apparatus begins to skid:

- **Stay off the brake.** Until the vehicle slows, your brakes will not work and could cause you to skid more.
- **Steer.** Turn the steering wheel in the direction you want the vehicle to go. As soon as the vehicle begins to straighten out, turn the steering wheel back the other way. If you do not do so, your vehicle may swing around in the other direction and you could start a new skid.

- **Continue to steer.** Continue to correct your steering, left and right, until the vehicle is again moving down the road under your control.

Intersections

During an emergency response, intersections present the greatest potential danger to fire apparatus. When approaching and crossing an intersection, you should not exceed the posted speed limit. If you are traveling too fast, you will not be able to stop in time to avoid a collision. Also, do not assume you have the right-of-way simply because you are using the emergency warning lights and sirens: Other drivers may not see or hear the fire apparatus as you approach the intersection. Scan the intersection for possible hazards such as right turns on red, pedestrians, or vehicles traveling fast or weaving through traffic. Observe the traffic in all four directions: left, right, front, and rear. Proceed through the intersection only when you can account for all lanes of traffic at the intersection. In multilane intersections, you should clear each lane individually before proceeding. In some cases, other vehicles may block your view or the view of other drivers. You may think the lane is cleared and completely miss the other vehicle that is proceeding through the intersection. In some cases, you may have to slow down to less than 5 mi/h (8.04 km/h) while in the intersection and visually clear each lane of traffic before proceeding.

At all intersections with green lights or where the fire apparatus has the right-of-way, you should slow down as necessary to ensure safe operation by disengaging the throttle and pressing the brake pedal. This strategy ensures that you are ready for any potential hazard—that you are driving defensively. When approaching an intersection where all lanes of traffic are blocked with other vehicles, you should turn off all sirens and horns at least 200′ (60.96 m) before the stopped traffic is reached. You should leave the emergency lights on and bring the fire apparatus to a stop at least 100′ (30.48 m) from the nearest vehicle in traffic. This stance will let the civilian drivers know that you are still there and responding to an emergency, but will not "push" them into the intersection. Never encourage or force traffic to proceed against red lights or to advance into dangerous traffic conditions. Instead, stay in the far left lane so that when the light turns green or the left green arrow is activated, you can proceed with lights and sirens through the intersection. The fire apparatus should come to a complete stop at all uncontrolled intersections, stop signs, and yellow and red lights. Proceed through the intersection only after it has been determined that the other vehicles have stopped and that it is safe to do so.

Many jurisdictions use **traffic signal preemption systems** at intersections to assist fire apparatus during the emergency response. These systems can change the signal from red to green for an emergency vehicle as it approaches the intersection. A flashing light on the fire apparatus, also known as an **emitter**, will trigger a **receiver** on the traffic signal and change the light to allow the fire apparatus to secure the right-of-way through these intersections in as safe and efficient a manner as possible. In jurisdictions using this type of traffic control system, each emergency vehicle is equipped with an emitter, which is usually mounted next to the emergency pump lights, and which emits visible flashes of light or invisible infrared pulses at a specified frequency. A receiver device is mounted on or near intersection traffic control devices to recognize the signal and preempt the normal cycle of traffic lights. Once the emergency vehicle passes through the intersection and the receiving device no longer senses the remote triggering device, normal operation resumes. Such a system provides the fire apparatus with the benefit of a quicker response. You should be alert to the fact that these intersections will require a heightened level of awareness and understanding. Other fire apparatus that are responding to the same emergency scene may also trigger this system from another direction and proceed through the intersection at the same time.

You should also be aware that an encounter by the public with an emergency response vehicle en route to an emergency is not a normal, everyday occurrence. This fact is clearly evident from the confusion that is often witnessed when fire apparatus approach civilian traffic at intersections controlled by traffic signal preemption systems. For example, while the traffic signal preemption system cycles the light through the normal sequence of yellow and then red, an approaching fire apparatus is typically unable to determine how long the light was green for cross traffic. An unusually short green light for cross traffic may appear to that traffic as a malfunctioning light, encouraging the civilian drivers to run a stale yellow light. Proceed only after ensuring that it is safe to do so: A green light given as a result of traffic signal preemption system activation should not be assumed to be safe until verified visually, in all directions.

When a green light is given to the responding fire apparatus due to a traffic signal preemption system's activation well in advance, the driver/operator must still slow down as necessary to ensure safe operation. If the system is not granted a green light on approach, you should come to a complete stop just before the intersection to ensure safe operation. This delay will provide time for the system to activate the traffic signal, allow the intersection to clear completely, and give those civilian drivers who are either inattentive or prone to running stale yellow lights a chance to clear through.

As is the case with all intersections, when approaching a traffic signal preemption system controlled intersection where all lanes are blocked, turn off all sirens and horns but leave the emergency lights on. The system is only operational when the vehicle's emergency lights are activated and the **parking brake** is in the off position. When activation has been granted and the green light delay occurs, activate sirens and horns for a few seconds to give the civilian vehicles time to clear the intersection completely.

Whenever you approach an unguarded railroad crossing, you should bring the fire apparatus to a complete stop before entering the grade crossing. Do not assume that the track is devoid of any trains in an attempt to continue the response. While stopped at the crossing, turn off all sirens, roll down the windows, operate the fire apparatus at an idle, and listen for the sound of an oncoming train. Railroad crossings must be treated with the same caution as any other intersection. Remember that with the siren activated, it may be difficult to hear a train horn or crossing bells. At intersections with railroad crossbars, you should never proceed between the crossbars in an attempt to continue the response. In this situation, you may have to alert the dispatch center or other responding units of a delayed response due to an inability to cross the railroad tracks safely.

Approaching the Scene of an Emergency

All emergency scenes have one thing in common: They are all dynamic in nature. A scene can go from bad to worse in a matter of seconds. For this reason, fire fighters should always be cautious. As a driver/operator, as you approach the scene, you should slow down, identify the correct address/location, and recognize any potential hazards. If you take all three of these actions, you can make the scene safer for the members of your crew and anyone else in the immediate area.

Slow Down

When fire apparatus are responding to an emergency, it usually attracts a lot of attention from the public. People are naturally curious about what is going on and may want to help. This is especially true in residential neighborhoods. Children may dart out into traffic, others may try to flag down fire apparatus, or people may stop their cars in the middle of the street. Some individuals even become so preoccupied with watching the emergency that they do not pay attention to their surroundings. This is why it is so essential for you to slow down and proceed with the utmost caution, especially the last few blocks to the emergency Figure 8-3.

Identify the Address/Location

Responding to the wrong location cannot only be embarrassing, but may also lead to loss of property and even lives. This type of mistake may be avoided by writing down the address when the call is dispatched. If you forget the address, you can reference the information that you wrote down earlier. Although some fire departments have onboard computers that can be used to locate the incident, it is always better to have a hard copy as a backup.

Once the apparatus is in the vicinity of the emergency, all crew members aboard the fire apparatus should help in locating the address. In some response areas, this may be difficult to do even during daylight hours. During the night, the task is usually even harder. That is why everyone on the fire apparatus should be looking for the correct location. Firefighting is a team concept. Although the driver/operator is responsible for delivering the members of the company to the scene, you may need some assistance along the way.

In large apartment complexes, especially garden-style apartments, finding the location can be even more difficult. Unfortunately, not every apartment complex is numbered the same way. While some may have a letter to identify the building and a number for the actual apartment, others use the exact opposite addressing system Figure 8-4. Knowing the response district and having access to current maps can make all the difference. This information is best obtained before the emergency.

In rural areas, many fire departments rely on the knowledge of the fire department members. They may use local landmarks and old terminology to locate an emergency site. While this approach is not the preferred method to find emergency locations, in some jurisdictions it is the only way. In these situations, it is critical to have someone guide responders from other outside agencies to the scene as they seek the correct location. This may involve staging a unit on a main street to direct others down a dirt road that is not labeled.

Locating the correct location may involve more than just looking for a physical address. Some calls may be dispatched to an area and not a specific address. When trying to locate an emergency scene, the entire crew should look for some of the following signs:

Figure 8-3 Slow down as you approach the scene, you never know who else is on the road.

Figure 8-4 Not every apartment complex is numbered the same way.

Figure 8-5 Smoke can help you in locating the incident.

- Civilians attempting to wave them down
- Smoke in the area Figure 8-5
- Police cars with their emergency lights on
- Large crowds of people
- Congested streets that normally are devoid of traffic
- Residential lights that are flickering on and off during the night
- Headlights from a car that is off the roadway

Recognize Potential Hazards

Recognizing potential hazards while approaching the incident can be difficult sometimes. You may respond to the same types of calls over and over. This routine may become monotonous, and it is easy for some members to become complacent. By constantly being on the lookout for hazards, you can ensure the safety of your crew. Some of the hazards that you may encounter include fallen power lines in the area, violent scenes, large amounts of pedestrian traffic, and debris in the roadway from a motor vehicle accident (MVA). While these hazards may be easily mitigated, failure to recognize them could put the crew's safety at risk. Do not become complacent; always be aware of your surroundings.

Fire Scene Positioning

While approaching a fire scene, you should attempt to view at least three sides of the structure. To do so, you may have to drive slightly past the structure. This positioning gives the crew on the fire apparatus a better size-up of the building, including the company officer who may have to share initial size-up information with other responding units.

Every fire department should train its members on proper staging procedures. Always follow your fire department's staging procedures. Some fire departments require that the first engine, ladder, and battalion chief go directly to the fire scene. With only a few necessary units at the scene, the incident commander (IC) can begin to orchestrate strategies and tactics without being overwhelmed. Eventually, specific tasks will be assigned to each unit by the IC. If tasks have not been delegated within a reasonable amount of time, the responding fire apparatus should attempt to contact the IC for an assignment.

Nothing Showing

When nothing is showing as the first-arriving units are approaching the incident, the emphasis should be on the proper positioning of the fire apparatus. This initial placement should allow members to effect an investigation and allow for future operations. In most cases, it requires parking near the main entrance to the building. Stay outside with your fire apparatus, monitor all radio traffic, and await orders from the IC. While the other members of the crew are completing an investigation on the building's interior, you should further size up the incident from the exterior and prepare for any needed operations. This effort may include the following steps:

- Determining the best position to spot other incoming fire apparatus
- Locating the nearest hydrant
- Locating the sprinkler system or standpipe connection Figure 8-6
- Preparing equipment that may be requested by the officer or IC, such as carbon monoxide detectors, thermal imaging cameras, and positive-pressure ventilation fans
- Observing any potential hazards on the scene

Always follow your department's standard operating procedures (SOPs) and chain of command. Do not make decisions or orders when you do not have the authority to do so. Obtain information about the incident and relay it through the chain of command.

Working Fire

Once the fire apparatus arrives at a working fire, the driver/operator must position it for maximum potential benefit. Unfortunately, there is usually a natural inclination to drive the fire apparatus as close to the fire as possible, which often results in positioning of fire apparatus that is both dysfunctional and

Figure 8-6 A fire department connection.

dangerous. The placement of all fire apparatus on the fire ground should take into account the following considerations:
- Your fire department's SOP
- A direct order from the IC
- A conscious decision on the part of the company officer based on existing or predictable conditions

Efficient fire apparatus placement must begin with the arrival of the first-responding fire apparatus. This vehicle will set the tone for the entire incident. If the initial fire apparatus is placed in the optimal position, then other fire apparatus will follow suit. In contrast, improper placement of the first-arriving units will create a problem that will have to be tolerated for the remainder of the incident. First-arriving fire apparatus should park at maximum advantage and go to work; later-arriving fire apparatus should be placed in a manner that builds on the initial plan and allows for expansion of the operation. Remember that firefighting is a team concept. Driver/operators who are already on the scene should communicate to those who are en route to the emergency and let them know in advance the most advantageous positions in which to place their vehicles. While this decision is usually the responsibility of the IC, all personnel will be affected if later-arriving units are positioned erroneously.

> **Driver/Operator Tip**
>
> Once at the scene of the incident, do not place the front of the fire apparatus too close to the rear of another fire apparatus. This will block in fire apparatus and may prohibit some equipment from being accessed, such as ground ladders carried on an aerial apparatus. Most modern aerial apparatus carry their ground ladders in the rear compartment. To remove the ladders, there must be a clear area behind them greater than the nested length of the ground ladder; for a 35' (10.66 m) extension ladder, this can be as much as 20' (6.096 m). Likewise, poor positioning of fire apparatus can cause problems in accessing the hose that is carried on the apparatus. If too many units are confined into a tight space together, the fire fighters will have a difficult time deploying the attack lines around other apparatus. Do not block other operations with your fire apparatus. Remember, hose lines can be extended; ladders cannot.

Later-arriving companies should follow their fire departments' staging procedures. Everyone must maintain an awareness of which site access provides the best tactical options and ensure that the immediate fire area does not become congested with fire apparatus. When possible, an access lane should be maintained down the center of the street. Park unneeded units out of the way. Fire apparatus that is not working should be left in the staging area or parked where it will not compromise access. Take advantage of the equipment on fire apparatus already in the fire area instead of bringing in more fire apparatus—many times only additional personnel are needed to address the incident.

When fire apparatus are on a fire scene, they are classified into one of two categories:
- **Working.** These fire apparatus are actually in use on the fire scene. The fire apparatus or the equipment that they carry is actively being used.
- **Parked.** These fire apparatus are not being used on the fire scene. Their only reason for being on scene, for the time being, is to transport personnel to the scene of the fire. No significant amount of tools or equipment from such fire apparatus is actively being used. Given these facts, the fire apparatus should remain in the appropriate staging area or be positioned so that they do not compromise the access route of other incoming units.

Staging

NFPA 1561, *Standard on Emergency Services Incident Management System*, identifies the need to provide a standard system to manage reserves of responders and other resources at or near the scene of the incident. If too many units arrive at the incident without direction, their presence may lead to possible freelancing and confusion. The IC will have great difficulty controlling the personnel and maintaining discipline at such a chaotic scene. Accountability of all personnel on scene would also be complicated if it is not known when units arrived or who was given an assignment.

To avoid this dilemma, later-arriving fire apparatus should stay in an uncommitted position and await orders from the IC, a practice commonly referred to as **staging**. Staging is the standard procedure used to manage uncommitted resources at the scene of an incident. The objective of staging is to provide a standard system of initial placement for responding fire apparatus, personnel, and equipment prior to their assignment at tactical incidents **Figure 8-7**. Staging also provides a way to control and record the arrival of subsequent resources and eventually their assignment to specific locations or functions within the Incident Management System (IMS).

When used successfully, proper staging of fire apparatus can accomplish the following goals:

Figure 8-7 The objective of staging procedures is to provide a standard system of initial placement for responding fire apparatus, personnel, and equipment prior to their assignment at tactical incidents.

- **Reduce unnecessary radio traffic.** During the first few critical moments of a structure fire, vital information is disseminated to all members who are responding. The IC will give a size-up of the incident. First-arriving units may request a water supply or additional resources to be sent to a specific location. Now is not the time to clutter the radio channel with needless radio traffic.
- **Reduce excessive apparatus congestion at the scene.** When too many fire apparatus are cluttering the scene, matters can become both dangerous and confusing. By placing fire apparatus in an uncommitted location close to the immediate scene, the IC can evaluate conditions prior to assigning companies. This will allow time for command personnel to formulate and implement a plan without undue confusion and pressure.
- **Provide the IC with a resource pool.** The IC may then assign units and resources at his or her leisure. These units should be ready to deploy as required by the IC.

Level I Staging

Many fire departments use two levels of staging. Generally, **Level I staging** is in effect for all first alarm assignments, or incidents involving three or more units. For example, during a working structure fire response, all units continue responding to the scene until the first-in unit reports its arrival on the scene. Following its arrival and assumption of command, the first-in unit begins the assignment of the remainder of the dispatch. During Level I staging, the only units that proceed directly to the scene of a working structure fire are the first-due engine, ladder, and chief **Figure 8-8**. These units respond directly to the scene and initiate appropriate operations as directed by their department's SOPs or by the IC. All other units stage in their direction of travel, uncommitted, approximately one block from the scene. A position that supports the maximum number of tactical options with regard to access, direction of travel, water supply, and other considerations is preferred. Staged units announce their arrival, report their company designation, and identify their staged location/direction (e.g., "Engine Five, South"). These units remain in position until they receive an assignment from the IC.

Level II Staging

Level II staging is utilized on second or greater alarms or when mutual aid units report to an incident. This type of staging places all reserve resources in a central location and automatically requires the implementation of a staging area manager. The **staging area manager** is the person responsible for maintaining the operations of the staging area. The **staging area** is an area away from the incident where units will park until requested to enter the emergency scene **Figure 8-9**. This area should allow staged apparatus to access any geographic point of the incident without delay or vehicle congestion.

First alarm units that are already in Level I staging or en route to Level I staging will stay in Level I unless otherwise directed by the IC. All other responding units will proceed to the Level II staging area.

When activating Level II staging, the IC should give an approximate location for the staging area. This area should be some distance away from the emergency scene to reduce site congestion, yet close enough to perform a prompt response to the incident site. Large parking lots or empty fields are excellent choices for a staging area.

The IC may designate a member of the department to serve as the staging area manager. When this position is not designated, the first fire officer to the Level II staging area will assume this role. The staging area manager takes a position that is both visible and accessible to incoming and staged companies. This is accomplished by leaving the emergency lights operating on the fire apparatus. No other fire apparatus positioned in the Level II staging area should leave its emergency lights on. All incoming resources into the staging area are logged and their availability reported to the IC. All subsequent arriving units should report in person to the staging area manager and await assignment, a strategy that reduces unnecessary radio traffic. Within the staging area, like units should be positioned next to one another—for example, all ladders next to each other, and all engines next to each other. This makes the staging area manager's job of identifying resources easier. Only the staging area manager may release resources from the staging area.

Figure 8-8 Level I staging.

Figure 8-9 Level II staging.

Engines and Ladders

Engine and ladder companies are the quintessential bread-and-butter apparatus of the fire service. The proper placement of ladders and engines should complement one another so as to rescue civilians and effect total fire extinguishment. Never assume that these two types of units can be positioned in the same way; instead, each unit has a specific function that should be taken into account when positioning the apparatus at a fire scene. Some fire departments have driver/operators who operate both types of fire apparatus on a regular basis. If this is the case, position the fire apparatus according to its function—not in the order in which it arrives on the fire ground.

The majority of fire departments in the United States have both engines and ladders in their fleets. This section focuses primarily on these two indispensable pieces of equipment. Although other fire apparatus are also discussed, their presence on a fire scene is usually not as critical to its outcome.

The placement of the first-arriving engine or ladder should be based on the initial size-up and the fire department's SOPs. If the engine driver arrives on scene first, he or she must choose a setup position that provides for efficient operation of the engine company as well as the first-arriving ladder apparatus. Just because your vehicle is the first fire apparatus to arrive at the scene, it does not mean that you are entitled to the best position on the fire scene. Instead, you must look at the big picture and imagine how the scene may unfold. An engine company that occupies a spot that would be better suited for a ladder apparatus is an example of poor fire apparatus positioning and inexperience.

Usually, only the engine company responds to the fire scene with hose, water, and a fire pump on board the fire apparatus. Generally, the engine's placement depends on the conditions encountered upon arrival. While other responding fire apparatus may assist the engine company, it is ultimately the engine crew's responsibility to provide the water flow for fire extinguishment. This water may be supplied through handlines or master streams. It is imperative that the engine company establish a water supply that will last for the duration of the incident.

The ladder company should be positioned to support the engine company's operations. When positioning a ladder apparatus on the scene of a fire, the driver/operator must take into account all of the hazards that may plague the engine company as well as some additional hazards that are specific to an **aerial device** (a generic term used to describe the hydraulically powered aerial ladder or platform that is operated from the top of an aerial apparatus). Aerial ladders and aerial platforms may be used as launching points for rescue, entry, search, and ventilation operations. They may also be used to stretch hose lines to upper floors or the roof, bridge a gap, perform ladder pipe operations, and serve as observation posts from which to assess conditions. When their need is evident upon arrival, aerial devices should be raised immediately. When need for them is anticipated to arise later, these devices should be positioned for rapid setup and future use. In these situations, the driver/operator of the ladder apparatus should remain in the vicinity of the turntable until the fire is under control.

> **Driver/Operator Tip**
>
> Each fire department should develop a procedure for the placement of initial fire apparatus that driver/operators should follow upon their arrival at an incident. Always follow your fire department's procedures, and learn the procedures for other jurisdictions covered by mutual aid agreements.

The IC may give specific instructions regarding the placement of fire apparatus and the operations to be performed. As a driver/operator, you must base your decision about placement of the fire apparatus on the following conditions: rescue potential and exposures.

Rescue Potential

Rescue is the first priority for every responding fire apparatus. If civilians are potentially trapped inside the structure upon arrival, you should ensure that the front of the structure is available for the ladder apparatus. Position the fire apparatus to assist in or perform this operation. In this scenario, engine apparatus should be out of the way but preparing to go to work. For efficient and safe control, all apparatus drivers must work together to ensure that the aerial apparatus is positioned for maximum use and minimum stress to the aerial device. Usually, the front of the building is left available for the first-arriving ladder apparatus regardless of the conditions. It may be necessary or advantageous for ladder apparatus to circle the block and come in from the opposite end of the street, if such action will improve the fire apparatus placement.

Exposures

When exposure protection is necessary upon arrival, do not position the fire apparatus between the fire and the exposure. Doing so may cause the fire apparatus to become an exposure problem itself. Instead, position the vehicle far enough away from the exposure to remain safe, but close enough to deliver fire streams for exposure protection. Usually, a distance of 30′ (9.144 m) is sufficient to allow the unit to operate, yet remain a safe distance from the fire. Regardless of the initial placement, if conditions change (i.e., rapid fire spread that exposes the fire apparatus) and lead to collapse potential, then repositioning may be required and must be accomplished quickly and safely. Proper training and planning are essential in such cases.

As a driver/operator, you must always think ahead and position the fire apparatus with the idea that it might have to be repositioned at some point during the incident. Do not get blocked in or out of the fire area. Consider the collapse potential based on severe fire conditions and building construction. Many fire fighters have been killed and fire apparatus damaged because of failure to recognize a building's collapse potential. To ward off this threat, most fire departments specify a **collapse zone**—that is, is a distance of 1½ times the height of the building in which fire fighters and fire apparatus must not be located in case of a building collapse. In buildings with bowstring trusses, identification of an even larger area as the collapse zone may be required. As such a building's walls and bowstring roof assembly fail, they may propel outward greater than 1½ times the height of the building. When tall buildings are involved with fire and a danger of collapse is present, a collapse zone of 1½ times the building's height may not be practical, as it would require positioning fire apparatus so far away that they may not be effective.

In some circumstances, the fire apparatus should be positioned in one of the **corner safe areas** of the fire ground. In his book *Safety and Survival on the Fireground*, retired Fire Chief Vincent Dunn describes studying this area by looking at the structure from a bird's-eye view: There are four areas of the fire ground that may not be covered by collapsing walls. As the fire progresses, fire fighters must continually reevaluate the potential for building collapse. Positioning the ladder apparatus at the corner safe areas of a building affords coverage on two fronts. This strategy enables coverage of a much wider area, permitting greater access and providing observation points from which to check the stability of the building and other issues.

When positioning the fire apparatus, note the locations of street lights, traffic signals, trees, utility poles, and wires at street corners or other parts of the site. Placement of the aerial device should be oriented toward providing as much effective operating area for the basket/tip as possible on both fronts of the building.

Fire Conditions
The initial engine company needs to be positioned for the efficient deployment of the first attack line—the most important attack line on the fire ground. You should not position the fire apparatus with the pre-connect attack line directly in front of the building's entrance. While this positioning may simplify the attack line deployment, it will place the engine in a position that compromises the operation of other incoming fire apparatus, especially the ladder apparatus. Instead, the engine should be placed past the structure, with the front of the building being left open for the ladder company. Most engines have an excess of fire hose, whereas aerial devices have a fixed length. Do not render the aerial device useless by blocking it out just to make stretching the attack lines easier.

Water Supply
First-due engine companies approaching the scene with any evidence of a working fire in a structure should secure a water supply. The next-in engine company may be too far away or encounter a delay while responding to perform this task. Exceptions to securing the water supply as the first-arriving engine company's primary task may sometimes arise, however—for example, when there is an obvious critical rescue requiring the entire crew or when the exact location of the fire in a multiple-unit occupancy is unknown. Whenever possible, the supply line should be large-diameter hose (LDH), as this equipment reduces the friction loss and provides an adequate water supply. Always notify other responding apparatus when laying LDH across streets and intersections, as this supply line may block other fire apparatus from reaching the scene of the fire. When laying the hose from the hose bed, always attempt to position it to the same side of the street as the hydrant.

Slope
Positioning uphill from the incident may prevent future problems, such as those caused by water runoff from the fire scene. The slope of the area will usually not affect normal engine operations. NFPA 1901, *Standard for Automotive Fire Apparatus*, requires that fire apparatus have two wheel chocks mounted in readily accessible locations, each designed to hold the fire apparatus when loaded to its maximum in-service weight, on a 20 percent grade with the transmission in neutral and the parking brake released. When the pump is engaged, make sure that these wheel chocks are placed in the proper positions. If there is any doubt about the unit's ability to keep from rolling down the hill, chock it or move it!

Some sloping surfaces may not allow for adequate deployment of an aerial device. To operate with 100 percent capacity of the aerial device, some aerial apparatus stabilizers can correct for a slope and grade of only a few degrees. You may be able to park the fire apparatus in an optimal position, but if the aerial device is incapable of operating in that position, this placement is futile. You must know the limitations of the fire apparatus that you are operating and position it accordingly.

Terrain and Surface Conditions
The terrain is the landscape on which the apparatus is positioned. It may be an asphalt roadway or a concrete driveway, but sometimes may include areas off the roadway. In these circumstances, you should position the fire apparatus with future pump operations in mind. If possible, leave the fire apparatus on the asphalt roadway or a concrete driveway that can support the weight of the fire apparatus. If this positioning is not feasible, you should prevent significant amounts of water from flowing underneath the fire apparatus if it is positioned on top of soil that is unstable or may become unstable. Otherwise, the fire apparatus may become stuck and the ensuing mess may hamper future operations. Where the ground is of doubtful stability, as is sometimes the case with vacant lots or other unpaved areas that may have hidden voids, and if the terrain is deemed not substantial enough, ladder apparatus should be positioned elsewhere.

Wind Conditions
Upon arrival, note the wind conditions. Place the fire apparatus out of the oncoming smoke and heat. Too much smoke will starve the engine of fresh air and cause it to shut down **Figure 8-10**. Should the wind shift during the operation and compromise the engine operation, notify the IC. If possible, it may be necessary to reposition the engine at another location. While you are operating the fire apparatus, if you need SCBA because of the smoke conditions, move the fire apparatus: If you cannot breathe the air, neither can the fire apparatus. Remaining in the original position may cause the fire apparatus to stall and render its pumping operations useless.

The wind may also affect master stream operations. When deck guns are not reliable because of heavy wind conditions, the use of portable ground monitors may be a better alternative.

The wind may also limit the operations of some aerial devices on scene. Most aerial devices are designed to be operated in winds up to 50 mi/h (80.46 km/h) without any reduction in tip load. Refer to the operating manual to determine the recommended operational extremes.

Overhead Obstructions
Overhead wires may interfere with any aerial device operation. Do not be intimidated by overhead wires when the situation

Chapter 8 Approaching the Fire Ground 205

Figure 8-10 Too much smoke will starve the engine of fresh air and cause it to shut down.

Figure 8-11 Use caution when operating around trees.

clearly calls for the use of the aerial device; rather, exercise caution and be creative in your approach. The IC should have wires removed by the utility company when fire conditions warrant doing so. All aerial devices should remain a minimum of 10′ (3.048 m) from all overhead wires. Do not try to guess which wires may be energized; instead, consider all wires to be live until proven otherwise.

When trees obstruct operations, it is possible to extend or raise the aerial device through light branches. However, retraction or lowering of the boom through branches may present a problem, and some cutting may be required to overcome this obstacle. Use caution when operating around trees, as an electrical hazard should always be suspected Figure 8-11 ▶ . Before an incident occurs, you should practice positioning the fire apparatus for operation with overhead obstructions in the area to become familiar with the fire apparatus's limitations and functions under these conditions. Possible alternatives may include placing the fire apparatus on sidewalks, setting it up at corners, and extending the aerial device parallel with the front of a building. The crew should also practice using this device at intersections with light posts, traffic signs or signals, intersecting overhead wires, and other obstacles to enable personnel to judge where and how fire apparatus is to be positioned for maximum coverage under similar circumstances.

Auxiliary Appliances

An **auxiliary appliance** is a standpipe and/or sprinkler system. The determination of whether a building has one or both of these systems is best made during development of a preincident plan. A **preincident plan** is described by NFPA 1620, *Recommended Practice for Pre-incident Planning*, as a document developed by gathering general and detailed data used by responding personnel to determine the resources and actions necessary to mitigate anticipated emergencies at a specific facility. When creating such a preincident plan, you should locate the fire department connection, fire pump, standpipe system, and sprinkler system in the building. If this information is not readily available, you will have to rely on other members of the crew to help locate these systems, which may slow down the initial operations and result in more work for the fire fighters on scene. For example, if an engine company responds to a fire in a multistory apartment building, the crew may prefer to use the building's built-in standpipe system to establish a water supply. If you cannot identify the location of the fire department connection, the engine crew will not be able to use the standpipe to connect attack lines and, therefore, will have to stretch additional hose lines up to the fire floor. This delay in water application may cause the fire to progress past the capabilities of a single hose line.

Positioning of Other Fire Scene Apparatus

Engines and ladder trucks are not the only fire apparatus required at a fire scene. Depending on the size, construction, occupancy, and involvement of the structure, a multitude of other types of fire apparatus may be needed to mount an effective response.

Specialized Fire Apparatus

Specialized fire apparatus may be equipped with an assortment of specialized equipment, such as that needed for heavy technical rescue, hazardous materials response, or mobile air supply. The personnel assigned to these fire apparatus are required to have specialized training, and their role on the fire scene is usually to provide support operations.

Specialized fire apparatus may need to be positioned close to the incident depending on the conditions. The drivers of these vehicles may have to be creative in their approach to the fire scene. Approaching from the same direction as all the other units may not be effective, for example. Instead, these driver/operators must listen to the radio, observe scene conditions, and identify the most efficient position for the task assigned.

Command Vehicles

A **command vehicle** is one that the chief uses to respond to the fire scene. This vehicle should be positioned at a location that will allow maximum visibility of the fire building and surrounding area. It should be easily identified at the scene and placed in a logical position. The command vehicle should not restrict the movement or positioning of other apparatus at the fire scene.

Ambulances

An **ambulance** is a specially designed vehicle that is capable of transporting the sick and injured. It is usually staffed with trained emergency technicians and/or paramedics. These vehicles should be parked in a safe position that will provide the most effective treatment and transportation of fire victims and firefighting personnel, while not blocking other apparatus or interfering with firefighting operations. Ambulance drivers are also responsible for positioning their vehicles with a clear route of egress. During large fire scenes, these emergency vehicles may be staged with other apparatus in a Level II staging area. When requested, they will respond to the scene. Once on the fire scene, the driver may stay with the vehicle while other personnel load the patient into the vehicle. This practice ensures that the patient is picked up as close to the scene as possible without compromising fire scene operations if the ambulance needs to be relocated.

Positioning at an Intersection or on a Highway

Operating at an emergency scene that is located either on or adjacent to a highway or intersection is extremely dangerous. Personnel should understand and appreciate the high risk that fire fighters are exposed to when they are working in or near moving vehicles. According to the U.S. Fire Administration, the fifth leading cause of fatal injuries to fire fighters in 2005 was being struck by an object. Four fire fighters died after being struck by vehicles or other objects while on duty Table 8-1 . This type of fire fighter fatality is not uncommon; indeed, it happens every year. Each call near a roadway should be treated with caution. Always consider moving traffic as a threat to scene safety.

Each day, fire fighters are exposed to motorists of varying abilities, with or without licenses, with or without legal restrictions, and driving at speeds from creeping along to going well beyond the speed limit. Some of these motorists have visual impairments, and some are impaired because of the use of alcohol and/or drugs. On top of everything else, motorists often look at the scene and not the road. Their lack of attentiveness while passing a roadside emergency scene may affect fire fighters' safety as they work at the scene.

When the fire apparatus arrives at the scene, other members may have a desire to quickly dismount the apparatus and go to work. The driver should not allow personnel to exit the cab until the driver is satisfied with the position of the fire apparatus. To achieve this goal, it may be necessary to back into position or angle the fire apparatus off of a roadway. Only when the unit is parked and ready should the other members be allowed to dismount.

Table 8-1 Causes of Fatal Injuries to U.S. Fire Fighters, 2005

Cause	Number of Deaths
Stress/overexertion	62
Vehicle collision	25
Caught/trapped	9
Fall	5
Struck by object	4
Contact/exposure	3
Assault	1
Other	6
Total	**115**

Source: U.S. Fire Administration, "Fire Fighter Fatalities in the United States, 2005."

The following subsections describe four actions that all fire fighters can take to protect themselves and the other crew members while operating in traffic conditions.

■ Never Trust Traffic

Every fire fighter must have a healthy respect for all vehicles. Do not assume that because vehicles are moving around an emergency scene that the danger is gone. Anytime that the scene is near moving traffic, there is a potential danger to fire fighters. Fire fighters should exit the apparatus on the curb side or the nontraffic side whenever possible. Always look before stepping out of the fire apparatus or into any traffic areas on scene. When walking around fire apparatus parked adjacent to moving traffic, keep an eye on traffic, and walk as close to the fire apparatus as possible. Never turn your back to oncoming traffic for extended periods of time.

■ Engage in Proper Protective Parking

This aspect of safety relies on your ability as the driver/operator. As you position the fire apparatus, think about the consequences of your actions. Never allow convenience to compromise safety. Always position your apparatus to protect the scene, patients, and emergency personnel, and to provide a protected work area. When possible, position the fire apparatus at a 45-degree angle away from the curbside to direct motorists around the scene Figure 8-12 . Initial fire apparatus placement should always allow for adequate parking of other fire apparatus and a safe work area for emergency personnel. Allow enough distance between the fire apparatus and the scene to prevent a moving vehicle from knocking fire apparatus into the work areas.

■ Reduce Motorist Vision Impairment

During an emergency, the need for emergency lights is evident. At the incident, the use of emergency lighting may still be needed

Voices of Experience

I was recently promoted to driver/operator and I was assigned as a rover which meant that I called shift command to find out what truck I was assigned to for the day. On this particular day I was roved to an engine company in a small station with only one engine and an ambulance. The response area wasn't too tricky, mostly houses and apartment buildings with the occasional strip mall. The crew was led by a very capable captain with a lot of experience and included a senior fire fighter with a lot of experience, a probationary fire fighter, and myself, the new driver/operator.

It was a hot day in the middle of June. The temperature that afternoon ran around 115°F (46°C). We received a dispatch for a grass fire. Upon arrival, we had light smoke in a small area of grass, about 500 square feet (152 m) or so. I parked the engine directly to the west of the field, assuming that we would knock this grass fire down with our brush line. The senior fire fighter pulled the brush line and started to knock it down when a very hard gust of wind had hit the fire and ignited a bunch of oleander trees instantly.

The fire spread from tree to tree until it found an exposure to the south: a condominium complex. The senior fire fighter and captain came back to the engine to pull another hose line so that they could reach the fire and protect the exposure. I hand jacked about 300 feet (92 m) of 4 inch (102 mm) supply line to establish a water supply while they extended their attack line to the exposures. If I had parked closer to the hydrant, I would not have had to pull so much hose.

Additional resources arrived and they took care of the condominium complex as my crew contained the small grass fire that had spread over the whole five acres and consumed trees, power lines, and cars.

The biggest lesson that I learned that day is never assume that a little fire is just going to go out. Before then, I had seen lots of these types of fires and they all went out just like we wanted them to. However, with the help of wind and plenty of dry brush, this fire showed me how unpredictable fire really can be. Parking the engine is an integral part of any firefighting operation. The position of the first engine is critical to fire fighter safety, customer safety, protection of property, and to the overall effectiveness of the operation.

Mike Warriner
Phoenix Fire Department
Phoenix, Arizona

> "It is good to have an open mind, to think ahead, and to learn from the more experienced driver/operators on all types of apparatus."

Figure 8-12 An example of proper protective parking.

Driver/Operator Tip

Each colored emergency light on the fire apparatus serves a specific purpose and results in a different reaction from other motorists. The following list identifies the lights' color and reaction:

- **Red.** For most civilians, a red light identifies the need to stop. Unfortunately, it may also attract those drivers who are under the influence of drugs and/or alcohol as well as fatigued drivers.
- **Blue.** A blue light identifies the fire apparatus as being associated with either fire or police. States may have different laws dictating who can and cannot use blue lights on their vehicles. A blue light has good visibility during both daytime and nighttime operations.
- **Amber.** An amber light signals danger or caution. This color is widely used by other services to get drivers' attention. Many experts believe it to be the best warning light for the rear of emergency vehicles. It may deter those drivers who are fatigued or who are under the influence of drugs or alcohol. During foggy conditions, an amber-colored light is more readily visible than other colored lights.
- **Clear.** Clear light is associated with caution. Although it provides good visibility, such a light should normally be shut off at an emergency scene, to prevent blinding other drivers. For the same reason, it should not be used at the rear of the fire apparatus.

when the safety of personnel is otherwise compromised. Never hesitate to operate emergency lighting at a scene. However, understand that emergency vehicle lighting provides a warning only and does not ensure effective <u>traffic control</u>, which entails protecting the emergency scene from oncoming traffic by redirecting, blocking, or stopping all moving vehicles.

While most state laws require only one lighted lamp exhibiting red light visible under normal atmospheric conditions from a distance of 500′ (152.4 m), many fire apparatus exceed this requirement. Unfortunately, the use of too many lights at a scene may create a dangerous situation. An excess of emergency lights flashing can cause a carnival effect and create confusion for motorists. Limit the number of fire apparatus operating emergency lights to only those blocking oncoming traffic—and even these fire apparatus may not require all emergency lights to be on.

If provided, use directional arrows at the rear of the apparatus to direct any oncoming traffic Figure 8-13. It is safer to divert traffic with advanced placement of signs and traffic cones rather than to rely on warning lights on fire apparatus to reroute oncoming vehicles.

Wear High-Visibility Reflective Vests

Turnout gear does not adequately identify a fire fighter who is operating in or near traffic conditions. The reflective trim on the turnout gear may be dirty, covered with other equipment, or missing. To effectively identify themselves, fire fighters should wear ANSI-approved reflective vests. These vests should be retro-reflective and fluorescent. Each fire department should require all personnel operating under these conditions to wear such vests. Some manufacturers make reflective vests out of flame-retardant material that may be worn over turnout gear.

Manual on Uniform Traffic Control Devices

The U.S. Department of Transportation's Federal Highway Administration publishes the *Manual on Uniform Traffic Control Devices for Streets and Highways* (MUTCD). Under federal law, each state is required to adopt the provisions in this manual. Section 6I, "The Control of Traffic through Incident Management Areas," applies to all incidents that fire fighters might encounter on or near the roadway. It defines a <u>traffic incident</u> as an emergency traffic occurrence, a natural disaster, or other unplanned event that affects or impedes the normal flow of traffic. When traffic incidents occur, some form of traffic control must take place.

Figure 8-13 Use of directional arrows to divert the flow of traffic.

Safety Tip

The American National Standards Institute has identified three classes of safety vests:

- **Class 1.** This vest has the lowest level of visibility. It is generally worn in environments where speeds do not exceed 25 mi/h (40.2 km/h) by parking lot attendants, roadside maintenance workers, delivery vehicle drivers, and warehouse workers.
- **Class 2.** This category includes the most popular style of safety vest. It is commonly worn in environments where traffic is moving in excess of 25 mi/h (40.2 km/h) by construction workers, utility workers, school crossing guards, and emergency responders. A fire-retardant Class 2 safety vest that meets the NFPA 191A standard is also available for fire fighters; it is constructed of treated fluorescent polyester and carries reflective striping.
- **Class 3.** This vest offers the highest level of visibility. It is worn in environments where the traffic is moving in excess of 55 mi/h (88.51 km/h) by roadway construction workers, utility workers, survey crews, and emergency response personnel.

Figure 8-14 Traffic incident.

The goals of traffic control are fourfold:
- To improve responder safety while working at the incident
- To keep the traffic flowing as smoothly as possible around the incident
- To prevent the occurrence of secondary accidents at the scene
- To prevent unnecessary use of the surrounding road system

Within 15 minutes of arriving on the scene of a traffic incident, the IC should estimate the magnitude of the incident, the expected length of the queue of backed-up motorists on the highway or roadway, and the duration of the incident. According to the *MUTCD*, traffic incidents may be classified into one of three general classes of duration, each of which presents its own unique hazards and traffic control needs Figure 8-14 ▶ :

- **Major traffic incidents** include fatal crashes involving multiple vehicles, hazardous materials incidents on the highway, and other disasters. They usually require closing all or part of the highway for a period exceeding 2 hours. When this type of incident occurs, fire fighters must request assistance from traffic engineering and law enforcement to divert traffic around and past the incident.
- **Intermediate traffic incidents** are less severe in nature and usually affect the lanes of travel for 30 minutes to 2 hours. Traffic control is required to divert moving traffic around and past the incident. The highway may need to be closed for a short period to allow fire fighters to accomplish their task. Law enforcement personnel usually handle traffic control needs.
- **Minor traffic incidents** may involve minor crashes and disabled vehicles. Lane closures are kept to a minimum and are less than 30 minutes in duration. Traffic control is needed only briefly, if at all. The fire fighters may handle any minor traffic control needs.

MUTCD defines a **traffic incident management area (TIMA)** as an area of highway where temporary traffic controls are imposed by authorized officials in response to a traffic incident, natural disaster, hazardous material spill, or other unplanned incident. This area is further subdivided into the following sections:

- The **advance warning area** is the section of highway where motorists are informed about the upcoming situation ahead. It may be identified by an emergency vehicle with its lights activated or by warning signs. On highways, the advance warning signs should be positioned farther ahead of the actual site because of the high speeds at which vehicles are traveling. When roadways are smaller and have lower speed limits, the distance can be shortened.
- The **transition area** is where the vehicle is redirected from its normal path and where lane changes and closures are made.
- The **activity area** is the section where the work activity takes place. It may be stationary or may move as work progresses.
- The **buffer space** is a lateral and/or longitudinal area that separates motorist flow from the work space or an unsafe area. This area might also provide some recovery space for an errant vehicle.
- The incident space is the area where the actual incident is located.
- The **traffic space** is the portion of the highway in which traffic is routed through the activity area.
- The **termination area** is the area where the normal flow of traffic resumes.

By defining these areas, the fire fighters gain a better understanding of where fire apparatus should be positioned at the scene. Communicating to incoming apparatus and describing their placement becomes easier when the scene is divided into separate areas.

> ### Safety Tip
>
> On August 5, 1999, two career fire fighters of the Midwest City Fire Department (MCFD) were struck by a motor vehicle on a wet and busy interstate. MCFD Ladder Company 2 and Squad 2 had responded to a single motor vehicle accident on the interstate. The ladder apparatus was positioned approximately 150′ (45.72 meters) behind Squad 2, near the median wall, with its emergency lights left on.
>
> A few minutes after arriving on scene, Ladder Company 2 was hit from behind by a passenger vehicle. Fire fighters began to attend to the injuries of this driver while other members began to flag traffic away. A fire fighter who was watching the oncoming traffic situation noticed a vehicle coming toward the rear of Ladder Company 2 and the fire fighters. Two warnings were yelled out over the radio. This car hit several fire fighters as they were attempting to flee its path. The vehicle collided with the median wall and wedged into the space between Ladder Company 2 and the wall. Several company members were able to avoid the impact of the vehicle, but two fire fighters and a civilian were not so fortunate. The impact knocked the three individuals approximately 47′ (14.32 m), killing one fire fighter and severely injuring the other fire fighter and the civilian. While the fire fighters were attending to the injured, yet another vehicle spun out of control on the interstate and struck the vehicle that had impacted the rear of Ladder Company 2.

■ Motor Vehicle Accidents

A <u>motor vehicle accident (MVA)</u> may involve one or more vehicles, either on or off the roadway. This type of emergency is a very common reason for calling out fire fighters. In metropolitan areas, the majority of these incidents tend to result in only very minor damage to both the vehicles and the passengers. By comparison, MVAs in rural areas and highways may have quite different outcomes; they are usually very serious and result in great damage to both the vehicles and the passengers.

As the fire apparatus approaches the accident scene, traffic is usually backed up behind the MVA. This makes the approach to the scene slower than normal and may be frustrating for the fire fighters, who are ready to go to work. As the driver/operator, you must not get impatient and allow your emotions to get the best of you. Proceed with caution and remain calm. In this situation, you must have a consistent approach to the incident site. Do not weave the fire apparatus in and out of traffic to gain access to the scene, as such maneuvers confuse other motorist as to what your intentions are. Keep the fire apparatus in the far left lane while approaching the scene. If necessary, the fire officer may use the public address system on the fire apparatus to direct the backed-up traffic to the far right and allow the fire apparatus to reach the scene.

Positioning at an Intersection

Motor vehicle accidents are more likely to occur at an intersection than anywhere else. These accidents usually involve more than one vehicle. A unique hazard that is present with an accident in an intersection is large groups of people attempting to help the accident victims. Traffic control should be the first priority once the fire apparatus arrives at the scene. If the moving traffic is not controlled, then the scene will not be safe for fire fighters. For most incidents, the fire apparatus itself can be positioned to shield the work area for fire fighters and to protect fire fighters from moving traffic.

Whenever possible, police should be called for assistance with traffic control. Police officers are specifically trained to carry out this type of operation. The initial fire apparatus must assess the parking needs of later-arriving units and specifically direct the parking and placement of these vehicles as they arrive to provide protective blocking of the scene. When parking the fire apparatus to protect the scene, be sure to protect the work area as well. This area must be protected so that victims can be extricated, treated, moved about the scene, and loaded into ambulances safely. Do not position the fire apparatus exhaust in the direction of the victims who are entrapped in motor vehicles.

At intersections or at sites where the incident is near the middle of the street, two or more sides of the incident may need to be protected. Fire apparatus should block all exposed sides. Where fire apparatus are limited in numbers, prioritize the blocking scheme from the most critical sides to the least critical. Once enough fire apparatus have blocked the scene, park or stage unneeded vehicles off the street whenever possible. When ambulances are positioned at a scene, always protect the victim loading areas.

> ### Driver/Operator Tip
>
> If the fire apparatus is equipped with a traffic control device, it may be tied into the emergency lighting system. In such a case, when the emergency lights are activated, so is the traffic control device. While the fire apparatus is positioned at an intersection for the response to an emergency and operating its emergency lights, this may pose a problem. The intersection lights would continue to cycle through at the request of the apparatus's traffic control device. To overcome this problem, the traffic control device is designed to shut off when the fire apparatus's parking brake is set. This system ensures that the fire apparatus can operate its emergency lights while parked and not disrupt the directional lights at an intersection.

Positioning on a Highway

Because speeds are higher, traffic volume is more significant, and civilian motorists have little opportunity to slow, stop, or change lanes, fire fighters must be constantly aware of moving vehicles on highways. Although at times the scene may seem safe, matters can change at a moment's notice.

Near Miss REPORT

Report Number: 09-0000403
Report Date: 04/10/2009 10:40

Synopsis: Assume downed lines are energized.

Event Description: Crews were dispatched for a personal injury collision with a report of one trapped or pinned. Crews arrived on the scene to find a passenger vehicle on its side that had left the roadway. It hit a pole snapping it in half, and came to rest approximately thirty feet off the road. Power lines were down in the area immediately adjacent to the extrication scene. We had one head trauma patient still in the vehicle.

The squad and engine crews took a position next to the scene with downed wires between the apparatus and the vehicle. At several points during the call, crews walked over the lines with equipment, thinking that the lines were not energized or were not high-power lines due to the coloring and positioning of the lines. When the utilities arrived to confirm power shut off, they suggested that the lines could not be assumed to be de-energized and that, while normally neutral/grounding lines, they could be or become highly energized at any time.

Lessons Learned: Never assume that downed lines are not energized or could not become energized during a call. Work around these lines at all times until utilities confirm they are rendered safe. Don't let tunnel vision of a "good call" or priority patient cause you to cut corners. Position apparatus proactively at the beginning of the call using the assumption that power lines are significant hazards. Reinforce the need for situational awareness of all members of crew and encourage even junior members to speak up if they see something that is unsafe. Stop operations and reposition if a hazard is recognized in the middle of the incident.

Demographics

Department type: Combination, Mostly paid
Job or rank: Battalion Chief/District Chief
Department shift: 24 hours on—48 hours off
Age: 25–33
Years of fire service experience: 11–13
Region: FEMA Region III
Service Area: Suburban

Event Information

Event type: Non-fire emergency event: auto extrication, technical rescue, emergency medical call, service calls, etc.
Event date and time: 03/27/2009 21:00
Hours into the shift: Volunteer
Event participation: Witnessed event but not directly involved in the event
Weather at time of event: Cloudy and Rain
Do you think this will happen again? Yes
What were the contributing factors?
- Situational Awareness
- Decision Making
- Human Error
- Communication
- Weather

What do you believe is the loss potential?
- Life-threatening injury

When approaching an emergency scene on a highway, identify a position that will allow fire fighters to work in a safe area. Sometimes this may involve disrupting the normal flow of traffic or blocking it off completely. The safety of the fire fighters should always be the first priority—not the continuous flow of traffic. When in doubt about the proper positioning of the fire apparatus, always err on the side of caution: Block the highway to provide a safe working area for fire fighters. For emergencies on a highway, continue to block the scene with the first-arriving fire apparatus to provide a safe working area. Other companies may then be used to provide additional blocking if needed. If possible use the largest, heaviest fire apparatus (usually ladder apparatus) as the first blockers.

Vehicle Fires

Fire fighters respond to more vehicle fires than they do structure fires. In fact, approximately 25 percent of all reported fires in the United States involve vehicles. These types of fires should not be taken lightly. Given the various amounts of plastics, foams, and synthetic materials that modern-day vehicles are constructed from, a vehicle may be consumed quickly by an intense, fast-moving fire. Most often, the vehicle is usually so severely damaged before the fire department arrives on scene that it is a total loss. The majority of vehicle fires result in a total loss.

When responding to a vehicle fire, do not position the fire apparatus where it will become an exposure hazard. Always position the fire apparatus at least 30′ (9.144 m) away from all involved vehicles Figure 8-15.

Try to position the fire apparatus in a location that is uphill and upwind of the burning vehicle. While this may not always be the most advantageous position on the scene, you do not want smoke or flammable liquids to compromise the safe operation of the fire apparatus. If you have to position it downhill from the burning vehicle, a dike should be created in front of the fire apparatus to pool any flammable liquids. Be aware that the brakes on the burning vehicle may be compromised by the fire conditions and the vehicle may roll downhill; do not place the fire apparatus in a position that might allow the burning vehicle to roll into it.

You should position the fire apparatus at a 45-degree angle to the burning vehicle to protect the area near the pump panel as well as the scene. Most fire apparatus have the fire pump mounted on the driver's side in the middle of the vehicle; this is the location where the driver stands while operating the fire pump. Be aware of where the fire fighters are deploying hose lines at the scene. While the other members of the crew are extinguishing the fire, the driver/operator is responsible for providing adequate water from the pump and ensuring that the crew remains safe. This effort may involve positioning traffic cones or warning devices to alert oncoming traffic to the emergency scene.

Railroads

When positioning the fire apparatus near a railroad track, try to place it on the same side as the incident. This placement will ensure that fire fighters do not have to cross the tracks, thereby risking injury or death from oncoming trains. Every railroad track should be considered active. If possible, contact dispatch and request that the railroad tracks be shut down while on-scene operations continue. **Never** park the fire apparatus on top of the railroad tracks.

Positioning at the Emergency Medical Scene

An emergency medical scene can be just as dangerous as the other types of incidents to which fire fighters respond. Usually, the danger is not associated with the emergency scene itself, but rather with the people involved. Every day fire fighters are surprised by a scene that becomes violent when initially it appeared safe. At a moment's notice, what looks like a routine call can turn into a deadly encounter. Thus the first priority when arriving at an EMS scene is to provide a protected environment for fire fighters to work in. If the fire fighters are not safe, then they cannot provide adequate care. Ideally, the driver/operator should position the fire apparatus either 100′ (30.48 m) before or after the address. This placement will allow the entire crew to size up the situation and recognize any potential hazards. Do not position the fire apparatus directly in front of the address, as such placement does not allow the fire fighters adequate time to identify or react to any potential hazards.

Some incidents have a greater potential for violence than others, including assaults, fights, and domestic disputes. When fire fighters are requested to respond to these types of calls, you should turn off the emergency lights and sirens a few blocks away from the physical address of the incident. Fire fighters should enter the scene on their own terms and not rush into an unknown situation. For such calls, you should drive slowly past the address and spot the fire apparatus at least 100′ (30.48 m) from the building/location.

At EMS scenes, the fire apparatus should be positioned for a quick exit. If necessary, turn it around. This may involve backing the fire apparatus into an alley or side street. NFPA 1002 requires that all driver/operators complete an exercise—called the alley dock exercise—that simulates the process of backing a fire apparatus into an alley or tight space. This exercise measures your ability to drive on a street past a simulated area, and then back up the fire apparatus into the dock provided.

During the alley dock exercise, a street may be simulated by arranging marker cones 40′ (12.19 m) from a boundary line. The marker cones should mark off an area 12′ (3.65 m) wide and

Figure 8-15 A vehicle fire.

20′ (6.096 m) long, indicating the "dock" the fire apparatus will back into. As part of the exercise, you will pass the marker cones with the dock on the left and then back up the fire apparatus, using a left turn into the dock. This exercise should then be completed with the dock on the right side of the fire apparatus. The minimum depth of the apparatus bay is determined by the length of the fire apparatus. During the entire alley dock exercise, the fire apparatus must remain within the marked boundary and move in a continuous motion, except when required to change direction of travel. A spotter is necessary for this exercise.

To perform the alley dock exercise, follow the steps in **Skill Drill 8-1** ▶ :

1. Position the rear of the fire apparatus past the dock's opening and at a 90-degree angle to the marker cones. (**Step 1**)
2. Ensure that a spotter is correctly positioned behind the fire apparatus. (**Step 2**)
3. Activate the emergency lights. (**Step 3**)
4. Roll down the windows. (**Step 4**)
5. Turn off any mounted stereo equipment. (**Step 5**)
6. Disengage the parking brake, if set. (**Step 6**)
7. Shift the transmission into reverse. (**Step 7**)
8. Proceed in a reverse mode and turn the fire apparatus to align it with the objective. (**Step 8**)
9. Continue backing the fire apparatus until it has reached the desired objective or the spotter signals "stop." (**Step 9**)

Special Emergency Scene Positioning

Some types of emergencies are rarely encountered by most fire fighters. Although these incidents occur with less frequency than others, the need for proper scene positioning is always paramount. Special emergency scenes may include anything from a building collapse to a hazardous materials incident. During these incidents, fire fighters should be very cautious and resist the urge to rush into the scene and mitigate the situation. Usually, these incidents unfold slowly at first, until enough information has been gathered to determine the appropriate course of action.

When responding to these incidents and preparing to position the apparatus, the driver/operator must consider the **control zones**—areas at an incident that are considered hot, warm, or cold, based on the severity of the incident—that surround the incident:

- **Hot zone:** the area for entry teams and rescue teams only. This zone immediately surrounds the dangers of the site (e.g., hazardous materials release) and is demarcated to protect personnel outside the zone.
- **Warm zone:** the area for properly trained and equipped personnel only. This zone is where personnel and equipment decontamination and hot zone support take place.
- **Cold zone:** the area for staging vehicles and equipment until requested by the IC. The command post is located in this zone. The public and the media should be clear of the cold zone at all times.

The following list identifies some ideas for proper apparatus positioning during these incidents:

- **Hazardous materials incident.** The first course of action at any hazardous materials incident is to isolate the area and prevent anyone from entering it. The first-arriving company may have to position its apparatus to block a highway to prevent anyone from entering the scene. The proper apparatus position for these incidents is uphill and upwind. The material should be identified; once the scene is deemed safe, and if needed, the apparatus may then be driven closer to the scene.
- **Building collapse.** During this type of incident, the primary danger to fire fighters and fire apparatus is secondary collapse. Position the apparatus out of the collapse zone. Heavy equipment such as bulldozers and cranes may be required for on-scene operations. Do not block this equipment from reaching the incident; always leave a path for its entrance and exit.
- **Trench collapse.** Once a trench fails, the probability of secondary collapse is quite high. For this reason, first-arriving units should be positioned no closer than 150′ from the trench. All other incoming nonessential apparatus should stage at least 200′ from the trench. Only equipment that is needed for a rescue should be brought any closer than the first-arriving units.
- **Terrorism.** These types of incidents may present as an explosion, a building collapse, release of radioactive material, or any other potential hazard. Terrorist incidents have the potential to injure and kill large numbers of people. During a possible terrorist incident, fire fighters should position the apparatus based on the demands of the emergency. Be aware of the potential for future hazards and the possible need for rapid escape, but position the apparatus to best accomplish the tasks at hand.

Skill Drill 8-1

Performing the Alley Dock Exercise

1. Position the rear of the fire apparatus past the dock's opening and at a 90-degree angle to the marker cones.

2. Ensure that a spotter is correctly positioned behind the fire apparatus.

3. Activate the emergency lights.

4. Roll down the windows.

5. Turn off any mounted stereo equipment.

6. Disengage the parking brake, if set.

7. Shift the transmission into reverse.

8. Proceed in a reverse mode and turn the fire apparatus to align it with the objective.

9. Continue backing the fire apparatus until it has reached the desired objective or the spotter signals "stop."

Wrap-Up

Chief Concepts

- No member of any fire department should be allowed to drive an emergency vehicle or fire apparatus until he or she has completed a training course approved by the fire department.
- Fire fighters should always be cautious. As the fire apparatus approaches the scene, the driver/operator should slow down, identify the correct address/location, and recognize any potential hazards.
- While approaching a fire scene, the driver/operator should attempt to view at least three sides of the structure.
- Operating at an emergency scene that is located either on or adjacent to a highway or an intersection is extremely dangerous.
- The first priority upon arrival at an EMS scene is to provide a protected environment for fire fighters to work in.
- When responding to a special emergency scene, the driver/operator must consider the control zones (areas at the incident that are considered hot, warm, or cold, based on the severity of the incident) that surround the incident.

Hot Terms

activity area Area of the incident scene where the work activity takes place; it may be stationary or may move as work progresses.

advance warning area The section of highway where drivers are informed about an upcoming situation ahead.

aerial device An aerial ladder, elevating platform, aerial ladder platform, or water tower that is designed to position personnel, handle materials, provide continuous egress, or discharge water. (NFPA 1901)

ambulance A vehicle designed, equipped, and operated for the treatment and transport of ill and injured persons. (NFPA 450)

auxiliary appliance A standpipe and/or sprinkler system.

brake fade Reduction in stopping power that can occur after repeated application of the brakes, especially in high-load or high-speed conditions.

braking distance The distance that the fire apparatus travels from the time the brakes are activated until the fire apparatus makes a complete stop.

buffer space Lateral and/or longitudinal area that separates traffic flow from a work space or an unsafe area; it might also provide some recovery space for an errant vehicle.

centrifugal force The outward force that is exerted away from the center of rotation or the tendency for objects to be pulled outward when rotating around a center.

Code 1 response Response in a fire apparatus in which no emergency lights or sirens are activated.

Code 2 response Response in a fire apparatus in which only the emergency lights are activated; no audible devices are activated.

Code 3 response Response in a fire apparatus in which both the emergency lights and the sirens are activated.

collapse zone An area encompassing a distance of 1½ times the height of a building, in which fire fighters and fire apparatus must not be located in case of a building collapse.

command vehicle A vehicle that the chief uses to respond to the fire scene.

control zones A series of areas at hazardous materials incidents that are designated based on safety concerns and the degree of hazard present.

corner safe areas Areas outside a building where two walls intersect; these areas are less likely to receive any damage during a building collapse.

critical speed Maximum speed that a fire apparatus can safely travel around a curve.

emitter A device that emits a visible flashing light at a specified frequency, thereby activating the receiver on a traffic signal.

intermediate traffic incident A traffic incident that affects the lanes of travel for 30 minutes to 2 hours.

Level I staging Initial staging of fire apparatus in which three or more units are dispatched to an emergency incident.

Level II staging Placement of all reserve resources in a central location until requested to the scene.

liquid surge The force imposed upon a fire apparatus by the contents of a partially filled water or foam concentrate tank when the vehicle is accelerated, decelerated, or turned. (NFPA 1002)

major traffic incident A traffic incident that involves a fatal crash, a multiple-vehicle incident, a hazardous materials incident on the highway, or other disaster.

minor traffic incident A traffic incident that involves a minor crash and/or disabled vehicles.

motor vehicle accident (MVA) An incident that involves one vehicle colliding with another vehicle or another object and that may result in injury, property damage, and possibly death.

parking brake The main brake that prevents a fire apparatus from moving even when it is turned off and there is no one operating it.

preincident plan A document developed by gathering general and detailed data that are used by responding personnel to determine the resources and actions necessary to mitigate anticipated emergencies at a specific facility. (NFPA 1620)

reaction distance The distance that the fire apparatus travels after the driver/operator recognizes the hazard, removes his or her foot from the accelerator, and applies the brakes.

receiver A device placed on or near a traffic signal to recognize a signal from the emitter on an emergency vehicle and preempt the normal cycle of the traffic light.

staging A specific function whereby resources are assembled in an area at or near the incident scene to await instructions or assignments. (NFPA 1561)

staging area A prearranged, strategically placed area, where support response personnel, vehicles, and other equipment can be held in an organized state of readiness for use during an emergency. (NFPA 424)

staging area manager The person responsible for maintaining the operations of the staging area.

termination area The area where the normal flow of traffic resumes after a traffic incident.

total stopping distance The distance that it takes for the driver/operator to recognize a hazard, process the need to stop the fire apparatus, apply the brakes, and then come to a complete stop.

traffic control The direction or management of vehicle traffic such that scene safety is maintained and rescue operations can proceed without interruption. (NFPA 1006)

traffic incident A natural disaster or other unplanned event that affects or impedes the normal flow of traffic.

traffic incident management area (TIMA) An area of highway where temporary traffic controls are imposed by authorized officials in response to an accident, natural disaster, hazardous materials spill, or other unplanned incident.

traffic signal preemption system A system that allows the normal operation of a traffic signal to be changed so as to assist emergency vehicles in responding to an emergency.

traffic space The portion of the highway where traffic is routed through the activity area of a traffic incident.

transition area The area where vehicles are redirected from their normal path and where lane changes and closures are made in a traffic incident.

Driver/Operator in Action

During a weekly training drill, you and your crew discuss some problems with the last few working fires to which your engine company has been called. Everyone is eager to learn and accept criticism from one another. An open discussion is held, and some interesting observations concerning fire apparatus placement are made by other members of the crew. How will you respond?

1. What is staging?
   A. A procedure to manage uncommitted resources at an emergency scene
   B. A way to determine which responders will arrive first at an incident
   C. A procedure to determine which tasks will be accomplished first at an incident
   D. A place for fire fighters to go when they need some rehabilitation

2. To make the scene safer for all involved, when approaching the incident, the driver/operator should do all of the following except
   A. Slow down.
   B. Identify the correct address/location.
   C. Give a size-up of the incident.
   D. Recognize any potential hazards.

3. Engines and _____ are the quintessential "bread and butter" fire apparatus of the fire service.
   A. ambulances
   B. rescue vehicles
   C. ladder apparatus
   D. command vehicles

4. Which of the following is *not* one of the four actions that all fire fighters can take to protect themselves and the other crew members while operating in traffic conditions?
   A. Never trust the traffic.
   B. Wear a high-visibility reflective vest.
   C. Engage in protective parking.
   D. Move quickly between moving vehicles.

Responding on the Fire Ground

CHAPTER 9

NFPA 1002 Standards

4.3.7* Operate all fixed systems and equipment on the vehicle not specifically addressed elsewhere in this standard, given systems and equipment, manufacturer's specifications and instructions, and departmental policies and procedures for the systems and equipment, so that each system or piece of equipment is operated in accordance with the applicable instructions and policies. [p. 220–250]

(A) Requisite Knowledge. Manufacturer's specifications and operating procedures, and policies and procedures of the jurisdiction. [p. 220–250]

(B) Requisite Skills. The ability to deploy, energize, and monitor the system or equipment and to recognize and correct system problems. [p. 237–250]

5.2 Operations. [p. 220–250].

5.2.1 Produce effective hand or master streams, given the sources specified in the following list, so that the pump is engaged, all pressure control and vehicle safety devices are set, the rated flow of the nozzle is achieved and maintained, and the apparatus is continuously monitored for potential problems:

(1) Internal tank
(2)* Pressurized source
(3) Static source
(4) Transfer from internal tank to external source
[p. 220–250]

(A) Requisite Knowledge. Hydraulic calculations for friction loss and flow using both written formulas and estimation methods, safe operation of the pump, problems related to small-diameter or dead-end mains, low-pressure and private water supply systems, hydrant coding systems, and reliability of static sources. [p. 232–250]

(B) Requisite Skills. The ability to position a fire department pumper to operate at a fire hydrant and at a static water source, power transfer from vehicle engine to pump, draft, operate pumper pressure control systems, operate the volume/pressure transfer valve (multistage pumps only), operate auxiliary cooling systems, make the transition between internal and external water sources, and assemble hose lines, nozzles, valves, and appliances. [p. 232–250]

5.2.4 Supply water to fire sprinkler and standpipe systems, given specific system information and a fire department pumper, so that water is supplied to the system at the correct volume and pressure. [p. 232–250]

(A) Requisite Knowledge. Calculation of pump discharge pressure; hose layouts; location of fire department connection; alternative supply procedures if fire department connection is not usable; operating principles of sprinkler systems as defined in NFPA 13, NFPA 13D, and NFPA 13R; fire department operations in sprinklered properties as defined in NFPA 13E; and operating principles of standpipe systems as defined in NFPA 14. [p. 232–250]

(B) Requisite Skills. The ability to position a fire department pumper to operate at a fire hydrant and at a static water source, power transfer from vehicle engine to pump, draft, operate pumper pressure control systems, operate the volume/pressure transfer valve (multistage pumps only), operate auxiliary cooling systems, make the transition between internal and external water sources, and assemble hose line, nozzles, valves, and appliances. [p. 232–250]

Additional NFPA Standards

NFPA 1961 *Standard on Fire Hose* (2007)
NFPA 1962 *Standard for the Inspection, Care, and Use of Fire Hose, Couplings, and Nozzles and the Service Testing of Fire Hose* (2008)
NFPA 1963 *Standard for Fire Hose Connections* (2009)
NFPA 1964 *Standard for Spray Nozzles* (2008)
NFPA 1965 *Standard for Fire Hose Appliances* (2009)

Knowledge Objectives

After studying this chapter, you will be able to:

- Describe securing a water source after arriving on scene.
- Describe the driver/operator's responsibility with proper hose layouts.
- Describe cab procedures when positioning the fire apparatus at the fire ground.
- Describe the driver/operator's responsibilities prior to exiting the cab of the fire apparatus.
- Describe the driver/operator's responsibilities after exiting the fire apparatus.
- Describe the driver/operator's responsibility to make connections to a fire department sprinkler and/or standpipe connection.
- Describe the driver/operator's role in troubleshooting problems on scene with the fire apparatus or its equipment.
- Describe the driver/operator's role in the safe operation of the pump.

Skill Objectives

After studying this chapter, you will be able to:

- Assemble a hose line from a hydrant to a pump.
- Inspect a solid-stream nozzle.
- Inspect a fog nozzle.
- Engage the fire pump.
- Hand-lay a supply line.
- Connect a hose to a fire department connection (FDC).
- Perform a changeover operation.
- Operate an auxiliary cooling system.
- Disengage the fire pump.

You Are the Driver/Operator

It's 11:00 P.M. and a heavy rain is falling. You are the driver/operator on Engine 2. A fire call comes in, and your crew responds. Your apparatus is the first to arrive at the scene of a large single-family structure with fire and smoke showing. The incident commander (IC) calls for a 1¾" (45 mm) preconnect with a fog nozzle to be deployed. The second-due engine is in the process of laying a supply line to your engine from a hydrant. The IC calls for the primary attack line to be charged and flowed at 150 GPM (570 L/min). Your apparatus carries 600 gallons (2270 L) of water. The attack crew checks the nozzle for proper operation, makes entry, and starts to flow water. Additional fire fighters now pull a second 1¾" (45 mm) line with a fog nozzle as a backup attack line and are looking back at you, calling for water. Your supply line just arrived.

1. What do you do first?
2. Which hose should you use?
3. Which nozzles should you employ?

Fire Hose, Appliances, and Nozzles Overview

Functions of Fire Hose

Fire hose is used for two main purposes: as supply hose and as attack hose.

Supply hose (also known as **supply lines**) Figure 9-1 is used to deliver water from a static source or from a fire hydrant to an attack pumper. The water can come directly from a hydrant, or it can come from another engine that is being used to provide a water supply for the attack pumper. Supply line sizes include 2½" (64 mm), 3" (76 mm), 4" (102 mm), 5" (127 mm), and 6" (152 mm) hoses. Supply hose is designed to carry larger volumes of water at lower pressures compared to attack lines.

Attack hose (also known as **attack lines**) Figure 9-2 is used to discharge water from an attack pumper onto the fire. Most attack lines carry water directly from the attack pumper to a nozzle that is used to direct the water onto the fire. In some cases, an attack line is attached to a deck gun, an aerial device, or some other type of master stream appliance. Attack lines usually operate at higher pressures than do supply lines. These lines can also be attached to the outlet of a standpipe system inside a building.

Sizes of Hose

Fire hose ranges in size from 1" (25 mm) to 6" (152 mm) in diameter Figure 9-3. The nominal hose size refers to the inside diameter of the hose when it is filled with water. Smaller-diameter hose is used as attack lines; larger-diameter hose is almost always used as supply lines. Medium-diameter hose can be used as either attack lines or supply lines.

Small-diameter hose (SDH) Figure 9-4 ranges in size from 1" (25 mm) to 2" (51 mm) in diameter. Many fire department vehicles are equipped with a reel of ¾" (19 mm) or 1" (25 mm) hard rubber hose called a **booster hose** (or **booster line**), which is used for extinguishing small outdoor fires.

The hose that is most commonly used to attack interior residential structure fires is either 1½" (38 mm) or 1¾" (45 mm) in diameter. These lines are usually preconnected directly to a discharge and ready for a quick deployment. Some fire departments also use 2" (51 mm) attack lines. Each length of attack hose is usually 50' (15 m).

Medium-diameter hose (MDH) Figure 9-5 has a diameter of 2½" (64 mm) or 3" (76 mm). Hose in this size range can be used as either supply lines or attack lines. Large handline nozzles are often used with 2½" (64 mm) hose to attack larger fires. When used as an attack line, the 3" (76 mm) size is more often used to deliver water to a **master stream device** or a fire department connection. These hose sizes also come in 50' (15 m) lengths.

Large-diameter hose (LDH) has a diameter of 3½" (89 mm) or more. Standard LDH sizes include 4" (102 mm) and 5" (127 mm) diameters, which are used as supply lines by many fire departments. The largest LDH size is 6" (152 mm) in diameter. Standard lengths of either 50' (15 m) or 100' (30 m) are available for LDH.

Figure 9-1 Supply hose brings the water from a water source to the pump.

Figure 9-3 Fire hose comes in a wide range of sizes for different uses and situations.

Figure 9-2 Attack hose is used to discharge water from the pump to the fire.

Figure 9-4 Small-diameter hose is used mostly for attacking vehicle fires and dumpster fires.

Fire hose is typically designed to be used as either attack hose or supply hose, although some hose is designed to serve both purposes. Attack hose must withstand higher pressures and is designed to be used in a fire environment where it can be subjected to high temperatures, sharp surfaces, abrasion, and other potentially damaging conditions. LDH supply hose is constructed to operate at lower pressures than attack hose and in less severe operating conditions; however, it must still be durable and resistant to external damage. Attack hose can be used as supply hose, but LDH supply hose must never be used as attack hose unless that usage is recommended by its manufacturer.

Attack hose must be tested annually at a pressure of at least 300 pounds per square inch (psi) (2070 kilopascal [kPa]) and is intended to be used at pressures up to 275 psi (1900 kPa). Supply hose must be tested annually at a pressure of at least 200 psi (1380 kPa) and is intended to be used at pressures up to 185 psi (1275 kPa). Most LDH is constructed for use as supply lines; however, some fire departments use special LDH that can withstand higher pressures. This hose may be used in water supply operations in case of high-rise fires or large industrial fires.

Figure 9-5 Medium-diameter hose has multiple uses, including serving as attack lines for commercial buildings, supplying master stream devices, and in some cases serving as supply hose for small fires.

Attack Hose

Attack hose is designed to be used for fire suppression. During this activity, it may be exposed to heat and flames, hot embers, broken glass, sharp objects, and many other potentially damaging conditions. For this reason, attack hose must be tough, yet flexible and light in weight.

Most fire departments use two sizes of hose as attack lines for fire suppression. The smaller size is usually either 1½″ (38 mm) or 1¾″ (45 mm) in diameter. Medium-diameter (2½″ [64 mm]) hose is most often used for heavy interior attack lines and for certain types of exterior attacks. These lines can be either double-jacket or rubber-covered construction.

1½″ and 1¾″ Attack Hose

Most fire departments use either 1½″ (38 mm) or 1¾″ (45 mm) hose as the primary attack line for most fires. Both sizes of hose use the same 1½″ (38 mm) couplings. This attack hose is the type used most often during basic fire training. Handlines of this size can usually be operated by one fire fighter, although having a second person on the line makes it much easier to advance and control the hose. This hose is often stored on fire apparatus as a preconnected attack line in lengths ranging from 150′ (46 m) to 250′ (76 m), ready for immediate use.

The primary difference between 1½″ (38 mm) and 1¾″ (45 mm) hose is the amount of water that can flow though the hose. Depending on the pressure in the hose and the type of nozzle used, a 1½″ (38 mm) hose can generally flow water at a rate between 60 GPM (225 L/min) and 125 GPM (475 L/min). An equivalent 1¾″ (45 mm) hose can flow between 120 GPM (455 L/min) and 180 GPM (680 L/min). This difference is important, because the fire-extinguishing capability is directly related to the amount of water that is applied to it. A 1¾″ (45 mm) hose can deliver much more water and is only slightly heavier and more difficult to advance than a 1½″ (38 mm) hose line. As a consequence, the 1¾″ (45 mm) hose is one of the most widely used primary attack lines in the fire service.

2½″ Attack Hose

A 2½″ (64 mm) hose is used as an attack line for fires that are too large to be controlled by a 1½″ (38 mm) or 1¾″ (45 mm) hose line. A 2½″ (64 mm) handline is generally considered to deliver a flow of approximately 250 GPM (945 L/min). It takes at least two fire fighters to safely control this size of handline owing to the weight of the hose and the water and the nozzle reaction force. A 50′ (15 m) length of dry 2½″ (64 mm) hose weighs about 30 lb (14 kg). When the hose is charged and filled with water, however, it may weigh as much as 200 lb (90 kg) per 100′ (30 m) length.

Higher flows—up to approximately 350 GPM (1325 L/min)—can be achieved with higher pressures and larger nozzles. It is difficult to operate a handline at these high flow rates, however. For this reason, such flows are more likely to be used to supply a master stream device.

Booster Hose

A booster hose is usually carried on a hose reel that holds 150′ (46 m) or 200′ (76 m) of rubber hose. Booster hose contains a steel wire that gives it a rigid shape. This rigid shape allows the hose to flow water without pulling all of the hose off the reel. Booster hose is light in weight and can be advanced quickly by one person.

The disadvantage of booster hose is its limited flow. The normal flow from a 1″ (25 mm) booster hose is in the range of 40 GPM (150 L/min) to 50 GPM (190 L/min), which is not an adequate flow for extinguishing structure fires. As a consequence, the use of booster hose is typically limited to small outdoor fires and trash dumpsters. This type of hose should not be used for structural firefighting.

Supply Hose

Supply hose is used to deliver water to an attack pumper from a pressurized source, which could be a hydrant or another engine working in a relay operation. Supply lines range from 2½″ (64 mm) to 6″ (152 mm) in diameter. The choice of diameter is based on the preferences and operating requirements of each fire department. It also depends on the amount of water needed to supply the attack pumper, the distance from the source to the attack pumper, and the pressure that is available at the source.

Fire department engines are normally loaded with at least one bed of hose that can be laid out as a supply line. When **threaded hose couplings** are used, this hose can be laid out from the hydrant to the fire (known as a forward lay) or from the fire to the hydrant (known as a reverse lay). Sometimes engines are loaded with two beds of hose so they can easily drop a supply line in either direction. If Storz-type couplings are used or the necessary adapters are provided, hose from the same bed can be laid in either direction.

When 2½″ (64 mm) hose is used as a supply line, it typically comprises the same type of hose that is used for attack lines. This size of hose has a limited flow capacity, but it can be effective at low to moderate flow rates and over short distances. Sometimes two parallel lines of 2½″ (64 mm) hose are used to provide a more effective water supply.

Large-diameter supply lines are much more efficient than 2½″ (64 mm) hose for moving larger volumes of water over longer distances. Given this fact, many fire departments use 4″ (102 mm) or 5″ (127 mm) hose as their standard supply line. A single 5″ supply line can deliver flows exceeding 1500 GPM (56,980 L/min) under some conditions. LDH is heavy and difficult to move after it has been charged with water, however. This hose comes in 50′ (46 m) and 100′ (76 m) lengths. A typical fire engine may carry anywhere from 600′ (183 m) to 1250′ (381 m) of supply hose.

Soft Suction Hose

A **soft suction hose** is a short section of LDH that is used to connect a fire department pumper directly to the large streamer outlet on a hydrant **Figure 9-6**. Use of this type of hose allows as much water as possible to flow from the hydrant to the pump through a single hose. A soft suction hose may have Storz connections on both ends or a female connection on each end. If it has two Storz connections, an adapter is required to connect the large-diameter inlet to the engine and the hydrant. Many Storz connections have locks to prevent the couplings from coming loose; as the driver/operator, you should always make sure the coupling is locked before charging the line. If the hose uses two female threaded connections, one end should match the local

Figure 9-6 Soft suction hose.

hydrant threads and the other end should match the threads on a large-diameter inlet to the engine. The couplings may have large handles to allow for quick tightening by hand. The hose can range from 4″ (102 mm) to 6″ (152 mm) in diameter and is usually between 10′ (3 m) and 25′ (8 m) in length.

Hard Suction Hose

A **hard suction hose** is a special type of supply hose that is used to draft water from a static source such as a river, lake, or portable drafting basin Figure 9-7. The water is drawn through this hose into the pump on a fire department engine or into a portable pump. This type of line is called a hard suction hose because it is designed to remain rigid and will not collapse when a vacuum is created in the hose to draft the water into the pump.

Hard suction hose normally comes in 10′ (3 m) or 20′ (6 m) sections. The diameter of this type of hose is based on the capacity of the pump and can be as large as 6″ (152 mm). Hard suction hose can be made from either rubber or plastic; however, the newer plastic versions are much lighter and more flexible than the older rubber hoses.

Figure 9-7 Hard suction hose.

Long handles are provided on the female couplings of hard suction hose to assist in tightening the hose. To draft water, it is essential to have an airtight connection at each coupling. Sometimes it may be necessary to gently tap these handles with a rubber mallet to tighten the hose or to disconnect it. Tapping these handles with anything metal, however, could cause damage to the handles or the coupling—so always use the right tool for the job.

How to Assemble a Hose Line

Follow the steps in **Skill Drill 9-1** to connect a Storz coupling soft suction hose from a hydrant to a pump:

1. Using a hydrant wrench, remove the large-diameter discharge cap on the hydrant. (**Step 1**)
2. Open the hydrant and flush the hydrant until any debris is cleared. (**Step 2**)
3. Unroll the 15′ (4.4 m) to 25′ (7.6 m) length of soft suction hose. (**Step 3**)
4. Connect the matching threaded side of the hydrant adapter to the hydrant. (**Step 4**)
5. Connect the hose's Storz coupling to the Storz end of the hydrant adapter. (**Step 5**)
6. Connect the Storz coupling at the other end of the hose to the large-diameter inlet on the pump, using another adapter if necessary. (**Step 6**)

■ Fire Hose Appliances

A **fire hose appliance** is any device used in conjunction with a fire hose for the purpose of delivering water. As a driver/operator, you should be familiar with wyes, water thiefs, Siamese connections, double-male and double-female adapters, reducers, hose clamps, hose jackets, and hose rollers. It is important for you to learn how to use the various hose appliances and tools required by your fire department; that is, you should understand the purpose of each device and be able to use each appliance correctly. Some hose appliances are used primarily with supply lines, whereas others are most often used with attack lines. Many hose appliances can be employed with both supply lines and attack lines.

Wyes

A **wye** is a device that splits one hose stream into two hose streams. The word "wye" refers to a Y-shaped part or object. When threaded couplings are used, a wye has one female connection and two male connections. Unless the wye is preassembled to the hose line, you should use the foot-tilt method to connect the wye to the hoseline. This method helps to prevent cross-threading the two couplings.

The wye that is most commonly used in the fire service has one 2½″ (64 mm) inlet and splits into two 1½″ (38 mm) outlets. It is used primarily for 1¾″ (45 mm) attack lines that stretch for long distances from the fire apparatus. A **gated wye** is equipped with two quarter-turn ball valves so that the flow of water to each of the split lines can be controlled independently Figure 9-8. A gated wye enables fire fighters to initially attach and operate one hose line, and then to add a second hose later if necessary. The use of a gated wye avoids the need to shut down the hose line supplying the wye while attaching this second

Skill Drill 9-1

Connecting a Storz Coupling Soft Suction Hose from a Hydrant to a Pump

1. Remove the large-diameter cap on the hydrant.

2. Flush the hydrant.

3. Unroll the hose.

4. Connect the hydrant adapter to the hydrant.

5. Connect the hose to the hydrant adapter.

6. Connect the hose to the large-diameter inlet on the pump.

Figure 9-8 A gated wye is used to split one 2½" (64 mm) hose line into two 1¾" (45 mm) or 1½" (38 mm) lines.

Figure 9-9 A water thief.

line. To use this appliance correctly, you must know where it is positioned relative to the hoseline. As the driver/operator, you are responsible for providing the correct water flow to the nozzles. Using this appliance may change the required pressure for the attack line.

Water Thief

A **water thief** is similar to a gated wye, but includes an additional 2½" (64 mm) outlet Figure 9-9 . It is used primarily on attack lines. With this appliance, the water that comes from a single 2½" (64 mm) inlet can be directed to two 1½" (38 mm) outlets and one 2½" (64 mm) outlet. Under most conditions, it will not be possible to supply all three outlets at the same time because the capacity of the supply hose is limited. You should use the foot-tilt method to assemble the water thief to the 2½" (64 mm) hose line and then connect the other lines to the appliance as needed. When using different-size hoses that require different operating pressures, you may adjust the pressure by opening or closing the valves.

A water thief can be placed near the entrance to a building to provide the water for interior attack lines. One or two 1½" (38 mm) attack lines can be used in such cases. If necessary, they can then be shut down and a 2½" (64 mm) line substituted for them. Sometimes the 2½" (64 mm) line is used to knock down a fire, while the two 1½" (38 mm) lines are used during overhaul.

Siamese Connection

A **Siamese connection** is a hose appliance that combines two hose lines into one. The most commonly encountered type of Siamese connection combines two 2½" (64 mm) hose lines into a single 2½" (64 mm) hose line Figure 9-10 . This scheme increases the flow of water on the outlet side of the Siamese connection. A Siamese connection that is used with threaded couplings has two female connections on the inlets and one male connection on the outlet. It can be used with supply lines and with some attack lines. For example, you may connect a Siamese appliance to the 2½" (64 mm) inlet on the engine's pump, which allows for more water and pressure to be delivered from the pump. When making this type of connection, hold the Siamese connection level with

Figure 9-10 A typical Siamese connection has two female inlets and a single outlet.

the inlet and make sure you do not cross-thread the couplings. Once the Siamese connection is attached to the pump, connect first one hose line and then the other. Most Siamese appliances contain a built-in clapper valve to allow for one hose line to be charged before the other one is connected.

A Siamese connection is sometimes used on an engine inlet to allow water to be received from two different supply lines. This type of connection is also used to supply master stream devices and ladder pipes. Siamese connections are commonly installed on the fire department connections that are used to supply water to standpipe and sprinkler systems in buildings. Fire fighters should be familiar with their department's policies on the correct procedures for supplying fire department Siamese connections.

Adapters

Adapters are used for connecting hose couplings that have the same diameter but dissimilar threads. Dissimilar threads might be encountered when different fire departments are working together or in industrial settings where the hose threads of the building's equipment do not match the threads of the municipal

Figure 9-11 Double-male and double-female adapters are used to join two couplings of the same sex.

Figure 9-12 A reducer is used to connect a smaller-diameter hose line to the end of a larger-diameter line.

fire department. Adapters are also used to connect threaded couplings to Storz-type couplings. They are useful for both supply lines and attack lines.

Adapters can also be used when it is necessary to connect two female couplings or two male couplings. A **double-female adapter** is used to join two male hose couplings. A **double-male adapter** is used to join two female hose couplings. Double-male and double-female adapters are often used when performing a reverse hose lay Figure 9-11. With the reverse lay, the fire apparatus stops at the incident site and deploys hose; it then drives to the water source. If the engine company has its hose set up for a forward hose lay (water source to the fire scene), the first coupling to come off will be a female. The fire fighters may then connect a double-male adapter to the hose, followed by connection of a nozzle to the double-male adapter. After the engine company is positioned at the water source and a sufficient amount of hose is pulled, you would first connect a double-female adapter to the end of the hose and then connect it to a discharge on the pump.

Reducers

A **reducer** is used to attach a smaller-diameter hose to a larger-diameter hose Figure 9-12. Usually the larger end has a female connection and the smaller end has a male connection. One type of reducer is used to attach 2½″ (64 mm) couplings to 1½″ (38 mm) couplings. Many 2½″ (64 mm) nozzles are constructed with a built-in reducer, so that a 1½″ (38 mm) line can be attached for overhaul. For example, after the main fire is knocked down, fire fighters operating a 2½″ (64 mm) attack line would shut off the water flow at the nozzle's bale (the handle that controls the quarter-turn valve) and remove the tips on the nozzle, exposing the 1½″ (38 mm) male threads. The driver/operator would then bring them a smaller-diameter hose line with 1½″ (38 mm) couplings to the end of the hose line and attach the smaller hose with a smaller nozzle to continue fire suppression activities. Use of a smaller hose allows the crew to move around more easily and requires less personnel to operate the attack line. Reducers may also be used to attach a 2½″ (64 mm) supply line to a larger suction inlet on a fire pumper.

Hose Jacket

A **hose jacket** (also called a burst hose jacket) is a device that is placed over a section of hose to stop a leak Figure 9-13. Of course, the best way to handle a leak in a section of hose is to replace the defective section of hose. A hose jacket can provide a temporary fix until the section of hose can be replaced. This device should be used only in cases where it is not possible to quickly replace the leaking section of hose. It can be used for both supply lines and attack lines. As the driver/operator, you will often be in the best position to deploy a hose jacket if necessary. The hose jacket is hinged on one side and has a lock on the opposite side. Two thick rubber gaskets at each end are provided to trap the water inside the appliance. To deploy the hose jacket, you place the section of hose with the leak between the two rubber gaskets and securely lock the device in position.

The hose jacket consists of a split metal cylinder that fits tightly over the outside of a hose line. This cylinder is hinged on one side to allow it to be positioned over the leak. A fastener is then used to clamp the cylinder tightly around the hose.

Figure 9-13 A hose jacket is used to repair a leaking hose line.

Hose Clamp

A <u>hose clamp</u> is used to temporarily stop the flow of water in a hose line. Hose clamps are often applied to supply lines so as to allow a hydrant to be opened before the line is hooked up to the intake of the attack pumper. In such a case, the fire fighter at the hydrant does not have to wait for the pump operator to connect the supply line to the pump intake before opening the hydrant. At the fire scene, one of the fire fighters or the driver/operator should immediately place the hose clamp on the dry supply line. Once you pull enough supply hose from the hose bed and connect it to the intake, the clamp can be released. Ensure that the clamp is opened slowly so it does not injure you or cause water hammer. A hose clamp can also be used to stop the flow in a line if a hose ruptures or if an attack line needs to be connected to a different appliance **Figure 9-14**.

Figure 9-14 A hose clamp is used to temporarily interrupt the flow of water in a hose line.

Valves

Valves are used to control the flow of water in a pipe or hose line. Several types of valves are used on fire hydrants, fire apparatus, standpipe and sprinkler systems, and attack hose lines. The important thing to remember when opening and closing any valve or nozzle is to do it s-l-o-w-l-y so as to prevent water hammer.

The following types of valves are commonly encountered on the fire ground:

- <u>Ball valves</u>: Ball valves are used on nozzles, gated wyes, and engine discharge gates. These valves consist of a ball with a hole in the middle. When the hole is lined up with the inlet and the outlet, water flows through it. As the ball is rotated, the flow of water is gradually reduced until it is shut off completely **Figure 9-15A**. Ball valves are the most common discharge valves found on fire pumps.
- <u>Gate valves</u>: Gate valves are found on hydrants and on sprinkler systems. With this type of valve, rotating a spindle causes a gate to move slowly across the opening for the water flow. The spindle is rotated by turning it with a wrench or a wheel-type handle **Figure 9-15B**.
- <u>Butterfly valves</u>: Butterfly valves are often found on the large pump intake connections where a hard suction hose or soft suction hose is connected. They are opened or closed by rotating a handle for one-quarter turn **Figure 9-15C**.
- <u>Four-way hydrant valves</u>: This appliance, which is attached to a fire hydrant, is used in conjunction with a pumper to increase water pressure in relaying operations. The use of a four-way hydrant valve is covered in detail later in this chapter.

A.

B.

C.

Figure 9-15 **A.** Ball valve. **B.** Gate valve. **C.** Butterfly valve.

Nozzles

Nozzles are attached to the discharge end of attack lines to give fire streams shape and direction. Without a nozzle, the water discharged from the end of a hose would reach only a short distance. Nozzles are used on all sizes of handlines as well as on master stream devices.

Nozzles can be classified into three groups:
- Low-volume nozzles flow 40 GPM (150 L/min) or less. They are primarily used for booster hoses; their use is limited to small outside fires.
- Handline nozzles are used on hose lines ranging from 1½" (38 mm) to 2½" (64 mm) in diameter. Handline streams usually flow water at a rate between 60 GPM (225 L/min) and 350 GPM (1325 L/min).
- Master stream nozzles are used on deck guns, portable monitors, and ladder pipes that flow more than 350 GPM (1325 L/min).

Low-volume and handline nozzles incorporate a shut-off valve that is used to control the flow of water. In contrast, the control valve for a master stream is usually separate from the nozzle itself. All nozzles have some type of device or mechanism to direct the water stream into a certain shape. Some nozzles also incorporate a mechanism that can automatically adjust the flow based on the water's volume and pressure.

Nozzle Shut-Offs

The nozzle shut-off enables the fire fighter at the nozzle to start or stop the flow of water. The most commonly encountered nozzle shut-off mechanism is a quarter-turn valve. The handle that controls this valve is called a bale. Some nozzles incorporate a rotary control valve that is operated by rotating the nozzle in one direction to open it and in the opposite direction to shut off the flow of water.

Two types of nozzles are manufactured for the fire service: smooth-bore nozzles and fog-stream nozzles. Smooth-bore nozzles produce a solid column of water, whereas fog-stream nozzles separate the water into droplets. With a fog nozzle, the fire fighter can change the size of the water droplets and the discharge pattern by adjusting the nozzle setting. Nozzles must have an adequate volume of water and an adequate pressure to produce a good fire stream. The volume and pressure requirements vary according to the type and size of nozzle.

Smooth-Bore Nozzles

The simplest smooth-bore nozzle consists of a shut-off valve and a smooth-bore tip that gradually decreases the diameter of the stream to a size smaller than the hose diameter **Figure 9-16**. Smooth-bore nozzles are manufactured to fit both handlines and master stream devices. Those used for master streams and ladder pipes often consist of a set of stacked tips, where each successive tip in the stack has an increasingly smaller-diameter opening. Tips can be quickly added or removed to provide the desired stream size, which allows different sizes of streams to be produced under different conditions.

A smooth-bore nozzle has several advantages over a fog-stream nozzle. For example, this type of nozzle has a longer reach than a combination fog nozzle operating at a straight stream setting. In addition, a smooth-bore nozzle is capable of

Figure 9-16 Smooth-bore nozzle.

deeper penetration into burning materials, resulting in quicker fire knockdown and extinguishment. Smooth-bore nozzles also operate at lower pressures than many adjustable-stream nozzles: Most smooth-bore nozzles are designed to operate at 50 psi (345 kPa), whereas many adjustable-stream nozzles generally require pressures of 75 psi (520 kPa) to 100 psi (690 kPa). Lower nozzle pressures result in less nozzle reaction, which makes it easier for a fire fighter to handle the nozzle.

A solid stream extinguishes a fire with less air movement and less disturbance of the thermal layering than does a fog stream; this, in turn, renders the heat conditions less intense for fire fighters during an interior attack. It is also easier for the hose operator to see the pathway of a solid stream than a fog stream.

There are also some disadvantages associated with smooth-bore nozzles. Specifically, the streams from these nozzles do not absorb heat as readily as fog streams and are not as effective for hydraulic ventilation. A fire fighter cannot change the setting of a smooth-bore nozzle to produce a fog pattern; in contrast, a fog nozzle can be set to produce a straight stream.

To inspect a solid-stream nozzle, follow the steps in **Skill Drill 9-2**:

1. Inspect the nozzle for any damage. Remove the nozzle from the hose, and look at the exterior as well as the inside for any dents, broken parts, or missing pieces. (**Step 1**)
2. Ensure that the gasket located in the female swivel is in place and in good condition. (**Step 2**)

Chapter 9 Responding on the Fire Ground

Skill Drill 9-2

Inspecting a Solid-Stream Nozzle

5.2.4

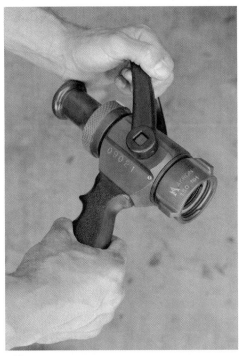
(1) Inspect the nozzle for damage.

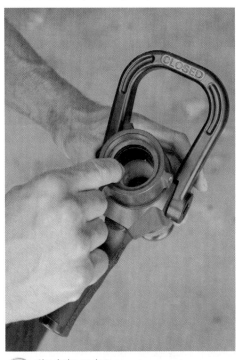
(2) Check the gasket.

(3) Inspect the nozzle tips.

(4) Operate the shut-off valve.

3. Ensure that the nozzle tips are properly labeled with their size and are attached. (**Step 3**)
4. Operate the shut-off valve from the closed position to the open position. The valve should not stick or be difficult to operate. (**Step 4**)

Fog-Stream Nozzles

Fog-stream nozzles produce fine droplets of water **Figure 9-17**. The advantage of creating these droplets of water is that they absorb heat much more quickly and efficiently than does a solid column of water. This characteristic is important when immediate reduction of room temperature is needed to avoid a flashover. Discharging 1 gallon (3.8 L) of water in 100 ft^3 (2.8 m^3) of involved interior space will produce enough steam to extinguish a fire in 30 seconds. Fog nozzles can produce a variety of stream patterns, ranging from a straight stream, to a narrow fog cone of less than 45 degrees, to a wide-angle fog pattern that is close to 90 degrees.

The straight streams produced by fog nozzles have openings in the center; in other words, a fog nozzle cannot produce a solid stream. The straight stream from a fog-stream nozzle will break up more quickly and will not have the reach of a solid stream delivered by a smooth-bore nozzle. In addition, this straight stream will be affected more dramatically by wind than will a solid stream.

The use of fog-stream nozzles offers several advantages over the use of smooth-bore nozzles, however. First, fog nozzles can be used to produce a variety of stream patterns by rotating the tip of the nozzle. In addition, fog streams are effective at absorbing heat and can be used to create a water curtain to protect fire fighters from extreme heat.

Fog nozzles move large volumes of air along with the water, which can be an advantage or a disadvantage, depending on the situation. For example, a fog stream can be used to exhaust smoke and gases through hydraulic ventilation. Unfortunately, this air movement can also result in sudden heat inversion in a room, which then pushes hot steam and gases down onto the fire fighters. If used incorrectly, a fog pattern can push a fire into unaffected areas of a building.

To produce an effective stream, nozzles must be operated at the pressure recommended by the manufacturer. For many years, the standard operating pressure for fog-stream nozzles was 100 psi (690 kPa). In recent years, some manufacturers have produced low-pressure nozzles that are designed to operate at 50 psi (345 kPa) or 75 psi (520 kPa). The advantage of low-pressure nozzles is that they produce less reaction force, which makes them easier to control and advance. Lower nozzle pressure also decreases the risk that the nozzle will get out of control.

Three types of fog-stream nozzles are available, with the difference between the types being related to the water delivery capability:

- A <u>fixed-gallonage fog nozzle</u> delivers a preset flow at the rated discharge pressure. The nozzle could be designed to flow 30, 60, or 100 GPM (115, 225, or 380 L/min).
- An <u>adjustable-gallonage fog nozzle</u> allows the operator to select a desired flow from several settings by rotating a selector bezel to adjust the size of the opening. For example, a nozzle could have the options of flowing 60, 95, or 125 GPM (225, 360, or 475 L/min). Once a setting is chosen, the nozzle will deliver the rated flow only as long as the rated pressure is provided at the nozzle.
- An <u>automatic-adjusting fog nozzle</u> can deliver a wide range of flows. As the pressure at the nozzle increases or decreases, its internal spring-loaded piston moves in or out to adjust the size of the opening. The amount of water flowing through the nozzle is adjusted to maintain the rated pressure and produce a good stream. A typical automatic nozzle could have an operating range of 90 to 225 GPM (340 to 850 L/min) while maintaining a discharge pressure of 100 psi (690 kPa).

To inspect a fog-stream nozzle, follow the steps in **Skill Drill 9-3**:

1. Inspect the nozzle for any damage. Remove the nozzle from the hose, and look at the exterior as well as the inside for any dents, broken parts, or missing pieces. (**Step 1**)
2. Ensure that the gasket located in the female swivel is in place and in good condition. (**Step 2**)
3. Ensure that the nozzle tip is operating properly. Rotate the nozzle tip through all positions several times; it should not bind or stick. (**Step 3**)
4. Operate the shut-off valve from the closed position to the open position. The valve should not stick or be difficult to operate. (**Step 4**)

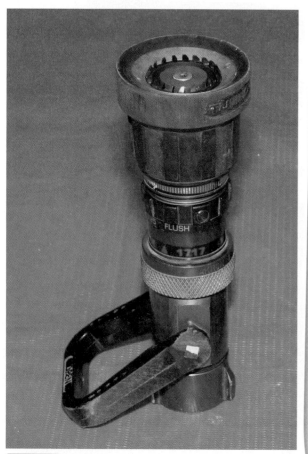

Figure 9-17 Fog-stream nozzle.

Skill Drill 9-3

Inspecting a Fog-Stream Nozzle

5.2.1

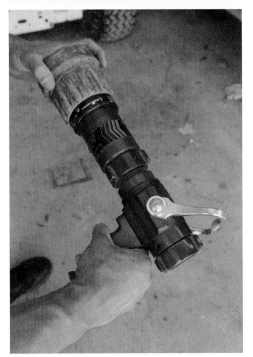
1. Inspect the nozzle for damage.

2. Check the gasket.

3. Inspect the nozzle tip.

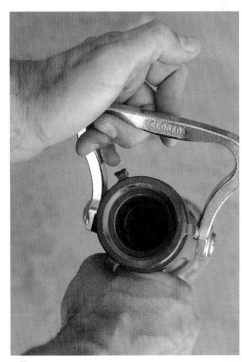
4. Operate the shut-off valve.

Other Types of Nozzles

Several other types of nozzles are used for special purposes. **Piercing nozzles**, for example, are used to make a hole in automobile sheet metal, aircraft, or building walls or ceilings, so as to extinguish fires behind these surfaces **Figure 9-18**.

Cellar nozzles and **Bresnan distributor nozzles** are used to fight fires in cellars and other inaccessible places such as attics and cocklofts **Figure 9-19**. These nozzles discharge water in a wide circular pattern as the nozzle is lowered vertically through a hole into the cellar. They work much like a large sprinkler head.

Water curtain nozzles deliver a flat screen of water that then forms a protective sheet ("curtain") of water on the surface of an exposed building **Figure 9-20**. The water curtains must be directed onto the exposed building because radiant heat can pass through the water curtain.

If your fire department has other types of specialty nozzles, you need to become proficient in their use and operation. Nozzles that are used for applying foam are described later in this chapter.

Nozzle Maintenance and Inspection

Nozzles should be inspected on a regular basis, along with all of the equipment on every fire department vehicle. In particular, nozzles should be checked after each use at an incident before they are placed back on the apparatus. They should be kept clean and clear of debris. Debris inside the nozzle will affect the performance of the nozzle, including possibly reducing flow. Dirt and grit can also interfere with the valve operation and prevent the nozzle from opening and closing fully. Applying a light grease to the valve ball will keep it operating smoothly.

On fog nozzles, inspect the fingers on the face of the nozzle. Make sure all fingers are present and that the finger ring can spin freely. Any missing fingers or failure of the ring to spin will drastically affect the fog pattern. Any problems noted should be referred to a competent technician for repair.

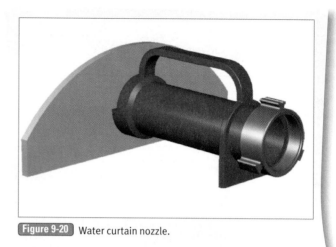

Figure 9-20 Water curtain nozzle.

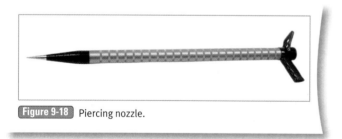

Figure 9-18 Piercing nozzle.

A. B.

Figure 9-19 A. Cellar nozzle. B. Bresnan distributor nozzle.

Fire Hose Evolutions

It is the driver/operator's responsibility to secure a water source upon arrival on the scene. Fire hose evolutions are critical in the driver/operator's success. Fire hose evolutions are standard methods of working with fire hose to accomplish different objectives in a variety of situations. Most fire departments set up their equipment and conduct regular training so that fire fighters will be prepared to perform a set of standard hose evolutions. As part of hose evolutions, specific actions are assigned to specific members of a crew, depending on their riding positions on the fire apparatus. Every fire fighter should know how to perform all of the standard evolutions quickly and proficiently; that is, when an officer calls for a particular evolution to be performed, each crew member should know exactly what to do.

As the driver/operator, you should know which pressures the lines need to be supplied at and should ensure that they are properly flaked out before they are charged with water pressure. Consequently, you need to not only know how to deploy the hose lines, but also which size, length, and nozzle configurations are required. You will need all of this information to supply the attack line with the correct water flow for fire extinguishment.

Hose evolutions are divided into supply line operations and attack line operations. Supply line operations involve laying hose lines and making connections between a water supply source and an attack pumper. Attack line operations involve advancing hose lines from an attack pumper to apply water onto the fire.

Supply Line Evolutions

The objective of laying a supply line is to deliver water from a hydrant or an alternative water source to an **attack pumper**. In most cases, this operation involves laying a hose line with a moving vehicle or dropping a continuous line of hose out of a bed as the fire apparatus moves forward. It can be done using either a forward lay or a reverse lay. A **forward lay** starts at the hydrant and proceeds toward the fire; the hose is laid in the same direction as the water flows—from the hydrant to the fire. A **reverse lay** involves laying the hose from the fire to the hydrant; the hose is laid in the opposite direction to the water flow. Each fire department will determine its own preferred methods and procedures for supply line operations based on available apparatus, water supply, and regional considerations.

Forward Hose Lay

The forward hose lay is most often used by the first-arriving engine company at the scene of a fire Figure 9-21 ▶. While en route to the fire scene, the driver/operator stops the engine at a hydrant close to the fire scene. A fire fighter dismounts the fire apparatus and either wraps the hose around the hydrant or places a strap around it to secure the hose in place. When this step is complete, the fire fighter signals the driver/operator to continue on to the fire by stating, "Driver, go." You then proceed to the fire scene no faster than 15 mi/h (24.14 km/h). At the fire scene, you disconnect the supply line from the hose bed, connect it to the pump, and call for the supply line to be charged with water. The forward hose lay allows the engine company to establish a water supply without assistance from an additional company. It also places the attack pumper close to the fire, allowing access to additional hose, tools, and equipment that are carried on the fire apparatus.

Figure 9-21 A forward hose lay is made from the hydrant to the fire.

Safety Tip

When performing a forward lay, the fire fighter who is connecting the hose to the supply hydrant must not stand between the hose and the hydrant. When the fire apparatus starts to move away, the hose could become tangled and suddenly be pulled taut—and anyone standing between the hose and the hydrant could be seriously injured. As the driver/operator, you should not move the apparatus unless you are sure all fire fighters are clear of the fire apparatus and in a safe position. Keep your window rolled down so you can effectively communicate with the fire fighter pulling the hose.

A forward hose lay can be performed using MDH (2½" [64 mm] or 3" [76 mm]) or LDH (3½" [89 mm] and larger). The larger the diameter of the hose, the more water that can be delivered to the attack pumper through a single supply line. When MDH is used and the beds are arranged to lay dual lines, a company can lay two parallel lines from the hydrant to the fire.

If the fire hydrant is close to the fire, it may supply a sufficient quantity of water from the hydrant alone (rather than using an interim supply engine to boost the flow). A 5" (127 mm) hose can supply 700 GPM (2650 L/min) over a distance of 500' (152 m) and lose only about 20–25 psi (138–172 kPa) of pressure due to friction loss.

Four-Way Hydrant Valve

In situations where long supply lines are needed or when MDH is used, it is often necessary to place a supply engine at the hydrant. The supply engine pumps water through the supply line so as to increase the flow to the attack pumper. Some departments use a four-way hydrant valve to connect the supply line to the hydrant, thereby ensuring that the supply line can be charged with water immediately, yet still allowing for a supply engine to connect to the line later.

When a four-way valve is placed on the hydrant, the water flows initially from the hydrant through the valve to the supply line, which delivers the water to the attack pumper Figure 9-22 ▼. The second engine can then hook up to the four-way valve and redirect the flow by changing the position of the valve. At this point, the water flows from the hydrant to the supply engine. The supply engine can then increase the pressure and discharge the water into the supply line, boosting the flow of water to the attack pumper. This operation can be accomplished without uncoupling any lines or interrupting the flow.

Figure 9-22 The four-way valve.

Safety Tip

When the fire fighter is connecting a supply line to a hydrant, he or she should wait to get the appropriate signal from the driver/operator before attempting to charge the line. If the hydrant is opened prematurely, the hose bed could become charged with water or a loose hose line could discharge water at the fire scene. Either situation will disrupt the firefighting operation and could cause serious injuries. Make sure that you know your department's signal to charge a hose line, and do not become so excited or rushed that you make a mistake. Confirm that all of the hose connections are in place before calling for water from the hydrant.

Driver/Operator Tip

When laying out supply hose with threaded couplings, you may find that the wrong end of the hose is on top of the hose bed. *Double-male connectors* allow you to attach a male coupling to a male coupling. *Double-female connectors* allow you to attach a female coupling to a female coupling. A set of adapters (one double-male and one double-female) should be easily accessible for these situations. Some fire departments place a set of adapters on the end of the supply hose for precisely this purpose.

As the driver/operator, you should know the contents of your fire apparatus as well as — if not better than — the rest of the crew. Think about potential situations that might require the use of the adapters before you have to use them on a fire scene. For example, if your fire apparatus is set up for a forward lay and it is used to deploy hose in a reverse lay at a fire, a double-male adapter will be needed at the nozzle and a double-female adapter will be needed at the pump discharge.

Reverse Hose Lay

The reverse hose lay is the opposite of the forward lay **Figure 9-23**. In the reverse lay, the hose is laid out from the fire to the hydrant, in the direction opposite to the flow of the water. This evolution can be used when the attack pumper arrives at the fire scene without a supply line. It is also used to supply the attack pumper directly through the hydrant.

The reverse hose lay may be a standard tactic in areas where sufficient hydrants are available and additional companies that can assist in establishing a water supply will arrive quickly. In this scenario, one company is assigned to lay a supply line from the attack pumper to a hydrant. The first engine arrives on the scene of a fire and begins a fire attack. The second engine arrives at the scene and drops off the fire fighters and the fire officer. At this point, the second engine deploys its LDH supply line to the first engine, where it is connected to that engine's intake. The second engine then proceeds to a water source, laying LDH as it drives

no faster than 15 mi/h (24.14 km/h). When this engine arrives at the water source, the driver/operator of the second engine will be operating alone. He or she secures a water supply, connects the supply line to a LDH discharge, and pumps water to the first engine through the supply line. The attack pumper focuses on immediately attacking the fire using water from the onboard tank. The supply engine stops close to the attack pumper, and hose is pulled from the bed of the supply engine to an intake on the attack pumper. The supply engine then drives to the hydrant (or alternative water source) and pumps water back to

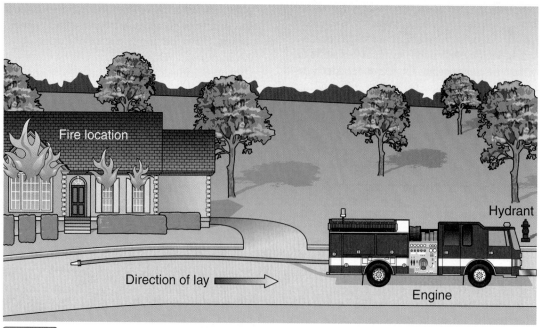

Figure 9-23 A reverse hose lay is made from the fire to a fire hydrant.

Driver/Operator Tip

A <u>split hose bed</u> is a hose bed that is divided into two or more sections. This division is made for several purposes:

- One compartment in a split hose bed can be loaded for a forward lay (female coupling out), and the other side can be loaded for a reverse lay (male coupling out). This arrangement allows a line to be laid in either direction without the use of adapters.
- Two parallel hose lines can be laid at the same time (called "laying dual lines"). Dual lines are beneficial if the situation requires more water than one hose line can supply.
- The split beds can be used to store hoses of different sizes. For example, one side of the hose bed could be loaded with 2½" (64 mm) hose that can be used as a supply line or as an attack line; the other side of the hose bed could be loaded with 5" (127 mm) hose for use as a supply line. This setup enables the use of the most appropriate-size hose for a given situation.
- All of the hose from both sides of the hose bed can be laid out as a single hose line. This is done by coupling the end of the hose in one bed to the beginning of the hose in the other bed.

In a variation of the split hose bed known as a combination load, the last coupling in one bed is connected to the first coupling in the other bed. When one long line is needed, all of the hose plays out of one bed first, and then the hose continues to play out from the second bed. To lay dual lines, the connection between the two hose beds is uncoupled, and the two hose lines can play out of both beds simultaneously. When the two sides of a split bed are loaded with the hose arranged in opposite directions, either a double-female or double-male adapter is used to make the connection between the two hose beds.

---

the attack pumper. Usually the supply engine parks in such a way that hose can be pulled from the supply engine to the inlet to the attack pumper.

Split Hose Lay

A <u>split hose lay</u> (also called an alley lay) is performed by two engine companies in situations where hose must be laid in two different directions to establish a water supply **Figure 9-24**. This evolution could be used when the attack pumper must approach a fire either along a dead-end street with no hydrant or down a long driveway.

To perform a split hose lay, the attack pumper drops the end of its supply hose at the corner of the street and performs a forward lay toward the fire. Normally, the fire fighter would get out of the fire apparatus and secure the supply hose to the hydrant. In this case, however, there is no hydrant in the immediate area. Instead, the fire fighter has to secure the end of the supply line to a fixed object such as a street sign or tree to keep it from dragging behind the attack pumper as the apparatus drives away. When the supply engine gets to the intersection where the end of the attack pumper's supply hose is anchored, it will stop and pull off enough hose to connect to the end of this supply line,

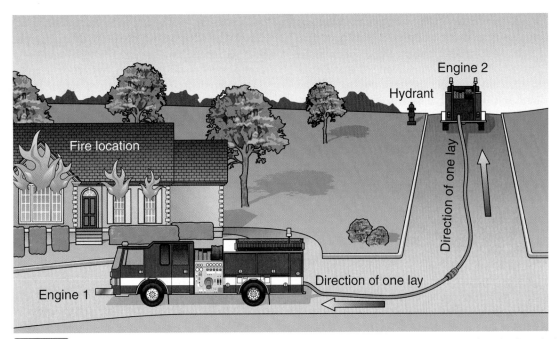

Figure 9-24 Engine 1 performs a forward hose lay from the corner to the fire. Engine 2 performs a reverse hose lay from the hose at the corner to a fire hydrant.

and then perform a reverse lay to the hydrant or water source. When the two lines are connected together, the supply engine can provide water to the attack pumper.

A split lay often requires coordination by two-way radio, because the attack pumper must advise the supply engine of the plan and indicate where the end of the supply line is being dropped and anchored. In many cases, the attack pumper is out of sight when the supply engine arrives at the split point.

A split hose lay does not necessarily require split hose beds. It can be performed with or without split beds if the necessary adapters are used.

Connecting a Fire Department Engine to a Water Supply

When an engine is located at a hydrant, a supply hose must be used to deliver the water from the hydrant to the engine. This special type of supply line is intended to deliver as much water as possible over a short distance. In most cases, a soft suction hose is used to connect directly to a hydrant. Alternatively, the connection can be made with a hard suction hose or with a short length of large-diameter supply hose.

Cab Procedures

When approaching the scene, you as the driver/operator should position the fire apparatus according to your fire department's policies and procedures and turn the front wheels toward the curb on a 45-degree angle Figure 9-25. If the fire apparatus then should move for any reason, it will travel in the direction of the curb and stop, making the scene a little safer for all involved. If no curb is present, you should still position the wheels in this manner so that the fire apparatus is not pushed straight ahead or into the other side of the street.

At this point, all unnecessary emergency lighting should be turned off except what is needed for the protection of personnel working in and around the fire apparatus on scene. For example, if you are the driver/operator of an engine on scene and your fire apparatus is surrounded at both the front and the back by other fire apparatus, it is probably safe to turn off the light bar and leave on the hazard lights.

Leaving all of the emergency lighting on can drain the electrical system of the fire apparatus. Newer fire apparatus have features that automatically change emergency response lighting to "on scene" modes; this is a feature that you would use when

Driver/Operator Tip

Radios play an important role in communications on the fire ground. Most fire apparatus have radio communications both inside the cab and outside the vehicle near the pump panel. If the fire apparatus is equipped with both, it may be necessary to transfer the radio communications headset from the cab to the outside radio.

Figure 9-25 Position the wheels of the apparatus at a 45-degree angle toward the curb to prevent the apparatus from moving if the parking brake should fail.

operating at the scene of the emergency. On older fire apparatus, you must explicitly choose which emergency lights are used.

Once you have arrived on scene and prior to exiting the cab of the fire apparatus, there are several steps that you need to complete. Exiting the cab prematurely can waste valuable time and cause you to overlook an important function in the cab. For example, if the pump is not properly engaged before you exit, you may spend time trying to troubleshoot the problem at the pump panel instead of getting water to the attack lines. When you are content with placement of the fire apparatus, the first procedure that you must perform is placing the transmission into neutral or park, based on the fire apparatus. Next, apply the parking brake. Failure to apply the parking brake on today's modern fire apparatus will not only cause an unsafe situation, but will also prevent the pump's throttle from working—thus making the pump inoperable.

The next step to be done prior to exiting the cab is to place the fire apparatus transmission from the "road" position to the "pump" position. Basically, you are transferring the motor's power from the drivetrain to the pump, which allows the pump to operate. A pump shift control switch can be found inside the cab Figure 9-26. This switch can be an electronic, mechanical, or pneumatic switch. Place your foot on the brake, and move the pump switch from the "road" position to the "pump" position. Look for the "Pump engaged" indicator light to ensure that this switch has actually taken place. Next, place the apparatus transmission into drive, which allows the motor to power the pump instead of the drivetrain of the fire apparatus. Upon proper engagement of the pump, a light near the pump shift control switch or in the cab will become illuminated, indicating that the pump is engaged with the transmission and ready to pump. This is the "Okay to pump" light. In conjunction with the light, you should hear the revolutions per minute (rpm) of the fire apparatus increase slightly and see a reading of 10 to 15 mi/h (16 to 24 km/h) on the speedometer. This occurs because the transmission is now spinning the impeller of the pump.

> **Driver/Operator Tip**
>
> If the fire apparatus does not go into the "pump" position after you complete the correct procedures, reverse the procedure to take the fire apparatus out of "pump" mode and try again.

Exiting the Cab

Once the pump is engaged, upon exiting the cab of the fire apparatus your first task is to chock the wheels of the fire apparatus. Follow your department's standard operating procedure (SOP) as to which wheels are to be chocked. Usually, two wheel chocks are placed against the rear driver's side tires. This step is necessary in case the fire apparatus moves during the incident—it will be stopped by the wheel's chocks.

Next, you need to circulate water into the pump. Walk to the pump panel on the fire apparatus and open the "tank to pump" valve. This valve allows water to flow from the onboard tank into the pump. Depending on the manufacturer of the fire apparatus, the label for this valve may use different terminology—for example, "Water" or "Tank to Pump." Becoming familiar with the layout of the pump panel and knowing how the valves operate will greatly enhance your efficiency in operating the pump.

Water is now flowing into the pump, but you need to remove any air inside the centrifugal pump for it to operate properly. To do so, you will need to operate the primer. This small positive-displacement pump draws air and water from the top of the centrifugal pump. Remember, a centrifugal pump can pump only fluids—not air.

Once the pump is primed (no air inside the pump), open the "tank refill" valve. This valve allows water to flow from the pump back into the onboard tank; it is basically a small discharge that directs water only back inside the onboard water tank. The label for this valve can have several names, such as "Recirculation Valve" or "Tank Refill." By completing this procedure, you ensure that water flows from the onboard tank into the pump, and then out from the pump back into the tank. This will help to keep the pump from overheating.

The steps for engaging the fire pump are summarized in **Skill Drill 9-4**:

1. Shift the transmission into neutral. **(Step 1)**
2. Set the parking brake. **(Step 2)**
3. Operate the pump shift control switch.
4. Ensure that the "Pump engaged" light is on. **(Step 3)**
5. Shift the transmission into the drive position. **(Step 4)**
6. Identify indicators of pump engagement inside the cab; the "Okay to pump" light should be on, and the speedometer should have a reading of 10–15 mi/h (16–24 km/h). **(Step 5)**
7. Exit the cab. **(Step 6)**
8. Chock the wheels of the apparatus. **(Step 7)**
9. Open the tank-to-pump valve to allow water to enter the pump. **(Step 8)**
10. Operate the priming pump to ensure all of the air is out of the pump. **(Step 9)**
11. Open the tank fill valve to recirculate water until attack lines are ready to be supplied.
12. Charge the attack lines. **(Step 10)**

Figure 9-26 The pump shift control switch transfers the engine's power from the "road" position to the "pump" position.

> **Driver/Operator Tip**
>
> Prior to opening the first discharge valve and charging the attack line, remember to close the tank fill valve. Failure to do so will cause the pump to treat the tank fill as a discharge, causing your pump discharge pressure settings to be incorrect.

Skill Drill 9-4

Engaging the Fire Pump

1. Shift the transmission into neutral.

2. Set the parking brake.

3. Operate the pump shift control switch. Ensure that the "Pump engaged" light is illuminated.

4. Shift the transmission into the drive position.

5. Identify indicators of pump engagement.

6. Exit the cab.

7. Chock the wheels of the apparatus.

8. Open the tank-to-pump valve to allow water to enter the pump.

9. Operate the priming pump to ensure all of the air is out of the pump.

10. Open the tank fill valve to recirculate water until attack lines are ready to be supplied. Charge the attack lines.

Securing a Water Source

As the driver/operator, securing a water source upon your crew's arrival on the scene is one of your primary responsibilities. Before you even leave the fire station to respond to the scene, you should have a working knowledge of water sources at or close to the scene. This can be accomplished by using a map book with hydrant locations in it and by being familiar with water sources within the response area, including lakes, ponds, canals, and drafting hydrants.

> **Driver/Operator Tip**
>
> It is not wise to always rely on the fire officer to help you find or spot a hydrant while responding to the scene. Many times the fire officer may be focused on listening to or speaking on the radio, or obtaining information from the onboard computer. Securing a water source is the driver/operator's responsibility, whether that source is a hydrant, draft, or other source.

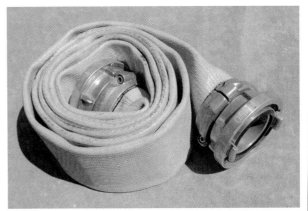

Figure 9-27 Large-diameter hose generally range from 15′ (5 m) to 25′ (8 m) in length and have names such as soft suction hose or pony lines.

Upon your arrival on scene, it is generally the fire officer's call whether you will lay a supply line to the incident. If the fire officer calls for a supply line, follow your fire department's SOP for laying this line. Many times if your crew is the first engine company on scene, the fire officer may forgo laying a supply line and go straight into the scene. This tactic is commonly employed when the second-due engine is immediately behind the first-due engine. Use of this tactic should be communicated by the officers on both engines so that a water supply is secured regardless of the order in which units arrive on scene.

Sometimes an available hydrant may be located in the close vicinity of the building on fire. In such a case, you may position the fire apparatus so that you can use a section of hose to connect to the hydrant instead of performing a forward lay. This is a typical **hand lay**. A hand lay occurs when you position the engine close to the fire scene and deploy the supply hose from the bed of the fire apparatus to the hydrant either yourself or with very little assistance from other fire fighters.

Hand Lays

If you find yourself having to create a hand lay to a hydrant, several considerations must be taken into account. First, how far is the fire apparatus to the closest hydrant? Second, what is the best hose to make the connection? Third, how long will this step take to complete?

To determine how far the fire apparatus is to the closest hydrant, either reference the map book in your fire apparatus or simply eyeball the estimated distance from your pump intake to the hydrant.

When determining the best hose to make the connection, you may or may not have several options for hose. Suppose the hydrant is approximately 15′ (5 m) away. If you use LDH in your fire department, it is common practice to carry short lengths of LDH for the purpose of making a connection between the fire apparatus intake and the hydrant. These lengths of hose generally range from 15′ (5 m) to 25′ (8 m) and have names such as soft suction hose or pony lines **Figure 9-27**. It is important for you to know how long each line found on the fire apparatus is and how many the fire apparatus carries. This can be accomplished during your morning checkout of the fire apparatus by removing the hose and laying it out for inspection.

Some fire departments may carry reduced-length 3″ (76 mm) or 2½″ (64 mm) hose for refilling the onboard tank from a hydrant. Caution should be used using hose with these diameters as supply lines during fire operations. You may find the need to pump more water than these lines can deliver.

If the hydrant is 60′ (18 m) feet away, using the shorter LDH lines most likely will not be an option. In most cases, pulling hose from the hose bed is the next choice. If you carry LDH, it is most often available in 100′ (30 m) lengths. You need to pull a full section from the hose bed and deploy it between the fire apparatus and the hydrant, keeping in mind while pulling it the direction in which you need to go; your goal is to minimize sharp turns and twists in the hose, thereby reducing kinks and severe bends that might cause a reduction of flow. Also, watch out for any oncoming traffic while operating in the roadway. Always wear your traffic safety vest as required by your department.

When considering the third question—How long will this hand lay take to accomplish?—you must take into account several factors:

- The distance from the fire apparatus to the hydrant
- The choice of hose
- Any possible obstacles in the way to deploy the hose, such as vehicles, trees, and bushes

An additional consideration when obtaining a water source is double-tapping a hydrant. Often when connecting to a dry-barrel hydrant during a hose lay or connecting straight in to the hydrant, fire departments use the largest (i.e., steamer) connection. This leaves two 2½″ (64 mm) port connections unused. Once the hydrant is opened, the use of these additional connections is gone, unless the hydrant is shut down. The hydrant is

Voices of Experience

While operating on the scene of a fire in an abandoned structure, a request came in for a mutual aid response in an adjacent county. Additional fire units, including mine, were dispatched to the scene. As we responded to the scene, I could already see the fire from the other side of the river that separated the two counties.

Upon arrival, we found a 5000 square foot (1524 m) structure with 99% involvement. The only portion that was not yet on fire was the basement level and garage. The home was located in a high-class neighborhood and was at the edge of the county—the furthest point on the water system.

The initial volunteer department on the scene was working with 2½ inch (51 mm) hose lines on each side of the structure. In order to quickly suppress this fire, we needed more water. The first responding mutual aid engine laid 1000 feet (305 m) of large diameter hose line (LDH) in a reverse hose lay to prepare for an aerial operation. An aerial ladder was set up in front of the burning structure in an attempt to cover the burning building and to protect the exposures: two houses less than 50 feet (15 m) from the burning building.

As the crews began to fight the fire, we learned that the hydrant was not going to be able to supply enough water to support the aerial operations. We had responded to the mutual aid request with our mostly urban response units that did not having drafting capabilities. However, the initial volunteer department had two units on the scene equipped with hard suction hose.

The next obstacle was obtaining access to a static water supply to draft from. The closest we could get the engines to a static water supply was around 25 feet (8 m). Not a problem. The driver/operators assembled 30 feet (9 m) of drafting hose. One fire fighter held the strainer down in the water until a weight could be found. The draft was established and the water supply was secure.

Now the fire ground operations were able to proceed in full. The aerial ladder was able to knock down the bulk of the fire to allow crews to save the homeowner's vehicles and property in the basement. The training and knowledge of the driver/operators, as well as the entire response's ability to adapt to the situation, made a difference in the ability to save some of the homeowner's belongings.

Brent Willis
Martinez-Columbia Fire Rescue
Martinez, Georgia

> "The training and knowledge of the driver/operators, as well as the ability to adapt to the situation made a difference in the ability to save some of the homeowner's belongings."

Driver/Operator Tip

When using the soft suction hose for the connection, knowing the length of the hose and the positioning of the fire apparatus is important. If the fire apparatus is too close to the hydrant for the length of the hose used, kinks in the hose can occur, which will cause a reduction of flow (GPM) to the pump.

By positioning the fire apparatus a proper distance from the hydrant based on the length of hose used, it should be possible to avoid having kinks in the hose. Knowing the intake locations on the fire apparatus will give you options for the hookup. Caution is needed when positioning the fire apparatus to ensure that you do not place the fire apparatus too far into the street. Doing so may make the road impassable for other vehicles — including additional fire apparatus. Understanding of the preferred positioning comes with practice by hooking up to the hydrant from all intakes on the fire apparatus, including the fire officer's side, front, and rear, if equipped.

Deploying the hose in an "S shape" is the best approach, as it will both decrease the chance of blocking the road and minimize or eliminate any kinks in the hose. To do so, you should keep the fire apparatus close to the side of the street that the hydrant is on. This will allow other fire apparatus to drive past your vehicle and position themselves on the fire scene. The last thing the engine apparatus wants to do is to block the scene for any ladder company that needs access.

Stop the engine apparatus with either the front bumper in line with the hydrant or the rear bumper in line with the hydrant. Do not position the intake directly even with the hydrant. Doing so will cause severe kinks in the short section of hose that is used to make this connection. Once the front or rear bumper is lined up with the hydrant, the short section of supply hose is connected to the intake, and the line is charged, it will have an "S shape" to it; one end of the "S" is connected to the hydrant and the other end is connected to the pump's intake.

---

capable of delivering additional water through these port connections. If you anticipate that additional water may be needed, an available appliance such as a gated ball valve should be placed on the dry-barrel hydrant prior to opening the hydrant. Then, if additional water is needed, a supply line can be connected from the valve to an intake on the fire apparatus **Figure 9-28**. Always follow your local SOP when obtaining a water supply.

Only with practice, practice, practice will you build confidence and become more proficient in securing a water source. Follow the steps in **Skill Drill 9-5** to hand lay a supply line:

1. Position the fire apparatus for the intake to be used. (**Step 1**)
2. Roll out or pull the supply line to the hydrant. (**Step 2**)
3. Connect the supply hose to the hydrant, after properly flushing the hydrant. (**Step 3**)
4. Connect the supply hose to the pump's intake. (**Step 4**)
5. Check for any kinks in the supply line. (**Step 5**)
6. Open the hydrant and charge the supply line. (**Step 6**)

Standpipe/Sprinkler Connecting

Connecting Supply Hose Lines to Standpipe and Sprinkler Systems

Another water supply evolution is furnishing water to standpipe and sprinkler systems. Fire department connections on buildings are provided so that the fire department can pump water into a standpipe and/or sprinkler systems. This setup is considered to be a supply line because it supplies water to standpipe and sprinkler systems. Such a supply line is connected to the discharge side of the attack pumper. The function of the hose line in this case is to provide either a primary or secondary water supply for the sprinkler or standpipe system. The same basic techniques are used to connect the hose lines to either type of system.

Standpipe systems are used to provide a water supply for attack lines that will be operated inside a building. Outlets are provided inside the building where fire fighters can connect attack lines. The fire fighters inside the building must then depend on fire fighters outside the building to supply the water to the fire department connection **Figure 9-29**.

Two types of standpipe systems exist:

- A dry standpipe system depends on the fire department to provide all of the water.
- A wet standpipe system has a built-in water supply, but the fire department connection is provided to deliver a higher flow or to boost the pressure.

The pressure requirements for standpipe systems depend on the height at which the water will be used inside the building. The fire department connection for a sprinkler system is also used to supplement the normal water supply. The required pressures and flows for different types of sprinkler systems can vary significantly. As a guideline, sprinkler systems should be fed at a pressure of 150 psi (1035 kPa) unless more specific information is available.

Due to certain building characteristics within a response area as well as the height and construction materials used in the building, private fire protection systems may be present in

Figure 9-28 The more outlets that are used on a dry-barrel hydrant, the more water you will have available to attack the fire.

Skill Drill 9-5

Hand Laying a Supply Line

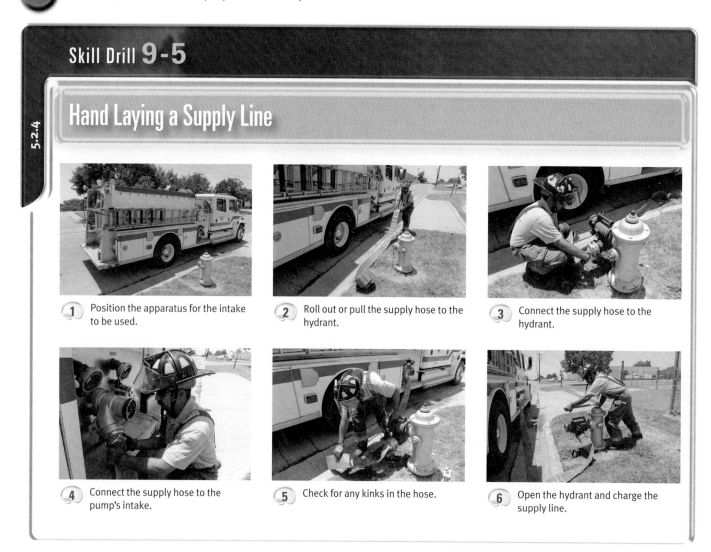

1. Position the apparatus for the intake to be used.
2. Roll out or pull the supply hose to the hydrant.
3. Connect the supply hose to the hydrant.
4. Connect the supply hose to the pump's intake.
5. Check for any kinks in the hose.
6. Open the hydrant and charge the supply line.

---

a fire building. As a driver/operator, it is your responsibility to have a complete and thorough understanding of these systems. Preincident planning as well as familiarization with your response area will help you to identify those buildings that have sprinkler systems and standpipes. When performing this planning or during routine driving through your response area, you should be looking for the fire department connections (FDCs), hydrant locations, and sprinkler valves associated with the building Figure 9-30 . If you are unaware of the locations or cannot find the FDC connection on scene, you become the owner of the problem, with all eyes looking toward you.

The fire department connection can be either free standing or wall mounted. It generally consists of a Siamese connection with two 2½″ (64 mm) female connections Figure 9-31 . Depending on the occupancy, multiple-connection standpipes may be present. Local codes may dictate the number and size of standpipes provided Figure 9-32 .

Each female connection should have a clapper valve inside the Siamese connection that will swing to the closed position on the connection that is not in use Figure 9-33 . This allows one side of the Siamese connection to be charged without water discharging from the other open intake. Depending on the fire department's SOP, it is generally recommended that both and/or all connections to the FDC be used when supplying a standpipe.

If the exterior FDC connection is damaged and connecting to it cannot be accomplished, an accepted practice is to hook up to the standpipe on the first floor of the building. This may

Figure 9-29 A standpipe connection.

Chapter 9 Responding on the Fire Ground

allow you to connect only one line into the standpipe, but better one line than none at all.

After the engine is connected to a sprinkler/standpipe, the fire department's SOP may specify when you will supply the standpipe. One factor that may affect this decision is whether the standpipe is a wet or dry pipe system. Whether a system is wet or dry may be dictated by state or local codes. A wet system has water under pressure in the pipe at all times; it is typically supplied by the water utility department having jurisdiction. A dry system has no water in the pipe and depends on the fire department to supply the water. Regardless of which system may be present, a supply line must be established.

Follow the steps in **Skill Drill 9-6** to connect hose to a fire department connection:

1. Secure a water supply using LDH to ensure the maximum available pressure and volume of water. (**Step 1**)
2. Position the apparatus in a safe and effective position for the given situation. (**Step 2**)
3. Engage the pump. (**Step 3**)
4. Pull enough 2½″ (64 mm) or 3″ (76 mm) hose to reach from the pump discharge to the FDC. (**Step 4**)
5. Remove or break the protective cover for the FDC. (**Step 5**)
6. Inspect the FDC for debris inside and any signs of damage. (**Step 6**)
7. Connect the male end to the FDC. (**Step 7**)
8. Connect the female end of the hose to the pump discharge.
9. Use a double-male adapter if you pull a female coupling off the hose bed and intend to connect it to a FDC. (**Step 8**)

Figure 9-30 Signs indicating the location of the FDC can be very helpful.

A.

B.

Figure 9-31 **A.** A free-standing FDC is usually located away from the building **B.** A wall-mounted FDC is located directly on the building itself. Usually, the sprinkler control room is located directly behind this connection inside the building.

Near Miss REPORT

Report Number: 09-0000479
Report Date: 05/08/2009 20:16

Synopsis: Seatbelts averted fatality while responding to medical emergency.

Event Description: An engine company staffed with 5 personnel was dispatched to a BLS medical emergency. The engine left the station operating with lights and sirens activated. The engine made a left hand turn out of the station onto a paved public road and proceeded north for approximately 100 feet until it reached an intersection controlled by a traffic signal. The traffic signal indicated a green light for the responding engine. As the engine made a left hand turn into the intersection and accelerated out of the turn, the pump operator lost control of the engine. As a result of this, the engine left the road and struck a ground pad transformer. The engine continued moving and crashed into a telephone pole, shearing it in half and ripping the officer's door off. The engine then proceeded approximately 300 additional feet up the middle of a paved public road before coming to a stop in the middle of the road.

After coming to a stop, one of the fire fighters exited the rear of the engine and found the pump operator with an altered level of consciousness and the officer trapped. The firefighter notified the Communications Center of the incident and requested additional resources to assist.

It is important to note that all five members of the apparatus were seated and using seatbelts at the time of the collision. After extrication, the pump operator and officer were transported to a Level I trauma center for evaluation. Both were treated and released within 24 hours of the incident. The officer has not returned to duty and is not expected to return to duty for an extended period of time. The pump operator has returned to work.

The three fire fighters in the crew compartment of the engine sustained injuries and were transported to a local community hospital for evaluation. All were treated, released, and have since returned to duty. The engine sustained severe damage and was destroyed in the collision. The original call for service had to be handled by other units from the department.

I believe that this incident could occur again in our department. Anyone reading this report should view it as an opportunity to save their life and the lives of the crew members by demanding compliance with all safety rules including the mandatory use of seatbelts.

Lessons Learned: First and most important seatbelts save lives. A policy of demanding that personnel wear their seatbelt is as useless as words on a page if it is not enforced. Relentless demand for compliance with this policy is why fire fighters wear their seatbelts. There is no doubt that the officer would have been ejected from the rig and likely killed if he had not been wearing his seatbelt.

The department already has a continual process in place to evaluate the use of lights and sirens to calls. For over three years, the department has followed a stringent set of guidelines for emergency responses involving lights and sirens. Until recently, the department did not apply this policy to medical calls. A limited number of low priority calls are now being dispatched no lights and no sirens. We will continue to evaluate this program and increase the number of low priority responses when the benefit of doing so outweighs the risk associated with a high priority response.

Our department responds to over 70,000 calls for service annually. The law of averages indicates that we will continue to experience apparatus crashes. We should do everything in our power to reduce the opportunity for a collision where possible. The department was recently awarded an AFG grant to purchase an apparatus driving simulator (similar to a flight simulator) that will allow driver operators to learn the skills necessary to safely operate large vehicles under a variety of conditions. It is expected that all driver operators will be required to re-certify annually in the simulator once the program is in place.

The role of human error in this case cannot be avoided. The pump operator failed to understand the fundamental physical features of the apparatus while driving. Although nationally certified as a fire apparatus driver operator under NFPA standards, certification alone is not enough to ensure competency. Periodic testing, training, and on-going evaluation are crucial to any program.

Near Miss REPORT (CONTINUED)

Report Number: 09-0000479
Report Date: 05/08/2009 20:16

Demographics

Department type: Combination, Mostly paid

Job or rank: Battalion Chief/District Chief

Department shift: 24 hours on—72 hours off

Age: 43–51

Years of fire service experience: 21–23

Region: FEMA Region III

Service Area: Suburban

Event Information

Event type: Vehicle event: responding to, returning from, routine driving, etc.

Event date and time: 04/21/2009 22:49

Event participation: Involved in the event

Weather at time of event: Clear with Wet Surfaces

What were the contributing factors?
- Decision Making
- Individual Action
- Human Error
- Situational Awareness

What do you believe is the loss potential?
- Minor injury
- Lost time injury
- Property damage
- Environmental

Performing a Changeover

When you first arrive at the scene, you may initially need to supply attack lines from your onboard tank, then have a supply line laid to your pump from another fire apparatus or a hand-laid supply line. When supplying attack lines from the onboard water tank, you are limited by the amount of water in the tank. For example, if you are supplying one attack line with 100 GPM (380 L/min) and your water tank contains 500 gal (1890 L), your tank will run dry in about 5 minutes. The tank water will last only so long, so a supply line will be needed to sustain a fire attack for a longer period. It is wise to recognize early on that the amount of water flowing through the attack lines is or will

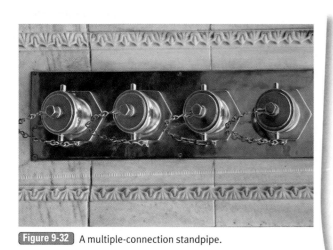

Figure 9-32 A multiple-connection standpipe.

Figure 9-33 A clapper on a Siamese connection allows for one hose to be charged while the other side is still being connected.

Skill Drill 9-6

Connecting Hose to a Fire Department Connection

1. Secure a water supply with LDH.

2. Position the apparatus appropriately.

3. Engage the pump.

4. Pull the hose to the fire department connection.

5. Remove or break the protective cover from the fire department connection.

6. Inspect the fire department connection.

7. Connect the male end of the hose to the fire department connection.

8. Connect the female end of the hose to the pump.

be far greater than the amount of water carried in the onboard tank. For this reason, it is important to establish a supply line early, and then perform a changeover operation.

During a changeover operation, you switch from the onboard water tank supply to an external water source. The goal with this task is simple: to make the changeover prior to running out of water in the tank and with the least amount of pressure fluctuation for the firefighting crew on the nozzle.

Over-pressurizing the attack lines can cause fire fighters to lose control of the handlines, causing an unsafe situation.

As the driver/operator, you need to constantly supply the attack lines with the correct pressure to ensure a safe operation. Two devices that will assist you in this task are the pressure relief valve and the pressure governor Figure 9-34 and Figure 9-35 . In a multistage pump, you will also need to use the transfer valve Figure 9-36 .

Chapter 9 Responding on the Fire Ground 247

Figure 9-34 Pressure relief valve.

Figure 9-35 Pressure governor.

Figure 9-36 Transfer valve.

There are several factors you need to keep in mind prior to the changeover. How much water is currently in the tank? What is the flow rate (GPM) and pressure (psi) currently used on the attack line by the firefighting crew? What is the incoming pressure from the supply line? What is the setting of the pressure control device?

At the beginning of the changeover, keep your eyes on the pressure gauge that the attack crew is using. The goal is to have a fluctuation that does not exceed 10 psi (70 kPa). Slowly open the valve from the external water source, which will introduce water into the pump. Once this valve has been fully opened, close the tank-to-pump valve. If the pressure-relieving device is set for a lower pressure than the incoming supply pressure, you will need to increase the setting on the valve above the incoming pressure from the supply line.

Once the changeover has been accomplished, recheck the gauges to ensure that they have the proper settings. If no other additional lines are required for pumping, now is a good time to refill the onboard tank. Slowly open the tank fill valve to allow water to refill the tank. The tank fill valve should be opened just enough to let water into the tank, but not enough to permit fluctuations of pressure for the attack lines.

Follow the steps in Skill Drill 9-7 ▶ to perform a changeover operation:

1. Calculate and flow an attack line from a discharge using the onboard water tank. (**Step 1**)
2. Slowly open the intake valve to allow water into the pump. (**Step 2**)
3. Attempt to keep pressure fluctuation no greater than 10 psi while using either a pressure governor or pressure relief valve. (**Step 3**)
4. After opening intake valve fully, close the tank-to-pump valve. (**Step 4**)
5. Adjust the pump settings as needed. (**Step 5**)

Skill Drill 9-7

Performing a Changeover Operation

1. Calculate and flow an attack line from a discharge using the onboard water tank.

2. Slowly open the intake valve to allow water into the pump.

3. Attempt to keep pressure fluctuation no greater than 10 psi while using either a pressure governor or pressure relief valve.

4. After opening intake valve fully, close the tank-to-pump valve.

5. Adjust the pump settings as needed.

Duties on Scene

As the driver/operator, you have additional duties and responsibilities on scene other than just pumping water or foam. Once the pumping operation is under way, you must monitor specific gauges on the pump panel, such as the oil pressure and engine temperature. During pumping operations, if you discover that the engine temperature is increasing, you need to take steps to reduce this temperature. First, shut down or turn off all unnecessary loads on the engine, such as the air conditioning system and external lighting. Next, if needed, open the auxiliary cooler on the fire apparatus. Sometimes simply opening the cab and lifting the engine cover to allow more air to get to the engine will help keep the engine cool.

Follow the steps in **Skill Drill 9-8** to operate an auxiliary cooling system:

1. Ensure the pump is operating at 50 psi (345 kPa) pump discharge pressure or lower.
2. Switch the transfer valve switch from either pressure (series) or volume (parallel) mode.
3. Increase the throttle to the desired pressure.

Of course, your final duty as the driver/operator on the fire ground is to disengage the fire pump. Follow the steps in **Skill Drill 9-9** to disengage the fire pump:

Skill Drill 9-9

Disengaging the Fire Pump

1. Reduce the throttle to idle speed.
2. Close the discharge valves.
3. Step into the apparatus's cab.
4. Place your foot on the brake pedal and move the transmission selector into neutral.
5. Ensure that the speedometer no longer has a reading of 10–15 mi/h (16–24 km/h).
6. Move the pump's shift control switch into the "road" position.
7. Confirm that the indicator lights ("Pump engaged" and "Okay to pump") are in the off positions.
8. Exit the cab, and close and drain all valves as necessary.

1. Reduce the throttle to idle speed. (**Step 1**)
2. Close the discharge valves. (**Step 2**)
3. Step into the apparatus's cab. (**Step 3**)
4. Place your foot on the brake pedal and move the transmission selector into neutral. (**Step 4**)
5. Ensure that the speedometer no longer has a reading of 10–15 mi/h (16–24 km/h). (**Step 5**)
6. Move the pump's shift control switch to the "road" position. (**Step 6**)
7. Confirm that the indicator lights ("Pump engaged" and "Okay to pump") are in the off positions. (**Step 7**)
8. Exit the cab, and close and drain all valves as necessary. (**Step 8**)

Wrap-Up

Chief Concepts

- Fire hoses are used for two main purposes: as supply hoses and as attack hoses.
- Fire hose evolutions are critical to the success of the fire-fighting operation.
- As the driver/operator, securing a water source upon arrival on scene is one of your primary responsibilities.
- Before you leave the fire station to respond to the scene, you should have a working knowledge of water sources at or close to the scene. This can be accomplished by using a map book with hydrant locations in it and by becoming familiar with water sources within the response area, including lakes, ponds, canals, and drafting hydrants.
- When your crew first arrives at the scene, you may initially need to supply attack lines from your onboard tank, but then have a supply line laid to your pump from another fire apparatus or via a hand-laid supply line.
- Once the pumping operation is under way, you must monitor specific gauges on the pump panel to ensure that an adequate supply of water or foam is available for the duration of the incident.

Hot Terms

adapter Any device that allows fire hose couplings to be safely interconnected with couplings of different sizes, threads, or mating surfaces, or that allows fire hose couplings to be safely connected to other appliances.

adjustable-gallonage fog nozzle A nozzle that allows the operator to select a desired flow from several settings.

attack hose (attack line) Hose designed to be used by trained fire fighters and fire brigade members to combat fires beyond the incipient stage. (NFPA 1961)

attack pumper An engine from which attack lines have been pulled.

automatic-adjusting fog nozzle A nozzle that can deliver a wide range of water stream flows. It operates by means of an internal spring-loaded piston.

ball valve A type of valve used on nozzles, gated wyes, and engine discharge gates. It consists of a ball with a hole in the middle of the ball.

booster hose (booster line) A non-collapsible hose that is used under positive pressure and that consists of an elastomeric or thermoplastic tube, a braided or spiraled reinforcement, and an outer protective cover. (NFPA 1962)

Bresnan distributor nozzle A nozzle that can be placed in confined spaces. The nozzle spins, spreading water over a large area.

butterfly valve A valve found on the large pump intake valve where the hard or soft suction hose connects to it.

cellar nozzle A nozzle used to fight fires in cellars and other inaccessible places.

double-female adapter A hose adapter that is used to join two male hose couplings.

double-male adapter A hose adapter that is used to join two female hose couplings.

fire hose appliance A piece of hardware (excluding nozzles) generally intended for connection to fire hose to control or convey water. (NFPA 1965)

fixed-gallonage fog nozzle A nozzle that delivers a set number of gallons per minute as per the nozzle's design, no matter what pressure is applied to the nozzle.

fog-stream nozzle A nozzle that is placed at the end of a fire hose and separates water into fine droplets to aid in heat absorption.

forward lay A method of laying a supply line where the line starts at the water source and ends at the attack pumper.

four-way hydrant valve A specialized type of valve that can be placed on a hydrant and is used in conjunction with a pumper to increase water pressure in relaying operations.

gated wye A valved device that splits a single hose into two separate hoses, allowing each hose to be turned on and off independently.

gate valve A type of valve found on hydrants and sprinkler systems.

hand lay A process in which the engine is positioned close to the fire scene and the driver/operator deploys the supply hose from the bed of the fire apparatus to the hydrant either alone or with very little assistance from other fire fighters.

handline nozzle A nozzle with a rated discharge of less than 350 GPM (1325 L/min). (NFPA 1964)

hard suction hose A hose used for drafting water from static supplies (e.g., lakes, rivers, wells). It can also be used for supplying pumpers from a hydrant if designed for that purpose. The hose contains a semi-rigid or rigid reinforcement designed to prevent collapse of the hose under vacuum. (NFPA 1963)

hose clamp A device used to compress a fire hose so as to stop water flow.

hose jacket A device used to stop a leak in a fire hose or to join hoses that have damaged couplings.

large-diameter hose (LDH) A hose of 3½″ (90 mm) size or larger. (NFPA 1961)

low-volume nozzle Nozzle that flows 40 GPM (150 L/min) or less.

master stream device A large-capacity nozzle supplied by two or more hose lines of fixed piping that can flow 300 GPM (1135 L/min). These devices include deck guns and portable ground monitors.

master stream nozzle A nozzle with a rated discharge of 350 GPM (1325 L/min) or greater. (NFPA 1964)

medium-diameter hose (MDH) Hose with a diameter of 2½″ (64 mm) or 3″ (76 mm).

nozzle A constricting appliance attached to the end of a fire hose or monitor to increase the water velocity and form a stream. (NFPA 1965)

nozzle shut-off A device that enables the fire fighter at the nozzle to start or stop the flow of water.

piercing nozzle A nozzle that can be driven through sheet metal or other material to deliver a water stream to that area.

reducer A device that connects two hoses with different couplings or threads together.

reverse lay A method of laying a supply line where the line starts at the attack pumper and ends at the water source.

Siamese connection A device that allows two hoses to be connected together and flow into a single hose.

small-diameter hose (SDH) Hose with a diameter ranging from 1″ (25 mm) to 2″ (51 mm).

smooth-bore nozzle A nozzle that produces a straight stream that consists of a solid column of water.

smooth-bore tip A nozzle device that is a smooth tube and is used to deliver a solid column of water.

soft suction hose A large-diameter hose that is designed to be connected to the large port on a hydrant (steamer connection) and into the engine.

split hose bed A hose bed arranged such that supply line can be laid out, or such that two supply lines can be laid out.

split hose lay A scenario in which the attack pumper lays a supply line from an intersection to the fire, and then the supply engine lays a supply line from the hose left by the attack pumper to the water source.

supply hose (supply line) Hose designed for the purpose of moving water between a pressurized water source and a pump that is supplying attack lines. (NFPA 1961)

threaded hose coupling A type of coupling that requires a male fitting and a female fitting to be screwed together.

water curtain nozzle A nozzle used to deliver a flat screen of water so as to form a protective sheet of water.

water thief A device that has a 2½″ (64-mm) inlet and a 2½″ (64 mm) outlet in addition to two 1½″ (38 mm) outlets. It is used to supply many hoses from one source.

wye A device used to split a single hose into two separate lines.

Driver/Operator *in Action*

When you are flowing water from the on-board water tank, you have to react quickly. From the moment that you pull the discharge for the first attack line, you will be losing water from the on-board water tank. Your next objective is to obtain a sustainable water supply before the on-board water supply runs out. Achieving this goal is not as easy as it might seem at first glance. You need to be an expert on how the fire pumper works and how the water supply in the on-board tank should be managed. Test your expertise by answering the following questions.

1. A 2½″ attack line is generally considered to deliver a flow of approximately _____?
   A. 150 GPM
   B. 200 GPM
   C. 250 GPM
   D. 300 GPM

2. Which type of hose lay starts at the hydrant and is laid to the fire scene?
   A. Forward lay
   B. Reverse lay
   C. Split lay
   D. Open lay

3. What is the first task that should be completed once the pump is engaged and the driver/operator exits the cab of the fire apparatus?
   A. Operate the primer
   B. Determine the amount of water in the tank
   C. Chock the wheels of the fire apparatus
   D. Open the tank-fill valve

4. When a short section of LDH is connected from the hydrant to the pump intake, what shape should the hose take?
   A. "L"
   B. "I"
   C. "U"
   D. "S"

Water Supply

CHAPTER 10

NFPA 1002 Standard

5.1.1 Perform the routine tests, inspections, and servicing functions specified in the following list in addition to those in 4.2.1, given a fire department pumper and its manufacturer's specifications, so that the operational status of the pumper is verified:

(1) Water tank and other extinguishing agent levels (if applicable)
(2) Pumping systems
(3) Foam systems [p. 258–260]

(A) Requisite Knowledge. Manufacturer's specifications and requirements, and policies and procedures of the jurisdiction. [p. 258–260]

(B) Requisite Skills. The ability to use hand tools, recognize system problems, and correct any deficiency noted according to policies and procedures. [p. 257–292]

5.2.1 Produce effective hand or master streams, given the sources specified in the following list, so that the pump is engaged, all pressure control and vehicle safety devices are set, the rated flow of the nozzle is achieved and maintained, and the apparatus is continuously monitored for potential problems:

(1) Internal tank
(2) Pressurized source
(3) Static source
(4) Transfer from internal tank to external source [p. 283–292]

(A) Requisite Knowledge. Hydraulic calculations for friction loss and flow using both written formulas and estimation methods, safe operation of the pump, problems related to small-diameter or dead-end mains, low-pressure and private water supply systems, hydrant coding systems, and reliability of static sources. [p. 283–292]

(B) Requisite Skills. The ability to position a fire department pumper to operate at a fire hydrant and at a static water source, power transfer from vehicle engine to pump, draft, operate pumper pressure control systems, operate the volume/pressure transfer valve (multistage pumps only), operate auxiliary cooling systems, make the transition between internal and external water sources, and assemble hose lines, nozzles, valves, and appliances. [p. 283–292]

5.2.2 Pump a supply line of 65 mm (2½ in.) or larger, given a relay pumping evolution the length and size of the line and the desired flow and intake pressure, so that the correct pressure and flow is provided to the next pumper in the relay. [p. 275–283]

(A) Requisite Knowledge. Hydraulic calculations for friction loss and flow using both written formulas and estimation methods, safe operation of the pump, problems related to small-diameter or dead-end mains, low-pressure and private water supply systems, hydrant coding systems, and reliability of static sources. [p. 257–292]

(B) Requisite Skills. The ability to position a fire department pumper to operate at a fire hydrant and at a static water source, power transfer from vehicle engine to pump, draft, operate pumper pressure control systems, operate the volume/pressure transfer valve (multistage pumps only), operate auxiliary cooling systems, make the transition between internal and external water sources, and assemble hose lines, nozzles, valves, and appliances. [p. 275–292]

5.2.4 Supply water to fire sprinkler and standpipe systems, given specific system information and a fire department pumper, so that water is supplied to the system at the correct volume and pressure.

(A) Requisite Knowledge. Calculation of pump discharge pressure, hose layouts; location of fire department connection; alternative supply procedures if fire department connection is not usable; operating principles of sprinkler systems as defined in NFPA 13, NFPA 13D, and NFPA 13R; fire department operations in sprinklered properties as defined in NFPA 13E; and operating principles of standpipe systems as defined in NFPA 14.

(B) Requisite Skills. The ability to position a fire department pumper to operate at a fire hydrant and at a static water source, power transfer from vehicle engine to pump, draft, operate pumper pressure control systems, operate the volume/pressure transfer valve (multistage pumps only), operate auxiliary cooling systems, make the transition between internal and external water sources, and assemble hose lines, nozzles, valves, and appliances. [p. 275–292]

10.1 **General.** The requirements of Fire Fighter I as specified in NFPA 1001 and the job performance requirements defined in Sections 10.1 and 10.2 shall be met prior to certification as a fire department driver/operator—mobile water supply apparatus. [p. 257–292]

10.1.1 Perform routine tests, inspections, and servicing functions specified in the following list, in addition to those specified in 4.2.1, given a fire department mobile water supply apparatus, so that the operational readiness of the mobile water supply apparatus is verified: [p. 258–263]

(A) Requisite Knowledge. Manufacturer's specifications and requirements, and policies and procedures of the jurisdiction. [p. 258–263]

(B) Requisite Skills. The ability to use hand tools, recognize system problems, and correct any deficiency noted according to policies and procedures. [p. 258–263]

10.2 **Operations.** [p. 257–292].

10.2.1 Maneuver and position a mobile water supply apparatus at a water shuttle fill site, given a fill site location and one or more supply hose, so that the apparatus is correctly positioned, supply hose are attached to the intake connections without having to stretch additional hose, and no objects are struck at the fill site. [p. 275–292]

(A) **Requisite Knowledge.** Local procedures for establishing a water shuttle fill site, method for marking the stopping position of the apparatus, and location of the water tank intakes on the apparatus. [p. 275–292]

(B) **Requisite Skills.** The ability to determine a correct position for the apparatus, maneuver apparatus into that position, and avoid obstacles to operations. [p. 275–292]

10.2.2 Maneuver and position a mobile water supply apparatus at a water shuttle dump site, given a dump site and a portable water tank, so that all of the water being discharged from the apparatus enters the portable tank and no objects are struck at the dump site. [p. 275–292]

(A) **Requisite Knowledge.** Local procedures for operating a water shuttle dump site and location of the water tank discharges on the apparatus. [p. 275–292]

(B) **Requisite Skills.** The ability to determine a correct position for the apparatus, maneuver apparatus into that position, avoid obstacles to operations, and operate the fire pump or rapid water dump system. [p. 275–292]

10.2.3 Establish a water shuttle dump site, given two or more portable water tanks, low-level strainers, water transfer equipment, fire hose, and a fire apparatus equipped with a fire pump, so that the tank being drafted from is kept full at all times, the tank being dumped into is emptied first, and the water is transferred efficiently from one tank to the next. [p. 283–292]

(A) **Requisite Knowledge.** Local procedures for establishing a water shuttle dump site and principles of water transfer between multiple portable water tanks. [p. 283–292]

(B) **Requisite Skills.** The ability to deploy portable water tanks, connect and operate water transfer equipment, and connect a strainer and suction hose to the fire pump. [p. 283–292]

Additional NFPA Standards

NFPA 1901, *Standard for Automotive Fire Apparatus* (2009)

Fire and Emergency Services Higher Education (FESHE) Model Curriculum

National Fire Science Degree Programs Committee (NFSDPC): Associate's Curriculum

Fire Protection Hydraulics and Water Supply

1. Apply the application of mathematics and physics to movement of water in fire suppression activities. [p. 257–258, 263–264]
2. Identify the design principles of fire service pumping apparatus.
3. Analyze community fire flow demand criteria.
4. Demonstrate, through problem solving, a thorough understanding of the principles of forces that affect water at rest and, in motion. [p. 257–258, 263–264]
5. List and describe the various types of water distribution systems.
6. Discuss the various types of fire pumps.

Knowledge Objectives

After studying this chapter, you will be able to:

- Describe the mechanics of drafting.
- Describe how to verify the operational readiness of the pump.
- Describe water supply management within the Incident Management System.
- Describe the process for selecting a suitable site for water drafting.
- Describe how to position a fire pumper for drafting.
- Describe the process for establishing a pumping operation from a draft.
- Describe how to perform drafting operations.
- Describe complications for drafting operations.
- Describe how to provide an uninterrupted water supply.
- Describe relay pumping operations.
- Describe water shuttle operations.
- Describe dump site operations.
- Describe water shuttle operations in the Incident Management System.

Skills Objectives

After completing this chapter, you will be able to:

- Perform the vacuum test.
- Position the hard suction hose into the water.
- Draft from a static water source.
- Provide water flow to handlines, master streams, and supply lines.

You Are the Driver/Operator

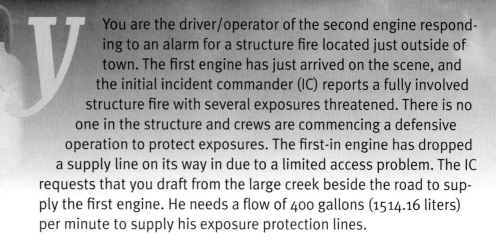

You are the driver/operator of the second engine responding to an alarm for a structure fire located just outside of town. The first engine has just arrived on the scene, and the initial incident commander (IC) reports a fully involved structure fire with several exposures threatened. There is no one in the structure and crews are commencing a defensive operation to protect exposures. The first-in engine has dropped a supply line on its way in due to a limited access problem. The IC requests that you draft from the large creek beside the road to supply the first engine. He needs a flow of 400 gallons (1514.16 liters) per minute to supply his exposure protection lines.

1. How can you be sure your fire pumper will draft water from the creek?
2. What must you consider as you select a site to draft from the creek?
3. How do you determine if there is enough water in the creek to meet your needs?

Introduction

Most of the time a driver/operator pumps water from the tank onboard the fire apparatus or from a pressurized water source such as a fire hydrant or a relay supply line. In many rural fire areas and other emergency situations or disasters, however, the usual water supplies either are not available or are inadequate. The ability to **draft** water from a static source makes your fire apparatus a versatile resource. These occasions—when you are called upon to draft and supply an uninterrupted flow of water from a static water source—present some of the most challenging operating conditions that you and your fire pumper will face. Success in these conditions will be achieved if you make sure that the pump on your fire apparatus is in good operating condition, if you understand and apply the principles involved in drafting, and if you can deliver water to the incident from a static water source.

The Mechanics of Drafting

Let's look closer at the drafting process to fully comprehend what is actually taking place. **Atmospheric pressure**, which is pressure caused by the weight of the atmosphere, is 14.7 pounds per square inch (psi) (equivalent to 101 kilopascals [kPa]) at sea level. As you move up in elevation from sea level, this supply pressure decreases because as the atmosphere thins or lessens, the pressure drops approximately half 1 psi (3.45 kPa) for every 1000 feet (304.8 meters [m]) of elevation above sea level. The atmospheric pressure at your drafting site is the maximum supply pressure for drafting. Thus, if the atmospheric pressure at your drafting site is 14.1 psi (101 kPa), then the vacuum pressure required to draft will not exceed 14.1 psi (101 kPa). Essentially, without atmospheric pressure, you would not be able to draft.

When you engage the primer, you begin pumping air out of the fire pump. This process creates a **vacuum**, which is any pressure less than atmospheric pressure. A vacuum must be contained inside some type of airtight vessel (the hard-sided supply hose, for example); otherwise, air would simply rush in to refill the vacuum. As the vacuum increases in the hose, the atmospheric pressure in it decreases, which results in greater pressure on the outside of the hose. This difference in pressure supports a column of liquid, and is commonly measured in units of inches or millimeters of mercury (Hg). Mercury was chosen as the liquid used to measure vacuum in the laboratory because its density makes it more convenient to use than water for this purpose. Table 10-1 lists the conversion factors used to convert

Table 10-1 Vacuum to Water Column and Pressure Conversions

U.S. Measurement System
1 inch of mercury = 1.13 feet of water
1 inch of mercury = 0.49 pound/square inch
1 pound/square inch = 2.31 feet of water

Metric System
1 millimeter of mercury = 1.36 centimeters of water
1 millimeter of mercury = 0.13 kilopascal
1 kilopascal = 10.20 centimeters of water

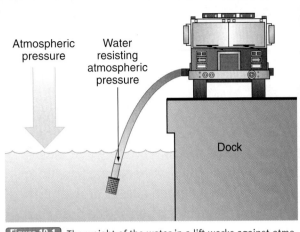

Figure 10-1 The weight of the water in a lift works against atmospheric pressure, reducing supply pressure.

inches (mm) of Hg to feet (cm) of water or pounds per square inch (kilopascals) of pressure.

A fire pumper in good condition should be able to develop a vacuum equal to about 22 inches Hg (50.78 cm Hg). This equates to the ability to lift water approximately 24.8′ (7.59 m) at sea level. While it is possible to lift water 20′ (6.10 m) or more, the discharge capacity of the fire pump is severely restricted due to the amount of vacuum being created on the supply side of the pump **Figure 10-1**. As you lift water to a greater height, more atmospheric pressure is required to support the weight of the water, which reduces the available pressure to move it. Stated simply, the higher you must lift the water, the less water you will be able to pump. **Dependable lift** is the height that a column of water can be lifted in a quantity considered sufficient to provide reliable fire flow; it is generally considered to be 15′ (4.57 m).

The pumps found in fire apparatus have their capacity ratings established by pumping from a draft. For example, a Class A pumper must lift water 10′ (3.05 m) through 20′ (6.10 m) of hard-sided supply hose and deliver its rated capacity at a discharge pressure of 150 psi (1034.2 kPa), as measured at the pump panel. The effect of lift on the discharge capacity of a typical 1000 gallons per minute (GPM) (3785.4 L/min) fire pumper is illustrated in **Table 10-2**.

Inspections, Routine Maintenance, and Operational Testing

Fire departments generally have a prescribed process for performing inspections, routine maintenance, and operational testing. Many fire departments have created a written report form for this process **Figure 10-2**. The availability of such a form is intended to help you perform the equipment inspection and operational testing in a consistent manner and to prevent items on the checklist from being missed. Fire apparatus inspection and routine maintenance are discussed fully in Chapter 6. Nevertheless, a few items on the checklist must be given special attention to ensure the operational readiness of the fire apparatus for drafting from a static water source. For example, you need to check the operational readiness of the priming system and the overall condition of the fire pump on your fire apparatus on a periodic basis.

Inspecting the Priming System

Priming is the process of removing air from a fire pump and replacing it with water. The priming pump is a positive-displacement pump used to remove the air from the fire pump during priming. This pump must be lubricated and sealed as it turns if it is to efficiently pump air **Figure 10-3**. The **primer oil reservoir** is a small tank that holds the lubricant for the priming pump. It is generally located slightly higher than the pump and often is enclosed behind an access panel on the side or the top of a fire apparatus. Be sure to check the level of oil in the primer oil reservoir **Figure 10-4**. If the priming pump runs out of lubricating oil, you will not be able to prime the pump. You may also severely damage the priming pump by running it without lubrication. Refill the reservoir to the full mark if needed, using the correct type and weight of oil recommended by the manufacturer.

When you check the primer oil level in the reservoir, you should also confirm that the **anti-siphon hole** is not clogged. This very small hole is found in the fitting on top of the primer oil reservoir or in the top of a loop in the lubrication line leading to the priming pump **Figure 10-5**. If the anti-siphon hole becomes plugged, air will not be able to enter the lubrication line following operation of the priming pump. This can cause the oil in the primer reservoir to continue to be siphoned into

Table 10-2 Effect of Lift on the Discharge Capacity of a 1000 GPM (3785.4 L/min) Fire Pump

Lift	5′ (1.52 m)	10′ (3.05 m)	15′ (4.57 m)	20′ (6.10 m)
Discharge	1140 GPM (4315.4 L/min)	1000 GPM (3785.4 L/min)	835 GPM (3160.8 L/min)	585 GPM (2214.5 L/min)

Note: The actual delivery performance for any fire pump depends on that pump's condition and other variables not addressed here. Many fire pumps may be capable of exceeding this performance and, therefore, are capable of discharging more water.

Conditions for this test: A 1000 GPM (3785.4 L/min) fire pump capable of discharging its rated capacity at a discharge pressure of 150 psi (1034.2 kPa) measured at the pump panel; 150 psi (1034.2 kPa) discharge pressure must be maintained for all tests; 2–10′ (3.05 m) lengths of 5″ (127 mm) hard suction line for the 5′ (1.52 m) and 10′ (3.05 m) lifts; 3–10′ (3.05 m) lengths of 5″ (127 mm) hard suction line for the 15′ (4.57 m) and 20′ (6.10 m) lifts; altitude of 2700′ (822.96 m); barometric pressure of 29.22 in. Hg (742.19 mm Hg); water temperature of 55°F (12.78°C).

Chapter 10 Water Supply

DAILY ENGINE INSPECTION SHEET

Week of _____   Unit I.D. _____
                                            Shop # _____

Date:	Mon.	Tues.	Wed.	Thurs.	Fri.	Sat.	Sun.
Name:							
Knox box serial number							
Fuel level							
Motor oil							
Radiator							
Wipers							
Gauges							
Brakes							
Starter							
Lights/siren							
Generator							
Mirrors							
Body condition							
Water level							
Pump controls/gauges							
Press control device							
Hydrant tools							
Hose/nozzles							
Appliances							
Tools/ladders							
SCBA—PPE							
Radios							
Box-lights							
Map-books/computer							
Keys							
Accountability							
Clipboard							
Tire pressure		Comments:					
Batteries		_____					
Transmission fluid		_____					
Bleed air tanks		_____					
Primer fluid		_____					
Drain valves		_____					
Tool box		_____					
Power tools		_____					

Figure 10-2 Following a printed inspection form ensures that nothing is overlooked on your fire apparatus.

Figure 10-3 The priming pump is lubricated by oil drawn from the primer oil reservoir.

Figure 10-5 The anti-siphon hole location found on modern fire apparatus.

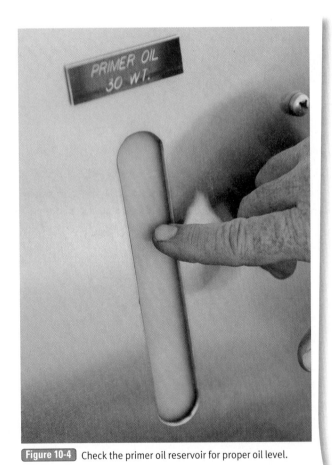

Figure 10-4 Check the primer oil reservoir for proper oil level.

Next, you will need to check the operation of the primer valve. It is located on the pump panel and is usually labeled "Primer" **Figure 10-6**. It is not necessary to start the engine for this test, but you will have to turn on the battery switch. Pull out on the primer valve handle firmly until it stops, and hold it in that position for several seconds. You should hear a whirring sound and see a small amount of water discharged onto the floor under the fire apparatus (if the pump is primed). Return the valve handle to its closed position. The valve should move freely and the primer motor should start and stop as you move the valve handle out and back in.

Safety Tip

Always chock the wheels of the fire apparatus when working around it.

the priming pump even after the priming pump stops running. The excessive flow of oil may create an oil spill on the floor of the fire apparatus bay, which can present a serious safety hazard. It will also drain the primer oil reservoir, which may result in no oil being available the next time you attempt to prime the pump.

Performing a Vacuum Test

The best way to assess the operational readiness of the priming system is to conduct a vacuum test periodically. This test also gives you a good indication of the general condition of the fire pump and all its valves and plumbing. NFPA Standard 1901,

Chapter 10 Water Supply

Figure 10-6 The primer control handle operates the priming pump and the primer valve at the same time.

4. Remove the caps from all discharge fittings. (**Step 4**)
5. Turn the battery isolation switch to the "on" position. It is not necessary to start the engine. (**Step 5**)
6. Pull the primer control handle out firmly until it stops, and hold it in that position. This will engage the primer motor and open the primer valve. (**Step 6**)
7. Continue to hold the primer control handle in the fully engaged position until the maximum vacuum is achieved (usually 15 to 20 seconds). You should see a vacuum reading of approximately 20–22 in. Hg (50.8–55.8 cm Hg) indicated by the supply master gauge on the pump panel. (**Step 7**)
8. Return the primer control handle to the fully closed position. (**Step 8**)
9. Record the vacuum reading from the supply master gauge on your inspection report form. Wait 5 minutes and record the reading again. The reading should not drop more than 10 in. Hg (25.4 cm Hg) in 5 minutes. If the reading drops excessively, you should note this fact on your inspection report form and have the fire apparatus scheduled for maintenance to repair the vacuum leak. (**Step 9**)

Driver/Operator Tip

Do not forget to refill the booster tank and prime the fire pump before placing the fire apparatus back in service!

Finding a Vacuum Leak

A fire pump that retains any vacuum for 5 minutes should prime and draft water easily in most situations. Vacuum leakage in excess of the amount specified by NFPA Standard 1901 may cause problems during the process of obtaining or maintaining a prime when you are attempting to lift water more than 10′ (3.05 m), or when you are pumping a relatively small volume of water.

To find a suspected vacuum leak, close all valves and possible vacuum leak sources, just as you did during the vacuum test. Remember to remove the caps from all discharges. Connect a fire hose between an intake port on the fire pump and a pressurized water source such as a hydrant. Shut off the fire apparatus engine, open the intake valve for the intake port to which you just connected the hose, and slowly open the hydrant to pressurize the fire pump and its components. Any leak that affected your vacuum test will become a water leak once the fire pump is pressurized with water. At this point, you will be able to see and/or hear the leak and locate its source.

Water Supply Management in the Incident Management System

Water supply management is a vital part of any fire incident and can become a major part of the Incident Management System (IMS) in scenarios where a reliable public or private water system is not available. Drafting operations, relay pumping, water

Standard for Automotive Fire Apparatus, lists the very specific conditions for this test when performed for the certification of a new fire pump. While the most important of these conditions are presented here, it is not necessary to exactly replicate the certification test conditions during a periodic vacuum test.

Take the time to remove all discharge caps and drain the booster tank prior to performing the vacuum test. Failure to do so will mask the effects of possible vacuum leaks due to worn or damaged valves in those components. You may be called upon to draft from a static source on the fire ground if you have attack lines deployed that have emptied your booster tank. If your fire pump has damaged or leaking valves in any of the discharges or tank lines, you will be unable to create the vacuum needed to lift water from the static source.

Follow the steps in **Skill Drill 10-1** to perform the vacuum test on a pump:

1. Chock the wheels of the fire apparatus to prevent movement while you are working on it. (**Step 1**)
2. Drain the water from the fire pump and the booster tank. (**Step 2**)
3. Close all drains, discharge valves, intake valves, cooler lines, and any other possible sources of vacuum leaks. Place caps on all suction or external intake fittings. (**Step 3**)

Skill Drill 10-1

Performing a Vacuum Test

1 Chock the wheels.

2 Drain the fire pump and booster tank.

3 Close all drains, valves, and other possible vacuum leaks to isolate the fire pump.

4 Remove the caps from all discharge fittings.

5 Turn the battery switch to the "on" position.

6 Firmly pull the primer control handle to the fully open position, and hold it in that position.

7 Continue to operate the priming pump until the maximum vacuum is achieved. The master supply gauge should show a reading in the range of 20–22 in. Hg (50.8–55.8 cm Hg).

8 Close the primer control handle.

9 Record vacuum reading from master supply gauge, wait 5 minutes, and record the reading again. The vacuum should not drop more than 10 in. Hg (25.4 cm Hg). Report excessive leakage following policy because fire pump maintenance or repair is indicated.

shuttles, and nurse tanker operations will become the responsibility of the water supply officer in these situations. The water supply officer position reports to the operations section chief and is responsible for ensuring that an adequate supply of water is available to mitigate the incident. In smaller incidents, if no operations section chief is designated, then this officer reports directly to the IC. Water supply officers should be given a priority on any scene, but particularly if a water shuttle is established. In those circumstances, a dump site and a fill site will be incorporated into the IMS organization.

Selecting a Drafting Site

Ensuring that the incident has a steady supply of water is critical to the success and safety of the incident response. When drafting water is the chosen method for providing that supply, the process of selecting an appropriate drafting site must involve determining the reliability of a static water source. Once the reliability has been established, operational requirements such as accessibility, purpose of the site, and positioning of the fire apparatus may influence site selection.

Determining the Reliability of Static Water Sources

It is very important to evaluate the reliability of a static water source before committing a fire pumper to draft from it. Failure to determine the reliability of a static water source can seriously affect your ability to support fire-ground operations and may even result in damage to the fire pumper. Several factors must be considered when you are determining the reliability of a static water source—namely, accessibility, quantity of water available, and quality of the water in the static water source.

Estimating the Quantity of Water Available

The first consideration in determining the reliability of a static water source is to estimate the quantity of water available. When considering a lake or large river, there is usually no question that an adequate supply of water will be available. In contrast, smaller static sources such as small rivers, streams, cisterns, and ponds require careful evaluation. These supplies may also be affected more dramatically by seasonal fluctuations in stream flows. You must be able to calculate the estimated amount of water available in both moving and nonmoving water sources.

Calculating Available Water in a Nonmoving Source

To calculate the amount of water available in a nonmoving source, you must determine the dimensions of the pond, cistern, or swimming pool. The length, width, and depth must be known or estimated. Multiplying the length (L) times the width (W), times the depth (D), times a constant (C), yields the quantity of water (Q) that is available. The constant is 7.5 for the U.S. system of measurement, reflecting the fact that there is 7.48 gallons in one cubic foot of water. The constant is 1000 for the metric system of measurement, reflecting the fact that there is 1000 liters in one cubic meter of water.

Formula 1: Calculating Available Water in a Nonmoving Source (U.S. Measurement System)

$$Q = L \times W \times D \times C$$

Where Q = gallons of water available
L = length of static source in feet
W = width of static source in feet
D = depth of static source in feet
C = a constant, 7.5 (7.48 gal in 1 cubic foot of water)

Example
Calculate the amount of water available in a small pond that is 100′ long and 80′ wide. You estimate the depth of the pond to be 4′.

$$Q = L \times W \times D \times C$$
$$Q = 100' \times 80' \times 4' \times 7.5$$
$$Q = 240{,}000 \text{ gal water available}$$

Formula 2: Calculating Available Water in a Nonmoving Source (Metric System)

$$Q = L \times W \times D \times C$$

Where Q = liters of water available
L = length of static source in meters
W = width of static source in meters
D = depth of static source in meters
C = a constant, 1000 (1000 L in one cubic meter of water)

Example
Calculate the amount of water available in a small pond that is 30 m long and 10 m wide. You estimate the depth of the pond to be 2 m.

$$Q = L \times W \times D \times C$$
$$Q = 30 \text{ m} \times 10 \text{ m} \times 2 \text{ m} \times 1000$$
$$Q = 600{,}000 \text{ L water available}$$

Calculating Available Water from a Moving Source

To calculate the amount of water available in a moving source, you must determine the dimensions of the stream, canal, or river. The width (W) and depth (D) must be known or estimated. Because this water source is moving, use velocity (V) in place of length in the formula. Velocity can be determined by measuring the distance the water travels in one minute. Multiplying the width (W), times the depth (D), times the velocity (V), times a constant (C), yields the quantity of water (Q) that is available. The constants are the same ones used in the previous examples.

Formula 3: Calculating Available Water from a Moving Source (U.S. Measurement System)

$$Q = W \times D \times V \times C$$

Where Q = gallons of water available
W = width of static source in feet
D = depth of static source in feet
V = velocity (distance traveled in one minute) in feet
C = a constant, 7.5 (7.48 gal in one cubic foot of water)

Example
Calculate the amount of water available in a stream that is 10 ft wide and 2 ft deep. The water is traveling 15 ft/min.

$$Q = W \times D \times V \times C$$
$$Q = 10 \text{ ft} \times 2 \text{ ft} \times 15 \text{ ft/min} \times 7.5$$
$$Q = 2250 \text{ GPM}$$

Formula 4: Calculating Available Water from a Moving Source (Metric System)

$$Q = W \times D \times V \times C$$

Where Q = liters of water available
W = width of static source in meters
D = depth of static source in meters
V = velocity (distance traveled in one minute) in meters
C = a constant, 1000 (1000 L in one cubic meter of water)

Example
Calculate the amount of water available in an irrigation ditch that is 2 m wide and 1 m deep. The water is traveling at a speed of 6 m/min.

$$Q = W \times D \times V \times C$$
$$Q = 2 \text{ m} \times 1 \text{ m} \times 6 \text{ m/min} \times 1000$$
$$Q = 12{,}000 \text{ L/min}$$

> **Driver/Operator Tip**
>
> When calculating available water from a moving water source, remember that you must replace the length measurement with a velocity measurement in your formula. Velocity is a distance per minute, so your answer will always be a quantity per minute—that is, either GPM or liters per minute (L/min).

Evaluating Water Quality of a Static Source
The next step in determining the reliability of a static water source is to consider the quality of the water in the static source. The water should be free of aquatic weeds, moss, algae, and other trash or debris. Whether floating or submerged, this type of foreign material tends to plug the screen on the supply hose and restrict your ability to deliver a reliable water supply for

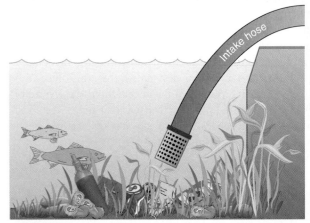

Figure 10-7 Weeds, trash, and other debris can clog the supply strainer and attack line nozzles.

firefighting operations **Figure 10-7**. Even small pieces of debris that may pass through the pickup screen can cause problems by plugging the nozzles on attack lines. Obviously, clean water is preferred over muddy or murky water, which contains sand or silt that can clog nozzles, damage the wear rings inside the pump, and even cause pump failure.

> **Driver/Operator Tip**
>
> Always thoroughly flush and refill the fire pump and booster tank with clean water following any drafting operation. Leaving contaminated water in the pump can lead to a variety of maintenance problems, including premature pump failure.

Accessibility to the Static Water Source
You must be able to safely position your fire apparatus close enough to the water source to completely submerge the strainer of the hard-sided supply hose in the water once it is connected to the fire pump. Soil conditions and site specific obstructions can affect your ability to position the fire apparatus close enough to draft.

In preparing for drafting, you must check the soil near the edges of the water source, especially a lake, river, or stream. Make sure that it is dry and solid enough to support the weight of the fire apparatus, especially for extended wet operations. Unstable soils and conditions may cause the vehicle to tip, slide, or roll over. Often when the soil is saturated, it will be too soft or muddy to support the fire apparatus. You must avoid driving a fire apparatus on soil in this condition because the vehicle will sink into the ground and may become stuck. This would render the fire apparatus unusable, causing a delay to the operation. In addition, a large tow truck may be required to free the fire apparatus, and the towing operation may itself damage components on the pump or the chassis.

Safety Tip

Be especially careful when considering driving a fire apparatus on a dirt-fill bank, levee, dike, or dam. These surfaces often appear to be firm and dry, but may hide a deeper layer of saturated soil. This layer of saturated soil may give way under the weight of a modern fire apparatus, causing the fire apparatus to slide or roll into the water.

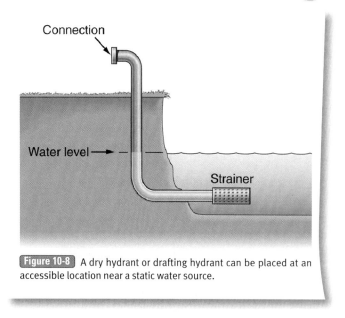

Figure 10-8 A dry hydrant or drafting hydrant can be placed at an accessible location near a static water source.

Once you have determined that the soil will support your fire apparatus, you should identify a drafting site that is relatively level and free of trees, bushes, rocks, fences, and other obstructions that may limit access to the static water source. Obstructions not only limit your ability to access the site, but also serve as a source of safety concerns. Rocks and other ground debris can cause trip hazards when you try to get through them to access the water. Operating on too steep of a slope can make the fire apparatus prone to shift or roll. It may also affect the cooling and lubrication systems of the fire apparatus, possibly causing premature mechanical failures.

Access to static water sources can be difficult during winter weather conditions. Snow can hide possible hazardous conditions, so be careful as you approach the water. Ice covering the water presents an access problem and a potential life-safety situation. You may have to chop a hole in the ice with an axe, a chainsaw, or an ice auger to gain access to the water source. Always firmly tap or "sound" the ice carefully to determine whether it is safe to walk on. Never walk on ice without a personal flotation device and safety line unless you know for certain that the water is less than waist deep.

Driver/Operator Tip

Paved or concrete boat launching ramps make good drafting sites on which to position fire apparatus. The ground is fairly level and is free of obstructions, allowing your crew to easily connect the hard-sided supply hoses to the pump and move them into place as you position the fire apparatus for drafting.

Special Accessibility Considerations

Drafting from a bridge is one possible solution to accessibility problems. You must be sure that the bridge is designed to support the weight of the fire apparatus, however. You should also evaluate the height of the **lift**—that is, the vertical distance from the water level to the center of the fire pumper. Lifting water more than 15′ (4.57 m) is not practical for most fire pumpers and requires additional lengths of hard-sided supply hose. The effects of restricting or blocking traffic also should be considered, along with the obvious safety hazards for personnel and equipment when they are operating near moving traffic.

It is beneficial to identify drafting sites during preplanning efforts rather than waiting for a fire to force you to find one in an emergency situation. Because calculating water availability from both static and moving sources can be complicated, it is best to perform these calculations when you are not under the pressure that comes from an emergency incident.

During preplanning, you will be able to determine which sources will and will not work for your operation. Problem areas can be identified and noted before the emergency operation, and predetermined drafting sites can be located. These drafting sites can be improved by clearing obstructions and installing gravel or pavement, if necessary, to provide a solid surface on which to drive the fire apparatus.

The installation of dry hydrants should be considered in areas where access is difficult or when a location would make an ideal drafting site that could be used frequently. Dry hydrants can be required, in local fire prevention codes, to be installed by the property owner to provide quick access to static water supplies **Figure 10-8**. Once installed, their maintenance can be required as part of the code to ensure that they are accessible and in good working condition. Otherwise, maintenance may be the responsibility of the fire department that installed them.

Consider placing **portable pumps** near static water sources whenever accessibility with fire pumper is not possible. These pumps can deliver water, through a hose, to another spot with better accessibility for fire apparatus. For example, if there is a deep spring pond that is located 20′ (6.10 m) off the roadway but cannot be directly accessed by a vehicle, a portable pump can be placed on the shore of the pond to supply a fire pumper using a section of hose on the discharge side of the pump. While

Figure 10-9 Portable pumps can offer solutions to accessibility problems.

Figure 10-10 Placing a dam in a flowing stream can increase the water depth to make it more suitable for drafting.

this drafting option is not ideal, multiple pumps may be used to supply larger volumes of water when needed.

Floating pumps may be used when the water is not deep enough to allow the use of conventional hard-sided drafting supply lines **Figure 10-9**. Floating pumps have intake ports and strainers that go below the surface of the water and do not draw air into the pump. The use and operation of portable and floating pumps are discussed in more detail later in this chapter.

Another option for drafting from relatively shallow flowing water sources is to place a dam in the streambed **Figure 10-10**. You can use fallen trees, rocks, soil, a ladder wrapped in a salvage cover, or anything else that will block the flow of water in the flowing stream. This should be considered an option of last resort, because extreme caution must be used to ensure that the following conditions are met:

- The water source, once dammed, will provide adequate quantities of water.
- Creating a dam will not have a negative effect downstream.
- The dam is strong enough to remain in place for the duration of the operation.
- Once the operation is completed, releasing the water will not have a negative effect downstream.

Operational Considerations for Site Selection

Whenever you have multiple, strong options for possible drafting sites, the best choice is determined by the location and the purpose of the drafting operation. If you are directly supplying water for firefighting operations that are nearby, then the closest available drafting site is usually the preferred option. Its use generally requires little coordination with other companies, but you must be mindful of fire apparatus movement in and out of the scene because hose lines will be extended from the pumper to the incident site.

As the distance from the incident increases, your water supply options become somewhat limited. You may have to consider other possible methods for obtaining water to supply your pumper. Consider a water shuttle, relay pumping, or a nurse tanker operation as a source for water. You may have to lay a dry supply line from an access point that can easily be extended from or connected to by another pumper without interfering with the operations of the incident. The other pumper(s) will finish establishing the supply line for you. In such cases, a water supply officer can be beneficial in coordinating the supply operation.

If your purpose for drafting is to supply a water shuttle or to be the source pumper in a relay pumping operation, operational considerations will apply to the site selection, in addition to the other considerations that have already been discussed. Try to select a drafting site whose use will not block access to the fire ground for any additional equipment that may respond to the incident. It is also wise to select a drafting site that allows water tenders in a shuttle operation to be filled without having to back up or turn around. Ultimately, these water sources should be as close as possible to the incident to aid in the quick delivery of water and eventual fire extinguishment.

Pumping from a Draft

The next step in the process of obtaining a water supply from a static water source is establishing a pumping operation from a draft. Once you have found a suitable location with an adequate water source that is close enough to reach, you are ready to set up your pump to draft. This task requires teamwork, and you should be assisted by two or more fire fighters. Hard-sided supply hose (also called hard suction hose or suction hose) is heavy and can be very awkward to handle when fewer than two fire fighters are available. Becoming proficient in this skill through regular drills will help your company provide a water supply quickly, which may mean reduced property damage and saved lives.

Making the Connection

The first step in establishing a drafting operation is to determine when you can connect the hard suction hose to the pump. If the location allows, you should place the fire apparatus in its final position before making any connections, while ensuring that you have adequate space to work around the pump panel and to connect the hose to the intake valve of the pump. If the space is limited, you may have to connect the hard suction hose away from the final position and then slowly move the fire apparatus with fire fighters carrying the hose as you advance toward your final position. This procedure can be very dangerous for the fire fighters as they walk alongside a moving fire apparatus and should not be the first choice for making suction hose connections.

Safety Tip

Be very careful when moving the fire apparatus into the final position for drafting. Fire fighters will be very close to the fire apparatus, carrying the heavy suction hose, and the footing may be very slippery. It is a good idea to have a spotter available to stop you at the first sign of a problem.

Driver/Operator Tip

It is not necessary to engage the pump transmission until after all hose connections have been made. Doing so too early will result in undue wear on the pump, which may lead to damage to the pump, seals, and bearings as a result of water not flowing.

Before any connections are made, the gaskets will need to be inspected for proper placement, cracks, and any debris that might affect your ability to create a vacuum and obtain a draft. Many times a hand-tight connection will be sufficient. Using a rubber mallet to tighten the connections, however, will provide a strong seal and ensure that they do not loosen when they are being moved around during placement. Regardless of the positioning constraints, the hard suction hose must be connected to the pump intake; in fact, it may take two sections to properly submerge the hose into the water.

Once you have determined the length of hose that is needed, a strainer must be placed on the end. The strainer will prevent large debris such as trash, rocks, weeds, small twigs, and animals/fish from entering the pump and causing pump failure and possibly permanent damage. It is important to ensure that the connection between the strainer and the hose is just as tight as the connection between the hard suction hose and the pump intake valve.

Three types of strainers are typically used for drafting operations, each with a specific function. The barrel strainer **Figure 10-11** is most commonly used for deep water sources in which you are confident that the strainer will not be able to contact the bottom of the source or large debris fields. Barrel

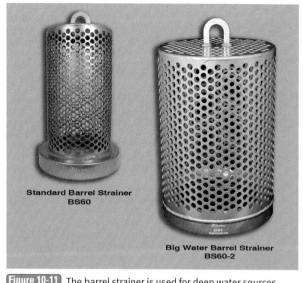

Figure 10-11 The barrel strainer is used for deep water sources.

strainers are cylindrically shaped and generally 10″ (254 mm) to 16″ (406 mm) long and made of aluminum.

The low-level strainer **Figure 10-12** is designed for use in clean, shallow water sources. This square-shaped, flat strainer has an opening between two flat plates of aluminum, typically with a 2″ (51 mm) wide gap between the 16″ (406 mm) plates on the top and bottom of the strainer. It is generally considered a lower-flow strainer because the intake opening is slightly smaller that the diameter of the hose, so as to avoid creating a strong draft that can draw in debris. A low-level strainer is used primarily when drafting from portable tanks because the water is free from debris and the strainer can touch the bottom without fear of dirt or silt entering the pump.

Figure 10-12 The low-level strainer is used primarily for drafting operations from a portable tank.

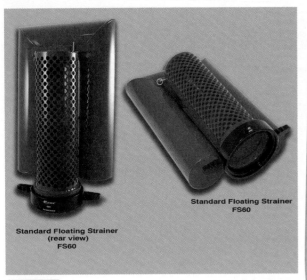

Figure 10-13 The floating strainer is used to avoid debris on the surface and from the bottom of the water source.

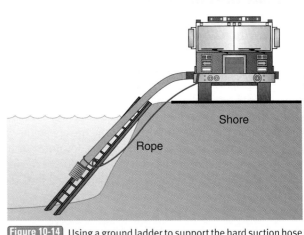

Figure 10-14 Using a ground ladder to support the hard suction hose will keep it off the shore line.

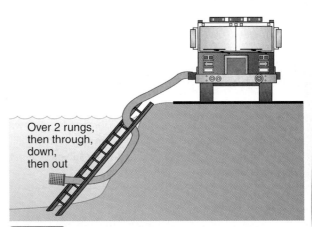

Figure 10-15 Weaving the strainer over and under two ladder rungs will keep it secure in moving water.

The floating strainer **Figure 10-13** is designed to operate below the surface scum, and above the weeds, dirt, and silt in the water source. This type of strainer is ideal for water sources in which the depth or the bottom quality is unknown. Floating strainers can be large and will take up a significant amount of space in fire apparatus cabinets. They may be made out of aluminum or polyethylene and contain a large, hollow chamber that allows them to float. As with the floating pump, the intake port is equipped with a strainer that is placed below the surface of the water and will not draw air into the pump.

Once all of the connections are made, you should be ready to place the hard suction hose and strainer into the water. Before doing so, however, you should tie a rope to the strainer to help you move the hose if needed once it is in the water and to secure the hose once you start drafting water. It is important to ensure that the strainer is completely immersed in the water so that air will not be drawn into the pump. When you are using a barrel strainer, there should be at least 24" (610 mm) of water in all directions around the strainer. The low-level and floating strainers are designed to operate in shallow conditions and require only 24" (610 mm) of water depth. You may wish to use a straight ground ladder to support the hard suction hose as it enters the water and to keep it off the shore line **Figure 10-14**. If you are drafting from a moving water source, you can use a straight ground ladder to weave the strainer over and under two rungs to prevent it from drifting due to the current of the water **Figure 10-15**.

Now that you have the connections made and the strainer securely placed, you are ready to beginning the drafting operation. Follow the steps in **Skill Drill 10-2** to position the hard suction hose in the water:

1. Stop the fire apparatus just short of where you expect to draft from. Leave plenty of room to connect the suction hoses together and then to the fire pump. Remember that these hoses are heavy and do not bend very easily, so allow enough room for your hose team to work. (**Step 1**)
2. Inspect the gasket in the female coupling for damage or debris. The gasket should be pliable enough to make an air-tight seal. (**Step 2**)
3. Connect the sections of suction hose together and install the strainer on the end that is placed in the water. The couplings should be made as tight as possible by hand and then tightened slightly more by striking the tightening lugs smartly two or three times with a rubber hammer. This action ensures an air-tight seal, and it prevents the couplings from working loose as the hoses are moved into position. Do not over-tighten the connections, as this can cut the gasket in the coupling, causing an air leak and preventing you from priming the fire pump. (**Step 3**)

Skill Drill 10-2

Positioning the Fire Apparatus for Drafting Operations

1. Allow plenty of room at the draft site to connect the suction hoses.

2. Inspect the gasket.

3. Assemble the suction hose and strainer.

4. Connect the suction hose to the fire pump. Connect the suction hose assembly and prepare for advancement if necessary.

5. Advance the suction hose assembly while moving the fire apparatus into position. Use extreme caution when operating around moving fire apparatus.

6. Once the fire apparatus is in its final position, set the parking brake and chock the wheels.

7. Ensure that the strainer assembly does not touch the bottom of the water source. There should be at least 24″ (0.61 m) of water in all directions around the strainer for drafting.

8. The suction hose assembly is ready to draft. Engage the pump at this time.

4. Connect the suction hose to the fire pump. If necessary, connect the suction hose assembly and prepare for advancement. **(Step 4)**
5. Advance the suction hose assembly while moving the fire apparatus into position. Use extreme caution when operating around moving fire apparatus. **(Step 5)**
6. Once the fire apparatus is in its finishing position, set the parking brake and chock the wheels. **(Step 6)**
7. Ensure that the strainer assembly does not touch the bottom of the water source. There should be 24″ (0.61 m) of water in all directions around the strainer for drafting. **(Step 7)**
8. The suction hose assembly is ready to draft. Engage the pump at this time **(Step 8)**

Driver/Operator Tip

Some situations may require ingenuity on your part to create a solution that will allow you to draft successfully. For example, in shallow water you may need to dig a hole in the bottom of the static water source or dam up the stream with a ladder and a tarp to provide sufficient water depth for drafting. Securely fastening a scoop shovel, pointed downward, underneath the strainer will prevent it from lying directly on the bottom of the static water source.

Preparing to Operate at Draft by Priming the Pump

The first step in preparing to draft from a water source is to prime the pump. Once the suction hose has been connected and placed into the water source, air is drawn from the pump using the priming pump. Activation of the priming pump reduces the atmospheric pressure inside the fire pump and the hard suction supply hose. The atmospheric pressure outside the pump presses down on the water source, which in turn causes water to rise inside the hard suction supply hose and replace the air being pumped out. This process continues until all of the air is removed from the pump and is replaced with water.

The steps involved in priming a fire pump are very similar to those used to test the priming pump or conduct a vacuum test. Before you begin, ensure that all of the drains and valves are closed. This will prevent air from entering the pump while you are trying to remove the air already in it. Firmly pull the priming pump handle out until it stops and hold it out; you will hear a loud whirring sound as the air is pumped out. Water and air will also be discharged under the fire apparatus; this is residual water in the pump that is being mixed with the air being pumped out. As you hold the handle out, a vacuum reading will appear on the supply master gauge, indicating that a draft is starting. More water will start to flow out of the priming pump and be discharged under the apparatus when the air is pumped out. Do not stop operating the priming pump until you note pressure on the discharge side; at that point, you can slightly advance the throttle.

Driver/Operator Tip

Most modern fire apparatus in use today is equipped with single-stage pumps. If you are drafting with a two-stage pump, it is recommended that you switch to the "volume" or "parallel" stage before you begin any drafting operation. This will allow the priming pump to completely evacuate all of the air that may become trapped in the "pressure" or "series" stage.

Drafting Operations

Once you have primed the pump, you are ready to start drafting water and producing discharge pressures **Figure 10-16**. Before you start supplying water to handlines, master streams, or a relay pumping operation, you will need to establish a dump line. A **dump line** is a small-diameter hose line (booster line, 1¾″ hose, or 1½″ hose) that remains in the open position and continues flowing water during the entire drafting operation. This line will help to maintain the draft even if your discharge pressures are low because it is constantly flowing water that has been drafted from the water source to start the flow of water. *Slowly* open the corresponding discharge valve for the dump line and begin flowing water. If the valve is opened too fast, you may cause a loss of vacuum and break the draft of the pump. You will have to slightly increase the engine speed (rpm) to maintain the draft.

The water from the dump line should be discharged back into the water source, away from the strainer, to avoid introducing air into the hard suction hose; an inflow of air may lead to loss of

Figure 10-16 Prepare the pump for drafting.

the vacuum and end the pump's draft. If you cannot discharge the dump line back into the water source, you will have to discharge the water onto the ground, away from your drafting site.

> **Driver/Operator Tip**
>
> If you must flow water onto the ground, make sure you discharge the water far enough away from the fire apparatus and drafting site that it will not saturate the soil under the fire apparatus. Otherwise, the instability may cause the fire apparatus to sink, slide, or tip over.

Once you sustain a flow of water through the dump line, you can begin to attach the hose lines that you will be supplying; these lines may differ depending on the purpose of your drafting operation. It will speed up your operation if other crew members can make these connections while you are establishing the draft and dump line. Once water is flowing from the dump line and the lines being supplied are attached, you are ready to increase the amount of water that will flow.

To provide water to the operation or to master streams and handlines from a static water source, follow the steps in **Skill Drill 10-3**:

1. **Prepare the pump for drafting.** Close all discharge valves and cap any suction fittings not being used.
2. Close all drain valves and tank valves.
3. Close the auxiliary cooler line (also called a booster cooler line).
4. Move the transfer valve to the "volume," "parallel," or "capacity" position if the pump is a two-stage pump.
5. Establish a dump line. This line will be used to flow a small amount of water to cool the pump while you are drafting but not otherwise flowing water. Make sure the water from this line will not run under the fire apparatus, as it may soften the ground and cause the fire apparatus to sink. **(Step 1)**
6. **Place the pump in gear.** Shift the pump into gear by following the pump manufacturer's recommended procedure. Confirm that the pump is correctly engaged by checking the shift indicator light or checking the reading on the speedometer.
7. Engage the shift lock mechanism (if the pump is equipped with one). **(Step 2)**
8. **Prime the pump.** Advance the hand throttle to the manufacturer's recommended setting (generally 1200 to 1500 rpm).
9. Pull the primer control handle out firmly until it stops, and hold it in that position. You will hear a loud whirring sound as the priming pump begins pumping air from the fire pump.
10. Notice that the suction hose begin to settle as it fills with water and gets heavier.
11. Look at the master supply gauge on the pump panel, which should begin to show a pressure reading on the negative side of zero. This measurement, which is given in inches of mercury, is another indication that the priming pump is developing a vacuum within the fire pump.
12. Continue holding the primer control handle in the fully engaged position until the whirring sound changes to a gurgling sound, and you see water being discharged from the priming pump onto the ground. You should also see some indication on the master discharge gauge of pressure being developed by the fire pump.
13. Push the primer control handle all the way in to the fully closed position. The fire pump is now primed and ready to begin delivering water. **(Step 3)**

    Note: Some manufacturers recommend priming the fire pump before placing the pump into gear. This step is taken because some newer pumps have a much stronger seal on the pump shaft; this seal will not tolerate operating without water to cool and lubricate it. If your equipment is of this type, simply reverse the steps described here. The only difference is you will not see a pressure indication on the master discharge gauge when the pump is primed because the fire pump is not turning; all other indicators will appear the same during the priming process.
14. **Adjust the discharge pressure.** Adjust the hand throttle or computer governor to obtain the desired discharge pressure.
15. **Slowly** open the discharge valve for the hose line that has been identified as the dump line. Opening the valve quickly may lead to a loss of the vacuum and cause the pump to lose draft.
16. Begin flowing a small amount of water back into the water source through the dump line for pump cooling purposes. **(Step 4)**
17. **Begin the water delivery.** The fire pump is now ready to begin delivering water to attack handlines, supply lines for an attack pumper, a relay pumping operation, a master stream appliance, a sprinkler system, or a standpipe system. You may begin supplying water according to your fire department's standard operating procedures (SOPs) as soon as the hoses are connected and you receive the order to do so. **(Step 5)**

    Note: Operation of the fire pump from this point will be very similar to pumping operations utilizing a pressurized source. These pumping operations are covered in detail in Chapter 7. Be sure to continuously monitor the fire apparatus for potential problems during the drafting operation.

Producing the Flow of Water

Whether you are supplying attack lines or are supplying water for the fire attack pumper, make sure the crews know that you are ready to start flowing water *before* you open the discharge valve. This notification can be made over the radio, through face-to-face contact with the crew, or through hand signals as established by the organization.

Slowly open the corresponding discharge valve for the hose line and begin flowing water. As with the dump line, if the

Skill Drill 10-3

Drafting from a Static Water Source

1. Prepare the pump for drafting.

2. Place the pump in gear.

3. Prime the pump. (Note: Steps 2 and 3 may be reversed if the pump manufacturer recommends priming the pump prior to shifting it into pump gear.)

4. Adjust the discharge pressure.

5. Begin the water delivery.

hose line is opened too fast, you may cause a loss of vacuum and end the pump's draft. You will have to increase the engine speed (rpm) by increasing the throttle to maintain the draft and provide adequate discharge pressure. Opening the discharge valve and increasing the engine speed are generally done simultaneously, with the emphasis being placed on increasing the engine speed over flowing water. The flow of water into the pump is usually slower than the discharge flow because the energy that it takes to create a vacuum is greater than the energy required to discharge the water. Continue to increase the throttle until you have reached the desired discharge pressure; then set the pressure relief valve or governor to prevent sudden spikes in pressure.

Once the draft has been accomplished; don't drop the draft until the operation is completed *and* either the water supply officer or the IC gives the order to do so. Establishing a drafting operation takes time, and the incident should not be left without

Skill Drill 10-4

Providing Water Flow for Handlines and Master Streams

1. Ensure that hose lines or pumpers are ready for water. Open the corresponding discharge valve slowly.

2. Increase the engine's speed (rpm) to achieve desired pressure.

3. Set the pressure-regulating mechanism.

---

a water supply unless directed through the incident command structure.

Following the steps in **Skill Drill 10-4** will help you provide the needed flow to handlines, master streams, and supply lines:

1. Ensure the receiving hose lines or pumpers are ready to receive water.
2. **Slowly open the corresponding discharge valve** for the hose line and begin flowing water. If it is opened too fast, you may cause a loss of vacuum and end the pump's draft. **(Step 1)**
3. **Increase the engine's speed (rpm)** to maintain the draft and provide adequate discharge pressure. Open the discharge valve and increase the throttle simultaneously until you have reached the desired discharge pressure. **(Step 2)**
4. **Set the pressure-regulating mechanism.** Adjust the pressure governor or pressure relief valve to the desired pressure. This is the initial setting—recheck it once water delivery commences. **(Step 3)**

Complications during Drafting Operations

You must be able to recognize some of the common problems that you may encounter while attempting to draft or during the operation itself. While operating on the shore of a static water source during an emergency incident is not the time to have to call in a mechanic to help diagnose a common problem. As a driver/operator, it is your responsibility to know what the problem is and how to correct it. This section highlights a few of the more common problems encountered during drafting operations and describes the actions needed to identify and correct them.

The driver/operator must continuously observe the in-take pressure while drafting. If you see any drop in pressure, check the strainer for the presence of debris. If you are flowing large volumes of water, you will be drafting (sucking) large volumes of water from your water source—and this action may draw debris from areas that were not visualized when you initially set up the operation. Simply wipe the strainer clean while it is underwater to clear the debris; be sure to remove the debris from the water source.

You must also continuously observe the engine and pump temperature gauges. Drafting operations require these components to work hard; sometimes, not enough water will flow to keep them cool. Remember, if you are simply supplying a dump line while you wait to fill a tanker shuttle operation, your flow will be less than 100 GPM. The decrease in flow will result in an increase in component temperature. Open and close the cooling lines as recommended by your fire apparatus manufacturer to prevent engine or pump overheating conditions.

If not enough water covers the strainer or if the draft is strong as a result of flowing large volumes of water, a whirlpool may form as a vacuum draws in water—just like in your kitchen sink when you release the stopper to drain the water. This whirlpool allows the pump to suck in air, which can break the vacuum and cause a loss of water supply. To avoid this problem, you can place an inflatable ball into the whirlpool, which acts to seal the whirlpool and prevent air from being sucked into the pump.

Another problem you may encounter during drafting operations is <u>cavitation</u>, which occurs when you attempt to flow water faster than it is being supplied to the pump. Cavitation during drafting operations is usually caused by a vacuum leak, which introduces air into the intake water supply. As the air flows through the pump, it creates small water hammers along its path through to the discharge point. Simply correcting the vacuum leak will correct the cavitation problem.

A more technical explanation of cavitation requires us to follow the path of water through the fire pump. As water is being drawn into the pump, the supply pressure is reduced owing to lift, friction loss, and/or flow restrictions. When the supply pressure is reduced to the point it registers as a vacuum, the boiling point of the water is reduced. When the boiling point is reduced sufficiently to match the ambient temperature of the water, the water will actually vaporize and form pockets of steam in the water supply. As these steam pockets travel through the pump to the discharge side, they are subjected to the discharge pressures; as a consequence, the steam pockets instantly collapse and re-form as water, creating a slight water hammer. The more severe the cavitation becomes, the more severe each water hammer will be.

Cavitation affects a fire pump's ability to deliver water and, if allowed to continue, can actually damage the pump. This problem is fairly easy to diagnose because it causes the discharge pressure reading on the pump panel to fluctuate significantly. Ultimately, the fluctuations become so severe that you cannot determine what the discharge pressure reading is. You will also hear a rattling noise that sounds like the pump is pumping gravel. When a pump is cavitating, its discharge pressure does not respond to an increase in the throttle (rpm) as it does under normal operations. In fact, increasing the throttle setting during cavitation generally makes the cavitation worse. You may also hear a sputtering or popping sound from the nozzles on your discharge lines, which is most apparent on the dump line because it is the closest line to the pump panel.

If you suspect cavitation during a drafting operation, follow these steps:

1. **Check the water around the intake screen on the hard supply hose.** The number one cause of cavitation during drafting is formation of a whirlpool above the intake screen, which in turn draws air into the intake hose. This problem occurs when a sufficient water level is not maintained above the intake. To alleviate this problem, lower the intake deeper into the water, construct a dam to back up the water in a stream, use a low-level strainer in portable drop tanks, or place a beach ball or similar round object in the whirlpool to stop it from introducing air in the supply intake.
2. **Check the vacuum reading on the master supply gauge.** This reading should indicate approximately 1 in. Hg (2.54 cm Hg) of pressure for each foot of lift under normal operating conditions. If the vacuum reading is at or near normal, suspect that a vacuum leak is allowing air to enter the water supply, causing cavitation. If the vacuum reading is significantly higher than normal, suspect cavitation due to supply restrictions.
3. **If you have a near-normal vacuum reading**, check for vacuum leaks by confirming the tightness of all supply hose couplings. Next, recheck all valves at the pump panel. Also check the water level in the booster tank, and gradually refill the booster tank if it is not full. If signs of cavitation persist, there may be a problem requiring the fire apparatus to be taken out of service for repairs. It is possible to continue drafting in most cases even with a vacuum leak, if it is necessary to continue the support of fire-ground operations. The key is to keep water moving through the pump at all times, which prevents the accumulation of air in the pump housing and avoids loss of the vacuum (priming). You can accomplish this by flowing water through the dump line whenever you are not supplying water through your discharge lines. You need to advise your IC and your hose teams of the problem if you are supplying attack lines. Also, you need to advise the attack pumper if a relay operation is under way.
4. **If you have a higher than normal vacuum reading**, check the intake screen: The second leading cause of cavitation is a plugged intake screen. An obstruction will restrict the ability of water to pass through into the intake hose and cause the vacuum on the supply side of the pump to increase. In such a case, you must clean the screen to allow the drafting operation to return to normal. If the screen is clean and the supply vacuum reading is still high, then confirm that the suction valve is fully open (if one is being used). If everything checks out and the vacuum reading is still high, try reducing the throttle slightly or gating down the discharges slightly; you are probably attempting to pump water faster than it can be supplied to the pump. You should notice an immediate improvement in the cavitation situation once the pump's discharge is lowered slightly.

The failure of the pump to prime is one of the most common problems that occurs during drafting operations. Many times the cause of the problem is driver/operator error—in many areas, pumping from draft is not a routine operation. Make sure that you are not the cause of the problem by double-checking the steps required for priming a pump. The fire pump on your fire apparatus should prime in 30 seconds or less in most cases.

An inability to prime the pump may be caused by any of several factors. Following these diagnostic steps will help you identify the problem and quickly find a solution:

1. Note the vacuum reading on the supply master gauge after holding the primer valve fully open for 20 to 30 seconds.
2. If the vacuum reading is less than 12–15 in. Hg (30.48–38.1 cm Hg), recheck all valves on the pump control panel to verify that they are in the closed position. This includes the supply, discharge, recirculation, and drain valves. The only valve that should be open is the one to which the hard suction line is connected (if you are using a gated suction). Recheck the couplings on the hard suction line to make sure they are tight. Make sure the suction strainer is completely submerged. Try priming the pump again. If you still cannot prime the pump and the booster tank is empty, try adding one or two 5-gal (19.93-L) buckets of water to the booster tank. Often a worn tank-to-pump valve will be the source of a vacuum leak. If you can cover it with water, you can generally obtain a prime. You should also check the primer oil

reservoir to make sure sufficient lubrication oil is available for the efficient operation of the priming pump. If you are still unable to obtain a prime, you may have other vacuum leaks, or mechanical problems with the pump or the primer system. These problems are generally not correctable on the fire ground and require that the fire apparatus be taken out of service for repairs.

3. If the vacuum reading is less than 12–15 in. Hg (30.48–38.1 cm Hg), make sure the intake screen is clean and you are not attempting to lift water too high. You should not attempt to lift water more than 15′ (4.57 m). Reposition the fire pump so the lift is 10–15′ feet (3.05–4.57 m), which is a more practical lift. Make sure the suction valve is open (if one is being used). Try to prime the pump again. If you are still unsuccessful and the vacuum reading is high, remove the suction hose from the pump and check both the inlet strainer on the fire pump and the inside of the suction hose for an obstruction. Reconnect the hard suction hose, and try to prime the pump again. If you are still unsuccessful and the vacuum reading is high, replace the hard suction hose with another hose and try again. It is possible you have a defective suction hose, which is collapsing internally when vacuum is applied.

Uninterrupted Water Supply

Establishing an uninterrupted water supply is one of the most important objectives for any IC. In areas where municipal water supplies are not readily available or static water supplies are not close to the incident, the IC must establish a plan to obtain an uninterrupted water supply through a relay pumping operation, a tanker shuttle, or a nurse tanker operation. These operations can be labor and equipment intensive. Appointing a water supply officer to oversee them will ease the burden on the IC.

Relay Pumping Operations

A <u>relay pumping operation</u> can be one of the most complex because it requires at least two pumpers, and sometimes as many as four, to form a relay. In this process, water is pumped from a water source (hydrant or static) through hose under pressure to the apparatus engaged in fire suppression efforts Figure 10-17 ▶ . Relay operations can be as simple as one fire pumper at a pond or lake and the other fire pumper at the fire scene, or as complex as the use of multiple pumpers to supply water over a long distance to the fire operation.

Components of a Relay Pumping Operation

To establish any type of relay, you must have a minimum of two fire pumpers, hose lines, and personnel. One fire pumper is located at a water source; it is called the source pumper Figure 10-18 ▶ . The other fire pumper is at the fire scene; it is called the attack pumper Figure 10-19 ▶ . Any other fire pumpers that are between these two apparatus and are positioned to maintain flow pressure in the relay are called relay pumpers. Two medium-size or large-diameter hose lines are used to supply water in this scenario. These operations will demand the use of personnel as they are being established, especially at the source pumper. Once the relay is established, generally the only crew member who needs to remain with the fire apparatus is the driver/operator, as long as the area around the fire apparatus is safe from traffic flow. If it is not, a second member should be left behind to assist the driver/operator; all other crew members could be used as resources if the IC chooses to do so.

Source Pumper

The source pumper is the most important pumper in a relay because it is located at a water source and supplies the water to the entire incident. The source of the water is either a fire hydrant or a static water source. The source pumper should consist of the largest fire pump of the units assigned to the relay pumping operation, thereby ensuring that it can provide the maximum amount of water. For example, if the source pumper has a 1500-GPM pump, it should supply the 1000-GPM pump on the relay pumper; this setup ensures that the 1000-GPM

Figure 10-17 A relay pumping operation.

Figure 10-18 Source pumper.

Figure 10-19 Attack pumper.

Figure 10-20 A relay pumper.

pump can be supplied with enough volume of water to meet its capacity if needed. If the reverse were the case, then the larger pump would not be supplied its capacity because the incoming flow would be less than 1500 GPM.

Attack Pumper

The attack pumper is usually the first unit on the scene of an incident, and its operation will dictate how much water will be needed, based on how many GPM the fire attack requires. If the attack pumper cannot establish its own water supply, it may lay a dry supply line into the scene to be filled later by the source pumper or a relay pumper. If the attack pumper is being positioned in an area that has limited access, it is generally recommended that a supply line be laid down from an open area to the limited access point. Another responding engine should then complete the connection and provide a water supply to the attack pumper.

Upon its arrival at the incident, the attack pumper will start supplying hose lines while operating from its water tank; it will then switch over to the relay supply line when it is ready. This two-step process provides a challenge to both the driver/operator and the initial IC, as the fire attack is initially limited by the capacity of the attack pumper's water tank. The attack pumper will be able to support additional hose lines only if the incoming water supply from the relay operation will support them. For example, if the relay operation is supplying 300 GPM of water, the attack pumper will be able to provide slightly less than 300 GPM of discharge flow to the hose lines. The amount of flow available is valuable information for the IC and must be identified so that the IC can make tactical adjustments to the overall strategy until the water supply can be improved. Ultimately, the attack pumper is the "workhorse" at the incident: It provides water to all of the hose lines that are attacking the fire. If the attack pumper does not have water, then there is no attack of the fire taking place.

Relay Pumpers

Relay pumpers are the fire apparatus that are placed in the middle of the relay pumping operation Figure 10-20. They obtain their water from the source pumper and increase the pressure to the next fire pumper in the relay. In essence, these pumpers act like booster pumps to maintain pressure throughout the length of the relay. A relay pumper cannot increase the volume of water (GPM) being pumped; it can only increase the pressure because the source pumper provides the flow (quantity) based on the volume being discharged by the attack pumper. If the source pumper is flowing 800 GPM at 80 psi, the relay pumper will receive a flow of 800 GPM but, due to friction loss in the hose lines, the water will be at a pressure of only 50 psi. The relay pumper will increase the pressure back to 80 psi and continue to flow 800 GPM to the next pumper, which will then increase the pressure again to overcome the friction loss until, ultimately, the 800 GPM flow reaches the attack pumper.

Equipment for Relay Pumping Operations

Fire hose is the primary path of travel for the water to get from one fire pumper to another in a relay operation. The hose can consist of a single large-diameter hose (LDH), which is 4″ or 5″ in diameter, or multiple medium-size hoses, which are 2½″ or 3″ in diameter. Single lines of LDH, by their design, have a very low friction loss variable—7 psi when flowing 1000 GPM in a 5″ hose line. Compare this to a single 3″ medium-size hose, which has a friction loss variable of 21 psi for just 500 GPM. To overcome the higher friction loss associated with use of medium-size hose lines, it is recommended to use two hoses for a water supply. To avoid this issue altogether, it is highly recommended that LDH be used whenever possible. To determine the friction loss for a relay pumping operation, you must know how much fire hose is in the part of the relay that you are supporting once it is established. When you determine how much hose will be flowing water, you can calculate the friction loss for your part of the relay operation and adjust your discharge pressure accordingly.

> **Driver/Operator Tip**
>
> How can you quickly determine the length of your supply line? Color the middle sections of the hose in your hose bed. Assuming your pumper carries 1000' (305 m) of supply hose, color-coding the hose at the halfway point will make it easier to calculate the length of the hose lay. If you have three 100' sections of hose past the colored hose, you can quickly determine that you have 800' of hose out. Color-coding the last section or two will help you recognize that you are reaching the end of your supply hose before you get too far and then realize that you ran out.

Relay pumping operations may require some other standard equipment, such as hose adapters to connect hoses with different types of threads, a Siamese hose adapter (which merges two medium-size hose lines into a single LDH), a wye (which splits a single LDH into two or three medium-size hose lines), and reducers (which connect larger single hose lines with smaller single hose lines). No matter which size hose you are using to supply water from the source pumper, each pumper in the relay operation and the attack pumper must be equipped with an intake relief valve. Most modern pumpers are equipped with these valves at the time they are built, but older pumpers may need to be retrofitted with an external intake relief valve. Such a valve will prevent the incoming water supply from reaching an excessively high pressure, which may damage the pump and possibly interrupt the relay operation. Relief valves discharge the excessive pressure through a port located on the valve.

■ Personnel for Relay Pumping Operations

A safe relay pumping operation requires that you have adequate personnel for the operation. A minimum of two crew members should be assigned to each fire pumper if the pumper will be exposed to moving traffic and other hazards around the fire apparatus. As the driver/operator, it is your responsibility to operate the pump and manage the intake and discharge pressures. The second fire fighter is responsible for managing the area around the fire apparatus so as to ensure the safety of the members by monitoring vehicle movement around the fire pumper and warning you of possible dangers.

■ Preparing for a Relay Pumping Operation

The length of a relay pumping operation is primarily determined by the distance that the water source is from the incident. The

Figure 10-21 Forward lay.

amount of water required by the attack pumper and the size of the supply line can also affect whether a relay pumper is required and, if so, how many are needed. As discussed earlier, smaller-diameter supply hose lines decrease the flow capacity and increase the friction loss. Generally, LDH represents the best option for relay pumping operations.

Given that supply hose is critical for a successful relay pumping operation, it must be in place before water can be flowed. A **forward lay** Figure 10-21 ▲ is a supply hose line that is brought to the scene by a fire pumper from the water source. An engine that connects to a fire hydrant and then proceeds to the fire scene is an example of a forward lay. Hose can also be laid backward from the attack pumper to a water source or to another fire pumper in the relay line; both are examples of a **reverse lay** Figure 10-22 ▶ . Alternatively, one fire pumper may lay hose into the scene while another fire pumper lays hose out from the first engine's coupling to the water source; this approach is called a **split lay** Figure 10-23 ▶ . Another type of split lay involves laying a line from an open area to the pumper when there is limited access and having a second pumper connect to that hose and move to the water source.

Once you have the hose in place for the relay, you will need to determine whether a relay pumper is needed by calculating the amount of friction loss for the length of the hose lay. If the discharge pressure will exceed 80 psi for LDH or 175 psi for medium-size hoses, a relay pumper is needed to boost the water pressure. A good rule of thumb is that no relay operation should extend for more than 800'. Maintaining relay operations over a shorter distance will keep the pump discharge pressures manageable while ensuring the proper quantity of water flows to the attack pumper. Keep in mind that the attack pumper's needs are the deciding factor in terms of the pressure and flow needed.

Adapters, reducers, manifolds, and appliances may also be used in a relay pumping operation. When two fire pumpers use different threads on their couplings for hose, **adapters** allow for both pumpers to connect to the same hose. When the sizes of the hose lines are different, **reducers** are needed to connect the mismatched lines together. **Manifolds** greatly increase

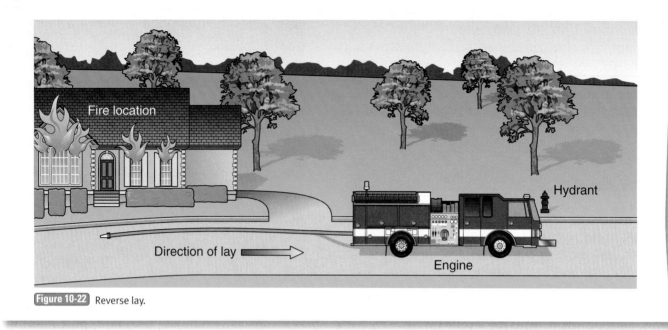

Figure 10-22 Reverse lay.

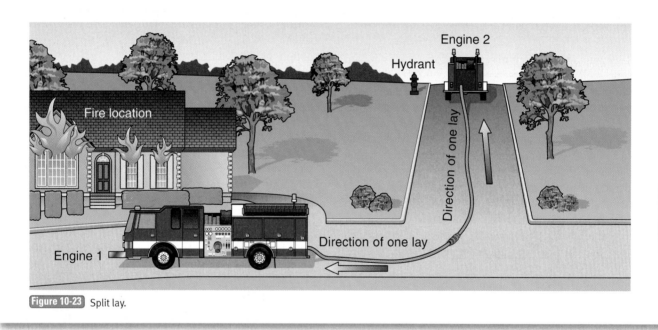

Figure 10-23 Split lay.

the effectiveness of relays by allowing one LDH to supply two to four medium-size lines with water. When a LDH is split into multiple hose lines, be aware that the overall friction loss increases due to the compression of the water back into smaller vessels in addition to the flow restrictions within the appliance itself. An appliance is any other component that you can flow water through. The Siamese and gated wye are examples of appliances that, when used in a relay pumping operation, must be taken into account when calculating the friction loss.

Calculating Friction Loss

As stated earlier, the attack pumper's needs drive the relay pumping operation. For this reason, the attack pumper's driver/ operator must communicate to the source pumper's driver/operator how much water (GPM) is flowing now or is planned to flow. The attack pumper driver/operator can determine his or her flow quickly by adding together the flow of each handline or master stream that the attack pumper is discharging. For example, if the attack pumper were flowing 500 GPM from three handlines, the driver/operator would communicate to the source pumper that he or she needs 500 GPM.

The driver/operator of the source pumper would then calculate the distance between the two pumpers in hose. For this example, assume it is two lines of 3″ hose laid out in 400′. To calculate the friction loss, you will divide the flow in GPM by the number of hose lines being used—two lines, in this case. Each line will flow half of the water required; 500

GPM divided by 2 equals 250 GPM. A 3″ hose line at that flow has a friction loss factor of 6 psi per 100 feet, which is multiplied times the number of feet in hundreds (4, in this example), which gives us a total friction loss of 24 psi. Once we have calculated the friction loss, we must add 20 psi for residual pressure to prevent cavitation, for a total discharge pressure of 44 psi.

Attack pumper is flowing 500 GPM; source pumper is 400′ away and has connected to the attack pumper with two 3″ lines:

500 GPM ÷ 2 (number of lines) = 250 GPM
Friction loss for 3″ hose line for every 100′ = 6 psi
6 psi × 4 (number of feet in hundreds) = 24 psi + 20 psi
(residual pressure) = 44 psi pump discharge pressure

■ Types of Relay Pumping Operations

Two basic types of relay pumping operations are performed. The first type of relay pumping operation is called a calculated flow relay. It requires the driver/operator of the source pumper to obtain from the attack pumper the flow (GPM) required for the fire operations; the driver/operator must then calculate the friction loss for each hose layout in the relay and add 20 psi for residual pressure at the attack pumper. This type of relay is the most common, as it uses actual flow calculations to determine the quantity of water required.

The second type of relay pumping operation is called a constant pressure relay. In this type of operation, the source pumper and the relay pumpers supply a constant pressure and flow for the operation regardless of the flow (GPM) being discharged by the attack pumper. Many fire departments have standardized the supply pressures that they use for constant pressure relays, based on prior experience. For example, pressures of 175 psi and 80 psi for medium-size hose lines and LDH, respectively, work best in most situations. Use of a constant pressure relay eliminates the need for pump calculations unless excessive elevation changes are present, in which case you can simply calculate the changes and add (or subtract) the required pressure. The pumper driver/operator in this type of relay sets the pump discharge pressure to 175 psi or 80 psi, depending on the size of the supply line, and maintains this pressure for the duration of the relay pumping operation. Setting the pressure relief valve or pressure governor will maintain the desired pressure in the event of any changes at the attack pumper and will prevent over-pressurizing of hose lines or damage to the pump.

■ Operating the Source Pumper

The relay pumping operation starts at the source pumper and is where the largest fire pump should be located. The hose between the pumpers should be placed through either a forward, reverse, or split hose lay. Remember, LDH provides the highest efficiency for relay pumping operations. Once at the water source, the source pumper will begin a drafting operation from a static water source (see Skill Drill 10-3) or will connect to a fire hydrant.

When setting up a source pumper at a static water source, you must consider the maximum amount of water available for the operation. Once you have established the draft and are flowing water through the dump line, connect the hose lines to be supplied and notify the receiving driver/operator that you are ready to start flowing water. To determine the flow rate, you will need to calculate the friction loss for the supply hose based on the GPM rate to be supplied, the length of the relay, the size of the hose being used, plus a residual pressure of 20 psi. If the next pumper is not ready, then wait until it is ready before initiating the flow. If you are supplying the attack pumper, advise it that you are ready; the attack pumper will then tell you when to send the water.

Whether you are supplying the next pumper in a relay operation or the attack pumper, the order of the steps remains the same. Open the corresponding discharge valve and slowly begin the flow of water while simultaneously advancing the throttle as you are opening the discharge valve; you do not want to lose the draft of the pump (i.e., interfere with the pump's vacuum strength) by flowing more water out than is being supplied to the pump. When you reach the desired pressure (calculated or constant), set the pressure relief valve or the pressure governor to prevent damage to your pump if the flow is stopped or interrupted by a relay pumper or the attack pumper.

Establishing a relay pumping operation from a hydrant works in the same way as using a hydrant as a source of water for your own supply; the only difference is that you are supplying another pumper and not the handlines attacking the fire. Select a suitable hydrant that is as close to the incident site as possible. Flush the hydrant before connecting to it with a soft-sided supply hose line or LDH. If neither of these types of hose is available, you can connect to the hydrant using two medium-size hose lines to establish a water supply. Contact the receiving driver/operator that you are ready to start supplying water. If the next pumper is not ready, wait until it is ready before you begin to flow water. If you are supplying the attack pumper, advise its driver/operator that you are ready to begin the flow; the attack pumper crew will then tell you when to send the water. Fill the supply lines slowly by opening the discharge valves. Advance the throttle to the desired pressure (calculated or constant) and monitor the residual pressure from the hydrant. You will need to set the pressure relief valve or the pressure governor to prevent damage to your pump if the flow is stopped or interrupted by a relay pumper or the attack pumper.

■ Operating the Attack Pumper

The attack pumper is generally the first pumping apparatus to arrive on the scene of a fire. It establishes the hose lines that attack the fire: two 1¾″ (4.4 cm) hose lines flowing at 100 GPM each, requiring a total water flow of 200 GPM from the pumper. If you start and continue flowing water from the tank on the fire pumper until the relay operation is set up, you will have to monitor your tank level closely to ensure that interior fire crews do not run out of water. Most fire pumpers have 500- to 1000-gal water tanks, which allow them to supply the two 1¾″ hose lines for 2½ to 5 minutes before running out of water.

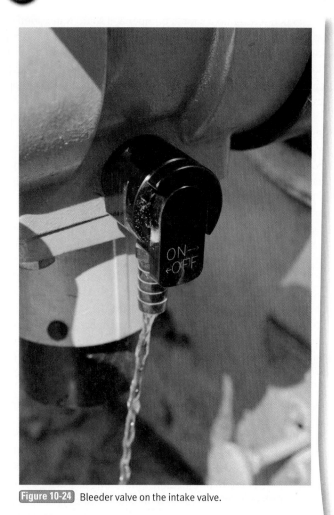

Figure 10-24 Bleeder valve on the intake valve.

attack pumper. When it is full, close the tank fill valve. From this point forward, the attack pumper will be running solely off the relay or source pumper(s).

Advise the source pumper of how much water you need and indicate whether the incoming pressure is sufficient to maintain at least 20 psi of residual pressure. If you need more flow, contact the personnel providing the supply quickly to communicate your needs, but *do not* try to get more water than the source pumper is supplying, as this will cause your own fire pump to cavitate. Until you receive the required pressure/flow, you must decrease the pressure to the handlines that the attack pumper is currently supporting. It is important to notify the crews operating those lines of the situation so that they may make tactical adjustments or retreat to a safe location until the water supply issue is resolved.

■ Operating the Relay Pumper

The operation of the relay pumper can be quite simple: All you are doing is laying out hose between the next pumper and your location and then supplying the required pressure and volume to the next pumper in line or to the attack pumper. A relay pumper may receive its supply from the source pumper or another relay pumper, depending on the distance between the water source and the attack pumper. If you participate in establishing the relay operation, you will need to provide hose to the receiving pumper, another relay pumper, or the attack pumper, through a forward, reverse, or split hose lay. Remember, LDH is the most efficient choice of hose for relay pumping operations. You will need to calculate the friction loss for the supply hose based on the flow rate (GPM) to be supplied, the length of the relay, the size of the hose being used, plus a residual pressure of 20 psi. Once the hose is in place, the source pumper will contact you to advise that the water flow is ready to start.

A relay pumper is considered ready when the supply hose to the pumper and the discharge hose to the next pumper are connected. When the water fills the supply hose, bleed off the air in the hose line from the bleeder valve located on your pumper's intake valve, until a steady stream of water comes out of the bleeder. Remember, your supply line will contain more air if the hose lay is long. Slowly open the intake valve to allow the water to enter the pump. Your tank-to-pump valve should be closed and the pump recirculating valve should be open, which will help prevent the pump from overheating until you start flowing larger quantities of water. Open the discharge valve(s) for the corresponding discharge port(s) to begin flowing water, and adjust the throttle to increase the flow to the desired pressure based on your calculations for friction loss or the predetermined pressure for the relay.

As the driver/operator of a relay pumper, you must monitor the incoming residual pressure from either the source pumper or another relay pumper; your goal is to maintain 20 psi of residual pressure. The residual pressure can be increased by contacting the source pumper to request an increase in pressure or by decreasing your flow pressure, which can negatively affect the lines supplied by the attack pumper. If the flow or pressure to

Getting the water from the source pumper, whether directly or through a relay is critical to continued operations and fire fighter safety. Once the supply lines are charged, bleed off the air in the hose line from the bleeder valve located on the pumper's intake valve, until a steady stream of water comes out of the bleeder **Figure 10-24**. The air is present as a result of the hose being filled with water; this air is displaced by the water as it flows into the hose, and is forced to the end of the line. It is always a good practice to bleed your supply lines; this action prevents air from entering the pump and causing it to cavitate. The longer the hose lay, the more air that will be displaced.

Slowly open the intake valve while backing down the throttle until the intake valve for the supply line is fully open. The pressure from the incoming water supply will increase the discharge pressure to the hose lines unless you decrease the throttle rate. Close the tank-to-pump valve and slowly open the tank fill valve; this action will prevent the pump from creating a vacuum in the tank and drawing air into the pump through the open valve into a tank containing little or no water. Only slightly open the tank fill valve; too large an opening will take away most of the incoming pressure, so that the handlines will not receive the required pressure and flow. Fill the tank on the

the attack pumper is ever affected, you must communicate that fact to the driver/operator of the source pumper immediately.

Once the flow is stabilized, set the pressure relief valve or the pressure governor to prevent damage to the fire pump if the flow is stopped or interrupted by either the supply or discharge side of the pump. Remember that you are part of a relay operation and are affected by the actions of others supplying your pumper or those receiving your supply.

> **Driver/Operator Tip**
>
> If you will be operating for extended periods of time without flowing large volumes of water, consider establishing a dump line away from the fire apparatus so you can continue to flow water without overheating the pump.

Water Relay Delivery Options

Depending on the nature of the relay pumping operation, the last pumper in the series may supply two attack pumpers. If this is the case, then the last pumper will use a manifold—that is, one LDH that is split into four or five medium-size hose lines that provide the water supply for multiple attack pumpers **Figure 10-25**.

To use the manifold in a relay pumping operation, the last relay pumper provides a supply line to the manifold via a LDH from a discharge port. If two attack pumpers are being used at different locations within the incident, each can be supplied with two medium-size hose lines through the manifold. Once the medium-size hose lines have been attached, contact the attack pumpers to advise them that you are ready to flow water. At that point, the valves on the manifold can be opened to start the flow of water. The pumper supplying the manifold will need to account for the additional friction loss of the appliance plus the residual pressure for the attack pumpers to maintain the correct pump discharge pressure.

Joining an Existing Relay Pumping Operation

Most relay pumping operations are completely set up all at one time from the beginning of the incident, because adding a pumper to an existing relay operation usually requires that the flow of water be stopped. Depending on the size of the relay pumping operation, this step could involve a number of pumpers. If you are called to assist with a relay pumping operation and your fire apparatus is not part of the initial setup of the relay, then you need to be able to hook up your pumper to the hose lines that are already in use. Pumpers are generally added to an existing relay pumping operation because there is not enough pressure or flow to meet the demands of the attack pumper.

Many fire departments that regularly use relay pumping operations because of issues related to limited water supply will use a <u>relay valve</u> **Figure 10-26** in long hose lays. A relay valve lets you hook up into the relay without shutting down the entire relay operation. Simply hook up both the supply line and the discharge line to your fire apparatus before opening the relay valve **Figure 10-27**. Slowly open the intake valve to your fire apparatus to obtain a water supply. Once the valve is completely open, slowly open the discharge valve from your pump to the hose line that is connected to the relay valve. Observe the incoming supply pressure, and increase the discharge pressure to provide the GPM and pressure required by the attack pumper. Monitor the incoming pressure to maintain 20 psi of residual pressure, and set the pressure relief valve or the pressure governor to prevent damage to your pump if the flow is stopped or interrupted.

If a relay valve is not present or has not been used, then the task of hooking into an existing relay operation is more complex. If you must hook up to an existing relay, get all of your equipment ready and place your fire apparatus in the proper position. Once you have calculated the required discharge pres-

Figure 10-25 Manifold.

Figure 10-26 Relay valve.

Figure 10-27 If you are hooking into a relay valve, hook up both the supply line and the discharge line to your fire apparatus before opening the valve.

sure based on the length of your hose lay, advise command and the attack pumper that you are ready to hook into the relay. The source pumper and any other relay pumpers in the supply line will then shut down their supply lines while the attack pumper switches from the supply line of the relay to tank water. Attach the incoming line from the source or other relay pumper to your intake valve first, and attach the discharge line to the discharge valve.

Contact the source pumper that you are ready for water, and bleed off the air in the line as this flow enters the intake valve. Contact the next pumper in the relay or the attack pumper to advise it that you are flowing water again, and increase the discharge pressure to meet the required GPM and pressure flow. After you have water flowing, set your relief valve or governor to prevent any sudden surges.

Driver/Operator Tip

You need to be quick when hooking into a relay pumping operation that is already flowing water. The attack pumper may have only 2½ to 5 minutes of water in its tank to supply the hose lines and crews operating at the fire. Training is the key to success when it comes to establishing relay pumping operations, especially if you are joining an existing relay operation. Practice your skills in all positions of this operation—source, relay, and attack pumper—to maintain your proficiency as a member of a relay pumping team.

Pressure Fluctuations in a Relay Pumping Operation

As mentioned previously, the needs and capacity of the attack pumper affect all of the other fire pumpers in the relay operation. The driver/operator of the attack pumper determines the GPM needed and notifies the driver/operators of the relay and/or source pumpers of its supply requirements. Unfortunately, should something change suddenly, there may not be time to communicate the need to modify the operation. Mechanical failures or an increase or decrease in hose line flows, for example, can cause pressure fluctuations. These generally do not have a large effect on the operation and can be handled by the pressure relief valves or the pressure governors.

At other times, you must manually adjust the pump pressures. During an operation, if the attack pumper were flowing 500 GPM and a 200-GPM nozzle was shut down, all of the pumpers in the relay would see an increase in pressure. This event should not be a concern unless the pressure fluctuation is greater than 200 psi—a magnitude of change that could cause hose line failure or pump damage. In such a scenario, the driver/operator of the attack pumper would need to decrease the pump discharge pressure and each pumper in the relay would need to do the same.

Should the attack pumper open an additional line that results in an increase of flow (GPM), all pumpers in the relay will need to ensure that they do not drop below 20 psi residual pressure to avoid pump cavitation by simply advancing the pump throttle to increase the discharge pressure.

Shutting Down a Relay Pumping Operation

The order to terminate or shut down a relay pumping operation is given by the IC or the water supply officer only after careful consideration: Relays are difficult to establish and, once in place, should be maintained as long as necessary. Shutting down a relay pumping operation is a simple process, but all pumpers participating in the operation must act in a coordinated fashion.

When shutting down the relay pumping operation, the attack pumper acts first. The driver/operator of the attack pumper must ensure that the tank of the fire apparatus is full and then slowly throttle down. It is important to keep the hose lines flowing water while you shut down a relay, as this will prevent fluctuations in water pressure throughout the entire relay pumping operation. The next pumper in the relay makes sure that its tank is full and slowly throttles down, while still flowing water. This sequence continues until all pumpers have completed this step, ensuring that their tanks are full. When the source pumper is reached, it contacts the attack pumper to begin closing the intake valve; all pumpers in the relay then begin closing the intake and discharge lines until all pumpers are no longer flowing water, including the dump line from the source pumper. The hose lines can then be disconnected and drained before being reloaded or rolled up. This systematic approach to terminating a relay pumping operation prevents pump damage by keeping the water flowing at low levels before it is finally shut down completely.

The source pumper can then close the hydrant to stop the flow of water into this pump. Alternatively, if it is providing water from a static water source, simply stopping the flow of water through the pump will break the vacuum and draft of the pump. This action will stop the flow of water through the hard suction hose into the pump. If that does not occur immediately, disengage the pump, remove a cap from an unused discharge port, and open the valve. This move should allow the water to drain back through the hard suction hose line and back into the static water source. The setup steps are simply reversed, all equipment and hose are placed back onto the fire apparatus, and the pumper is ready to return to service.

Safety for Relay Pumping Operations

Safety is the utmost concern in any operation and cannot be overlooked when conducting relay pumping operations. It is an understatement to say that water is critical to the success of any firefighting operation. In areas where an uninterrupted water source is not available and must be established, knowing where and how you will obtain a water supply starts with planning before the incident begins. Identifying large static water sources, dry hydrants, or potential drafting sites before an incident occurs will help improve the efficiency of the incident response.

Incident safety is improved when crew members have the appropriate training, knowledge of how the equipment works, and practice in using it. Which equipment does your company have available? Which appliances are needed for the operation? Do your crew members know how to use them? Make sure that you fully understand the role played by each position within a relay pumping operation, and practice those roles regularly. An effective driver/operator will know which equipment he or she will need or use to provide water to the scene of a fire through a relay pumping operation.

Communication is another area that must be addressed as part of any relay pumping operation. Contact the IC for a tactical channel to which all pump operators can be assigned. With this communications setup, all pump operators can talk to one another directly when making requests for additional water or other pressure adjustments without distracting the fire attack operations personnel who are communicating on the main tactical channel. In addition, constant communication between the pump driver/operators will help prevent situations that might otherwise cause pump damage.

Fire hose is another source of safety concerns during a relay pumping operation. Watch for any signs of damage to or possible trouble with the hose. If a catastrophic failure were to occur in the hose lines of a relay pumping operation, the safety of the fire fighters on the handlines at the fire scene will be compromised. If you can identify defects and replace the hose section before it is used, you will save time later and have a safer operation. If you do see a defect after the operation has begun, simply get the replacement hose in place and follow the shutdown procedures used for entering an existing relay pumping operation, in which the attack pumper switches to tank water before stopping the flow of water from the relay.

The last area of concern regarding safety focuses on the personnel working around a relay pumping operation and their use of personal protective equipment (PPE). The pump driver/operator is at risk for injury if he or she is not wearing some type of protective gear. At a minimum, turnout pants and boots, a helmet, and gloves are required. This PPE will protect the driver/operator's hands, feet, and head from injury caused by falling equipment or catastrophic failure of a hose line. If crew members will be operating in or around the street and moving traffic, it is recommended that they wear reflective vests as well, even during daytime operations.

Driver/Operator Tip

The safety of both firefighting personnel and the public should be on your mind at all times. When you have laid hose down in a roadway, have the police close the road to vehicle traffic. If the police are not present at the time, then have a support vehicle block the road.

Water Shuttle

Although an uninterrupted water supply is essential for all firefighting operations, a water source may not always be close enough that a relay operation is a practical solution. In many rural and remote areas of a jurisdiction, water must be carried to the fire scene in an operation called a **water shuttle**. A specialized vehicle called a tanker or a tender is used for this operation: While the National Incident Management System (NIMS) uses the term "tender," the "tanker" label is still widely used in the fire service on a national basis. For our purposes, we will refer to these specialized vehicles as tankers for the remainder of this chapter.

A water shuttle does not just consist of the tankers, but also includes **fill sites** and **dump sites**. A fill site is a location where the tankers can get their water tanks filled; a dump site is where the tanker can offload the water. Water shuttles can become very large and complex operations. In such cases, a water supply officer is generally assigned to coordinate the required activities. In contrast, if the incident is small or quickly mitigated, the water shuttle may be quite simple and short in duration.

Fill Sites

A fill site is any location at which a tanker can be filled, such as a hydrant, a pond, a lake, or a river. Fill sites are not always readily available, however, and they can often be great distances from the incident and beyond the limits for a practical relay operation. It is wise to select a drafting site that, if possible, allows tankers in the shuttle operation to enter and exit without having to back up or turn around. This will improve safety at the fill site, as personnel will not need to walk in the area as backing spotters while vehicles are entering and exiting. A fill site utilizing a static water source has the same characteristics and hazards as the locations used for a drafting operation. In fact, the only difference between a fill site and a static water source for drafting is the distance from the incident. Fill sites utilizing a hydrant are often the most reliable due to their unobstructed access and the quantity and quality of the water that they offer.

Driver/Operator Tip

Preplanning in regard to fill site locations is highly important as they are likely to be very few and far between. The time to be looking for suitable fill sites is not when a fire needs to be extinguished. Consider making the preplanning excursion serve as a departmental training exercise.

Voices of Experience

We had just sat down to lunch for a bowl of firehouse chili when a report of a structure fire in the rear of a furniture store came in. Approximately six blocks out, I noticed a large black column of smoke in the vicinity of the fire. As we approached, I told the driver/operator that we had "good water" (aka high water pressure in the hydrants) in the area and that there was a hydrant just south of the address.

As the acting lieutenant, I assumed command at the scene. After the size-up, I transmitted a second alarm due to the size of the building and the amount of fire showing. This fire would have to be fought defensively.

We secured an interrupted supply of water using 200′ (60 m) of 5″ (127 mm) hose line. The first fire attack line placed into operation was a 2½″ (64 mm) hose. The second fire attack line placed into operation was an elevated master stream supplying 3″ (76 mm) hose. Finally, I ordered the use of the deck gun. The defensive suppression operations began flooding the structure with water.

Suddenly, the driver/operator appeared with a fretful look on his face. I immediately enquired, "What's wrong?"

"The pumper is making a weird nose and it is throttling up and down," he said. "And now the 5″ (127 mm) supply is bouncing violently!"

> "Suddenly, the driver/operator appeared with a fretful look on his face."

I ran with great haste to the fire apparatus and began shutting all nonessential fire attack lines down. There was no residual water left in the single water main that was supplying this entire firefighting operation. We were attempting to supply more water to fight the fire than the domestic water system could afford.

That day I learned these lessons:

- Know the water main size within your response area.
- Driver/operators must watch their gauges vigilantly.
- Be mindful of fire apparatus using the same water main for their water supply.
- Consider notifying the water department to increase the water flow to the water grid, if necessary.
- High water pressure does not equate to high water flow.

Michael Washington
Cincinnati Fire Department
Cincinnati, Ohio

Establish a Fill Site Using a Hydrant

Using a hydrant as a fill site can be a very simple operation, especially if you find a suitable hydrant that has a flow pressure of at least 50 psi. This pressure will allow the tanker to be filled quickly and without the help of a pumper. There are only a couple of ways that you can use a hydrant as a fill site. Regardless of how you use the hydrant, however, you must flush the hydrant until the water runs clear before connecting any hose lines.

If you choose to use the 2½" (6.4 cm) outlets on the hydrant, first connect a control valve—either a quarter-turn hydrant valve or a screw-type gate valve **Figure 10-28** —to the outlets. You can then connect two 2½" or 3" (6.4 cm or 7.6 cm) medium-size hose lines to the control valves.

Using a control valve on each outlet allows you to supply two tankers simultaneously without having to shut down the hydrant between filling operations. It also allows you to use both outlets at the same time to fill one tanker more quickly. Both options improve your flexibility to adapt to whatever circumstances may arise. Be sure that the control valves are closed, and then open the hydrant fully; your fill site is now ready. Having the hydrant open and the hoses already connected will speed up the process of filling a tanker in a shuttle operation. If control valves are not available, consider using hose clamps on the hose lines to stop the flow of water. As a last result, you can simply shut the hydrant down between filling each tanker.

LDH can also be used at a fill site from a hydrant. If the hydrant is located far off the road or another hard surface, attach a short section of LDH from the steamer connection of the hydrant to a manifold. You can attach as many as four medium-size hose lines to a manifold, thereby enabling you to fill either one or two tankers at a time with ease and at the speed that you prefer. If the hydrant is located close to the road or another hard surface, you can simply connect the LDH to the steamer connection and supply the tanker directly if it has a large-diameter intake connection. If you do not have a manifold and the tanker is not equipped with a large-diameter intake connection, you can use a wye to split the LDH into two medium-size hose lines and supply the tanker. Using LDH will increase the flow (GPM) and fill the tanker faster, but this approach requires that the hydrant be closed between filling operations unless you are using a manifold.

If possible, leave a fire fighter (or two) at the hydrant fill site to fill incoming tankers and to monitor the site. While this may be a good task for junior personnel, an experienced and knowledgeable crew member may be able to manage the entire fill site, make the connections, and complete the filling operations without the tanker driver/operator ever having to leave the cab of the truck. If fire personnel cannot be spared for this task, leave the hose next to the hydrant for the next incoming tanker to speed up the operation. In this circumstance, you should call for the police to monitor the site, as you will have a lot of equipment left unattended and the apparatus movement in the area could become a traffic hazard or a danger to bystanders.

Establishing a Fill Site at a Static Water Source

Establishing a fill site from a static water source entails following the same procedures discussed earlier in this chapter. The location of the fill site should be chosen after considering if it relates to how other fire apparatus will access the incident; ideally, you should select a site that will not block egress to the fire scene. Other factors that must be considered during the process of determining the reliability of a static water source include accessibility of the supply, quantity of available water, and quality of the water.

Once you have selected the best location, simply follow the steps outlined in Skill Drills 10-2, 10-3, and 10-4 to start flowing water through the dump line. Remember, water must be constantly flowing during a drafting operation to prevent the loss of draft (vacuum) in the pump and to prevent the pump from overheating. You can discharge the dump line back into the static water source.

After establishing a flow of water, you can use your pumper as a sort of "hydrant": Use either two medium-size hose lines to supply the tankers, a LDH line if the tanker is equipped with one, or both methods if you need to supply multiple tankers simultaneously. As with all drafting operations, start your flow through the discharge valves slowly, while increasing the throttle at the same time, to avoid breaking the vacuum and the draft of the pump. When the tanker is full, slowly throttle down and close the discharge valves, while keeping the dump line open during the entire operation.

The benefit of establishing a fill site at a static water source is that the pumper drafting the water can supply the tankers being filled at a remote location that is better suited for tanker movement. For example, the pumper might be set up at the shore of

Figure 10-28 Screw-type gate valve in place on a hydrant.

a small lake, but provide fill hoses to a location on a roadside, several hundred feet away. This type of operation allows the tankers to drive up, fill their tanks, and drive away, all without turning around or backing.

Fill Sites in Inaccessible Areas

Sometimes you may not be able to get a pumper close to the static water source because the area is blocked by trees or surrounded by undeveloped roadway or pathways. In these situations, you may have to consider using portable pumps as a way of getting the water to the tankers. Establishing a suitable drafting site for a portable pump requires that you have 24" (61 cm) of water in all directions around the strainer and enough hose to get the water to the fill site.

As when you establish a drafting site for a pumper, you must ensure that the quantity of water available is large enough to support drafting with the portable pump. If the source is large but not deep enough, you can use a floating strainer for the portable pump hard suction hose; this strainer is exactly like one used with the hard-sided hoses used by a pumper, only smaller. If a floating strainer is not available, you may need to dig a hole in the water to provide the depth you need for the strainer or create a dam in a creek using a ladder wrapped in a salvage cover. Before undertaking any such damming operation, consider how changing the stream might affect the surrounding areas.

> **Driver/Operator Tip**
>
> Consider putting a washtub in the bottom of the hole to prevent any sediment from getting into the strainer. This measure will keep the water relatively clean and may help to prevent damage to the pumps that use the water from this source. (This tip has saved many buildings.)

Once you have the area ready for the portable pump to start drafting, connect the hard-sided suction lines to the pump, and place a strainer on the intake end of the hose. You can use a ladder to secure these smaller hard-sided suction hoses in moving water sources, just as you can with the hoses from a pumper. Connect the discharge lines to the pump and determine where you will discharge the water once it starts flowing. Most of the portable pumps will flow at a maximum rate of only 250 GPM, so using a dump line will not be possible. You will not be able to use control valves to stop the flow of water at any point while the pump is running, so you should consider putting the hose lines into a **portable tank**; otherwise, you can supply a pumper directly. Start the portable pump and obtain a draft by following the manufacturer's guidelines to start flowing water. Once water is in the portable tank, a pumper can draft from the portable tank and supply the tankers without having to shut down the portable pump. You may want to use multiple portable pumps to supply multiple portable tanks in an operation, given that portable pumps cannot provide large quantities of water at any time. Set one portable pump up at a time to increase the effectiveness of the operation.

Portable floating pumps have become a very popular option for obtaining water in inaccessible areas in recent years. These devices are very good choices for use at a fill site at a static water source. Depending on the flow (GPM) you need to supply, you might consider using multiple floating pumps. When placing a floating pump into service, follow the manufacturer's guidelines for operating the pump, and make sure the fuel tank is full of fuel.

Once in the water, portable floating pumps are difficult to turn on and off and reposition, so make sure that they are ready for a long operation before launching them. Attach the hose to the discharge port of the portable floating pump, and tie off the pump with a rope to a secure place on the bank or shore of the static source. You must first place the floating pump in the water *before* you start it, because it will automatically prime, begin to draft, and start pumping the hose line very quickly. Given this fact, it is important to have personnel in place to direct the flow into a portable tank or have the hose attached to an engine to serve as its water supply. Consider using multiple floating portable pumps to fill multiple portable tanks because, like portable pumps, these devices have limited discharge capabilities.

Whether you are using a floating or stationary portable pump, you must monitor the pump to ensure that it is functioning properly and flowing water. Check for clogged strainers if the flow is reduced, as this is the most common source of flow restrictions.

When an inaccessible area will serve as a fill site, the use of some type of portable pump that discharges the water into a portable tank is a highly practical means of obtaining a water supply. This portable tank operation must be supported with a pumper so that incoming tankers do not have to establish a draft each time they arrive to obtain water, which would significantly slow the operation down. Once the portable tank is filled, the pumper drafting from that static water source can flow water through a dump line while waiting to fill the tankers as they arrive. The key is for the pumper to slowly fill the tankers so that it does not drain the portable tank faster than it can be supplied by the portable pumps.

> **Driver/Operator Tip**
>
> At no time should portable tanks and pumping operations be left unattended. The fill area can pose a danger to bystanders owing to the constant flow of water and the attractive nuisance that a full portable tank can present to children as a drowning hazard.

▮ Filling Tankers

Whether the fill site is being supplied by a hydrant or a pumper (i.e., a static water source), the procedure for filling the tanker (tender) is the same. When filling tankers, you should fill only one at a time when you are using medium-size hose lines. Trying to fill more than one tanker at a time will certainly slow down the water supply operation, although you have the flexibility to fill two tankers at one time if the situation dictates it. For example, if you are filling a 1500-gal tanker and a 3000-gal tanker, you can get the smaller tanker back to the scene more quickly if you fill both tankers simultaneously. You will have to judge whether sacrificing speed is worth the increased efficiency. Without question, you can fill more than one tanker at a time when you are using a LDH hydrant manifold or a pumper from draft.

Having two hose lines ready with control valves and personnel at the ends ready to connect to the tankers will increase the speed of the operation and reduce turnaround times for the operation. When you arrive to fill your tanker, simply attach one hose at a time from the control point—that is, the control valves on the hydrant, the hose clamps, the manifold from a LDH connection at the hydrant, or the pumper being used to supply the fill site from a draft—to the tank fill valves on the tanker. Be sure that the tank fill valves are open on the tanker before you open the control point valves to fill the tank. It is easy to overlook the step of opening these valves, but by doing so you will waste valuable time.

Once the tank is completely full on the tanker, close the control point valve before closing the tank fill valve on the tanker. Following this sequence allows you to disconnect the hose from the tanker, because it will have less pressure in it. Remove the hose lines and advise the driver/operator of the tanker that it is safe to drive away from the fill site. Repeat this operation for each tanker until the operation is terminated through IMS by either the IC or the water supply officer.

Safety for Water Shuttle Operations

Water shuttle operations present the same risks that are encountered in relay pumping operations. The purpose of both operations is to provide an uninterrupted water supply for the firefighting efforts. The water shuttle operation relies on a fill site to obtain the water and a dump site to deliver it to the attack pumper; a relay pumping operation uses a source pumper to provide the water to the attack pumper. A water shuttle, however, requires the frequent movement of apparatus. Thus the safety concerns associated with this type of water supply should be identified before we discuss the components of the operation.

Steps that can be taken to improve incident safety include preplanning locations for obtaining the water, being trained in the use and operation of the equipment, and knowing which equipment you have to use. Communication between the tankers and the fill sites may require assignment of a separate radio frequency, which avoids problems on the main tactical channel as tankers report their locations or indicate that they are returning for refilling. Personnel operating at the fill sites will need to wear the same level of PPE as is worn during a relay operation to protect these crew members from falling equipment, catastrophic equipment failures, and traffic in and around the fill site.

A water shuttle operation involves moving tankers full of water from the fill site to the dump site and returning with an empty tanker. Tankers are specialized types of fire apparatus that carry several hundred or several thousand gallons of water, which can weigh more than 80,000 pounds. Many of the vehicles being used as tankers may have previously been used in another capacity before being converted to fire department use. Put simply, they were never intended to be driven under emergency conditions and at high rates of speed. Unfortunately, most personnel are not familiar with the handling characteristics of these large vehicles and do not make adjustments to their driving habits when at the wheel of these apparatus. These two circumstances alone have caused too many fire fighters to lose their lives.

Extreme caution must be emphasized while driving tankers, because they do not handle in the same manner as a standard pumper. The ability to maneuver is greatly reduced with tankers, due to their size and weight. Driver/operators should be well trained in driving these vehicles both full and empty before they attempt to operate one during an emergency incident. *At no time* should these vehicles be operated beyond the posted speed limit, regardless of what is allowed by your state motor vehicle laws or codes.

> **Safety Tip**
>
> Never drive a tanker that is not completely full or completely empty. Driving such a vehicle with a partially full tank can cause dangerous weight shifting, which can result in the tanker rolling over.

Establishing Dump Site Operations

Establishing the dump site location to create a static water source for the firefighting operation takes practice and forethought. The location at which the shuttled water will be offloaded is called a dump site. This site should be on relatively level ground, be firm, and not be susceptible to significant changes as a result of fire apparatus movement or getting wet from the water. The location must be large and strong enough to be able to support the weight of the water and the weight of the incoming tankers, and it should provide enough room for the movement of the tankers in and out of the site. Large parking lots and fields that are flat and smooth can make excellent locations for dump sites.

The ideal dump site location is close enough that the source pumper can supply the attack pumper(s) without a relay pumper, but not so close that the tankers cannot access it. For example, if the attack pumper is positioned just past the building and a ladder truck is in the front, the dump site might be located down the street at an intersection that would allow for the tankers to approach and depart in different directions. While there may be space in the street that is closer, the efficiency of the operation dictates that the dump site be located farther away. As in the case of a long driveway, the attack pumper may be located near the building but is being supplied by a source pumper located on the main roadway.

In cases when there are no intersections or large parking lots, care must be taken not to obstruct the flow of additional responding apparatus or the tanker shuttle by placing the portable tanks in the roadway. Thought should be given as to which way a tanker will offload its water: Does the tanker offload from the rear or can it dump from the side? Once you have determined that you have a good site, you are ready to set up the portable tanks and create a static water source for a source pumper.

Near Miss REPORT

Report Number: 08-0000344
Report Date: 07/17/2008 00:34

Synopsis: Fire fighter falls off back step while laying supply line.

Event Description: Our engine company was alerted that it would be second arriving to a dwelling fire at 4 am in the morning. Four inches of snow fell since our last response at midnight and our ride was rather slow. The first engine gave a size up of a two story dwelling with fire showing Division 2 on the Charlie side. Cars were in the driveway and no one was outside, so the first engine took a tankline into the house to hit the fire and search for life. The pump operator reported he had a half tank as we arrived and searched for a hydrant. We needed to do a reverse lay and I was assigned to hook up to the first engine with our 4 inch hose. I dropped the line with the first engine operator and mounted the back step to pull hose from the bed. I assumed we were at the water supply and was pulling another fold of hose and stepping down when the operator accelerated and pulled away. Holding the hose, I found myself face to the street and the water supply was some 100 feet away. I stood up, dusted myself off, and walked to assist in making the supply. The back step buzzer had been broken for years and we had no communication. He stopped to let a police car move.

Lessons Learned: Riding the back step is still dangerous whether you are going 10 feet or 100 feet. You shouldn't be on the back step without a partner (in communications) watching us. The back step buzzer was fixed within a week. If the hose lay is less then 100 feet, we walk. If it's longer, the fire fighter rides in the cab and the officer observes the lay for any snags and radios the engine operator.

Demographics

Department type: Paid Municipal
Job or rank: Fire Fighter
Department shift: 24 hours on—24 hours off
Age: 43–51
Years of fire service experience: 14–16
Region: FEMA Region V
Service Area: Suburban

Event Information

Event type: Fire emergency event: structure fire, vehicle fire, wildland fire, etc.
Event date and time: 12/28/2007 04:00
Event participation: Involved in the event
Weather at time of event: Cloudy and Snow

What were the contributing factors?
- Equipment
- Human Error
- Decision Making

What do you believe is the loss potential?
- Minor injury

Using Portable Tanks

Portable tanks are usually carried on tankers and should hold as much water as the capacity of the tanker will allow. For example, a 3000-gal tanker usually carries a portable tank that can hold 3000 gal. If the tanker cannot offload its entire tank at one time, it will slow the operation because the tank's contents must be offloaded incrementally or the tanker will have to dump any water that it has left before leaving for the fill site. Using portable tanks at a dump site can be very easy and should be able to be set up quickly and safely. When using portable tanks, thought should be given to the possibility of expanding the operation. The area should be big enough to hold two to three portable tanks connected together in addition to the source pumper, which will be drafting from these tanks. Portable tanks provide a source of water that is free from debris and not subjected to the effects of stream currents and unknown shore soil conditions.

Before deploying any portable tank, place a salvage cover or tarp on the ground in the area where the tank will be positioned to prevent any damage to the tank by rocks, sticks, or glass that may be present on the ground. Once the salvage covers are in place, remove the portable tank from the fire apparatus and place it on the salvage cover. A self-expanding type of portable tank can be laid flat on the salvage cover and is ready for water. If you are using a metal frame type of portable tank, open the portable tank by completely unfolding the frame outward and pushing the lining into place until it touches the frame all around the edges. If you are setting up multiple portable tanks, start with the tank at the high point of the ground and work downward. Ensure that the portable tank's drain is on the low side of the tank, and check that the drain is closed or secure before filling the tank.

The tank is initially filled by a tanker before it enters the water shuttle operation. Once the tanks are in place, the tanker positions itself to offload the water through a **dump valve** on the tank. The dump valve is a large opening, which can be as large as 12″ to 16″, that is connected directly to the tank that allows it to be emptied quickly, in some cases within 1 minute **Figure 10-29**. It is important that the portable tank not overflow while it is being filled, so initially fill the tank slowly. If multiple portable tanks are being used, once the first tanker has emptied its tank and moves out of the way, the next tanker can get into position to repeat the steps until all of the portable tanks are full.

Offloading Tankers

When offloading a tanker, the driver/operator must review the dump site characteristics upon arrival to ensure that his or her specific tanker can access the location. Despite the best of intentions, not all of the personnel who set up the dump site may know what your tanker requires or how it maneuvers. As the driver/operator, you should know your vehicle better than anyone. Once again, consider how much room you have to maneuver, whether the area will support the weight of your tanker, how you will offload the tanker, and in which direction you must travel to exit the site. If a dirt area is being used as the dump site, consider what happens when the water is splashed

A.

B.

Figure 10-29 (A) A manually operated dump valve. (B) A dump valve from a pneumatic valve.

on the ground: If the dirt turns into mud, the tankers may sink and become stuck.

Once the tanker is positioned at the dump site, unloading it through a dump valve is the easiest function you can perform **Figure 10-30**. These valves may be located on each side of the vehicle as well as the rear; they are found directly on the tank and allow the water to exit or dump from the tank very quickly. Some newer tankers have been equipped with in-cab pneumatic controls to open the dump valves, so you do not have to get out of the cab to operate the valves. Once the tanker is in position and has been given approval to start dumping water, simply activate the switch and the valve opens, which releases the water from the tank.

In older tankers, the water is released manually by personnel at the dump tank. This is one of the most common methods used to offload a tanker. Using these valves does not require any changes to the procedure: The tanker must still be positioned at the portable tank, but the dump valve (5–6 in.^{2}) is opened manually by a fire fighter at the portable tank.

Some tankers are equipped with a jet dump as alternative means to offload their water. With this type of operation, the tank on the apparatus is built slightly differently—it consists of a vessel that can be pressurized rather than simply holding water. The action of "jet dumping" starts when the vehicle's tank is pressurized; once it reaches the predetermined pressure, the contents—water in this case—are forced out under pressure. This process can be dangerous, as you are forcing the water out under pressure in a large horizontal column that is discharged quickly. Driver/operators must be highly skilled and well trained before they attempt to operate this type of fire apparatus. If a jet dumping process is used, personnel at the dump site must exercise caution, as the pressurized column of water coming out of the discharge could lead to the collapse of the portable tank.

Another method of offloading a tanker when the large dump valves are not installed on the fire apparatus is to attach medium-size lines (2½" to 3") to the pump of the tanker and place the other end of the lines into the portable tank. Either the hose must be held in place by personnel at the dump site or a commercially made device must be attached to the tank to free up personnel in the dump site from performing this task. The driver/operator of the tanker should pump water through these hose lines at a maximum pressure of 50 psi.

Use of Multiple Portable Tanks

The use of multiple portable tanks has become a standard practice in rural firefighting operations. When the tanks are set up, they are positioned starting at the high point of the ground and then going down slope. Remember, the main objective is to keep the water moving toward the tank being used for drafting. Even if the ground appears flat, there is usually a slightly noticeable slope. If there is not, simply hook the tanks together using the rings on the drains if they are so equipped. Many tanks are equipped with two "drains" so that they can be hooked together; otherwise, a **jet siphon** may be used to move the water from one tank to another. Jet siphon adapters use the hard suction hose as a pathway for flowing water that is being forced through the hose by Venturi forces. The jet siphon has a connection for a 1½" or 1¾" hose to flow water through the hose, which will create the Venturi forces and draw the water from the tank up through the hose and be discharged into the second tank. The jet siphon is connected to one end of a loose hard suction hose in the tank from which you wish to remove water; the other end of the hose is secured to the top of the tank that you want water to flow into. A 1½" or 1¾" hose is attached to the jet siphon device and placed into the tank once it is full. The source pumper connects the hose to a discharge port and pumps it at a pressure of 100 to 125 psi. Do not use the dump line to supply the jet siphon, as it may not always be in operation and the dump line must constantly flow water. Multiple jet siphons can be used to move the water between tanks, depending on how many tanks you have in operation.

Source Pumper Considerations for Portable Tank Operations

The main objective in moving water from one tank to another when using multiple tanks is to keep the main portable tank full. Obviously, when only one tank is being used, it represents the main tank—the one being used by the source pumper for its water supply.

After the source pumper has established a draft from this static water supply, all efforts should be taken to prevent air from being introduced into the pump—which explains why it is preferred that dumping operations are done in one tank while the drafting is done from another tank. When only one tank is being used, the tanker operator must use care not to dump the

Figure 10-30 Unloading water through a dump valve.

water on or near the strainer, as introduction of the water creates air turbulence when it enters the tank. Using a blow-up ball is very important in portable tank operations because the size of the water "container" is much smaller and whirlpools are created easily in shallow water.

The main tank should not run low on water during the operation if the tanker shuttle is working as designed. If it does run out, the incident will lose its water supply and put fire fighters at risk if they are operating in interior positions at the fire scenario and lose water. If the main tank continues to run low, the incident commander or the water supply officer must be notified of the situation and may consider adding more tankers to the shuttle operation to keep up with the water demands.

Driver/Operator Tip

Use a low-level strainer when drafting from portable tanks. This approach will allow you to get the maximum volume out of the tank without worrying about drawing in debris or silt as might occur in a pond, stream, or lake.

■ Traffic Flow within a Dump Site

The movement of tankers into and out of the dump site can cause significant congestion, which may result in slow water delivery to the portable tanks. This congestion can be prevented with a little foresight during the layout of the dump site and the portable tanks. Consideration must be given to how the tankers will offload their contents: Will two tankers offload their water at the same time? How will they approach the portable tanks to dump the water? In which direction will the tankers depart the dump site to return to the fill site? Establishing a smooth traffic pattern will improve efficiency and increase safety around the dump site—the tankers will need to make fewer movements if they do not have to back up to reposition themselves for offloading.

Driver/Operator Tip

Using two members for a crew on a tanker will improve fire fighter safety by providing a fire fighter to act as a backing guide whenever the vehicle operates in reverse at the fill or dump site. The second crew member can act as a spotter as the tanker approaches the portable tank to offload its water.

Most modern tankers have the ability to offload from the sides and the rear of the vehicle, which gives the driver/operator greater freedom in deciding the direction of travel into the dump site and determining how to approach the portable tank. Some older tankers can offload only from the rear of the vehicle, which means that backing the vehicle into position is unavoidable.

Once you know how the tankers will be offloaded, you can determine the best direction of approach to the portable tanks.

The portable tanks should have been placed in as open an area as possible so that you can easily maneuver around the site. If both sides of the tanks are accessible and offloading into two tanks will not disrupt the source pumpers' drafting operation, then two tankers can offload their contents at the same time. This will significantly improve the efficiency of the operation.

All tankers should travel the same path to enter and then exit the dump site. These vehicles are large and difficult to maneuver, especially in scenarios characterized by heavy congestion. Once you determine the entry and exit directions to the site, be sure to communicate the traffic plan to all driver/operators of the tankers. A traffic plan that flows in a semi-circular direction will be the most efficient because the vehicles will not be traveling on the same roadway in opposite directions. For example, have the tankers approach the dump site from the north, enter from the east, and travel in a westerly direction to the portable tanks for offloading. Once completely unloaded, the tankers can exit the dump site on the west side and travel north again toward the fill site. If a circular traffic pattern is not possible, try to have the tankers back into position at the portable tank(s). The roads that lead into and out of the fill site should also be different to prevent congestion there Figure 10-31.

Driver/Operator Tip

Vehicles that are parked along the travel routes and fire apparatus that are parked and not being used increase the risk of an accident with a tanker. If police assistance is available, have the police department shut down the roads on which the tankers are traveling for added safety.

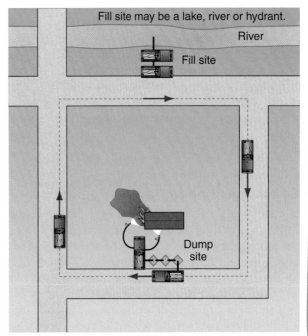

Figure 10-31 The traffic pattern of a dump site should flow in a semi-circular direction.

Nurse Tanker Operations

The operation of offloading a tanker into another tanker or into a pumper as a form of a water source is called a **nurse tanker operation**. Its name derives from what happens as a mother nurses her child: This type of operation eliminates the use of portable tanks (bottle), as the attack pumper is supplied directly by the nurse tanker (mother). A nurse tanker operation is set up when there is no room to establish a dump site, a relay operation, or a combination of the two. Such operations can also be used when the size of the fire is small enough that establishing such a detailed supportive water supply mechanism is not practical. The nurse tanker operation can be performed at the start of an incident as a way to provide water to the attack pumper while the relay or fill-and-dump operation is being established. In this sense, it represents a stop-gap measure that creates the needed time to set up a longer-duration water supply and provides water to support the fire attack until an uninterrupted water supply is established.

The ideal nurse tanker should be the largest tanker that is available because it will serve as the primary source of water for the fire attack operation. Tractor-trailer tankers, which can carry as much as 6000 gal of water, are very good choices for this operation because they can be supplied by smaller tankers through a second nurse tank operation. The nurse tanker must have a fire pump to supply the water to the attack pumper, but only needs to be big enough to maintain 20 psi of residual pressure.

Most tractor-trailer tankers have a small pump located on the trailer to assist with offloading. The larger the pump on the tanker, the more water it will be able to supply and with greater pressure. The nurse tanker essentially can function like a combination of a contained portable tank and source pumper if it positioned nearby the attack pumper, provided it can meet the demands of the attack pumper.

Other tankers can supply, via nursing, this new "source"; alternatively, it can be supplied by the source pumper at a dump site that was established farther away to keep the nurse tanker full. The availability of fire pumps on tankers can open many options for providing a reliable water supply.

When you are filling a nurse tanker, do not fill it too fast. Remember, you want to avoid or limit the introduction of air into the water. Also, because tankers are filled from the bottom, air bubbles are trapped in the water until they can float to the top. Using two medium-size hose lines at 50 psi should be sufficient to fill the nurse tanker.

Communication for Tanker Operations

Communication plays a vital role on the fire ground; likewise, it needs to be a vital part of the water supply and tanker shuttle operations. A number of tankers may be operating around an emergency scene in addition to traveling to the fill site and back to the dump site. Driver/operators need to communicate with one another effectively, especially if the fill or dump sites are located off a narrow roadway or in a tight area. The water supply officer should request from the IC that all tanker operators be assigned to a tactical channel. This practice will eliminate important but nonemergent radio traffic on the main tactical channel, in which communications between the IC and fire attack crews must have priority. The tanker driver/operators must be able to communicate their locations within the shuttle flow and indicate whether they are filling or returning to the dump site so that command officers can make decisions based on the time it takes to sustain a steady flow of water to the attack pumper.

Staying in communication is also a way of maintaining accountability for each tanker and its personnel. If a tanker and crew have not been in contact with a member of the water supply team, having these communication capabilities in place will allow members of the command structure to reach them and determine if a problem exists. If a tanker and crew are out of contact, it would constitute an emergency for the IC and would be treated as a missing fire fighter mayday.

> **Driver/Operator Tip**
>
> Many fire departments operate on different radio frequencies, which can cause complications at an emergency. Planning an exercise with neighboring fire departments to work out any communication issues prior to the incident will help things run smoothly during an actual mutual aid call.

Water Shuttle Operations in the Incident Management System

Having a sustainable water supply is a vital part of any fire incident scene. Water supply officers should be given a top priority at any scene; if this position is not officially assigned, however, the IC accepts the responsibility for this function. Once a water shuttle operation is established, then a water supply officer should be established as part of the IMS, which may also include a dump site officer and a fill site officer.

The dump site officer is responsible for establishing the dump site, creating the traffic patterns, and ensuring that the source pumper at the site has a sustained water supply. If the incident is complex and includes multiple dump sites, these officers may be responsible for all of them and will have to coordinate the larger flow between the fill site and each dump site depending on the demands of each location. They will need to communicate with each dump site to ensure that the water is routed to where it is needed.

The fill site officer is responsible for establishing the fill site location(s), ensuring that the water source for filling the tankers will meet the demand, and creating the traffic patterns in and out of the fill site. Filling two tankers at a time requires a strong water source such as a pumper at draft or a hydrant with a LDH manifold. If these are not present and the fill operation is the cause of slower dumping cycles, fill site officers may consider establishing multiple fill sites. This will require the same level of communication between multiple fill sites as it does between multiple dump sites. The ultimate goal is to provide an uninterrupted water supply to the fire incident.

Wrap-Up

Chief Concepts

- The ability to draft water from a static source makes your fire apparatus a versatile resource.
- Fire departments generally have a prescribed process for performing inspections, routine maintenance, and operational testing of water supply apparatus and mechanisms.
- Water supply management is a vital part of any fire incident and can become a major part of the Incident Management System (IMS) in the event that a reliable public or private water system is not available.
- Drafting operations, relay pumping, water shuttles, and nurse tanker operations will become the responsibility of the water supply officer.
- Once you have found a suitable location with an adequate water source that is close enough to reach, you are ready to set up your pump to draft. This task requires teamwork, and the driver/operator should be assisted by two or more fire fighters.
- The first step in preparing to draft from a water source is to prime the pump.
- Once the suction hose has been connected and placed into the water source, air is drawn from the pump using the priming pump.
- Once the pump is primed, you are ready to start drafting water and producing discharge pressures.
- Whether you are supplying attack lines or supplying water for the fire attack pumper, be sure that the crews know that you are ready to start flowing water *before* you open the discharge valve—either over the radio, through face-to-face contact with the crew, or through hand signals as established by the organization.
- Slowly open the corresponding discharge valve for the hose line and begin flowing water.
- Establishing an uninterrupted water supply is one of the most important objectives for any IC.
- In many rural and remote areas of a jurisdiction, water must be carried to the fire scene in an operation called a water shuttle, in a specialized vehicle called a tanker (tender).
- Establishing the dump site location to create a static water source for the firefighting operation takes practice and forethought.
- The location at which the shuttled water will be offloaded is called a dump site.
- The process of offloading a tanker into another tanker or into a pumper as a form of a water source is called a nurse tanker operation.
- Having a sustainable water supply is a vital part of any fire incident scene.

Hot Terms

adapter Any device that allows fire hose couplings to be safely interconnected with couplings of different sizes, threads, or mating surfaces, or that allows fire hose couplings to be safely connected to other appliances.

anti-siphon hole A very small hole in the fitting on top of the primer oil reservoir or in the top of the loop on the lubrication line leading to the priming pump.

atmospheric pressure Pressure caused by the weight of the atmosphere; equal to 14.7 psi (101 kPa) at sea level.

cavitation A condition caused by attempting to move water faster than it is being supplied to the pump.

dependable lift The height that a column of water can be lifted in a quantity considered sufficient to provide reliable fire flow.

draft The pressure differential that causes water flow. (NFPA 54)

dump line A small-diameter hose line that remains in the open position and flowing water during the entire drafting operation.

dump site A location where a fire apparatus can offload the water in its tank.

dump valve A large opening from the water tank of a mobile water supply apparatus for unloading purposes. (NFPA 1901)

fill site A location where the fire apparatus can get water tanks filled.

forward lay A method of laying a supply line where the line starts at the water source and ends at the attack pumper.

jet siphon Appliance that connects a 1½″ or 1¾″ hose to a hard-sided hose coupling and creates a pathway for flowing water that is being forced through the hose by Venturi forces.

large-diameter hose (LDH) A hose of 3½″ (90 mm) size or larger. (NFPA 1961)

lift The vertical height that water must be raised during a drafting operation, measured from the surface of a static source of water to the center line of the pump intake. (NFPA 1911)

manifold A large-diameter hose line that is split into four or five medium-size hose lines.

medium-size hose Hose that is 2½″ or 3″ in diameter.

nurse tanker operation The process of supplying water to an attack engine directly by a supply engine.

portable floating pump A small fire pump that is equipped with a flotation device and built-in strainer, capable of being carried by hand and flowing various quantities of water depending on its size.

portable pump A type of pump that is typically carried by hand by two or more fire fighters to a water source and used to pump water from that source.

portable tank Any closed vessel having a liquid capacity in excess of 60 gal (227 L), but less than 1000 gal (3,785 L), and not intended for fixed installation. (NFPA 30)

primer oil reservoir A small tank that holds the lubricant for the priming pump.

priming The process of removing air from a fire pump and replacing it with water.

reducer A device that connects two hoses with different couplings or threads together.

relay pumping operation The process of moving water from a water source through hose to the place where it will be needed.

relay valve An appliance placed in a hose lay that allows an attack engine to connect into a relay pumper operation without interrupting the flow of water.

reverse lay A method of laying a supply line where the line starts at the attack pumper and ends at the water source.

split lay A scenario in which the attack engine lays a supply line from an intersection to the fire, and the supply engine lays a supply line from the hose left by the attack engine to the water source.

vacuum Any pressure less than atmospheric pressure.

water shuttle The process of moving water by using fire apparatus from a water source to a location where it can be used.

Driver/Operator in Action

While positioning the fire apparatus to begin drafting operations in a rural setting, you think to yourself that supplying attack lines from the on-board water tank or a hydrant is usually much easier than drafting water in a rural setting. At the rural setting, you will have to operate alone at the draft site and must rely on your ability to troubleshoot any problems that arise. These operations may be complicated by many variables, and you will have to make the operation work or the fire fighters at the scene will not have enough water to protect the lives and property in danger. As a driver/operator, there is quite a bit of weight resting on your shoulders in such a scenario.

1. What is a pressure that is less than atmospheric pressure called?
   A. Vacuum
   B. Draft
   C. Siphon
   D. Suction

2. Lifting water more than _____ is not practical for most fire pumpers and requires additional lengths of hard-sided supply hose.
   A. 3'
   B. 5'
   C. 10'
   D. 15'

3. The installation of _____ should be considered in areas where access to a water supply is difficult or when the installation would make an ideal drafting site that could be used frequently.
   A. dry-barrel hydrants
   B. dry hydrants
   C. wet-barrel hydrants
   D. fire department connections (FDC)

4. Which of the following is not one of the three types of strainers that are generally used for drafting operations?
   A. Barrel strainer
   B. Low-level strainer
   C. Floating strainer
   D. High-level strainer

Foam

CHAPTER 11

NFPA 1002 Standard for Fire Apparatus Driver/Operator Professional Qualifications

5.1.1 Perform the routine tests, inspections, and servicing functions specified in the following list in addition to those in 4.2.1, given a fire department pumper and its manufacturer's specifications, so that the operational status of the pumper is verified:
1. Water tank and other extinguishing agent levels (if applicable)
2. Pumping systems
3. Foam systems [p. 298–323]

5.2.3 Produce a foam fire stream, given foam-producing equipment, so that properly proportioned foam is provided. [p. 298–323]

(A) Requisite Knowledge. Proportioning rates and concentrations, equipment assembly procedures, foam system limitations, and manufacturer's specifications. [p. 298–323]

(B) Requisite Skills. The ability to operate foam proportioning equipment and connect foam stream equipment. [p. 298–323]

Additional NFPA Standards

NFPA 11 *Standard for Low-, Medium-, and High-Expansion Foam* (2005)
NFPA 30 *Flammable and Combustible Liquids Code* (2008)
NFPA 36 *Standard for Solvent Extraction Plants* (2009)
NFPA 101 *Life Safety Code®* (2009)
NFPA 402 *Guide for Aircraft Rescue and Fire-Fighting Operations* (2008)
NFPA 412 *Standard for Evaluating Aircraft Rescue and Fire-Fighting Foam Equipment* (2009)
NFPA 1150 *Standard on Foam Chemicals for Fires in Class A Fuels* (2004)
NFPA 1403 *Standard on Live Fire Training Evolutions* (2007)
NFPA 1901 *Standard for Automotive Fire Apparatus* (2009)

Knowledge Objectives

After studying this chapter, you will be able to:
- Describe how foam works.
- Describe the foam tetrahedron.
- Describe foam characteristics.
- Describe the different types of foam concentrates.
- Describe foam expansion rates.
- Describe foam percentages and their importance.
- Describe foam guidelines and limitations.
- Describe the different types of foam application systems.

Skill Objectives

After studying this chapter, you will be able to:
- Batch-mix foam.
- Operate an in-line eductor.
- Operate the around-the-pump proportioning system.
- Operate a balanced-pressure proportioning system.
- Operate an injection foam system.
- Operate a compressed-air foam system (CAPS).
- Apply Class A foam on a fire.
- Apply foam with the roll-on method.
- Apply foam with the bankdown method.
- Apply foam with the raindown method.

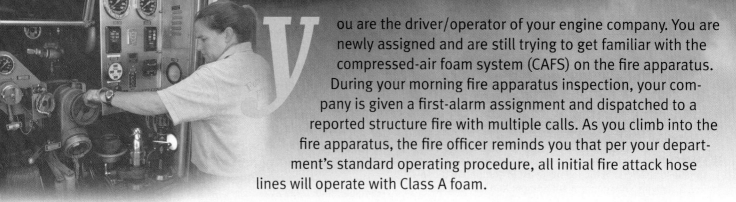

You Are the Driver/Operator

You are the driver/operator of your engine company. You are newly assigned and are still trying to get familiar with the compressed-air foam system (CAFS) on the fire apparatus. During your morning fire apparatus inspection, your company is given a first-alarm assignment and dispatched to a reported structure fire with multiple calls. As you climb into the fire apparatus, the fire officer reminds you that per your department's standard operating procedure, all initial fire attack hose lines will operate with Class A foam.

As you are driving to the fire scene, you see a large column of black smoke and realize that your fire apparatus will be the first on the scene. You start to think about the foam system, including how it operates. The following questions start to pop up in your mind:

1. What is the procedure to operate the foam system?
2. At what percentage should you supply Class A foam to the fire attack hose lines?

Introduction

Water has been the main means of suppressing fires for many years. Water is effective, bountiful, and relatively inexpensive to use. While water is certainly effective, foams have added a new dimension to fighting fires. Among the many hazards that fire fighters must contend with are the wide variety of **combustible and flammable liquid** incidents. Successful control and extinguishment of these incidents requires not only the proper application of **foam** on the fuel surface, but also an understanding of the physical characteristics of foam. A full understanding of foam and its application is imperative to a safe and successful suppression operation. Lack of familiarity with the chemical characteristics of foam and its application can cause severe problems for both you and your crew—so know your equipment and operate *safely*.

With improvements in, and greater simplicity of, application techniques and the versatility of foam concentrates, the use of foam for all types of fires is becoming more common. The National Institute of Standards and Technology (NIST) has determined that foam is three to five times more effective than plain water in extinguishing fires, which can justify its use and its added expense at an incident site.

History

Foam has been available for firefighting for many years. In the 1800s, it was introduced as an extinguishing agent for flammable liquid fires. This foam was produced by mixing two powders—aluminum sulfate and sodium bicarbonate—with water in a foam generator. This method was used for chemical foam extinguishers as well. Inside the fire extinguisher, the two chemicals were stored separately. To operate the fire extinguisher, a seal was broken and the extinguisher was inverted, allowing the aluminum sulfate and sodium bicarbonate to mix and create the foam.

In the 1940s, a foam concentrate based on liquid protein was introduced. This foam, which was made from natural animal protein by-products, was produced by mechanically mixing the protein foam concentrate with water in a **foam proportioner**. A foam proportioner is a device that mixes the foam concentrate into the fire stream in the proper percentage. Protein-based foam was primarily used to fight flammable liquid fires aboard Navy ships **Figure 11-1**.

In the 1960s, fluoroprotein foam (FP) and aqueous film-forming foam (AFFF) were introduced. These foams were more versatile and performed better than protein foam. In particular, they were able to knock down fires faster than protein foam and had a longer blanket life.

Driver/Operator Tip

Training with foam concentrates and foam systems is critical to being able to properly operate the system when you are at "the big one." Many fire departments do not like to use foam concentrate for training because of the cost factors involved. Even so, the only true way to make sure that you know how to operate the foam system and to make sure that it is operating properly is to train as if you are at a real incident. Training foams are available and offer an inexpensive training alternative to using regular foam concentrates.

Figure 11-1 The protein foam first used on Navy ships in the 1940s is still used in the fire service today.

In the 1970s, alcohol-resistant foams were introduced. Such foams could be used for both hydrocarbon and polar solvent fuels. The introduction of alcohol-resistant foams allowed flexibility in dealing with the many types of fuels that fire fighters encounter on a daily basis.

While the use of foam has been limited for many reasons, the technological improvements made to foams and the equipment associated with their application has made these agents' use in today's fire service more common and acceptable. As more fire apparatus have been equipped with foam systems, operation and maintenance of these systems have emerged as important skills for the driver/operator. The knowledge necessary to properly operate and maintain these systems can come only from learning about and training with the systems.

Overview

Why is foam used for firefighting? What are the reasons for equipping fire apparatus with foam systems? What are the benefits that make firefighting with foam so popular that fire departments are willing to spend thousands of dollars to add this equipment to their fire apparatus? Is it a fad, or does firefighting with foam truly make a difference?

With the introduction of the automobile and the use by industry of petroleum products, fire fighters soon realized that water was not effective for extinguishing these types of fires. Because water is heavier than petroleum, its application to petroleum-fueled fires caused more problems than it solved. In fact, water would actually spread the fire and make the situation worse. A method was needed to deal with the growing use of petroleum products and the resulting incidents that occurred when those products ignited. Firefighting foam was one of the methods available to deal with these issues.

What Is Foam?

The foam used for firefighting purposes is a stable mass of small, air-filled bubbles. It has a lower density than oil, gasoline, or water, meaning that it will float on top of the fuel. The way that foam works to help extinguish a fire is very simple **Figure 11-2**. Once applied to the burning fuel, the foam will float and form a blanket on the surface. This blanket, if applied correctly, will stop or prevent the burning process by separating the fuel from the air, lowering the temperature of the fuel, and/or suppressing the release of flammable vapors.

Foam is created through the application of four components: water, foam concentrate, mechanical agitation, and air. First, water is mixed with the foam concentrate in various ratios to produce a **foam solution**. This solution must then be mixed with air by some form of mechanical agitation. In firefighting, this mechanical agitation usually takes the form of a nozzle that mixes the air and foam solution to form the final product, which is referred to as **finished foam**. Too little foam concentrate in the water will produce a foam solution that is too thin (lean) to be effective and may quickly dissipate into the fuel. Too much concentrate will produce a foam that may be too thick (rich) to be properly expanded or aspirated when mixed with air.

The expansion of foam solution depends on good mechanical agitation and effective **aeration**, the process of introducing air into the foam solution. When an insufficient amount of air is introduced into the solution stream, the solution is poorly aerated. This results in foam with few bubbles; fewer bubbles cause the foam to break down quickly so that it does not suppress vapors effectively. Poorly aerated foam will also break down quickly when exposed to heat and flame.

Foam Tetrahedron

A **foam tetrahedron** **Figure 11-3** depicts the elements needed to produce finished foam. If any of the sides of the tetrahedron are missing or not at the proper mixture, foam production will be affected. The result can range from a poor-quality, ineffective foam to no foam produced at all. The foam tetrahedron includes four components:

1. Water
2. Foam concentrate
3. Air
4. Mechanical agitation

Foams come in two basic types: **chemical foams** and **mechanical foams**. Chemical foams are produced through a reaction between two chemicals, like the one that took place in the chemical foam extinguishers used in the 1800s **Figure 11-4**. Today, this type of foam is rarely used because it requires the combination of two different chemicals before the foam can be made, which may be difficult to carry out at a fire scene. Also, because of the large amount of powders required, this type of foam has become obsolete in today's fire service.

Mechanical foams are produced when water and foam concentrate are mixed in the appropriate amounts (**proportioned**). The ratio of foam concentrate to water must be the correct percentage if the foam-creation operation is to be effective. The class of materials involved in the spill or fire, which dictates the type of fire, determines the percentage of foam needed.

Hydrocarbon fuels, such as gasoline, jet fuel, and kerosene, have a lower **surface tension** than water. When these types of

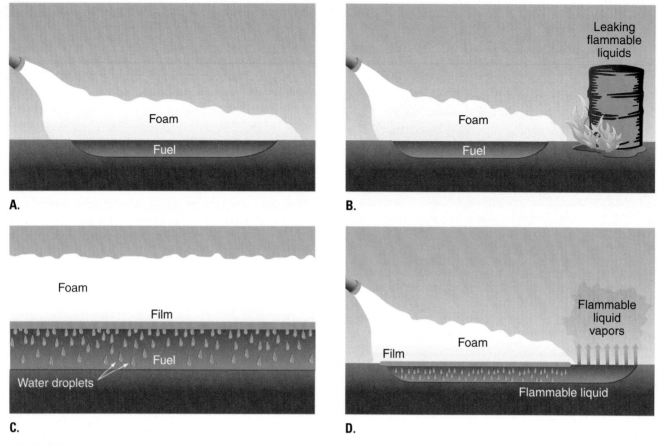

Figure 11-2 **A.** Foam forms a blanket over the fuel surface and *smothers* the fire. **B.** A foam blanket *separates* the flames from the fuel surface. **C.** A foam blanket *cools* the fuel and any adjacent heat and ignition sources by slowly releasing the water that forms a major portion of the foam. **D.** A foam blanket *suppresses* the release of flammable vapors that can mix with air.

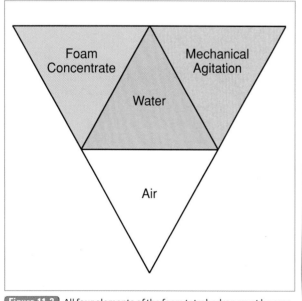

Figure 11-3 All four elements of the foam tetrahedron must be present to make foam.

fuels and water are mixed, the two fluids quickly separate; the fuel rises to the top and the water remains on the bottom. When **foam concentrate** is mixed with water, the surface tension is reduced, allowing the foam solution to float on the surface of the fuel. Its presence will suppress the vapors by separating the fuel from oxygen and the ignition source as well as by cooling the fuel below its ignition temperature.

Producing finished foam has become a very common practice for operators of fire apparatus. The modern foam systems are user friendly, and the easy setup and application steps have made foam an important tool for fire fighters to have at their disposal.

Foam Characteristics

Good foam must have the right combination of physical characteristics if it is to be effective. Specifically, it must have good knockdown speed and flow as well as good fuel resistance.

Knockdown speed and flow is the time required for a foam blanket to spread out across a fuel surface. The foam must also be able to flow around obstacles to achieve complete extinguishment and vapor suppression.

Foam must have good heat resistance so that it can avoid breakdown from the effects of direct flame contact with burning fuel vapors or the heat generated from metal objects. **Fuel**

Figure 11-4 Chemical foam extinguishers were first used in the 1800s.

Driver/Operator Tip

Operating foam lines that produce poor-quality foam or no foam at all can prevent the desired results of the operation from being achieved and endanger the fire fighters who are managing the hose lines. It is the driver/operator's responsibility to make sure that foam is being produced at the proper ratio. Know your foam system!

resistance is foam's ability to minimize **fuel pick-up**, which is the absorption of the burning fuel into the foam itself. This **oleophobic** (ability to shed hydrocarbons) quality reduces the amount of fuel saturation in the foam and prevents the foam blanket from burning.

Ideally, foam will produce a good vapor-suppressing blanket. A vapor-tight foam blanket reduces the generation of flammable vapors above the fuel surface and minimizes the chance of reignition Figure 11-5 . When used on polar solvent fuels, foam must also be alcohol resistant. Given the facts that alcohol readily mixes with water and that foam is mostly water, a foam blanket that is not alcohol resistant will quickly dissolve into the fuel and be destroyed.

Foam Classifications

Firefighting foams are classified as either Class A or Class B. Understanding the difference in classes of foam as well as the advantages and disadvantages of each class will help you to safely and effectively handle an incident where foam is needed.

Figure 11-5 A good vapor-suppressing foam blanket is needed to prevent reignition.

Class A Foams

<u>Class A foams</u> are used on ordinary combustible materials such as wood, textiles, and paper; they are also effective on organic materials such as straw and hay. Class A foams, which are often referred to as wetting agents, are very effective because they improve the penetrating effect of water and allow for greater heat absorption. Foam manufacturers state that these wetting agents can extinguish fire in Class A materials as much as 20 times faster than water Figure 11-6 .

Figure 11-6 Class A foam improves the penetrating effect of water.

Figure 11-7 A thick blanket of Class A foam applied to an exposure can provide adequate protection to that exposure, preventing the spread of fire.

Class A foam is particularly useful for protecting buildings in rural areas during forest and brush fires when the water supply is limited. It can also be used as a fire barrier—that is, an obstruction to the spread of fire. A thick blanket of Class A foam applied to an exposure can provide adequate protection to that exposure, thereby preventing the spread of fire Figure 11-7. Many fire departments use Class A foam while performing initial attack and overhaul of fires.

Class A foam increases the effectiveness of the water as an extinguishing agent by reducing the surface tension of the water. This allows the water to penetrate dense materials instead of running off the surface; it also allows more heat to be absorbed. In addition, the foam keeps water in contact with unburned fuel to prevent its ignition. Class A foam can be added to water streams and applied with various nozzles.

Class B Foams

Class B foams are used on hydrocarbon, combustible fuels, or polar solvent fires Table 11-1. A hydrocarbon is a chemical compound that contains the elements carbon and hydrogen; fuel oil is an example. Polar solvent and water-miscible fuels, such as acetone, mix readily with water and will degrade the effectiveness of ordinary Class B foam.

Class B foams are divided into the following categories:
- Protein foams
- Fluoroprotein foams
- Alcohol-resistant film-forming fluoroprotein foam (AR-FFFP)

Protein Foams

Protein foams are used for extinguishment of Class B fires involving hydrocarbons. These agents can protect flammable and combustible liquids where they are stored, transported, and processed. Protein foams contain naturally occurring proteins—that is, animal by-products—as the foaming agent. They are based on hydrolyzed keratin protein derived from sources such as chicken feathers or hooves. Protein foams also include stabilizers and inhibitors, which prevent corrosion and control viscosity. They produce highly stabilized mechanical foam with good expansion properties, excellent heat and burnback resistance, and good drainage characteristics.

Regular protein foams may be created using either fresh or salt water. These foams must be properly aspirated, however; thus they should not be used with non-aspirating structural fog nozzles. Regular protein foams have slower knockdown characteristics than other concentrates, but they provide a long-lasting foam blanket after the fire is extinguished Figure 11-8.

Fluoroprotein Foams

Fluoroprotein foam (FP) consists of hydrolyzed protein, stabilizers, preservatives, and synthetic fluorocarbon surfactants. This type of foam is used for hydrocarbon vapor suppression and extinguishment of fuel-in-depth fires. Notably, FP is effective for subsurface application to hydrocarbon fuel storage tanks.

Fluoroprotein foams are intended for use on hydrocarbon fuels and some oxygenated fuel additives. They have excellent heat and burnback resistance, and they maintain a good foam blanket after extinguishment of the fire. The addition of surfactants makes the foam more fluid, which increases the knockdown rate and provides better fuel tolerance than is possible with protein foam. Fluoroprotein foams may be created using either fresh or salt water. Because they must be properly aspirated, these foams should not be used with non-aspirating structural fog nozzles Figure 11-9.

Film-forming fluoroprotein foam (FFFP) use protein and fluorochemical surfactants. With these foams, the fluorochemical surfactants are generally present at a higher

Table 11-1 Class B Foams and Their Properties

Property	Protein	Fluoroprotein	AFFF	FFFP	AR-AFFF
Knockdown	Fair	Good	Excellent	Good	Excellent
Heat Resistance	Excellent	Excellent	Fair	Good	Good
Fuel Resistance	Fair	Excellent	Moderate	Good	Good
Vapor Suppression	Excellent	Excellent	Good	Good	Good
Alcohol Resistance	None	None	None	None	Excellent

Notes: AFFF = aqueous film-forming foam; FFFP = film-forming fluoroprotein foam; AR-AFFF = alcohol-resistant aqueous film-forming foam.

polysaccharide **polymer** additive. Whereas polar solvents will destroy FFFP, the polymer in AR-FFFP forms a membrane to separate the polar solvent from the foam blanket.

Synthetic Foams

Aqueous film-forming foam (AFFF) is based on combinations of fluorochemical surfactants, hydrocarbon surfactants, and solvents. AFFF requires a low-energy source to produce high-quality foam. This foam can be applied using a variety of foam application systems. Because of the versatility of AFFF, it is used by the majority of municipal and airport fire departments in the United States. AFFF is also used at refineries, manufacturing plants, and other operations involving flammable liquids.

AFFF is very fluid and quickly flows around obstacles and across the fuel surface. This ability to flow quickly allows AFFF to achieve a very fast knockdown of hydrocarbon fires. This foam can be used as a premixed solution, is compatible with dry chemical agents, and can be created using either fresh or salt water.

AFFF consists of a mixture of synthetic foaming agents and fluorochemical surfactants. It extinguishes fire by forming an aqueous film on the fuel surface. This film comprises a thin layer of foam solution, which quickly spreads across the surface of a hydrocarbon fuel, creating an extremely fast fire knockdown **Figure 11-10**. The surfactants found in AFFF reduce the surface tension of the foam solution, which allows it to remain on the surface of the hydrocarbon fuel. The aqueous film is formed by the action of the foam solution draining from the foam blanket.

Application of AFFF needs to be accomplished by using aspirating foam nozzles for maximum performance. Nozzles that will be used for foam application must be tested for compatibility with the foam system being used.

Figure 11-8 Regular protein foams have slower knockdown characteristics than other concentrates, but they provide a long-lasting foam blanket after the fire is extinguished.

concentration than is available with standard FP. FFFP is able to form a vapor-sealing film on nonpolar solvents. Knockdown performance is improved because the foam releases an aqueous film on the surface of the hydrocarbon fuel.

Alcohol-Resistant Film-Forming Fluoroprotein Foams

Alcohol-resistant film-forming fluoroprotein foam (AR-FFFP) can be used on both hydrocarbon and water-soluble fuels. On hydrocarbon fires, the film-forming ability of this foam allows for rapid fire knockdown and excellent burnback resistance. On water-soluble fuels, its resistance to water-soluble solvents occurs because of the formation of a cohesive polymeric membrane on the fuel surface, which protects the foam from contact with polar fuels. AR-FFFP is basically FFFP with a

Alcohol-Resistant Aqueous Film-Forming Foams

Alcohol-resistant AFFF (AR-AFFF) foam comprises a combination of synthetic detergents, fluorochemicals, and high-molecular-weight polymers. Neither polar solvents nor water-miscible fuels

Figure 11-9 Fluoroprotein foams must be properly aspirated and should not be used with non-aspirating structural fog nozzles.

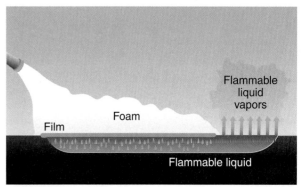

Figure 11-10 The film formed by application of AFFF comprises a thin layer of foam solution, which quickly spreads across the surface of a hydrocarbon fuel, creating an extremely fast fire knockdown.

are compatible with non-alcohol-resistant foams. Common polar solvents include solutions of the following compounds:

- Alcohols: isopropyl, methanol, ethanol
- Esters: butyl, acetate
- Ketones: methyl ethyl ketone
- Aldehydes: cinnamaldehyde, tolualdehyde

When non-alcohol-resistant foam is applied to the surface of a polar solvent, the foam blanket quickly breaks down into a liquid and mixes with the fuel. By comparison, AR-AFFF acts like a conventional AFFF on hydrocarbon fuels, forming an aqueous film on the fuel surface. When applied to polar solvents, however, AR-AFFF forms a polymeric membrane on the fuel surface. This membrane separates the fuel from the foam and prevents destruction of the foam blanket Figure 11-11.

AR-AFFF is one of the most versatile types of foam. It provides good knockdown and burnback resistance, and it has a high fuel tolerance on polar solvent and hydrocarbon fires.

Safety Tip

Today, alternative fuels are being used in a variety of vehicles. These fuels include ethanol, natural gas, propane, hydrogen, and methanol. Fuel mixtures composed of gasoline and ethanol have become quite common, for example. Gasoline mixtures containing more than 10 percent ethanol are polar solvents and will mix with water. Alcohol-resistant foams should be used for incidents involving these types of fuels.

Synthetic Detergent Foams (High-Expansion Foams)

High-expansion foams are highly effective in confined-space firefighting operations and in areas where access is limited or entry is dangerous to fire fighters, such as basements, shipboard compartments, warehouses, aircraft hangers, and mine shafts. These types of foams can be used in fixed generating systems and portable foam generators.

High-expansion foams can be used on either Class A or Class B fires. Each foam manufacturer offers a different product, however, and it is your responsibility as a driver/operator to determine which foam can be used on the fire. High-expansion foams are based on combinations of hydrocarbon surfactants and solvents. They are useful on fuels such as liquefied natural gas for vapor dispersion and control. In certain concentrations, these foams are effective extinguishing agents for flammable liquid spill fires of most types. The foam concentrates are normally proportioned between 1 percent to 3 percent for these uses Figure 11-12.

Safety Tip

Make sure that the foam concentrate that you are using is designed for the product and the application method being used. Serious injuries or death can occur if the appropriate foam is not applied properly.

Fire control and extinguishment are achieved by rapid smothering and cooling of the fire. High-expansion foams have a tremendous smothering and steam generation effect because the water contained in them is divided into such fine particles, which enhance the heat absorption quality of the water. This arrangement also presents a potential hazard: Care must be taken with regard to electrical power sources in the area when this type of foam is applied. Remember that foam is mostly water and presents the same electrical shock potential to fire fighters as does water application.

Figure 11-11 When applied to polar solvents, AR-AFFF forms a polymeric membrane on the fuel surface. This membrane separates the fuel from the foam and prevents destruction of the foam blanket.

Figure 11-12 Synthetic detergent foam is an effective extinguishing agent for flammable liquid spill fires of most types.

Foam Concentrates

Foam concentrates are designed to be mixed with water at specific ratios **Table 11-2**. Foam concentrate ratios vary from 1 percent to 6 percent (i.e., 1 to 6 percent foam concentrate-to-water ratio in the final foam product). The amount of concentrate varies depending on the manufacturer, the type of application, and the type of fuel. For example, when using Class A foam for initial fire attack on a vehicle fire, you might use 1 percent ratio of foam concentrate to water. This ratio allows the foam to help extinguish the fire without creating too much foam. If you were supplying a hose line that was spraying foam on a house to protect it during a brush fire, you might increase the ratio to 3 percent, thereby creating a thicker foam that would last longer and thus aid in protecting the house for a longer period.

The foam concentrate must be proportioned at the percentage listed by its manufacturer. Each foam is tested and approved for certain types of fires and at specific ratios—so you should always follow the manufacturer's guidelines when using this product. The foam concentrates are manufactured at different percentages for a variety of reasons. Thus the product's components include the concentrate (a unique chemical compound) plus any freeze protection additives. Military-use specifications and cost are some of the basic factors that determine the percentage of concentrate used in any particular situation.

The trend in the industry is to reduce foam concentrate percentages as low as possible, for several reasons. First, lower proportioning rates mean less bulk in storage for fire departments. Second, they mean that you can double your firefighting capacity by carrying the same volume of foam concentrate or you can cut your foam supply in half without reducing the company's fire suppression capabilities. Third, lower proportioning rates can reduce the cost of fixed foam system components and concentrate transportation costs. Historically, foam concentrates were manufactured for use at ratios between 3 percent and 6 percent. Today, however, foam concentrates are produced for uses at ratios as low as 1 percent and as high as 6 percent, depending on the liquid fuel and the manner in which the foam is to be used. Remember, the foam concentrate percentage must match the fuel to which the foam is being applied; if there is a mismatch, the foam may not be effective in controlling the fire.

As mentioned earlier, alcohol-resistant foams can be used effectively on both hydrocarbon and polar solvent fuel. AR-AFFF is a commonly used concentrate for this purpose; it is available in different percentages. AR-AFFF concentrate is available as a 3 percent/3 percent product, which means that it can be used at 3 percent for incidents involving either hydrocarbon fuels or polar solvents. It is also available as a 3 percent/6 percent mixture, which means that foams for hydrocarbon-based incidents are proportioned at a 3 percent concentrate ratio, and foams for incidents involving polar solvents are proportioned at a 6 percent concentrate ratio. Another option available is AR-AFFF 1 percent/3 percent, which means foams for hydrocarbon-based incidents are proportioned at 1 percent and foams for polar solvent–based incidents are proportioned at 3 percent.

Given that there are so many types of foam concentrates available, the selection of the right concentrate is critical to the safe and effective handling of an incident. Knowledge of the foam types and systems available will assist incident commanders (ICs) in their ability to mitigate an incident. The importance of preplanning and training for these types of incidents cannot be stressed enough.

Foam Expansion Rates

The foam expansion rate is the ratio of finished foam to foam solution after the concentrate is mixed with water, agitated, and aspirated through a foam-making appliance. For example, a low-expansion foam has an expansion ratio of 20:1, which means that there are 20 parts of finished foam to every one part of foam solution. The air inside the bubbles makes up the expanded part of the finished foam. NFPA 11 classifies foam concentrates into three expansion ranges:

- Low expansion
- Medium expansion
- High expansion

Low-Expansion Foam

<u>Low-expansion foam</u> has a foam expansion ratio of up to 20:1. This type of foam is primarily designed for use on flammable and combustible liquids. Low-expansion foam is effective in controlling and extinguishing most Class B fires. Special low-expansion foams are also used on Class A fires where the penetrating and cooling effect of the foam solution is important.

Medium-Expansion Foam

<u>Medium-expansion foam</u> has a foam expansion ratio in the range of 20:1 to 200:1. This kind of foam is used primarily to suppress vapors from hazardous chemicals. Foams with expansion ratios between 30:1 and 55:1 have been found to produce the optimal foam blanket for vapor mitigation of highly reactive chemicals and low-boiling-point organics.

High-Expansion Foam

<u>High-expansion foam</u> has a foam expansion ratio in the range of 200:1 to 1000:1. This type of foam is designed for confined-space firefighting. The foam concentrate consists of a synthetic, detergent-type foam used in confined spaces such as basements, mines, shipboard, and aircraft hangars.

Table 11-2 Foam Concentrate-to-Water Ratios

Protein foam	3–6%
Fluoroprotein foam	3–6%
Aqueous film-forming foam (AFFF)	1–6%
Film-forming fluoroprotein foam (FFFP)	3–6%
Alcohol-resistant aqueous film-forming foam (AR-AFFF)	3–6%
Class A foam	0.1–3%

Foam Proportioning

Foam cannot be produced if it is not proportioned properly. Several types of foam application systems are available for this purpose, ranging from basic operations to more advanced systems. With the proper training, these systems are user friendly and will produce good-quality finished foam.

■ Proportioning Foam Concentrate

The application of foams at the proper percentage depends on the foam concentrate being mixed at the proper percentage with water. This percentage of concentrate may be introduced into the water stream by any of several methods. It is imperative that you are familiar with the type of equipment that is used in your agency's foam operations.

As the driver/operator, you have the responsibility to produce effective foam streams. You must have knowledge of all aspects of foam operations so that the foam produced will do what it is intended to do. Foam that is not mixed at the proper percentage or is used on a material or product for which it was not intended will not only be ineffective, but could also put the fire fighters on the hose lines in danger.

To produce finished foam, the water, air, and foam concentrate must be mixed at the proper ratio. In other words, foam concentrate must be added to the water that is flowing from a discharge line to form the proper mixture. This mixture or percentage is based on the type of foam concentrate being used, the type of material involved in the incident, and the type of equipment used to produce the finished foam.

■ Foam Proportioning Systems

A foam proportioner is the device that mixes the foam concentrate into the fire stream in the proper percentage. The two types of proportioners—eductors and injectors—are available in a wide range of sizes and capabilities. Foam solution can also be produced by batch mixing or premixing.

Batch Mixing

<u>Batch mixing</u> is the process of pouring foam concentrate directly into the fire apparatus water tank, thereby mixing a large amount of foam at one time. The proper amount of foam concentrate must be poured into the onboard tank to produce the desired percentage in the finished foam **Figure 11-13**. While this method requires no special appliances, there are some problems associated with batch mixing:

- The foam solution is corrosive to the apparatus's pipes, pump, and water tank.
- It is difficult to adjust and maintain the correct application rate, especially if additional water is supplied for an external source.
- The addition of foam solution may cause the gauges to become inaccurate and overflow the water tank with foam when the pump is recirculating the mixture.

Class A foam concentrates that are batched mixed must be used within 24 hours to be effective. As with some of the other systems used to create foam, all discharges will be delivering foam when supplied from the onboard tank. Another issue is

Figure 11-13 Batch mixing is not the most effective way to mix foam concentrate and water.

that when the water tank is empty, there is no longer any foam available—a new batch has to be mixed. As long as you know the amount of water in the onboard water tank, then you can simply add the foam concentrate to achieve the desired foam solution percentage. For example, if the water tank on the fire apparatus holds 500 gallons of water and you want to use Class A foam at 1 percent, you would add 5 gallons of foam concentrate to the water tank.

To prepare foam through batch mixing, follow the steps in **Skill Drill 11-1**:

1. Make sure that you know the correct amount of water in the fire apparatus water tank. (**Step 1**)
2. Determine the correct percentage of foam required for the fire.
3. Add the correct amount of foam concentrate directly into the top of the water tank's fill port. (**Step 2**)
4. Discharge the foam through the attack lines. (**Step 3**)

Premixing

Premixing of foam solutions is a technique that is usually reserved for portable fire extinguishers. Many fire departments use these extinguishers filled with premixed foam solutions for small flammable liquid spills at the scene of a motor vehicle

Chapter 11 Foam

Skill Drill 11-1

Batch Mixing Foam

5.2.3

1. Make sure that you know the correct amount of water inside the fire apparatus water tank.

2. Determine the correct percentage of foam required. Add the correct amount of foam concentrate to the water tank.

3. Apply the foam on the fire.

accident. They are quick and easy to deploy but have a limited amount of foam and should be applied only to small fires or fuel spills. Always refer to the foam manufacturer's recommendations regarding how long the foam concentrate should be used after mixing it with water.

Foam Eductors

Induction involves the use of an eductor to introduce the appropriate amount of foam concentrate into a stream of water flowing from a discharge. An **eductor** is an appliance that uses the Venturi principle (i.e., suction effect) to induce foam concentrate into the water stream. An eductor can be built into the plumbing of an engine, or a portable eductor can be inserted in an attack hose line. A foam eductor is usually designed to work at a predetermined pressure and flow rate; its metering valve can be adjusted to set the percentage of foam concentrate that is educted into the stream. For example, a simple in-line foam eductor has a small pick-up tube that is submerged in foam concentrate. As the water travels through the in-line eductor, it draws foam concentrate into the water at the desired percentage using the Venturi principle.

Two types of eductors are used in the fire service today: in-line eductors and bypass eductors. In-line eductors **Figure 11-14** have long been used to proportion foam. These appliances may be mounted permanently on the apparatus, or they can be portable and connected anywhere along the hose lay. In-line eductors should be used only for the application of foam. Those mounted permanently to the fire pump are dedicated to one foam discharge; such **pump-mounted eductors** are in-line devices dedicated to producing foam only. By comparison, **bypass eductors** are permanently mounted appliances that can be used for water or foam application, depending on what is required at the incident scene.

As mentioned earlier, eductors use the Venturi effect to mix a specific amount of foam concentrate into the water stream. A suction effect is created at the narrow inlet to the eductor. This narrow passage increases water velocity, which reduces its pressure as it flows into the larger induction area. Foam concentrate is introduced into the eductor using a metering valve, which allows the driver/operator to adjust the percentage of foam concentrate being educted into the water stream.

Most eductors are calibrated to flow the rated capacity at 200 pounds per square inch (psi) (460 kilopascals [kPa]) inlet pressure. Eductor inlet pressure of 200 psi is necessary to overcome friction loss through the eductor as well as the friction loss between the eductor and the nozzle.

Eductors are available for delivering flow rates of 30, 60, 95, 125, and 250 gallons per minute (GPM) (114, 227, 360, 473, and 946 liters per minute [L/min]), respectively). The nozzle used with the system must match the flow rating of the eductor. It is important to follow the manufacturer's recommendations when selecting nozzles to use with foam eductors.

Metering Device

The **metering device** **Figure 11-15** controls the flow of concentrate into the eductor. The amount of foam concentrate introduced into the water stream is controlled by adjusting the metering valve. As the driver/operator, you should adjust the metering valve to the desired percentage to handle the situation at hand. Metering valves have adjustable settings that range from 0 (the closed position) to 6 percent.

The percentage set on the metering valve will be achieved only if the inlet pressure at the eductor matches the manufacturer's recommended inlet pressure. If the eductor is operated

Figure 11-14 In-line eductors are very simple to operate and are found in many fire departments.

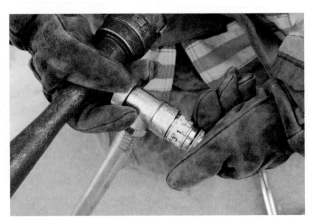

Figure 11-15 The metering device controls the amount of foam concentrate educted into the water.

at less than the recommended inlet pressure, a lower flow of water will go through the eductor. This lower flow of water will, in turn, result in a higher percentage of foam concentrate being introduced into the water stream. Depending on the available supply of foam concentrate, unnecessary wastage of foam concentrate will occur and the driver/operator could run out of foam before the fire is extinguished.

The opposite will occur if the inlet pressure to the eductor exceeds the pressure recommended by the manufacturer. A higher inlet pressure at the eductor will cause the foam solution to have a lower percentage of concentrate, which could affect the company's ability to handle the situation in which the foam is being applied.

Operating the In-line Eductor

The in-line eductor is a very simple, efficient, and inexpensive type of foam proportioner. Many fire departments still use it as standard equipment to create foam at their fire scenes. These devices are constructed out of rugged brass alloys and, if properly maintained, can last for many years. The eductor is attached to the hose line no more than 150' (48 m) from the nozzle—a placement that makes it easy to set up and take down. Although this type of device is simple to operate, some very specific instructions must be followed to ensure its proper use.

To operate an in-line eductor, follow the steps in **Skill Drill 11-2**:

1. Make sure that all necessary equipment is available, including an in-line foam eductor and air-aspirating nozzle. Ensure there is enough foam concentrate to suppress the fire and that it is the correct type of foam for the job. (**Step 1**)
2. Don all personal protective equipment (PPE). (**Step 2**)
3. Procure an attack line, remove the nozzle, and replace it with the air-aspirating nozzle if necessary. (**Step 3**)
4. Place the in-line eductor in the hose line no more than 150' (48 m) from the nozzle. (**Step 4**)
5. Place the foam concentrate container next to the eductor, check the percentage at which the foam concentrate should be used (this information will be found on the container label), and set the metering device on the eductor accordingly. (**Step 5**)
6. Place the pick-up tube from the eductor into the foam concentrate, keeping both items at similar elevations to ensure sufficient induction of foam concentrate. (**Step 6**)
7. Charge the hose line with water, ensuring there is a minimum pressure of 200 psi (460 kPa) at the eductor. Flow water through the hose line until foam starts to come out of the nozzle. The hose line is now ready to be advanced onto the burning material. (**Step 7**)
8. Apply the foam using one of the three application methods (sweep technique, bankshot technique, or raindown technique—discussed later in this chapter) depending on the situation. (**Step 8**)

Around-the-Pump Proportioning System

An **around-the-pump proportioning (AP) system** operates on the same principle as the in-line or bypass eductor system. The AP system diverts a portion of the water pump's output from the discharge side of the pump and sends it through an eductor. A vacuum is created at the eductor's foam concentrate inlet, which draws foam concentrate through the metering valve and into the eductor. The foam solution (water/concentrate) is then sent to the suction side of the water pump, where it mixes with the incoming water and is distributed throughout the discharge piping.

The AP system offers several advantages over other foam-creation methods:

- The process used for engaging the pump is the same for water or foam operations.
- The variable flow discharge rate allows for adjustment of the foam depending on the specific application.
- Variable pressure operations are possible, usually within a range of 125 psi (288 kPa) to 250 psi (576 kPa).
- There are no backpressure restrictions, because the system is not affected by hose length or elevation loss.
- There are no nozzle restrictions because the system will operate with any size or type of fixed-gallon nozzle.

As with any foam system available, AP systems have certain limitations with which you must be familiar. Knowing these limitations will enable you to use the system in an efficient and safe manner. All discharges will have *either* foam *or* water available at the same time. You will not be able to supply some lines with water and other lines with foam simultaneously: It's either one or the other.

> **Driver/Operator Tip**
>
> The foam solution is discharged from the eductor to the intake side of the water pump. Thus, when you are using foam, all outlets will be discharging foam when they are opened. Along with the discharges used for firefighting, the following sources will also discharge foam:
> - Tank fill
> - Pump cooler
> - Engine cooler
> - Operation of the primer pump
>
> Be aware of the potential problems with sending foam solution into these areas.

The maximum inlet pressure to the water pump cannot be more than 10 psi (23 kPa). Any pressure greater than 10 psi will affect the operation of the eductor. Some manufacturers' systems are designed to operate with an inlet pressure of as much as 40 psi (92 kPa), however, so check your system operating instructions for proper inlet pressure.

AP systems operate properly when water is supplied from the onboard tank or from draft. As a consequence, a pressurized source (hydrant or relay) will affect the operation of the system. Most fire apparatus with AP systems have a direct tank fill that does not go through the pump. In such systems, the tank is filled

Skill Drill 11-2

Operating an In-line Eductor

1. Ensure all equipment is available, including the air-aspirating nozzle, foam concentrate, and in-line eductor.

2. Don PPE.

3. Add the air-aspirating nozzle to the attack line.

4. Place the in-line eductor no more than 150′ (48 m) from the nozzle.

5. Place the foam concentrate next to the eductor, check the percentage for the foam, and set the metering device.

6. Place the pick-up tube into the foam concentrate.

7. Charge the hose with a minimum pressure of 200 psi (460 kPa) at the eductor. Flow water through the hose until foam begins and the hose is ready to advance on the fire.

8. Apply the foam.

Safety Tip

As the driver/operator, you must monitor the tank water level. Running out of water will cause firefighting operations to cease. Conversely, overfilling the tank can cause the water to overflow and possibly run down into the foam blanket.

directly from the pressurized source, and the water pump supply comes from the onboard tank.

All discharges must be flushed after the foam operation has ended, even if they were not used. When using automatic nozzles, the flow must be determined to ensure that you set the metering valve correctly. The appropriate setting for the metering valve depends on the type of concentrate being used, the percentage needed for the incident, and the rate of flow (in GPM). When the flow rate changes, as when a line is shut down, the metering valve needs to be adjusted to accommodate the new flow.

To operate an AP system, follow the steps in **Skill Drill 11-3**:

1. With the system engaged, operate the fire pump as you would during normal water pump operations. **(Step 1)**
2. Push the "on" button at the pump panel display to turn the foam system on. **(Step 2)**
3. Set the desired percentage of foam for the fire using the up/down arrow buttons. **(Step 3)**
4. Open the foam discharge valve, if required, and set the desired pressure.
5. Discharge the foam through the attack lines. **(Step 4)**

Driver/Operator Tip

Do not operate the AP system for extended periods of time with the discharges closed. Even though these outlets are closed, the system will continue sending water through the eductor, and foam concentrate will continue to be discharged into the water pump. The water pump will become rich with concentrate; as a result, concentrate will be wasted and the water pump may overheat. Some driver/operators have run out of foam concentrate because they operated the foam system with the discharges closed.

Balanced-Pressure Proportioning Systems

Balanced pressure systems are extremely versatile and accurate means to deliver foam. In fact, balanced-pressure proportioning is the most common method used for foam system application in today's fire service. This system uses a diaphragm-type pressure control valve to sense and balance the pressures in the

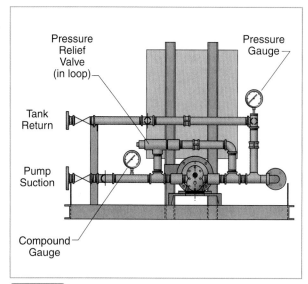

Figure 11-16 Foam concentrate and water pressure are kept in balance because the excess foam concentrate is allowed to return to the storage tank.

Courtesy of Williams Fire & Hazard Control, Inc.

foam concentrate and water lines to the proportioner. The valve keeps the foam concentrate and the water pressure in balance by allowing excess foam concentrate to return to the foam concentrate storage tank **Figure 11-16**. This proportioning occurs automatically for flows within the operating limits of the foam concentrate pump and the discharges.

The foam concentrate pump is a separate pump that supplies foam concentrate to the pressure control valve and the ratio controller. The pressure control valve is a balancing valve: It maintains equal foam concentrate and water pressures at the ratio controllers. Thus the pressure control valve automatically maintains equal pressure within the foam-creation system. Unused foam concentrate from the foam pump discharge is returned to the foam concentrate storage tank.

A ratio controller is a device required for each foam outlet to proportion the correct amount of foam concentrate into the water stream over a wide range of flows and with minimal pressure loss. This modified Venturi device provides a metered pressure drop for controlled injection of foam concentrate into a zone characterized by reduced pressure.

Metering valves receive foam concentrate from the foam pump and, in turn, discharge the concentrate to the individual ratio controllers. Put simply, these devices function as a concentrate shut-off. They also allow you to adjust the foam percentage for each discharge. The metering valve settings available usually range from "off" to 6 percent. Foam solution is produced at any ratio controller/discharge connection by opening the corresponding foam solution metering valve to the desired percentage rate. Plain water is simultaneously available at the remaining discharge connections.

Skill Drill 11-3

Operating an Around-the-Pump Proportioning System

1. With the system engaged, operate the fire pump as you would during normal water pump operations.

2. Push the "on" button at the pump panel display to turn the foam system on.

3. Set the desired percentage of foam for the fire using the up/down arrow buttons.

4. Open the foam discharge valve, if required, and set the desired pressure. Discharge the foam through the attack lines.

Injection systems depend on the water flow for their operation. They will adjust the amount of foam concentrate being injected into the discharge manifold depending on the flow rate (GPM) and the percentage set by the driver/operator. The foam concentrate pump will also adjust its operating speed (rpm) depending on the amount of water flowing.

Injection systems are unaffected by changes in suction and discharge pressure. The only limitation on such a system is the rated capacity of the foam concentrate pump. There are many different sizes of foam concentrate pumps available, so it is imperative that you know the limits of your foam concentrate pump.

Figure 11-17 A duplex gauge allows you to monitor the foam concentrate and water pressures simultaneously.

Driver/Operator Tip

A foam concentrate pump rated at a maximum output of 5 GPM would be able to supply concentrate at the following flow rates:

- 0.5 percent foam concentrate produces 1000 GPM (3785 L/min) foam solution.
- 1 percent foam concentrate produces 500 GPM (1893 L/min) foam solution.
- 3 percent foam concentrate produces 167 GPM (632 L/min) foam solution.
- 6 percent foam concentrate produces 84 GPM (318 L/min) foam solution.

A <u>duplex gauge</u> **Figure 11-17** at the pump panel allows the driver/operator to monitor the foam concentrate and water pressures. These pressures should be "balanced" (equal)—a feat that is achieved automatically by the balanced-pressure system. For example, if the flow of water into the pump increases, the system will balance itself to compensate for the additional pressure to maintain an accurate flow of correctly proportioned foam solution. A manual throttling control valve located at the pump panel enables the driver/operator to override the automatic system, by increasing or decreasing foam concentrate pressure, in case a problem occurs within the system.

Many balanced-pressure systems are equipped with a <u>foam heat exchanger</u>. The heat exchanger uses water as its cooling source and prevents the foam concentrate from overheating.

To operate a balanced-pressure proportioning system, follow the steps in **Skill Drill 11-4**:

1. With the system engaged, operate the fire pump as you would during normal water pump operations.
2. Engage the foam pump's PTO (power take-off).
3. Push the "on" button at the pump panel display to turn the foam system on.
4. Set the desired percentage of foam for the fire using the up/down arrow buttons.
5. Open the foam discharge valve, if required, and set the desired pressure.
6. Discharge the foam through the attack lines.

Injection Systems

<u>Injection systems</u> use an electrically operated, variable-speed foam concentrate pump that directly injects foam concentrate into the discharge side of the pump manifold. These systems are designed for use with Class A foam concentrates and with many Class B foam concentrates, depending on the viscosity of the concentrate.

Direct injection systems are very user friendly. A control unit located at the pump panel allows the driver/operator to operate the foam system. This control unit has an on/off button that is used to start and secure foam operations. It also allows the driver/operator to monitor the water flow rate, concentrate injection percentage, total amount of water flowed, and total amount of concentrate used.

With injection systems, the foam concentrate percentage can be adjusted while the system is in operation. Concentrate injection rates can be set anywhere in the range of 0.1 percent to 10 percent. These percentages are changed in 0.1-percent increments and are selected using a button on the control unit.

Injection foam systems can be used with standard nozzles, aspirating nozzles, and a CAFS. These systems can be set up to supply multiple discharges on the water pump. The foam concentrate is injected into the discharge manifold at only one injection point, so all discharges that are piped from the foam discharge manifold will be capable of delivering foam. Although other discharges not piped to the foam manifold will be capable of discharging water, all discharges from the foam manifold will produce foam.

To operate an injection foam system, follow the steps in **Skill Drill 11-5**:

1. With the system engaged, operate the fire pump as you would during normal water pump operations.
2. Push the "on" button at the pump panel display to turn the foam system on.

3. Set the desired percentage of foam for the fire using the up/down arrow buttons.
4. Open the foam discharge valve, if required, and set the desired pressure.
5. Discharge the foam through the attack lines.

Compressed-Air Foam System

A <u>compressed-air foam system (CAFS)</u> combines compressed air and a foam solution to create the finished foam. It is designed for a quick and effective fire attack and exposure protection. This type of system is generally used to generate Class A foam—specifically, to create a high-quality finished foam that clings to vertical surfaces Figure 11-18 ▼. This foam holds moisture on the fuel surfaces where it either evaporates or penetrates, thereby cooling and lowering the fuel's temperature.

The production of finished foam depends on the correct mixture of water, foam concentrate, and air within the CAFS. This kind of system uses a fire pump, a foam concentrate injection system, and an air compressor to produce high-quality finished-foam streams inside the fire apparatus piping. The foam aeration on the CAFS takes place at the pump discharge—that is, air is introduced at the pump. Thus the foam solution is mixed with air prior to reaching the end of the hose line. This type of system allows for a more uniform bubble structure and better finished foam.

CAFS has some key benefits for firefighting operations:

- The quality of the foam is greatly improved due to its consistently smaller bubble structure.
- The foam produced by CAFS works four to five times faster than water in suppressing fire.
- CAFS uses less water, so it reduces the extent of water damage.
- The quick knockdown, fast fire suppression, and less extensive water damage associated with this type of foam system may help fire investigators in determining the cause of the fire.

- The reach of the fire stream is improved. The foam discharged will be lighter than water because the discharging foam is a mixture of foam solution and *air*.
- The improved stream reach allows for an initial fire attack from a greater distance, which reduces the risk of injury to fire fighters.
- The weight of the attack line is less than that of a line using just water. CAFS hose lines are typically filled with approximately 30 percent compressed air. Fire fighters will be able to move the line more easily, which will reduce their fatigue.
- Because the hose is filled with air, it will float and can be used for water rescue or as a temporary barrier to contain spills.
- Because the hose is filled with air, friction loss is insignificant.
- There is not the normal pressure loss with elevation.

There are also some issues with the CAFS that driver/operators should be aware of:

- You need to know about the operation of the air compressor.
- Water and air are incompressible. Foam solution must be present in the water stream prior to injecting compressed air. If foam concentrate is not present, unmixed water and air will be discharged in an erratic manner, and the hose line could be "jerked" out of the hands of the fire fighters.
- A CAFS burst hose line will react more erratically due to the air in the line.
- Nozzle reaction is greater due to air in the hose, so nozzles should be opened *slowly*.

To operate a CAFS, follow the steps in Skill Drill 11-6 ▶:

1. With the system engaged, operate the fire pump as you would during normal water pump operations. (**Step 1**)
2. Set the desired discharge pressure for the attack line. (**Step 2**)
3. Push the "on" button at the pump panel display to turn the foam system on (if not already activated). (**Step 3**)
4. Select either "wet" or "dry" foam to be delivered to the discharge by setting the desired percentage of foam for the fire using the up/down arrow buttons. (**Step 4**)
5. Open the discharge valve. (**Step 5**)
6. Discharge the foam through the attack lines. (**Step 6**)

Figure 11-18 A compressed-air foam system (CAFS) creates foam that has the consistency of shaving cream, which enables it to cling to the building's surface.

Safety Tip

A CAFS should be operated at the water flow rates that would be used if just water was being applied. While this type of foam system will extinguish the fire more quickly than a water-based foam system, proper flow rates are still needed to protect fire fighters.

Skill Drill 11-6

Operating a Compressed-Air Foam System

1. With the system engaged, operate the fire pump as you would during normal water pump operations.

2. Set the desired discharge pressure for the attack line.

3. Push the "on" button at the pump panel display to turn the foam system on (if not already activated).

4. Select either "wet" or "dry" foam to be delivered to the discharge by setting the desired percentage of foam for the fire using the up/down arrow buttons.

5. Open the discharge valve.

6. Discharge the foam through the attack lines.

Voices of Experience

While working as a volunteer fire fighter in a rural area, our department used foam on several occasions. Our rural county had a large number of oil field pumping stations. These pumping stations would pump crude oil out of the ground and into holding tanks. These tanks averaged 30 feet (9 m) tall and 15 feet (5 m) in diameter. The tanks would fill and then pump oil into an oil refinery. The top of these tanks were only tack welded and the tanks had a single vent pipe to allow for pressure changes and liquid level changes.

The vent stack would constantly be giving off flammable vapors. When storms would come through the area, lighting would inevitably strike the vent stack. This would ignite the flammable vapors in the tank, blowing off the top of the tank, and setting the oil on fire. Our volunteer department would get the call of an "oil tank on fire."

> *"The better we got, the less foam we had to use."*

On arrival, we would be presented with a flammable liquid fire at the top of a tall tank. Our standard procedure was to set up a foam operation to extinguish the fire. We would set up a 95 gallons per minute (GPM) (360 L/min) foam eductor and a 200 foot (76 m) 1½″ (38 mm) attack line with a red foam tube. This tube was 4 foot (102 mm) long with air inlets near the valve. This tube mixed air into the foam solution to make the finished foam (aka: the big, bubbly, fluffy, white stuff.)

We would use the rain-down technique to apply foam to the burning oil surface fire. The fire officer would stand out away from the two crew members on the hose line and direct the foam stream onto target. With the rain-down technique, the foam is aimed up into the air, the stream arcs over, and the stream breaks up into large flakes of foam. The fire officer would give the command to the hose team to adjust the position of the nozzle to get the flakes of foam to land where he wanted them to. The fire officer would use the wind to help with the fall of the foam. As the foam flakes begin to hit the surface of the fire, the Aqueous Film Forming Foam (AFFF) foam properties would begin to start to suppress the flammable vapors. As the foam began to take effect, the fire officer would direct the line to move across the surface to complete extinguishment.

Depending on the wind and rain, it would take approximately 10 gallons (38 L) of foam concentrate at a 3% concentration to get full extinguishment. The more volatile the oil in the tank, the more foam concentrate would be required to put out the fire. We got very good at this operation, sometimes repeating it 3 to 4 times a month. The better we got, the less foam we had to use. Practice make prefect!

Roger Westhoff
Houston Fire Department
Houston, Texas

Nozzles

Nozzles are an important part of all foam operations. The proper nozzle is needed for fire fighters to be able to produce a good-quality foam blanket. The following types of nozzles are available:

- Medium- and high-expansion foam generators
- Master stream foam nozzles
- Air-aspirating foam nozzles
- Smooth-bore nozzles
- Fog nozzles

Medium- and High-Expansion Foam Generators

Foam generators produce medium- and high-expansion foams. These devices usually consist of either a mechanical blower or a water-aspirating appliance. A water-powered aspirating type of generator uses a water-motor-driven fan to produce the required air flow. Some water-powered high-expansion generators are designed to achieve expansion rates of as much as 1000 gal (3785 L) of finished foam for every 1 gal (3.785 L) of foam solution. The final expansion ratio achieved depends on the generator used, the solution flow rate, and the operating pressure.

Mechanical generators Figure 11-19 ▼ operate similarly to a water-aspirating generator. The difference is that the mechanical generator is electrically powered by a fan, which forces the air flow through the unit. High-expansion foam systems are designed for use in total flooding applications and are effective at the following locations:

- Mines
- Warehouses
- Aircraft hangars
- Basements
- Storage buildings
- Paper warehouses
- Machinery spaces
- Hazardous waste facilities
- Wildland fire breaks

High-expansion foams produce large volumes of foam which exclude oxygen from the incident area. In addition, the low water content of this type of foam reduces water damage.

Master Stream Foam Nozzles

Master stream foam nozzles allow operators to deal with large incidents where handline nozzles are not able to handle the demands for foam suppression. These master stream nozzles can be supplied from the fire apparatus's onboard systems (injection); alternatively, they can have an eductor as part of the nozzle. A pick-up tube draws the foam into the master stream nozzle; this tube is placed into the foam concentrate supply and operated as an eductor system. Master stream foam nozzles can be mounted permanently on the fire apparatus, can be removable so that they can be used as a portable unit, or can be strictly portable devices.

Air-Aspirating Foam Nozzles

Aspirating foam nozzles are designed to mix air with the foam solution as it is discharged. This effect is achieved without having to add a clamp-on tube to a fog nozzle. These types of foam nozzles are designed to aspirate the foam solution to produce good-quality finished foam.

Smooth-Bore Nozzles

Smooth-bore nozzles are the nozzle of choice when using CAFS. Because the aeration of the foam solution takes place in the piping and hose, the finished foam will be discharged at the nozzle. Application—not aeration—is the major concern at the discharge of a nozzle using CAFS.

Fog Nozzles

Fog nozzles can be used to produce finished foam. The problem with using these types of nozzles is that they generally do not provide the best aeration of the foam solution. When the incident involves polar solvent fuels, for example, these nozzles may not deliver a foam quality that is able to extinguish the fire. Foam aeration tubes are available that can be easily and quickly clamped onto the end of a fog nozzle to aerate the foam solution more efficiently.

Foam Supplies

Foam concentrate is stored in containers that range in size from 5-gal (19-L) pails Figure 11-20 ▶ to 55-gal (208-L) drums. Totes and trailers of foam concentrate are available in different sizes as well, ranging from 100 to 1000 gal (378.5 to 3785 L) of foam concentrate. Foam may be stored in its container without change in its original physical or chemical characteristics. Freezing and thawing usually will not have any effect on modern foam concentrates, but it is important to check with the concentrate manufacturer for technical information.

Figure 11-19 This mobile foam fire apparatus is one example of a mechanical generator.

Near Miss REPORT

Report Number: 08-0000044
Report Date: 01/25/2008 11:40

Synopsis: Problem with CAFS unit identified at live burn.

Event Description: Department wide training at a donated house. We burned for 2 days with great results. Good training and no injuries resulted for either day. Followed NFPA 1403 and had a great game plan.

Senior personnel (Deputy Chief, District Chief, and 2 Officers from a neighboring Department), planned an end of the day burn to evaluate the effects of CAF in a live burn situation. CAFS engine was set up and burn room was prepared with hay. Crew entered the burn room and activated hose line in the burn room. Foam was not directly aimed at fire, but towards the general area. During the initial stages, a burst of air exited the line (1¾"), and the fire rapidly increased. It was determined that this was not a flashover, but rapid fire intensification due to the disruption of the fire conditions. Nozzleman was on all 4's inside the room when this occurred. Nozzleman felt heat increasing, and exited the room and house per the exit plan. No injury resulted, but coat, helmet, and facepiece received heavy fire damage. We interviewed the FF and DC to get their interpretation of the incident. Upon investigation of the CAFS Engine the following day by the Quartermaster and maintenance personnel, the CAFS system was inspected and found that there were issues. The CAFS flow gauge was not working and the system did not work in the "Auto" mode. The CAFS system was put in manual or Tool mode and still did not work properly but did inject a high concentration of air in the systems that overloaded the hose line with air over water/foam. Station Officers were notified that the CAFS system was out of service until repairs could be completed.

Lessons Learned: CAFS should not be routinely used as an initial knockdown tool in a overly hot environment. It is similar to what happens when a direct attack is initiated on a fire that is near flashover or mushrooming. The burst of air coupled with the smaller droplets do not give a cooling effect, but instead displaces the heat which travels back to the hose team. Equipment issues can contribute to worsening conditions. FF's should always be prepared for changing conditions when CAF's is entered into a live fire scenario.

Demographics

Department type: Combination, Mostly volunteer

Job or rank: Safety Officer

Department shift: Other. Training Burn

Age: 43–51

Years of fire service experience: 17–20

Region: FEMA Region VI

Service Area: Urban

Event Information

Event type: Training activities: formal training classes, in-station drills, multi-company drills, etc.

Event date and time: 01/20/2008 13:00

Event participation: Witnessed event but not directly involved in the event

Weather at time of event: Clear with Wet Surfaces

What were the contributing factors?
- Training Issue
- Equipment
- Decision Making

What do you believe is the loss potential?
- Other

Figure 11-20 The standard size of foam concentrate container is a 5-gal (19-L) pail.

The shelf life of foam concentrate will vary depending on the type of concentrate. Typically, protein concentrate has a shelf life in the range of 7 to 10 years. Synthetic concentrates and high-expansion concentrates have a shelf life of 20 to 25 years.

The environmental impact when using foam has been a concern for many years. Many types of foam have undergone testing to assess their impact on the environment. Information regarding the toxicity of foam concentrates is usually available from product environmental data sheets, material safety data sheets, product technical bulletins, and toxicity summary sheets. The manufacturer/supplier of the foam concentrate is a primary resource for researching this information.

Foam Application

Knowing the accepted methods for foam application is important for driver/operators. After all, you may have to use a handline to apply foam or you may need to assist other fire fighters in the application method. Consequently, you need to be the expert on all aspects of foam operation.

Applying Foam

Class A Foam

The use of Class A foam for structural firefighting is becoming a more common practice. The benefits of using foam to extinguish Class A fires or in incidents involving three-dimensional fuels have been proven through testing done by manufacturers and fire departments. The use of Class A foam for firefighting is similar to the use of water. Although the application methods are the same, the results will probably be better with the foam. Training and experimentation with Class A foam in fire situations is not recommended due to safety concerns. The confidence and experience needed in operating foam systems can best be achieved with live fire training.

To apply a Class A foam on a fire, follow the steps in **Skill Drill 11-7**:

1. Open the nozzle and test to ensure that foam is being produced. (**Step 1**)
2. Move within a safe range of the target and open the nozzle. (**Step 2**)
3. Direct the stream of foam onto the burning surface or the exposure that you are protecting. (**Step 3**)
4. Be aware that the fire fighters may have to change positions around the structure so as to adequately cover the entire surface. (**Step 4**)

> **Driver/Operator Tip**
>
> All live fire training should be conducted in compliance with NFPA 1403, *Standard on Live Fire Training Evolutions*. While live burns can be an effective teaching tool, the safety of the personnel involved must be the number one priority. **Safety First.**

Class B Foam

The methods for using Class B foam on Class B fires differ from those used when applying Class A foam on Class A fires. Directing a Class B foam stream into a Class B fire can disrupt the fuel and cause the fire to spread. Plunging the stream into an existing foam blanket will allow vapors to escape, which can result in spreading, reignition, or flare-up of the fire. Three methods are used to produce foam blankets on Class B fires:

- Roll-on method
- Bankdown method
- Raindown method

Sweep (Roll-on) Method

The **sweep (roll-on) method** should be used only on a pool of flammable product that is on open ground. With this technique, foam is placed on the fuel surface by directing the foam stream onto the ground in front of the product involved. The foam will then roll onto the surface of the material involved. It may be necessary to move the hose line to a different position to achieve total coverage. It may also be necessary to use multiple lines to cover the material. If multiple hose lines are used, be aware of the position of the other fire fighters operating in the area. A coordinated attack should be conducted by the IC, and a safety officer should be on the scene.

To perform the roll-on method, follow the steps in **Skill Drill 11-8**:

Skill Drill 11-7

Applying Class A Foam on a Fire

1. Open the nozzle and test to ensure that foam is being produced.

2. Move within a safe range of the target and open the nozzle.

3. Direct the stream of foam onto the burning surface or the exposure that you are protecting.

4. Be aware that the fire fighters may have to change positions around the structure so as to adequately cover the entire surface.

Skill Drill 11-8

Applying Foam with the Roll-on Method

1. Open the nozzle and test to ensure that foam is being produced.

2. Move within a safe range of the target and open the nozzle.

3. Direct the stream of foam onto the ground just in front of the pool of product.

4. Allow the foam to roll across the top of the pool of product until it is completely covered. Be aware that the fire fighters may have to change positions along the spill so as to adequately cover the entire pool.

1. Open the nozzle and test to ensure that foam is being produced. (**Step 1**)
2. Move within a safe range of the target and open the nozzle. (**Step 2**)
3. Direct the stream of foam onto the ground just in front of the pool of product. (**Step 3**)
4. Allow the foam to roll across the top of the pool of product until it is completely covered. (**Step 4**)
5. Be aware that the fire fighters may have to change positions along the spill so as to adequately cover the entire pool. (**Step 5**)

Bankshot (Bankdown) Method

The <u>bankshot (bankdown) method</u> is employed at fires where the fire fighter can use an object to deflect the foam stream and let it flow down the burning surface. This application of foam should be as gentle as possible to avoid agitating the material involved. If the material is located in an area where there is a wall, a tank, or other vertical object, then the foam can be directed at and allowed to run down that object. This approach will allow the foam to spread over the material and form a foam blanket.

To perform the bankdown method of applying foam, follow the steps in **Skill Drill 11-9**:

Skill Drill 11-9

Applying Foam with the Bankdown Method (FF2)

1. Open the nozzle and test to ensure that foam is being produced.

2. Move within a safe range of the target and open the nozzle. Direct the stream of foam onto a solid structure such as a wall or metal tank so that the foam is directed off the object and onto the pool of product. Allow the foam to flow across the top of the pool of product until it is completely covered. Be aware that the fire fighters may have to bank the foam off several areas of the solid object so as to extinguish the burning product.

---

1. Open the nozzle and test to ensure that foam is being produced. (**Step 1**)
2. Move within a safe range of the target and open the nozzle.
3. Direct the stream of foam onto a solid structure such as a wall or metal tank so that the foam is directed off the object and onto the pool of product.
4. Allow the foam to flow across the top of the pool of product until it is completely covered.
5. Be aware that the fire fighters may have to bank the foam off several areas of the solid object so as to extinguish the burning product. (**Step 2**)

Raindown Method

The **raindown method** can be used when there is no vertical object to use for a bankshot and it would be too dangerous to get close and use the roll-on method of foam application. This foam application technique consists of lofting the foam stream into the air above the material and letting it fall gently down onto the surface. It can be an effective method as long as the foam stream can completely cover the material involved. Two caveats apply: This method might not be effective when wind conditions are unfavorable, and the coverage of the foam needs to be monitored.

To perform the raindown method, follow the steps in **Skill Drill 11-10**:

1. Open the nozzle and test to ensure that foam is being produced. (**Step 1**)
2. Move within a safe range of the target and open the nozzle. (**Step 2**)
3. Direct the stream of foam into the air so that the foam breaks apart in the air and falls onto the pool of product. (**Step 3**)
4. Allow the foam to flow across the top of the pool of product until it is completely covered.
5. Be aware that the fire fighters may have to move to several locations and shoot the foam into the air so as to extinguish the burning product. (**Step 4**)

Foam Compatibility

Class A concentrates and Class B concentrates are not compatible. Mixing these different classes of concentrate may cause the concentrate to gel and hinder operation of foam proportioning equipment. Likewise, Class B foam concentrates may not be compatible with each other. Check with the foam manufacturer for information on compatibility of particular types of foam.

Also make sure that onboard tanks on fire apparatus are properly marked. Many fire apparatus carry onboard water, Class A foam concentrate, and Class B foam concentrate tanks. In some instances, foam concentrate has been poured into the wrong tank—with undesirable results.

Skill Drill 11-10

Applying Foam with the Raindown Method

1. Open the nozzle and test to ensure that foam is being produced.

2. Move within a safe range of the target and open the nozzle.

3. Direct the stream of foam into the air so that the foam breaks apart in the air and falls onto the pool of product.

4. Allow the foam to flow across the top of the pool of product until it is completely covered. Be aware that the fire fighters may have to move to several locations and shoot the foam into the air so as to extinguish the burning product.

Wrap-Up

Chief Concepts

- While water is effective in suppressing fires, foams have added a new dimension to firefighting strategies.
- Foam for firefighting is a stable mass of small air-filled bubbles. It has a lower density than oil, gasoline, or water and will float on top of the fuel.
- The way that foam works to help extinguish a fire is very simple:
  - Once applied to the burning fuel, the foam will float on top of it, forming a blanket on the surface.
  - This blanket, if applied correctly, will stop or prevent the burning process by separating the fuel from the air, lowering the temperature of the fuel, and/or suppressing the release of flammable vapors.
- Foam is made up of four components: water, foam concentrate, mechanical agitation, and air.
- Firefighting foams are classified as either Class A or Class B.
- Foam concentrates are designed to be mixed with water at specific ratios, ranging from 1 percent to 6 percent.
- The foam expansion rate quantifies the ratio of finished foam to foam solution after being agitated and aspirated through a foam-making appliance.
- Foam cannot be produced if it is not proportioned properly.
  - Several types of foam application systems are available, ranging from basic operations to more advanced systems.
  - With the proper training, these systems can be very user friendly and will produce good-quality finished foam.
- The proper nozzle is needed for fire fighters to be able to produce a good-quality foam blanket.
- Foam concentrate is stored in containers of various sizes, including 5-gal (19-L) pails, 55-gal (208-L) drums, totes, and trailers.
- Knowing the several accepted methods for applying foam is important for driver/operators because they may have to use a handline to apply foam or may need to assist other fire fighters in the application method.
- Class A concentrates and Class B concentrates are not compatible; mixing these different classes of concentrate may cause the concentrate to gel and hinder operation of foam proportioning equipment.

Hot Terms

aeration The process of introducing air into the foam solution, which expands and finishes the foam.

aqueous film-forming foam (AFFF) A concentrated aqueous solution of fluorinated surfactant(s) and foam stabilizers that is capable of producing an aqueous fluorocarbon film on the surface of hydrocarbon fuels to suppress vaporization. (NFPA 11)

around-the-pump proportioning (AP) system A fire apparatus–mounted foam system that diverts a portion of the water pump's output from the discharge side of the pump and sends it through an eductor to discharge foam.

balanced-pressure system A fire apparatus–mounted foam system that uses a diaphragm-type pressure control valve to sense and balance the pressures in the foam concentrate and water lines to the proportioner.

bankshot (bankdown) method A method that applies the foam stream onto a nearby object, such as a wall, instead of directly at the fire.

batch mixing Pouring foam concentrate directly into the fire apparatus water tank, thereby mixing a large amount of foam at one time.

burnback resistance The ability of a foam blanket to resist direct flame impingement.

bypass eductor A foam eductor that is mounted to the pump and can be used for either water application or foam application.

chemical foam Foam produced by mixing powders and water, where a chemical reaction of the materials creates the foam.

Class A foam Foam intended for use on Class A fires. (NFPA 1150)

Class B foam Foam intended for use on Class B fires. (NFPA 1901)

combustible liquid Any liquid that has a closed-cup flash point at or above 100°F (37.8°C) as determined by the test procedures and apparatus set forth in Section 4.4. Combustible liquids are classified according to Section 4.3. (NFPA 30)

compressed-air foam system (CAFS) A foam system that combines compressed air and a foam solution to create firefighting foam.

concentration The percentage of foam concentrate contained in a foam solution. (NFPA 11)

duplex gauge A gauge at the pump panel that simultaneously monitors both the foam concentrate and water pressures.

eductor A device that uses the Venturi principle to siphon a liquid in a water stream. The pressure at the throat is below atmospheric pressure, allowing liquid at atmospheric pressure to flow into the water stream. (NFPA 11)

film-forming fluoroprotein foam (FFFP) A protein-based foam concentrate incorporating fluorinated surfactants, which forms a foam capable of producing a vapor-suppressing, aqueous film on the surface of hydrocarbon fuels. This foam might show an acceptable level of compatibility to dry chemicals and might be suitable for use with those agents. (NFPA 412)

finished foam The homogeneous blanket of foam obtained by mixing water, foam concentrate, and air.

fire barrier A continuous membrane or a membrane with discontinuities created by protected openings with a specified fire protection rating, where such membrane is designed and constructed with a specified fire resistance rating to limit the spread of fire, that also restricts the movement of smoke. (NFPA 101)

flammable liquid Any liquid that has a closed-cup flash point below 100°F (37.8°C), as determined by the test procedures and apparatus set forth in Section 4.4, and a Reid vapor pressure that does not exceed an absolute pressure of 40 psi (276 kPa) at 100°F (37.8°C), as determined by ASTM D 323, *Standard Test Method for Vapor Pressure of Petroleum Products (Reid Method)*. Flammable liquids are classified according to Section 4.3. (NFPA 30)

fluoroprotein foam (FP) A protein-based foam concentrate to which fluorochemical surfactants have been added. The resulting product has a measurable degree of compatibility with dry chemical extinguishing agents and an increase in tolerance to contamination by fuel. (NFPA 402)

foam A stable aggregation of small bubbles of lower density than oil or water, which exhibits a tenacity for covering horizontal surfaces. Air foam is made by mixing air into a water solution containing a foam concentrate, by means of suitably designed equipment. It flows freely over a burning liquid surface and forms a tough, air-excluding, continuous blanket that seals volatile combustible vapors from access to air. It resists disruption from wind and draft or heat and flame attack and is capable of resealing in case of mechanical rupture. Firefighting foams retain these properties for relatively long periods of time. Foams also are defined by expansion and are arbitrarily subdivided into three ranges of: (1) low-expansion foam—expansion up to a ratio of 20:1; (2) medium-expansion foam—expansion in the range of 20:1 to 200:1; and (3) high-expansion foam—expansion in the range of 200:1 to 1000:1. (NFPA 11)

foam concentrate (foam liquid) A concentrated liquid foaming agent as received from the manufacturer. (NFPA 11)

foam heat exchanger A device that uses water as its cooling source and prevents the foam concentrate from overheating.

foam proportioner A device or method to add foam concentrate to water so as to make foam solution. (NFPA 1901)

foam solution A homogeneous mixture of water and foam concentrate in the proper proportions. (NFPA 11)

foam tetrahedron A geometric shape used to illustrate the four elements needed to produce finished foam: foam concentrate, water, air, and mechanical aeration.

fuel pick-up The absorption of the burning fuel into the foam itself.

fuel resistance A foam's ability to minimize fuel pick-up.

high-expansion foam Any foam with an expansion ratio in the range of 200:1 to approximately 1000:1. (NFPA 1145)

hydrocarbon A chemical substance consisting of only hydrogen and carbon atoms. (NFPA 36)

hydrolyzed Decomposition of a chemical compound by reaction with water.

induction The use of an eductor to introduce a proportionate amount of foam concentrate into a stream of water flowing from a discharge.

injection system A foam system installed on a fire apparatus that meters out foam by pumping or injecting it into the fire stream.

knockdown speed and flow The time required for a foam blanket to spread out across a fuel surface.

low-expansion foam Any foam with an expansion ratio up to 20:1. (NFPA 1145)

mechanical foam Foam produced by physical agitation of a mixture of water, air, and a foaming agent.

mechanical generator An electrically powered fan used to aerate certain types of foams.

medium-expansion foam Any foam with an expansion ratio in the range of 20:1 to 200:1. (NFPA 1145)

metering device A device that controls the flow of foam concentrate into the eductor.

miscible Readily mixes with water.

oleophobic Oil hating; having the ability to shed hydrocarbon liquids.

oxygenated To treat, combine, or infuse with oxygen.

pick-up tube The tube from an in-line foam eductor that is placed in the foam bucket and draws foam concentrate into the eductor using the Venturi principle.

polar solvent A combustible liquid that mixes with the water contained in foam. Such solvents include non-petroleum-based fuels such as alcohol, lacquers, acetone, and acids.

polymer A naturally occurring or synthetic compound consisting of large molecules made up of a linked series of repeated simple monomers.

proportioned A combination of foam concentrate and water used to form a foam solution.

protein foam Foam created from a concentrate that consists primarily of products from a protein hydrolysate, plus stabilizing additives and inhibitors to protect against freezing, to prevent corrosion of equipment and containers, to resist bacterial decomposition, to control viscosity, and to otherwise ensure readiness for use under emergency conditions. These concentrates are diluted with water to form 3 percent to 6 percent solutions depending on the type. These concentrates are compatible with certain dry chemicals. (NFPA 11)

pump-mounted eductor An in-line eductor dedicated to the production of foam.

raindown method A method of applying foam that directs the stream into the air above the fire and allows it to gently fall onto the surface.

ratio controller A device required for each foam outlet to proportion the correct amount of foam concentrate into the water stream over a wide range of flows and with minimal pressure loss.

surface tension The elastic-like force at the surface of a liquid, which tends to minimize the surface area, causing drops to form. (NPFA 1150)

sweep (roll-on) method A method of applying foam that involves sweeping the stream just in front of the target.

viscosity A measure of the resistance of a liquid to flow. (NFPA 1150)

Driver/Operator in Action

Once the fire apparatus is positioned at the scene of the fire, the pump is engaged, and the attack lines are deployed and charged, then you can start to supply foam for the initial attack operations. Most foam systems are very simple and do not require complex procedures to operate them. Once foam is flowing, you may need to contact the fire fighters at the nozzle to determine if any adjustments are necessary. Some operations may require foam that is very "wet," while others may need a "dry" foam that can stick to the surface of the burning material. Either way, a fire requiring foam for suppression will test your knowledge.

1. In the 1940s, a liquid protein-based foam concentrate was made from what material?
   A. Animal protein by-products
   B. Whey protein
   C. Sodium bicarbonate
   D. Seaweed

2. Which of the following is *not* one of the four elements of the foam tetrahedron?
   A. Water
   B. Air
   C. Mechanical agitation
   D. Foam concentrate
   E. All are correct

3. Which type of foam should be used on polar solvents?
   A. AFFF
   B. AR-AFFF
   C. FFFP
   D. Class A

4. Foam concentrates are produced for uses at rates from 1 percent to __ percent?
   A. 3
   B. 5
   C. 8
   D. 6

Performance Testing

CHAPTER 12

NFPA 1002 Standard

4.2.1* Perform routine tests, inspections, and servicing functions on the systems and components specified in the following list, given a fire department vehicle and its manufacturer's specifications, so that the operational status of the vehicle is verified:

(1) Battery(ies)
(2) Braking system
(3) Coolant system
(4) Electrical system
(5) Fuel
(6) Hydraulic fluids
(7) Oil
(8) Tires
(9) Steering system
(10) Belts
(11) Tools, appliances, and equipment [p. 330–358]

(A) Requisite Knowledge. Manufacturer specifications and requirements, policies, and procedures of the jurisdiction.

(B) Requisite Skills. The ability to use hand tools, recognize system problems, and correct any deficiency noted according to policies and procedures.

5.1.1 Perform the routine tests, inspections, and servicing functions specified in the following list in addition to those in 4.2.1, given a fire department pumper and its manufacturer's specifications, so that the operational status of the pumper is verified:

(1) Water tank and other extinguishing agent levels (if applicable)
(2) Pumping systems
(3) Foam systems

(A) Requisite Knowledge. Manufacturer's specifications and requirements, and policies and procedures of the jurisdiction. [p. 330–358]

(B) Requisite Skills. The ability to use hand tools, recognize system problems, and correct any deficiency noted according to policies and procedures. [p. 330–358]

Additional NFPA Standards

NFPA 1500 *Standard on Fire Department Occupational Safety and Health Program* (2007)
NFPA 1901 *Standard for Automotive Fire Apparatus* (2009)
NFPA 1911 *Standard for the Inspection, Maintenance, Testing and Retirement of In-Service Automotive Fire Apparatus* (2006)

Knowledge Objectives

After studying this chapter, you will be able to:

- Describe the requirements for the performance tests as required by NFPA 1911.
- Describe the criteria for re-rating fire pumps.

Skills Objectives

After studying this chapter, you will be able to:

- Perform the no-load governed engine speed test.
- Perform the pump shift indicator test.
- Perform the pump engine control interlock test.
- Perform a gauge meter test.
- Perform a flow meter test.
- Perform a tank-to-pump flow test.
- Perform a vacuum test.
- Perform an internal relief valve test.
- Perform a priming system test.
- Conduct the pumping and overload tests for fire pumps.
- Perform the pressure control tests.

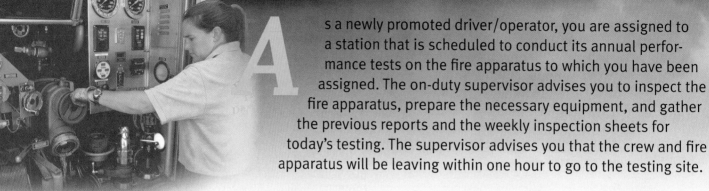

You Are the Driver/Operator

As a newly promoted driver/operator, you are assigned to a station that is scheduled to conduct its annual performance tests on the fire apparatus to which you have been assigned. The on-duty supervisor advises you to inspect the fire apparatus, prepare the necessary equipment, and gather the previous reports and the weekly inspection sheets for today's testing. The supervisor advises you that the crew and fire apparatus will be leaving within one hour to go to the testing site.

1. What do you need to inspect on the fire apparatus?
2. Which equipment do you need to inspect and bring to the testing site?
3. Which required forms do you need to bring to the testing site?
4. What should you do if you find any defects and/or deficiencies related to the fire apparatus prior to testing?

Introduction

After a fire apparatus has been accepted and placed in service, it has to be properly inspected, operated, maintained, and tested over its entire useful lifetime. On certain fire apparatus, the pumping system is designed to deliver a certain amount of water at a desired pressure and the hose lines and nozzles used are based on the pump's design. Failure of a fire pump to deliver the required amount of water at a certain pressure will impair the fire-suppression ability and the safety of fire fighters, possibly leading to injury and increased property damage.

Unlike the original acceptance test and the Underwriters Laboratory (UL) tests, which last three hours, fire apparatus are subject to annual performance testing. Today certification testing can be done by agencies other than UL, such as Factory Mutual or American Testing. The critical point is that the fire apparatus must be periodically tested by an independent, third-party certifying agency.

Performance tests are tests conducted after a fire apparatus has been put into service to determine if its performance meets predetermined specifications or standards. Performance testing of pumps is an integral and vital part of fire apparatus safety and maintenance. Imagine arriving on the scene of a multiple-alarm fire, laying hose, hooking up to the hydrant, and starting to pump water—only to have the pump fail after 10 minutes. One question that might be asked in this scenario is, Could the pump failure have been prevented by annual performance testing?

The Insurance Services Office (ISO) reports that fire is still the leading cause of loss cited in conjunction with personal and commercial property insurance policies. The ISO also reports that there is a definite correlation between improved fire protection and compliance with the NFPA standards. During the annual performance testing of fire pumps, the ISO Fire Suppression Rating Schedule assigns the highest point total to fire apparatus pumps.

NFPA 1911, *Standard for the Inspection, Maintenance, Testing, and Retirement of In-Service Automotive Fire Apparatus* calls for annual **service testing** of fire apparatus pumps. Fire pumps are water pumps with a **rated capacity** of 250 GPM (1000 L/min) or greater at 150 psi (1000 kPa) net pump pressure that are mounted on a fire apparatus and used for firefighting purposes. Fire pumps that are placed in service should be tested each year to determine if they are still capable of achieving their designed performance.

The performance tests for fire pumps do not just test the pump; rather, they test the entire pumping system—the engine, transmission, and pump, plus related accessories and devices used in operating the pumping system. You might consider the annual performance tests of fire pumps to be a "stress test" for the fire apparatus: To pass, the fire apparatus must deliver its original design flow and pressure; must show no signs of overheating, loss of power, or over-acceleration; and must not exhibit any other major defects during the test.

NFPA 1911, *Standard for the Inspection, Maintenance, Testing, and Retirement of In-Service Automotive Fire Apparatus* regulates the actual conduct of the performance testing of fire pumps. It states that such a test is to be conducted annually for all fire apparatus that have a fire pump with 250 GPM or larger capacity or if the pump and/or engine on the fire apparatus has been repaired or modified. Note that this test is not the same as a "New Apparatus Acceptance or UL (Underwriter's Laboratory) Test," which is conducted by an independent agency for the manufacturer and lasts approximately 3 hours. In addition to the annual performance test, NFPA 1911 calls for a fire pump test to be conducted whenever major repairs or modifications have been made to the pump or any component of the fire apparatus used in pump operations.

Fire Apparatus Requirements

Engine-driven accessories should not be functionally disconnected or otherwise rendered inoperative during the annual performance test. If the chassis engine drives the pump, all headlights, running lights, warning lights, and air conditioners, if provided, should continue to operate during the pumping portion of this test. The following devices are permitted to be turned off or at least not operated during the pump performance test:
- Aerial hydraulic pump
- Foam pump
- Hydraulic-driven equipment (other than a hydraulic-driven line voltage generator)
- Winch
- Windshield wipers
- Four-way hazard flashers
- Compressed-air foam system (CAFS) compressor

If any electrical loads are connected through an automatic electrical load management system, then that system should be permitted to automatically disconnect those loads during the course of the test.

When operating a pump, it is important that the engine temperature is kept within the proper range; neither a cold engine nor an excessively hot engine will give as good service as one that is run at the proper temperature. The oil pressure on the engine should be watched to see that the engine is being lubricated properly. The transmission gears should be monitored for overheating. Any unusual vibration of the engine or the pump, or any leak in the pump casing or connections, should be noted and managed. Centrifugal pumps are not self-priming, so they could lose their prime if a leak in the suction lines occurs.

When conducting performance tests on fire pumps, the fire apparatus should use the side intakes only. Although not always preferred, front- or rear-mounted intakes may be used in conjunction with side intake suction(s) for all pumps that provide flows of 1500 GPM or greater. All other intakes should remain in the closed position and be properly capped. Storz fittings are not recommended and should be removed prior to testing; they may be replaced with steamer caps during the performance test.

In case of any fire apparatus failure, time should be allotted to make the necessary repairs where applicable. **Failure** is defined as a cessation of proper functioning or performance. Other defects in the performance of the engine or the pump should be documented as well, and minor defects should be corrected immediately if possible. If such repairs cannot be done on site, then the fire apparatus should be rescheduled for testing at a later date.

Environmental Requirements

Prior to the commencement of any performance tests on fire apparatus pumps, the following environmental conditions must be determined and recorded prior to and immediately after the pump performance testing:

- Ambient air temperature
- Water temperature
- Atmospheric pressure

NFPA 1911 specifies that tests must be conducted when the ambient air temperature is between 0°F and 110°F (–18°C and 43°C). **Ambient air temperature** is the temperature of the surrounding medium and usually refers to the temperature of the air in which a structure is situated or a device operates. The water temperature should be between 35°F and 90°F (1.7°C and 32°C), and the air pressure (atmospheric pressure) at 29 in. Hg (736.6 mm Hg) or greater (corrected to sea level).

People who regularly conduct pump tests will tell you that those extremes are of little help if you want to maximize pump performance. For example, hot air temperatures can cause the engine cooling fan to run more and increase the parasitic power loss. Cold air temperatures can affect battery power. Although no one makes a specific recommendation, days with moderate temperatures are generally better choices for conducting performance tests than very hot or cold ones.

Water temperature extremes are even more important. Warm water, usually defined as water at a temperature greater than 90°F (32°C), is more likely to cavitate inside the pump and may result in a loss of up to 500 GPM (1900 L/min) in flow rate. It is particularly important that the water supply be non-aerated and that its temperature not exceed 90°F (32°C) when you are conducting a test pit, where repeated circulation of the water through the pump can increase water temperatures over time.

At the other end of the scale, cold water is more likely to freeze and foul test equipment, especially if the air temperature is also low. The general recommendation is a water temperature in the range of 35°F to 85°F (1.7°C to 29°C), with about 60°F (16°C) being ideal. If these criteria are not met, the pump performance could be affected dramatically.

Atmospheric pressure (sometimes called air pressure or barometric pressure) is also important. High atmospheric pressure pushes harder on the surface of the water being drafted and makes it easier to lift; low atmospheric pressure makes it harder to lift. Conducting the performance test on a day when a high-pressure area is parked over your test site will give the best results.

To correct local atmospheric pressure readings to sea level, add 1 in. Mercury (Hg) (25 mm Hg) for every 1000' (305 m) of elevation at the site. If the resulting corrected reading is less than 29 in. Hg (736.6 mm Hg) or if the atmospheric pressure readings are falling steadily, then postpone your test to a more favorable day.

If any environmental conditions are not within the specified limits, the test should be delayed until they are satisfactory. Otherwise, the results will need to be confirmed by another test at a later date.

Test Site

The test site should be located along an improved roadway or on solid ground where the water is from 4' to 8' (1.2 m to 2.4 m) below grade. It should be possible to reach the water from the pump intake with not more than 20' (6 m) of hard suction

Figure 12-1 A good test site is safe and will give accurate results for the fire apparatus's performance.

The theoretical values of lift and maximum lift must be reduced by the entrance and friction losses in the suction hose equipment to obtain the actual or measurable lift. The vacuum (i.e., negative pressure) on the intake side of a pump is measured in inches of Hg, usually written as "in. Hg" or "Hg." A vacuum of 1 in. Hg is equal to a negative pressure of 03.49 psi; that is, 1 in. Hg = 0.49 psi. A positive pressure of 0.49 psi at the bottom of a 1 in. (645 mm) container will support a column of water that is 1.13' (0.0344 m) high; therefore, a negative pressure of 0.49 psi at the top of the container will support the same column of water. Thus a reading of 16 in. Hg on a vacuum gauge will be equivalent to approximately 8 psi.

Fire departments that want to achieve the best performance from their pumps should consider several other factors when conducting performance tests. For example, they should select a test site with an adequate source of clear, fresh water. Salt water is denser than fresh water and should not be used in these periodic tests for two reasons: It reduces performance, and it accelerates the corrosion process within the apparatus's pump and piping. Muddy water also should be avoided because it often contains hidden debris and can clog the pump, valves, fittings, and gauge lines.

Make sure that the water source is configured to allow good performance and easy accessibility. Water sources should be accessible so that the fire apparatus can be parked on a level, hard surface. The fire apparatus should be positioned such that a 20' (6 m) length of hard suction hose can be connected to the pump inlet with the strainer submerged according to the NFPA 1911 requirements. Operating on a side slope can result in air pockets forming within the plumbing that might reduce performance. Operating on soft soil on a side slope can result in a dangerous tip-over condition and should also be avoided. Operating with the strainer too close to the surface of the water or in water that is too shallow can result in air entrainment caused by formation of a whirlpool.

Another option in conducting any performance testing is to use a properly designed pump test pit **Figure 12-2**. Hale Products' publication titled *Pump Test Pit Design Recommendations* suggests using a pit that is 13' to 15' (4 to 5 m) deep and approximately 2½ to 3 times longer than it is wide. For example, if the pit is 10' (3 m) wide, it should be 25' to 30' (7 to 9 m) long. A 10' (3 m) wide by 30' (9 m) long by 15' (5 m) deep pit would have a total volume of 4500 ft^3 (127 m^3). Multiply cubic feet by 7.5 to determine gallons. In this case, the pit has a total capacity of 33,750 gal (12,760 L), of which about 30,000 gal (113,560 L) is usable after allowing for partial fill. This would be sufficient to test pumps up to 3000 GPM (11,360 L/min) using the recommendation of 10 gal (38 L) of pit capacity for every 1 GPM (3.8 L/min) pump flow. The water should be drafted from one end of the pit and discharged back into the other end. Hale Products recommends that several removable baffles be placed within the pit between the two ends to control turbulence and help reduce aeration of the water

If you use a test pit, make sure that the usable pit capacity is at least 10 gal for every 1 GPM of the pump. For example, a 1500 GPM (5680 L/min) pump would require a pit with a usable capacity of at least 15,000 gal (56,780 L) of water. Pits with less capacity can result in air entrapment and excessive water

hose with the strainer submerged at least 2' (0.6 m) and with no humps in the hose.

The water should be at least 4' (1.2 m) deep where the strainer is located to provide clearance below the strainer and sufficient depth above it. A poor test site can result in as much as 150 to 500 GPM (570 to 1900 L/min) of lost flow rate. In some cases, a bad site can even force you to stop the test and start over **Figure 12-1**.

All tests requiring the flowing of water should be conducted whenever possible with the pump drafting. If it is impractical to provide all of the specified conditions, the authority having jurisdiction may authorize tests under other conditions. The **authority having jurisdiction (AHJ)** is an organization, office, or individual responsible for enforcing the requirements of a code or standard, or for approving equipment, materials, an installation, or a procedure.

When a suitable site for drafting is not available, the chosen site must provide a level area for stationing the fire apparatus, a source of hydrant water with sufficient flow, and an area that is suitable for discharging water.

Both the elevation of site and the lift should be recorded. The lift is the vertical height that water must be raised during a drafting operation, as measured from the surface of a static source of water to the center line of the pump intake. Depending on the rated capacity (GPM or L/min) of the pump that you are testing, a specific suction arrangement, diameter of suction hose, maximum number of suction lines, and maximum allowable lift will be required.

The maximum lift is the greatest difference in elevation at which the fire apparatus can draft the required quantity of water under the established physical characteristics of operation. These characteristics will take into account the following considerations:

- The design of the pump
- The adequacy of the engine
- The condition of the pump and the engine
- The size and condition of the suction hose and strainers
- The elevation of the pumping site above sea level
- The atmospheric conditions
- The temperature of the water

Figure 12-2 Use a properly designed pump test pit.

- Mechanical or digital hand-held tachometer for measuring the pump's shaft counter (rpm)
- Strobe-type tachometer and tape if the fire apparatus does not have a pump shaft (rpm)
- Suction hoses of appropriate sizes (diameter) and lengths with the appropriate type of strainer(s)
- One or more mallets to be used for tightening hard suction connections
- Two deluge appliances with stream straighteners
- Numerous lengths of 2½″ (64 mm) or 3″ (76 mm) hoses according to the recommended hose layouts
- A minimum of two wheel chocks
- Assorted wrenches and Allen keys
- Roll of plumber's tape
- Assorted spanner wrenches
- Hydrant wrench
- Calculator

Always keep the gauges used for testing purposes separate from the fire apparatus's gauges Figure 12-3 . The test gauges should themselves be regularly tested and calibrated once a year for accuracy. They are used to inspect the fire apparatus's gauges to see if they are in within acceptable ranges.

Driver/Operator Safety

Because of the significant pressures being applied when conducting pump testing, nozzles should be used with portable or mounted monitors. Hand-held nozzles should never be used during performance tests.

temperature rise, which can lead to loss of suction, cavitation, and significantly reduced pump performance.

Equipment Requirements

Prior to conducting performance (service) tests on fire apparatus, the AHJ must ensure that the proper equipment is available and in satisfactory working conditions. The following is a list of equipment that will be needed during the testing period:

- One 0–300 psi test gauge marked in 5-psi increments (QC-Test Kit)
- One 0–30 in. Hg vacuum gauge marked in 0.5″ increments (QC-Test Kit)
- Two 0–150 psi test gauges marked in 1-psi increments
- Two Pitot gauges, preferably the fix-mounted type or a hand-held unit with knife edge and air chamber
- One or more flow meters (optional) in lieu of the Pitot gauges.
- Assorted smooth-bore testing nozzles of appropriate diameter (size) for the required flow (GPM) for the various test points

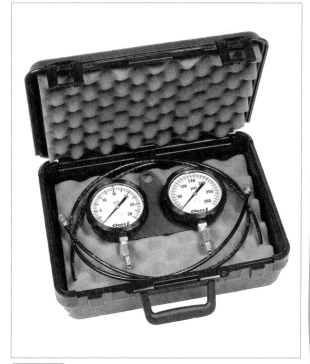

Figure 12-3 Keep testing gauges separate from the fire apparatus's gauges.

Nozzles that are suitable for testing usually can be found in the regular equipment of a fire department. However, the actual coefficient of discharge of each nozzle should be known; otherwise, the test results reported could be erroneous. The actual coefficient of discharge needs to be determined by a test conducted by a competent person using equipment such as weigh tanks or calibrated flow meters. Testing nozzles with affixed Pitot tubes and gauges is recommended, and should be conducted whenever possible with secured portable or mounted monitors Figure 12-4.

The size of the nozzle is usually chosen to give the desired discharge at a nozzle pressure between 60 psi and 70 psi (410 kPa and 48 kPa). This pressure is neither so high that the Pitot gauge is difficult to handle in the stream, nor so low that the normal inaccuracies of a gauge that is used at low pressure would come into play. Nozzle (Pitot) pressures less than 50 psi (350 kPa) or greater than 100 psi (690 kPa) should be avoided. The nozzle should always be used in conjunction with a securely placed monitor, and a test should never be conducted while any person holds the nozzle. Failure to abide by this recommendation can lead to serious injury.

Only smooth-bore nozzles should be used. Care should be taken that washers or gaskets do not protrude into the nozzle, because a perfectly smooth waterway is essential. Nozzle tips of 1½″ (38 mm) to 2¼″ (57 mm) in diameter are desired for use during various capacity and pressure tests. These tips should be free of nicks and scratches to ensure a smooth stream. Tips should be inspected, preferably prior to being attached and made ready for the test, to ensure that there is no mistake about the size of the tip being used.

A Pitot tube with an air chamber and pressure gauge is necessary for determining the velocity pressure of the water at the nozzle. The Pitot tube should be kept free of dirt and the air chamber free of water. Any water that accumulates in the air chamber should be removed after each test. The knife edges will inevitably get battered in service, but need to be kept sharp to reduce as much as possible the spray caused by inserting the Pitot tube into the stream. To ensure accurate and consistent readings, Pitot tubes should be fixed in the center of the

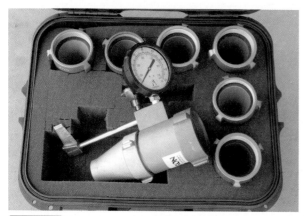

Figure 12-4 Testing nozzles with affixed Pitot tubes and gauges is recommended, and should be conducted whenever possible with secured portable or mounted monitors.

Figure 12-5 To ensure accurate and consistent readings, Pitot tubes should be fixed in the center of the stream, with the end of the tube located away from the end of the nozzle by a distance that is equal to half of the nozzle diameter.

stream, with the end of the tube located away from the end of the nozzle by a distance that is equal to half of the nozzle diameter Figure 12-5.

No-Load Governed Engine Speed Test

The first performance test is to check the governed engine speed. If the engine speed is not within ±50 rpm of the governed speed when the fire apparatus was brand new, this problem must be corrected before proceeding with the additional pump tests. This step is important because failure to have the proper governed engine speed will invalidate the results of the following tests. Check this speed while you are preparing the fire apparatus prior to the day of testing. Failure to operate at the correct governed speed during the performance tests is one of the most common problems that you will encounter.

To get the maximum power from the engine, run the engine up to the governed engine speed and make sure it stays there once the engine has reached normal operating temperature. At no time should you allow the engine to exceed its rated no-load governed speed. Check the air cleaner and fuel filter restrictions, and replace the filter elements if necessary to avoid power loss. Check the fan belt and adjust the tension to provide adequate engine cooling during the pump performance test. Likewise, check the alternator belt tension and the battery charge to ensure that they will provide enough electrical power to allow up to 45 seconds of uninterrupted primer operation during the vacuum and priming device tests. Finally, if you are conducting the test in an area where the diesel fuel is switched to a low-paraffin formulation for winter operations, make sure you fill the tank with the high-paraffin fuel used in the summer. Fuel with a high paraffin content burns hotter and provides more power.

Readings should equal the no-load governed engine speed as indicated when the fire apparatus was new. This information is listed on the original acceptance form(s) and should also be found on or near the pump panel on the UL plate Figure 12-6.

3. Record the readings from the tachometer inside the cab and the tachometer on the pump panel. Note any discrepancies between each tachometer.
4. Record all data on the performance test form. (**Step 3**)

> **Safety Tip**
>
> Ensure that the wheel chocks have been placed correctly prior to conducting the no-load governed engine speed test. Also verify that the air brake has been engaged.

Intake Relief Valve System Test

If the fire apparatus is equipped with an intake relief valve system or a combination intake/discharge system, an intake relief valve system test must be conducted to ensure that the system is operating in accordance with the manufacturer's specifications. A relief valve is a device that allows the bypass of fluids to limit the pressure in a system. One strategy for conducting this test is to use a second pumper to supply water to the pumper being tested. With this setup, pressure from the supply pump should be increased until the receiving pumper's intake relief valve system opens a dump valve. The pressure at which the system opens, dumps, or otherwise starts to operate should be recorded and reviewed against current operating procedures and the system adjusted accordingly. Record all data from this test on the performance test form.

Pump Shift Indicator Test

A test of the pump shift indicators should be made to verify that the pump shift indicators in the cab of the fire apparatus and on the driver/operator's panel indicate correct pump status when the pump is shifted from road mode to pump mode. Pump shift controls might include electrical, pneumatic, or mechanical components working individually or in combination to shift the pump drive system into and out of pump mode. Some pumps have manual backup shift controls available as well **Figure 12-7**. Pump shift indicators in the cab and on the driver/operator's pump panel on split shaft PTO pump drive systems typically require an electromechanical device, such as a switch mounted on the pump transmission, to sense pump shift status.

To perform the pump shift indicator test, follow the steps in **Skill Drill 12-2**:

1. Place the transmission in the neutral position. (**Step 1**)
2. Engage the parking brake. (**Step 2**)
3. Engage the pump shift. After the pump shift has been engaged, engage the transmission in the drive position. In a manual transmission, shift it into the highest gear or the indicated over-the-road gear. Note that some transmissions have an overdrive capability. (**Step 3**)

A.

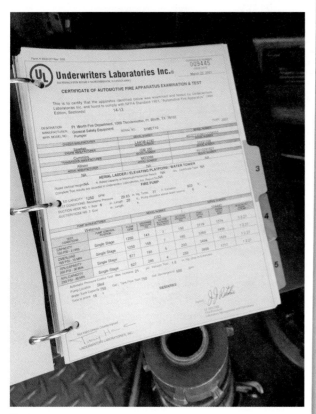

B.

Figure 12-6 The no-load governed engine speed can be found on or near the pump panel on the Underwriters Laboratory plate. **A.** UL plate. **B.** UL Certificate of Automotive Fire Apparatus Examination and Test.

To conduct the no-load governed engine speed test, follow the steps in **Skill Drill 12-1**:

1. With the assistance of another person, turn off all accessories powered by the engine, including headlights, emergency lights, siren activation switch, radio, and air conditioning. Place the drive train in neutral or park if the apparatus transmission is automatic. (**Step 1**)
2. Increase the engine speed (rpm) slowly, until the maximum governed engine speed is obtained. (**Step 2**)

Skill Drill 12-1

Conducting the No-Load Governed Engine Speed Test

1. With the assistance of another person, turn off all accessories powered by the engine, including headlights, emergency lights, siren activation switch, radio, and air conditioning. Place the drive train in neutral or park if the apparatus transmission is automatic.

2. Increase the engine speed (rpm) slowly, until the maximum governed engine speed is obtained.

3. Record the readings from the tachometer inside the cab and the tachometer on the pump panel. Note any discrepancies between each tachometer. Record all data on the performance test form.

4. An indicator light should come on inside the cab and on the pump panel labeled "Pump engaged" or "Okay to pump," indicating that the pump has been successfully engaged.
5. Record the information on the performance test form. (Step 4)

Pump Engine Control Interlock Test

Beginning with the 1991 edition of NFPA 1901, fire apparatus equipped with electronic or electric engine throttle controls were required to include an interlock system to prevent engine speed advancement, unless the chassis transmission is in neutral with the parking brake engaged; or unless the parking brake is engaged, the fire pump is engaged, and the chassis transmission is in pumping gear; or unless the fire apparatus is in the "Okay to pump" mode. This test is intended to inspect the proper operation of the interlock system **Figure 12-8**.

While the NFPA 1911 standard requires testing of the interlock in only two configurations, there are various combinations in which the chassis, transmission gear, the parking brake, and the pump shift in the driving compartment can be arranged. The engine speed control should also be adjustable at the driver/operator's panel when that combination is employed. You may wish to test whether the engine speed control is capable of being advanced in other configurations.

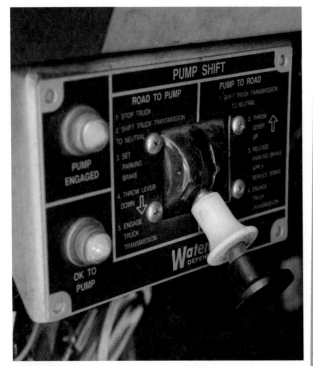

Figure 12-7 Pump shift controls might include electrical, pneumatic, or mechanical components working individually or in combination to shift the pump drive system into and out of pump mode.

Chapter 12 Performance Testing

Skill Drill 12-2

Performing the Pump Shift Indicator Test

(1) Place the transmission in the neutral position.

(2) Engage the parking brake.

(3) Engage the pump shift. After the pump shift has been engaged, engage the transmission in the drive position. In a manual transmission, shift it into the highest gear.

(4) An indicator light should come on inside the cab and on the pump panel labeled "Pump engaged" or "Okay to pump," indicating that the pump has been successfully engaged. Record the information on the performance test form.

Figure 12-8 The interlock system on the pump panel.

To perform the pump engine control interlock test, follow the steps in **Skill Drill 12-3**:

1. With the fire apparatus running, the pump gear in road mode with the transmission gear in the neutral position, and the air brake engaged, you should be able to advance the interlock system (throttle) located at the pump panel. (**Step 1**)
2. With the fire apparatus running, the pump gear in road mode with the transmission gear in the neutral position, and the air brake disengaged, you should not be able to advance the interlock system (throttle) located at the pump panel. (**Step 2**)
3. With the fire apparatus running, the pump gear engaged ("Pump engaged" or "Okay to pump"), the air brake engaged, and the transmission in pump gear, you should be able to advance the interlock system (throttle) located at the pump panel. (**Step 3**)
4. With the fire apparatus running, the pump gear engaged ("Pump engaged" or "Okay to pump"), the air brake disengaged, and the transmission in pump gear, you should not be able to advance the interlock system (throttle) located at the pump panel.
5. Record all information on the performance test form. (**Step 4**)

Gauge and Flow Meter Test

Discharge pressure gauges can be checked quickly against test gauges for accuracy. Individual discharge lines with gauges should be capped and the discharge valve opened slightly. The test gauge, master discharge gauge, and all discharge gauges should display the same reading. The test gauges are attached to the testing ports located on the pump panel.

Each water pressure gauge or flow meter must be checked for accuracy. Pressure gauges should be checked at a minimum of three points, including 150 psi (1035 kPa), 200 psi (1380 kPa), and 250 psi (1725 kPa). Any gauge that is off by more than 10 psi (69 kPa) must be recalibrated, repaired, or replaced.

To perform a gauge meter test, follow the steps in **Skill Drill 12-4**:

1. Connect the pressure test gauge (Si-Span QC Test kit) to the pressure testing port connection located on the pump panel. Do not connect the vacuum gauge at this time. (**Step 1**)
2. Engage the pump on the fire apparatus.
3. Ensure that all discharge gauges, including any preconnected lines, have been capped and that all discharge bleeder valves are in the closed position. (**Step 2**)
4. Ensure that the relief valve and governor have been turned off or set in the highest position. (**Step 3**)
5. Open all discharge valves. Note that discharge valves do not need to be in the fully opened position. (**Step 4**)
6. Slowly increase the pump throttle until the test pressure gauge is at 150 psi. Inspect all discharge gauges and the master compound gauge to verify that they are at 150 psi. (**Step 5**)
7. Increase the pump throttle until the test pressure gauge is at 200 psi. Inspect all discharge gauges and the master compound gauge to verify that they are at 200 psi. (**Step 6**)
8. Increase the pump throttle until the test pressure gauge is at 250 psi. Inspect all discharge gauges and the master compound gauge to verify that they are at 250 psi.
9. Decrease the pump throttle to idle and check the pump water temperature. If necessary, recirculate the pump water.
10. Note any discrepancies and record the information on the performance test form. (**Step 7**)

Driver/Operator Safety

Ensure that all discharges have been capped and are on securely before beginning the gauge meter test. Keep any personnel from standing in front of any discharge valves, as the pressures exerted on these valves during this test is extremely high and serious injury could occur in case of an unexpected discharge.

Flow meters (flow minders) need to be checked individually, using a hose stream with a smooth-bore tip and a Pitot tube to measure actual flow. Each flow meter must be checked for accuracy at the flows included in the test.

Any flow meter whose accuracy is off by more than 10 percent must be recalibrated, repaired, or replaced. Depending on the type of device, most flow meters can be recalibrated using either a small (eyeglass) screwdriver or a magnet during this test. Always follow the manufacturer's recommendations when recalibrating any flow meter **Figure 12-9**.

To perform a flow meter test, follow the steps in **Skill Drill 12-5**:

Skill Drill 12-3

Performing the Pump Engine Control Interlock Test

1. With the fire apparatus running, the pump gear in road mode with the transmission gear in the neutral position, and the air brake engaged, you should be able to advance the interlock system (throttle) located at the pump panel.

2. With the fire apparatus running, the pump gear in road mode with the transmission gear in the neutral position, and the air brake disengaged, you should not be able to advance the interlock system (throttle) located at the pump panel.

3. With the fire apparatus running, the pump gear engaged ("Pump engaged" or "Okay to pump"), the air brake engaged, and the transmission in pump gear, you should be able to advance the interlock system (throttle) located at the pump panel.

4. With the fire apparatus running, the pump gear engaged ("Pump engaged" or "Okay to pump"), the air brake disengaged, and the transmission in pump gear, you should not be able to advance the interlock system (throttle) located at the pump panel. Record all information on the performance test form.

Skill Drill 12-4

Performing a Gauge Meter Test

1 Connect the pressure test gauge (Si-Span QC Test kit) to the pressure testing port connection located on the pump panel. Do not connect the vacuum gauge at this time.

2 Engage the pump on the fire apparatus. Ensure that all discharge gauges, including any preconnected lines, have been capped and that all discharge bleeder valves are in the closed position.

3 Ensure that the relief valve and governor have been turned off or set in the highest position.

4 Open all discharge valves. Note that discharge valves do not need to be in the fully opened position.

5 Slowly increase the pump throttle until the test pressure gauge is at 150 psi. Inspect all discharge gauges and the master compound gauge to verify that they are at 150 psi.

6 Increase the pump throttle until the test pressure gauge is at 200 psi. Inspect all discharge gauges and the master compound gauge to verify that they are at 200 psi.

7 Increase the pump throttle until the test pressure gauge is at 250 psi. Inspect all discharge gauges and master compound gauge to verify that they are at 250 psi. Decrease the pump throttle to idle and check the pump water temperature. If necessary, recirculate the pump water. Note any discrepancies and record the information on the performance test form.

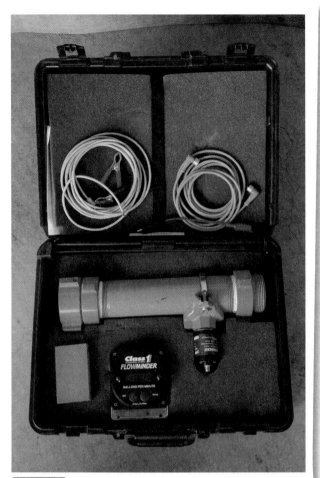

Figure 12-9 Any flow meter whose accuracy is off by more than 10 percent must be recalibrated, repaired, or replaced.

1. Connect the supply line to the intake valve on the fire apparatus. (**Step 1**)
2. Connect hose lines and nozzles that are suitable for discharging water at the anticipated flow rate to one or more of the discharge outlets. (**Step 2**)
3. Connect the flow meter to the nozzle or portable monitor. (**Step 3**)
4. Connect the hose line to the flow meter. (**Step 4**)
5. Put the fire apparatus into pump mode. (**Step 5**)
6. Make sure that the discharge valve(s) leading to the hose lines and nozzles are fully opened.
7. Adjust the engine throttle until the maximum consistent pressure reading on the Pitot gauge or the test flow on a flow meter is obtained for that specific pipe size.
8. Record the information on the performance test form. (**Step 6**)

Tank-to-Pump Flow Test

The flow rate should be compared with the rate designated by the manufacturer when the fire apparatus was new or with the rate established in previous testing. Rates less than the rate when the fire apparatus was new or as established in previous testing indicate problems in the tank-to-pump line or tank pump. For tanks with capacities greater than 750 gal (2840 L), the test is the same except that a flow of 500 GPM (1890 L/min) is used.

> **Safety Tip**
>
> During the tank-to-pump test, if you feel that the pump is beginning to run away, reduce the engine throttle quickly and close the appropriate discharge. This will prevent any unnecessary increase in pump temperature and cavitation.

If the fire apparatus is equipped with a water tank with a capacity of 300 gal (1135 L), up to and including 750 gal (2840), the tank-to-pump flow rate must be checked using the following procedures listed in **Skill Drill 12-6**:

1. Connect the supply line to the intake valve on the fire apparatus. The water tank should be filled until it overflows. (**Step 1**)
2. Close all intakes to the pump. (**Step 2**)
3. Close the tank fill and bypass cooling line. (**Step 3**)
4. Connect hose lines and nozzles that are suitable for discharging water at the anticipated flow rate to one or more of the discharge outlets. (**Step 4**)
5. Make sure that the tank-to-pump valve(s) and discharge valve(s) leading to the hose lines and nozzles are fully opened. (**Step 5**)
6. Adjust the engine throttle until the maximum consistent pressure reading on the discharge pressure gauge, Pitot gauge, or flow meter is obtained.
7. Close the discharge valve(s), and refill the water tank by opening the tank fill valve. The bypass line may be opened temporarily, if needed, to keep the water temperature in the pump within acceptable limits. (**Step 6**)
8. Fully reopen the discharge valve(s), and take a Pitot reading or other flow measurement while the water is being discharged. If necessary, adjust the engine throttle to maintain the discharge pressure.
9. Record the flow rate on the performance test form. (**Step 7**)

Vacuum Test

When a fire apparatus is new, the pump must be able to develop a vacuum of 22 in. Hg (75 kPa), unless the altitude is greater than 2000′ (610 m), in which case the vacuum attained is permitted to be less than 22 in. Hg by 1″ for each 1000′ (305 m) of altitude above 2000′ (610 m). The vacuum (i.e., reduction in atmospheric pressure) inside a pump or suction hose should not drop in excess of 10 in. Hg (34 kPa) in 5 minutes.

The vacuum test is basically a test of the priming system—specifically, the tightness of the pump, including its valves and fittings. It is not a test of the pump's ability to maintain a vacuum while pumping water. Remember to use the required vacuum based on your elevation above sea level as noted in the NFPA

Skill Drill 12-5

Performing a Flow Meter Test

1. Connect the supply line to the intake valve on the apparatus.

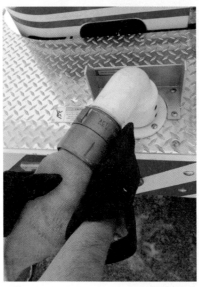
2. Connect the hose lines and nozzles that are suitable for discharging water at the anticipated flow rate to one or more of the discharge outlets.

3. Connect the flow meter to the nozzle or portable monitor.

4. Connect the hose line to the flow meter.

5. Put the fire apparatus into pump mode.

6. Make sure that the discharge valve(s) leading to the hose lines and nozzles are fully opened. Adjust the engine throttle until the maximum consistent pressure reading on the Pitot gauge or the test flow on a flow meter is obtained for that specific pipe size. Record the information on the performance test form.

Chapter 12 Performance Testing

Skill Drill 12-6

Testing the Tank-to-Pump Rate

5.1.1

1 Connect the supply line to the intake valve on the fire apparatus. Fill the water tank until it overflows.

2 Close all intakes to the pump.

3 Close the tank fill and bypass cooling line.

4 Connect hose lines and nozzles that are suitable for discharging water at the anticipated flow rate to one or more of the discharge outlets.

5 Make sure that the tank-to-pump valve(s) and discharge valve(s) leading to the hose lines and nozzles are fully opened.

6 Adjust the engine throttle until the maximum consistent pressure reading on the discharge pressure gauge, Pitot gauge, or flow meter is obtained. Close the discharge valve(s), and refill the water tank by opening the tank fill valve. The bypass line may be opened temporarily, if needed, to keep the water temperature in the pump within acceptable limits.

7 Fully reopen the discharge valve(s), and take a Pitot reading or other flow measurement while the water is being discharged. If necessary, adjust the engine throttle to maintain the discharge pressure. Record the flow rate on the performance test form.

1911. Leaking gaskets and improperly adjusted pump packing are two of the biggest sources of problems on this test, so preparation of the apparatus is important. If the primer device fails to produce a vacuum of at least 22 in. Hg (75 kPa) at sea level, the reason for the discrepancy must be determined and corrected prior to conducting any further testing.

To perform a vacuum test, follow the steps in **Skill Drill 12-7** ▶:

1. Drain the pump, all discharges, discharge drains, and intakes of water. **(Step 1)**
2. Inspect the priming oil reservoir. If applicable, replace the fluid as recommended by the pump manufacturer. **(Step 2)**
3. Install the vacuum-testing gauge (Si-Span QC Test kit) in the vacuum testing port connection located on the pump panel. **(Step 3)**
4. Remove any Storz fittings and piston intake valves on the intakes, and replace them with steamer caps. **(Step 4)**
5. Inspect all intake screens, and replace any damaged or corroded screens as necessary. **(Step 5)**
6. Make sure that all intake valves are open and either capped or plugged.
7. Remove the caps on all discharges, and close all discharge valves, all drains, and the pump drain. **(Step 6)**
8. With the fire apparatus running, operate the primer device in accordance with the manufacturer's instructions.
9. Ensure that the recirculating, engine or auxiliary cooling, and tank-to-fill valves are all closed.
10. With the fire apparatus running and all intakes and discharge valves closed and uncapped, confirm that a vacuum of 22 in. Hg (75 kPa) or higher is developed using the priming device. **(Step 7)**
11. Reduce the engine speed, turn off the fire apparatus engine, and listen for any air leaks. The vacuum should not drop more than 10 in. Hg (34 kPa) in 5 minutes. Do not use the primer after the 5-minute test period has begun.
12. Do not operate the engine at any speed greater than the governed speed during this test.
13. Close all intake valves, remove all caps and/or plugs from each valved intake, and perform inspections of valve parts.
14. Record all data on the performance test form. **(Step 8)**

Safety Tip

Prior to activating any device, you and the crew should have hearing protection on. Most priming devices can produce decibels that are high and can be harmful.

Priming System Test

At the start of the vacuum test, attention should be paid to the ease with which the pump can develop a vacuum. Before starting the priming process, close all discharges, drains, and water tank valves and petcocks; make sure that the gaskets in the suction line hose(s) are in place and free of foreign matter; close all intake valves; and tighten all intake caps and couplings.

For pumps that operate at less than 1500 GPM (5678 L/min), the priming device should be able to create the necessary vacuum in 30 seconds to lift water 10′ (3 m) through 20′ (6 m) of suction hose of the appropriate size. The priming device on pumps of 1500 GPM or larger should be able to accomplish this task in 45 seconds.

An additional 15 seconds might be needed where the pump system includes an auxiliary 4″ (100 mm) or larger intake pipe having a volume of 1 ft^3 (0.03 m^3) or more. Operate the controls as necessary to develop pressure, and then open one discharge valve to permit the flow of water. If the pump fails to pull a draft in a specific amount of time, note the cause and make any adjustments and/or repairs as necessary.

To perform a priming system test, follow the steps in **Skill Drill 12-8** ▶:

1. Attach suction hose(s) to the appropriate intake(s), and spot the fire apparatus. Place suction lines a minimum of 2′ (0.6 m) below the surface of the water; this should eliminate any whirlpools that might otherwise lead to cavitation.
2. Attach the appropriate length of hoses, flow meters, and Pitot gauges with testing nozzles to the fire apparatus and deluge sets based on the fire apparatus's rated capacity.
3. Make sure all the testing gauges and Pitot gauges are in a position where you can see them clearly. To prevent injury, make sure that all testing nozzles and appliances are secured. **(Step 1)**
4. Make sure the pump has been drained of all its water. Close and cap any intake valves and the pump drain. **(Step 2)**
5. Ensure that all discharge drains, valves, and engine and pump recirculating lines are in the closed position. **(Step 3)**
6. Put the fire apparatus in pump mode. **(Step 4)**
7. Slowly increase the pump throttle until a speed of 800 to 1200 rpm has been achieved. Always follow the manufacturer's recommendations for pump speed.
8. Activate the priming mechanism, noting the starting time and the time after the prime is obtained. The starting time is defined as the instant when the priming device begins to operate.
9. The pump is considered primed when water under pressure has entered a discharge hose or discharges onto the ground beneath the fire apparatus.
10. Record the time to prime on the performance test form. **(Step 5)**

Pumping Test Requirements

To avoid excessive parasitic power losses from lights and other electrical equipment, make sure that all loads not required for the pumping test are manually turned off or automatically shed by the electrical load manager. NFPA 1911 requires that if the pump is driven by the engine of the fire apparatus, then engine-driven accessories should not be functionally disconnected or otherwise rendered inoperable during the tests. If the chassis engine drives the pump, all headlights, running lights, warn-

Skill Drill 12-7

Performing a Vacuum Test

1. Drain the pump, all discharges, discharge drains, and intakes of water.

2. Inspect priming oil reservoir. If applicable, replace the fluid as recommended by the pump manufacturer.

3. Install the vacuum-testing gauge (Si-Span QC Test kit) in the vacuum testing port connection located on the pump panel.

4. Remove any Storz fittings and piston intake valves on the intakes, and replace them with steamer caps.

5. Inspect all intake screens, and replace any damaged or corroded screens as necessary.

6. Make sure that all intake valves are open and either capped or plugged. Remove caps on all discharges, and close all discharge valves, all drains, and the pump drain.

7. With the fire apparatus running, operate the primer device in accordance with the manufacturer's instructions. Ensure that the recirculating, engine or auxiliary cooling, and tank-to-fill valves are all closed. With the vehicle running and all intakes and discharge valves closed and uncapped, confirm that a vacuum of 22 in. Hg (75 kPa) or higher is developed using the priming device.

8. Reduce the engine speed, turn off the apparatus engine, and listen for any air leaks. The vacuum should not drop more than 10 in. Hg (34 kPa) in 5 minutes. Do not use the primer after the 5-minute test period has begun. Do not operate the engine at any speed greater than the governed speed during this test. Close all intake valves, remove all caps and/or plugs from each valved intake, and repeat the inspection of the valves. Record all data on the performance test form.

Skill Drill 12-8

Performing a Priming System Test

1. Attach suction hose(s) to the appropriate intake(s), and spot the apparatus. Place suction lines a minimum of 2′ (0.6 m) below the surface of the water; this should eliminate any whirlpools that might otherwise lead to cavitation. Attach the appropriate length of hoses, flow meters, and Pitot gauges with testing nozzles to the apparatus and deluge sets based on the apparatus's rated capacity. Make sure all the testing gauges and Pitot gauges are in a position where the operator can see them clearly. To prevent injury, make sure that all testing nozzles and appliances are secured.

2. Make sure the pump has been drained of all its water. Close and cap any intake valves and pump drain.

3. Ensure that all discharge drains, valves, and engine and pump recirculating lines are in the closed position.

4. Put the fire apparatus in pump mode.

5. Slowly increase the pump throttle until a speed of 800 to 1200 rpm has been achieved. Always follow the manufacturer's recommendations for pump speed. Activate the priming mechanism, noting the starting time and the time after the prime is obtained. The starting time is defined as the instant when the priming device begins to operate. The pump is considered primed when water under pressure has entered a discharge hose or discharges onto the ground beneath the fire apparatus. Record the time to prime on the performance test form.

Voices of Experience

Early February 2001, I was promoted to driver/operator. On my first day as a driver/operator I began dealing with the effects of a snow storm. Everything was shut down. Despite the weather, we had a structure fire that night on the south side of the city. Obviously nervous and driving in the snow, I took my time driving the apparatus and was extra careful.

Upon arrival, we found a 2-story house with fire coming from the A-side out the front door. I stopped where we thought a hydrant should be according to our hydrant maps. I dropped off a fire fighter and continued on past the house, so the Captain could get a good look at the other three sides of the structure. Unfortunately, the hydrant was not there. Fortunately, the fire fighter quickly discovered its actual location.

The Captain performed a size-up and called for additional engines. A member of our crew immediately pulled a 1¾″ (45 mm) attack hose line to start a defensive operation until the rest of the crews could arrive on scene. Before our attack line could be charged, the Captain noticed that a live electrical line had fallen onto the attack line, due to the snow and ice. Immediately we stopped the operation. We pulled another attack line out and put it in place.

When the second attack line was charged and operating, a crew member aimed it towards the fire coming out of the front door. The nozzle reaction pushed him backwards on the snow and ice.

When the other engines arrived on scene, two fire fighters entered the building to extinguish the fire so that a primary search could be started. Upon a quick primary search, a male victim was found next to the door, unconscious. The fire fighters immediately started CPR, but to no avail, the victim passed away.

> *The first day as a newly promoted driver/operator can be a difficult day.*

That was a day to remember. Snow, ice, finding a hydrant, live electric lines down on the attack line, a house destroyed, and a death. Many things happened that were not normal, many things could have happened to make the situation even worse. The main thing to remember is that you always need to be ready to adapt and overcome any situation that might be handed to you. If I did not have confidence in my abilities and the department did not have confidence in me, then the situation that night could have easily have become much worse.

Todd Dettman
Rocky Mount Fire Department
Rocky Mount, North Carolina

ing lights, and air conditioners should be operating during the pumping portion of this test.

Other equipment normally driven by the engine, such as the air compressor, engine fan, or power-steering pump, must remain in normal operation. Check for air system leaks, improper operation of the air compressor unloader, improper activation of the fan clutch, and other potential sources of power loss.

During the pumping test, the pump should not be stopped except when discharges are closed to permit changing hose layout or nozzle diameter. It is recommended that you continue to flow water from a discharge outlet when changing testing nozzles, as this action prevents the pump from overheating. The engine compartment should remain closed during the pumping test unless the fire apparatus was designed to meet an older standard that permitted testing with the compartment open.

When testing a pump at elevations up to 2000′ (610 m), 20′ (6.1 m) of suction hose of the appropriate size for the rated capacity of the pump should be used. A suction strainer and hose that will allow flow with total friction and entrance loss not greater than that specified should be used as well. When multiple suction lines are deployed, all suction lines are not required to be the same hose size. The number, length, and condition of suction hoses, as well as the altitude, water temperature, atmospheric pressure, and lift, are all factors that will affect the apparatus's performance while pumping from draft.

The test data for pumps of different capacities are outlined in NFPA 1911 and listed in **Table 12-1**. The test site should be arranged to meet those requirements. In addition, all gauges and flow measurement devices must be inspected and calibrated prior to the pump test.

Table 12-1 Test Data for Pumps of Different Capacities

Pump's Rated Capacity (GPM)	Nozzle Size (in.)	Nozzle Pressure (psi)	Flow (GPM)	Pump Pressure (psi)	Proportion of Capacity (%)
500	1½″	58	508	150	100
	1¼″	58	351	200	70
	1″	72	251	250	50
750	1¾″	68	750	150	100
	1½″	62	525	200	70
	1¼″	66	375	250	50
1000	2″	72	1008	150	100
	1¾″	60	704	200	70
	1½″	58	508	250	50
1250	2¼″	70	1260	150	100
	1⅞″	70	875	200	70
	1⅝″	64	627	250	50
1500	Two: 1¾″	68*	750*	150	100
	2″	78	1050	200	70
	1¾″	68	750	250	50
1750	Two: 1⅞″	70*	875*	150	100
	2¼″	67	1233	200	70
	1⅞″	70	875	250	50
2000	Two: 2″	71*	1001*	150	100
	Two: 1⅝″	80*	700*	200	70
	2″	72	1008	250	50

* Indicates each separate nozzle pressure and flow.

Pump Performance Test

In testing the pump, three variable factors come into play: pump speed, net pump pressure, and pump discharge rate. A change in any one of these factors will cause a change in at least one of the other factors. For example, any change in engine speed changes the pump speed. Any change in hose layout or valve position changes the pump pressure. Any change in the nozzle tip changes the discharge rate. Using these variables is the only way to reach the standard test condition desired.

During the pump performance test, the pump should be operated at reduced capacity and pressure for several minutes to allow the engine and transmission to warm up gradually. The pump speed should then be increased until the desired pressure at the pump is reached. If the desired pressure is not attained, one or more lengths of hose might have to be added, a smaller nozzle used, or a discharge valve throttled.

When the desired pressure is obtained at the pump, the Pitot gauge should be read to see if the required amount of water is being delivered. If the discharge is not as great as desired and it is believed that the pump will deliver a greater quantity of water, the discharge can be increased by further speeding up the pump. If speeding up the pump increases the pump pressure by more than 5 psi (34 kPa) or 10 psi (68 kPa), a length of hose should be taken out, a discharge valve should be opened slightly, or a large nozzle should be used.

A speed reading should be taken at the same time that the pressure readings are taken. Counting the revolutions for 1 minute generally ensures that readings will be sufficiently accurate **Figure 12-10**.

To achieve the maximum pump performance, check the suction screen on the pump inlet and remove any accumulated debris. Check the priming device fluid level and add more fluid if necessary. If the pump has packing seals, check and adjust the packing as required to minimize vacuum losses. Check and replace all cracked or missing suction and discharge hose and cap gaskets. Make sure that the hard suction hoses and suction strainer are free of soda cans, polishing rags, and other obstructions. You will need one 20′ (6.1 m) length of hard suction hose to conduct the tests. Pumps with rated capacities of 1500 GPM (5678 L/min) and greater may require two 10′ lengths of hose. The hose suction diameter, number of suction lines, and maximum lift for fire pumps are specified in NFPA 1911 **Table 12-2**.

In general, hard suction hoses with smooth interiors produce lower friction losses than do more flexible suction hoses with spiral-corrugated interiors. Basket-type suction strainers usually have lower entrance losses than do barrel-type strainers and are recommended for use with 1500 GPM (5678 L/min) or greater pumps. Float-type suction strainers should not be used during the pump performance test because the NFPA standard requires the strainer be submerged at least 2′ (0.6 m) below the surface of the water.

Figure 12-10 A speed reading should be taken at the same time that the pressure readings are taken.

Testing a pump at draft is preferable to testing it from a hydrant, because the true performance of the pump is easier to evaluate while pumping from a draft. If no suitable drafting locations are available, however, testing a pump from a hydrant is an acceptable method.

Discharge pressure is the water pressure on the discharge manifold of the fire pump at the point of gauge attachment. Gauge readings of this parameter reflect the pressure necessary for the pump to perform at the required net pump pressure. **Net pump pressure** is the sum of the discharge pressure and the suction lift converted to psi or kPa when pumping at draft, or the difference between the discharge pressure and the intake pressure when pumping at a hydrant or other source of water under positive pressure. For example, if the intake pressure gauge reads 30 psi (207 kPa) and the test requires a 150 psi (1034 kPa) net pump pressure, the discharge pressure gauge should read 180 psi (1241 kPa). Remember to reduce the net pump pressure by 1 for the 200 psi test and by 2 for the 250 psi test: With each successive test, less water is lifted during drafting, which also means less friction loss occurs in the suction hose.

When testing a pump from a hydrant, the intake hose should be of a size and length that will allow the necessary amount of water to reach the pump with a minimum intake gauge pressure of 20 psi (140 kPa) while flowing at the rated capacity. **Intake pressure** is the pressure on the intake passageway of the pump at the point of gauge attachment. Only the strainer (screen) at the pump intake connection will be required to test this component of the pump's performance.

Most fire apparatus are rated for operations at elevations up to 2000′ (610 m). Engine and pump performance may be reduced at higher elevations. If your location is more than 2000′ above sea level, contact your fire apparatus manufacturer for a corrected pump rating before you attempt to conduct the pump test.

The pump should be subjected to a pumping test of at least 45 minutes in duration. Allow the pump, transmission, and engine to warm up for approximately 10 minutes prior to beginning the pump tests. The pump should not be throttled down except when discharges are closed to permit changing the hose or a nozzle, or to change the position of a transfer valve **Figure 12-11**.

If the pump is a two-stage, parallel/series-type pump, the test at 100 percent capacity should be run with the pump in parallel mode; the test at 70 percent of capacity can be run with the pump in either series or parallel mode (as indicated on the UL plate); and the 50 percent capacity test should be run with the pump in series mode. The engine should not be throttled down except when the hose, a nozzle, or the position of the transfer valve is being changed.

A complete set of readings should be taken and recorded a minimum of five times during the 20-minute test for 100 percent rated capacity, a minimum of twice during the overload test, and a minimum of three times during each of the 10-minute tests for 70 percent capacity and 50 percent capacity. Document the time, pump speed counter (rpm), pump speed (rpm), pump tachom-

Table 12-2 Hose Suction Size, Number of Suction Lines, and Lift for Fire Pumps

Rated Capacity		Maximum Suction Hose Size		Maximum Number of Suction Lines	Maximum Lift	
gpm	L/min	in.	mm	1	ft	m
250	1000	3	75	1	10	3
300	1100	3	75	1	10	3
350	1300	4	100	1	10	3
500	2000	4.5	100	1	10	3
750	3000	6	110	1	10	3
1000	4000	6	150	1	10	3
1250	5000	6	150	2	10	3
1500	6000	6	150	2	10	3
1750	7000	8	150	2	8	2.4
2000	8000	6	150	1	6	1.8
2000	8000	8	200	3	6	1.8
2250	9000	6	150	1	6	1.8
2250	9000	8	200	3	6	1.8
2500	1000	6	150	1	6	1.8
2500	1000	8	200	4	6	1.8
3000	12000	6	150	2	6	1.8
3000	12000	8	200	4	6	1.8
3500	14000	6	150	2	6	1.8
3500	14000	8	200	4	6	1.8
4000	16000	6	150	2	6	1.8
4000	16000	8	200	4	6	1.8
4500	18000	6	150	2	6	1.8
4500	18000	8	200	4	6	1.8
5000	20000	6	150	2	6	1.8
5000	20000	8	200	4	6	1.8

Reproduced with permission from NFPA's 1911-07, *Inspection, Maintenance, Testing, and Retirement of In-Service Automotive Fire Apparatus*, Copyright © 2007, National Fire Protection Association. This reprinted material is not the complete and official position of the NFPA on the referenced subject, which is represented only by the standard in its entirety.

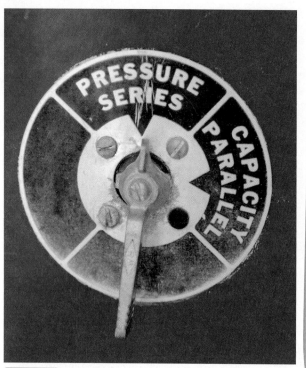

Figure 12-11 A transfer valve.

eter, pump intake test gauge (in. Hg), apparatus pump discharge, testing gauge pump discharge, nozzle Pitot readings, and flow meter(s) readings. During each testing period, record the engine water temperature, engine oil pressure (psi), transmission oil temperature, and voltage on the performance test form.

If the fire pump flow or pressure readings vary by more than 5 percent during a particular test, the reason for the fluctuation should be determined, the cause corrected, and the test continued or repeated. If a pump counter speed shaft (rpm) is not provided, the engine speed can be read with a photo-tachometer or strobe light off a rotating element.

The pump should be subjected to a pumping test consisting of the following tasks:

- Twenty minutes pumping at 100 percent of the rated capacity at 150 psi, net pump pressure. Readings should be taken a minimum of five times if they fluctuate. For a two-stage parallel/series-type pump, the test at 100 percent of capacity must be run with the pump in parallel mode.
- Ten minutes pumping at 70 percent of the rated capacity at 200 psi, net pump pressure. Readings should be conducted a minimum of three times if they fluctuate. For a two-stage parallel/series-type pump, the test at 70 percent of capacity can be run with the pump in either series/pressure or parallel mode.
- Ten minutes pumping at 50 percent of the rated capacity at 250 psi, net pump pressure. Readings should be conducted a minimum of three times if they fluctuate. For a two-stage parallel/series-type pump, the test at 50 percent of capacity can be run with the pump in series/pressure mode.

Capacity Test/150 psi Test (100 Percent Test)

The capacity test is conducted within several minutes after the priming test has been successfully completed. This sequence is intended to ensure that the engine, pump, and transmission have had an adequate amount of time to warm up. Inspect the following fire apparatus gauges, the engine water temperature, engine oil pressure, transmission temperature, and pump temperature, if applicable, to see if they are within their normal operating ranges prior to subjecting the fire apparatus to any further testing.

To perform a capacity test or 150 psi test, follow the steps in **Skill Drill 12-9**:

1. Ensure that the appropriate hose lines and deluge appliances are fixed and secured, and that testing nozzles, Pitot gauges, and flow meters are all attached. (**Step 1**)
2. After opening the discharge valves to the appropriate hose layout, gradually increase the engine speed until the net pump pressure of 150 psi is achieved. Remember that the net pump pressure is the difference between discharge pressure and intake vacuum in inches of mercury (in. Hg). (**Step 2**)
3. Check nozzle pressures with the Pitot gauges, and check flow rates if using flow meters. If the pressure is too high, close the appropriate valve(s) and readjust the engine speed to the correct net pump pressure. If the flow is too low, open the appropriate valve(s) and increase the engine speed to the correct net pump pressure. (**Step 3**)
4. Begin testing once the current flows and net pump pressure are obtained.
5. The following readings should be made and recorded every 5 minutes during the 20-minute test. (**Step 4**)
   a. Time
   b. Pump counter (rpm; if used)
   c. Pump speed (rpm; based on the specific type of pump manufacturer's correlation)
   d. Engine tachometer
   e. Engine temperature
   f. Oil pressure
   g. Voltage
   h. Automatic transmission temperature (if equipped)
   i. Pump intake (vacuum) apparatus gauge
   j. Pump intake (vacuum) test gauge
   k. Pump master discharge apparatus gauge
   l. Pump discharge test gauge
   m. Pitot gauge(s) and/or flow meter readings

Overload Test/165 psi Test

If the pump has a rated capacity of 750 GPM (3000 L/min) or greater, the fire apparatus must be subjected to an overload test consisting of pumping the rated capacity at 165 psi (1100 kPa) net pump pressure for at least 5 minutes without exceeding the maximum no-load governed speed. The overload test should be performed immediately following the test of pumping the rated

Skill Drill 12-9

Performing a Capacity Test (150 psi Test)

1. Ensure that the appropriate hose lines and deluge appliances are fixed and secured, and that testing nozzles, Pitot gauges, and flow meters are all attached.

2. After opening the discharge valves to the appropriate hose layout, gradually increase the engine speed until the net pump pressure of 150 psi is achieved. The net pump pressure is the difference between discharge pressure and intake vacuum in inches of mercury (in. Hg).

3. Check nozzle pressures with the Pitot gauges, and check flow rates if using flow meters. If the pressure is too high, close the appropriate valve(s) and readjust the engine speed to the correct net pump pressure. If the flow is too low, open the appropriate valve(s) and increase the engine speed to the correct net pump pressure.

4. Begin testing once the current flows and net pump pressure are obtained. The following readings should be made and recorded every 5 minutes during the 20-minute test.
   a. Time
   b. Pump counter (rpm; if used)
   c. Pump speed (rpm; based on the specific type of pump manufacturer's correlation)
   d. Engine tachometer
   e. Engine temperature
   f. Oil pressure
   g. Voltage
   h. Automatic transmission temperature (if equipped)
   i. Pump intake (vacuum) apparatus gauge
   j. Pump intake (vacuum) test gauge
   k. Pump master discharge apparatus gauge
   l. Pump discharge test gauge
   m. Pitot gauge(s) and/or flow meter readings

capacity at 150 psi (1000 kPa) net pump pressure. The pumping tests should not be started until the pump pressure and the discharge quantity are satisfactory.

To perform an overload test, follow the steps in **Skill Drill 12-10**:

1. Gradually increase the throttle speed to reach a net pump pressure of 165 psi, while at the same time closing down a valve enough so that the flow (GPM) remains at capacity.
2. Check nozzle pressures with the Pitot gauges, and check flow rates if using flow meters. If the pressure is too high, close the appropriate valve(s) and readjust the engine speed to the correct net pump pressure. If the flow is too low, open the appropriate valve(s) and increase the engine speed to the correct net pump pressure.
3. Begin testing once the current flows and net pump pressure are obtained.
4. The following readings should be made and recorded at least two times during this 5-minute test.
   a. Time
   b. Pump counter (rpm; if used)
   c. Pump speed (rpm; based on the specific type of pump manufacturer's correlation)
   d. Engine tachometer
   e. Engine temperature
   f. Oil pressure
   g. Voltage
   h. Automatic transmission temperature (if equipped)
   i. Pump intake (vacuum) apparatus gauge
   j. Pump intake (vacuum) test gauge
   k. Pump master discharge apparatus gauge
   l. Pump discharge test gauge
   m. Pitot gauge(s) and/or flow meter readings
5. Reduce the pump speed to make any necessary nozzle tip changes.

200 psi Test (70 Percent Test)

The 200 psi (1380 kPa) test is conducted immediately after the 165 psi overload test. There should be no delay in time between these tests unless a testing nozzle tip change is made.

To perform the 200 psi test, follow the steps in **Skill Drill 12-11**:

1. Use the same test procedures as for the capacity test, except use the proper nozzle sizes and hose layout to flow 70 percent of the pump's rated capacity.
2. For a two-stage pump, check the previous performance test sheets to see if the test should be run in the volume or pressure mode.
3. After opening the discharge valves to the appropriate hose layout and testing nozzles, gradually increase the engine speed until the net pump pressure of 200 psi is achieved.
4. Check nozzle pressures with the Pitot gauges, and check flow rates if using flow meters. If the pressure is too high, close the appropriate valve(s) and readjust the engine speed to the correct net pump pressure. If the flow is too low, open the appropriate valve(s) and increase the engine speed to the correct net pump pressure.
5. Begin testing once the current flows and net pump pressure are obtained.
6. Run the test for 10 minutes, and check and record the following readings at a minimum of three intervals during this test.
   a. Time
   b. Pump counter (rpm; if used)
   c. Pump speed (rpm; based on the specific type of pump manufacturer's correlation)
   d. Engine tachometer
   e. Engine temperature
   f. Oil pressure
   g. Voltage
   h. Automatic transmission temperature (if equipped)
   i. Pump intake (vacuum) apparatus gauge
   j. Pump intake (vacuum) test gauge
   k. Pump master discharge apparatus gauge
   l. Pump discharge test gauge
   m. Pitot gauge(s) and/or flow meter readings
7. Reduce the pump speed to make any necessary nozzle tip changes.

250 psi Test (50 Percent Test)

The 250 psi (1725 kPa) test is conducted immediately after the 200 psi test. There should be no delay in time between these tests unless a testing nozzle tip change is made.

To perform the 250 psi test, follow the steps in **Skill Drill 12-12**:

1. Use the same test procedures as for the capacity test, except use proper nozzle sizes to flow 50 percent of the pump's rated capacity.
2. For a two-stage pump, the test should be run in the pressure mode. This requires you to reduce the engine speed to idle and then change the transfer valve to pressure (if you have not already done so).
3. After opening the discharge valves to the appropriate hose layout, gradually increase the engine speed until a net pump pressure of 250 psi is achieved.
4. Check nozzle pressures with the Pitot gauges, and check flow rates if using flow meters. If the pressure is too high, close the appropriate valve(s) and readjust the engine speed to the correct net pump pressure. If the flow is too low, open the appropriate valve(s) and increase the engine speed to the correct net pump pressure.
5. Begin testing once the current flows and the net pump pressure are obtained.
6. Run the test for 10 minutes, and check and record the following readings at a minimum of three intervals during this test.
   a. Time
   b. Pump counter (rpm; if used)
   c. Pump speed (rpm; based on the specific type of pump manufacturer's correlation)
   d. Engine tachometer
   e. Engine temperature

Near Miss REPORT

Report Number: 05-0000174
Report Date: 05/27/2005 14:10

Synopsis: While enroute to a working auto extrication, the brakes failed on a large single axle rescue squad.

Event Description: While enroute to a working auto extrication, the brakes failed on a large single axle rescue squad while attempting to make a left turn onto a heavily traveled roadway. As the operator, I had overheated the brakes while slowing for a previous red traffic light at the base of a hill. As I approached the next intersection, I had downshifted the automatic transmission to 2nd gear, and applied the brakes. The rescue squad was traveling about 25-30 miles per hour. I immediately realized that I had no brakes and advised the officer of the same. The road that I was traveling came to a dead end at this intersection.

I turned to the left and felt the truck begin to lean to the right. The truck then rolled up on the right side tires. I steered slightly to the right and the truck started to right itself. However by steering to the right, we were now on a collision course with a utility pole and the guardrail. As if in slow motion, I turned slightly left again which caused the truck to roll back to the right on two wheels again, but missed both objects in our path. With the right side tires approaching the soft shoulder, the truck slowly rolled back to the left and landed on all six wheels. After checking with the officer on any possible injuries to him, we then communicated with the 2 fire fighters in the crew cab via the intercom, and found they too were uninjured. The weather was sunny, and road conditions were dry.

Lessons Learned:

1. KNOW the limitations of your equipment. I knew that this particular vehicle, which was not overloaded, but is close on its gross vehicle weight, could potentially have a brake problem if you were "hard" on the brakes several times without allowing any cooling off time.

2. Most importantly, REMEMBER THAT YOU ARE RESPONSIBLE FOR THE LIVES OF YOUR CREW AND OF THE MOTORISTS AROUND YOU!!!!! No rescue call is worth risking the lives of your crew and what good can you do if you don't arrive on the scene?

Demographics

Department type: Combination, Mostly volunteer
Job or rank: Driver/Engineer
Department shift: Respond from home
Age: 43–51
Years of fire service experience: 27–30
Region: FEMA Region III
Service Area: Suburban

Event Information

Event type: Non-fire emergency event: auto extrication, technical rescue, emergency medical call, service calls, etc.
Event date and time: 04/28/2005 14:15
Hours into the shift: 0–4
Event participation: Involved in the event
Do you think this will happen again? Uncertain
What were the contributing factors?
- Human Error
- Equipment

What do you believe is the loss potential?
- Property damage
- Life-threatening injury

f. Oil pressure
g. Voltage
h. Automatic transmission temperature (if equipped)
i. Pump intake (vacuum) apparatus gauge
j. Pump intake (vacuum) test gauge
k. Pump master discharge apparatus gauge
l. Pump discharge test gauge
m. Pitot gauge(s) and/or flow meter readings

7. Reduce the pump speed to make any necessary nozzle tip change.

Pressure Control Test

The pressure control device, a relief valve, or pressure governor located on the pump panel should be tested at 150 psi (1035 kPa), 90 psi (620 kPa), and 250 psi (1725 kPa) intervals. Care should be taken to perform the pressure control test using net pump pressure and net pressure rise readings. Some pressure control systems might not operate correctly if the hydrant pressure is too high; consult the system manufacturer's manual for further information (Figure 12-12▼ and Figure 12-13▶).

Closing all discharges in less than 3 seconds could cause instantaneous pressure rises, such that the pressure control device might not be able to respond rapidly enough to avoid damage to the pumping system. Taking more than 10 seconds to close the discharges is not a reasonable test of the pressure control device response capability. Controlling closure of the discharges can be performed manually or otherwise.

To perform a pressure control test, follow the steps in Skill Drill 12-13▶:

1. The pump should be delivering a flow at the rated capacity at a net pump pressure of 150 psi.
2. The pressure control device should be set in accordance with the manufacturer's instructions to maintain a discharge at a net pump pressure of 150 psi.

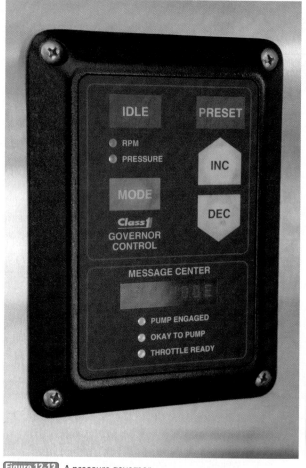

Figure 12-13 A pressure governor.

Figure 12-12 A relief valve.

3. All discharge valves should be closed no more rapidly than in 3 seconds and no more slowly than in 10 seconds. The rise in discharge pressure must not exceed 30 psi. (**Step 1**)
4. The original conditions of pumping the rated capacity at a net pump pressure of 150 psi should be reestablished. The discharge pressure should be reduced to a net pressure of 90 psi by throttling back the engine fuel supply with no change to the discharge valve setting, hose, or nozzles. (**Step 2**)
5. The pressure control device should be set in accordance with the manufacturer's instructions to maintain a discharge at a net pump pressure of 90 psi. (**Step 3**)
6. All discharge valves should be closed no more rapidly than in 3 seconds and no more slowly than in 10 seconds. The rise in discharge pressure must not exceed 30 psi. (**Step 4**)
7. Reduce the engine speed and make the necessary adjustments to the hose layout and nozzle tip size for the 250 psi test. (**Step 5**)
8. The pump should be delivering 50 percent of the rated capacity at a net pump pressure of 250 psi.
9. The pressure control device should be set in accordance with the manufacturer's instructions to maintain a discharge at a net pump pressure of 250 psi. (**Step 6**)

Skill Drill 12-13

Performing a Pressure Control Test

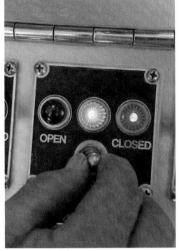

1. The pump should be delivering a flow at the rated capacity at a net pump pressure of 150 psi (1035 kPa). The pressure control device should be set in accordance with the manufacturer's instructions to maintain a discharge at a net pump pressure of 150 psi. All discharge valves should be closed no more rapidly than in 3 seconds and no more slowly than in 10 seconds. The rise in discharge pressure must not exceed 30 psi.

2. The original conditions of pumping the rated capacity at a net pump pressure of 150 psi should be reestablished. The discharge pressure should be reduced to a net pressure of 90 psi (620 kPa) by throttling back the engine fuel supply with no change to the discharge valve setting, hose, or nozzles.

3. The pressure control device should be set in accordance with the manufacturer's instructions to maintain a discharge at a net pump pressure of 90 psi.

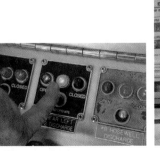

4. All discharge valves should be closed no more rapidly than in 3 seconds and no more slowly than in 10 seconds. The rise in discharge pressure must not exceed 30 psi.

5. Reduce engine speed and make the necessary adjustments in the hose layout and nozzle tip size for the 250 psi (1725 kPa) test.

6. The pump should be delivering 50 percent of the rated capacity at a net pump pressure of 250 psi. The pressure control device should be set in accordance with the manufacturer's instructions to maintain a discharge at a net pump pressure of 250 psi.

7. All discharge valves should be closed no more rapidly than in 3 seconds and no more slowly than in 10 seconds. The rise in discharge pressure must not exceed 30 psi. Record all data—including the rise in pressure at 150 psi, 90 psi, and 250 psi—on the performance test form.

10. All discharge valves should be closed no more rapidly than in 3 seconds and no more slowly than in 10 seconds. The rise in discharge pressure must not exceed 30 psi.
11. Record all data—including the rise in pressure at 150 psi, 90 psi, and 250 psi—on the performance test form. (**Step 7**)

Remember, for a relief valve to function properly, there must always exist a pressure differential of 30 psi or greater between the intake manifold and the discharge header of the pump. The excess flow (GPM) must have somewhere to go.

Post Performance Testing

After conducting all of the performance tests, it is recommended that you reduce the engine speed of the fire apparatus to idle. This will allow the engine, pump, and transmission to cool down for approximately 10 minutes. After the cool-down period, you can turn off all the engine-driven accessories that were turned on during the testing period.

After about 5 minutes, the fire apparatus can be switched from pump mode back into road mode. It is very important to keep the fire apparatus running at idle engine speed to allow the engine turbo (if applicable) time to cool down. The pump and suction lines will still have a vacuum, so it will be necessary to open a discharge valve to remove the vacuum that was created inside the pump. Remove all testing equipment, hoses, hard suction hoses, and nozzles, and place the fire apparatus back in service.

It is always recommended when operating from a draft of a static water source to backflush the pump and tank after pump testing. Backflushing requires the fire apparatus to be attached to a clean water source, such as a hydrant; all discharge and intake valves and the pump are then flushed. The tank also should be drained of water and then refilled. If you are testing a Darley fire pump, backflushing cannot be conducted by the standard method due to the Pitot valve located in the bottom of the discharge header. For a Darley pump, connect it to a high-volume water source and flush the system until all discharges are flowing clear.

The pumping system (i.e., engine, pump, and transmission) should exhibit no undue heating, loss of power, or other defect during the entire pump test. The average flow rate, discharge pressure, intake pressure, and engine speed should be calculated and recorded at the end of each phase of the performance test.

When the apparatus operates at or near full engine power while remaining stationary, the heat generated by its operation can raise the temperature of certain chassis and/or pumping system components above the level that can be touched without extreme discomfort or injury. However, as long as the fire apparatus can be operated and used satisfactorily for the required duration of the test under those conditions and the engine coolant temperature remains within normal range, its performance should be considered acceptable.

Normal wear in the pumping system can require speeds greater than those required at the time of delivery for the pumping test. Such variances are acceptable as long as the fire apparatus passes the performance tests without exceeding the no-load governed engine speed.

Final Test Results

Just imagine taking all the time and effort to conduct good performance tests on your fire apparatus—and then not getting full credit for them. Unfortunately, this outcome happens far too frequently.

If you do not record all the test conditions, all the readings, and all the information about the fire apparatus being tested, then you will not receive full credit for the performance tests. Even worse, you will not be able to compare the most recent test results with previous ones, which might otherwise enable you to notice long-term changes that could indicate hidden problems. Also, if you do not keep records of the tests in a permanent file, the ISO will not give you full credit on your next evaluation of the fire apparatus.

NFPA 1911 contains a test data form to be used with your annual performance tests. It contains headings and blank spaces to record all the data from the test. You can photocopy this form and use it as a permanent record that will satisfy both NFPA and ISO requirements. The form has spaces for the vehicle information, the weather conditions at the start and end of the tests (both are needed), and the readings and results of all tests. As a reminder that some tests require you to take several readings, the form even has spaces to record the required number of readings, plus one extra space for any additional reading. Record information neatly, using pen instead of pencil, if possible. Be sure to fill in the "Witnessed by" and "Date" lines at the bottom of the first page to make the test form more legally defensible **Figure 12-14**.

Pump testing should be conducted once per year, every year. If you test your pumps less often, you will not be able to notice any serious trends, and you will not receive full credit for your

---

PUMP PERFORMANCE TEST

Apparatus number or designation_____
Manufacturer_____
Serial no._____
Year manufactured_____
Model_____
Vehicle identification no._____
Model_____
Model_____
Engine make_____
Pump make_____
Ratio to engine_____
Pump rated capacity_____(gpm) (L/min)
 at_____(psi) (kPa)
Speed check taken from_____
Test site location_____
___
Test performed from   ❏ Draft   ❏ Hydrant
Suction hose size_____(in.) (mm)
Length_____(ft) (m)

Figure 12-14 The pump performance test form.
Reproduced with permission from NFPA's 1911-07, *Inspection, Maintenance, Testing, and Retirement of In-Service Automotive Fire Apparatus*, Copyright © 2007, National Fire Protection Association. This reprinted material is not the complete and official position of the NFPA on the referenced subject, which is represented only by the standard in its entirety.

irregular tests. For example, if the average time between the last three pump tests on an apparatus is three years, the ISO will give you only 50 percent of the test credit. If the average time between the last three tests is five years, you will get no credit, even if the vehicle passes the test.

If the test conditions are equivalent to those at the time of delivery of the fire apparatus and the speed of the engine increases by more than 10 percent of the original engine speed, the reason for decrease in performance should be determined and the deficiency corrected. Where test conditions differ significantly from the original test conditions at the time of the apparatus's delivery, results should be compared with those from the previous year's test. The test conditions should be maintained as consistently as possible from test period to test period.

Problem Solving

Most performance tests are conducted without incident. Nevertheless, trouble may develop during some tests, and an effort should be made to locate the source of trouble while the fire apparatus remains at the test site.

Failure to prime a centrifugal pump is a frequent source of trouble, with the usual reason for this failure being an air leak in the suction hose or pump. One way to trace this trouble is to remove all discharge hose lines, cap all discharge openings and the suction hose, and operate the priming mechanism in accordance with the manufacturer's recommendations. The intake gauge should be studied to determine the maximum vacuum that is developed, which should be at least 22 in. Hg (75 kPa) at altitudes of less than 1000′ (305 m). The primer should then be stopped. If the vacuum drops 10 in. Hg (34 kPa) or more in less than 5 minutes, there is a leak in the suction hose or pump assembly; it could be in a valve, drain cock, piping, casing, or pump packing.

You can also attempt to locate a leak by listening for air movement. Another method of checking for leaks is to connect the pump to a convenient hydrant, cap the pump discharge outlets, open the hydrant, and watch for water leaks. A leak can usually be corrected at the test site.

Two possible causes of failure of the pump to deliver the desired capacity, pressure, or both, are insufficient power and restrictions in the intake arrangement. Insufficient power is indicated by the inability of the engine to reach the required speed for the desired pumping condition.

Insufficient pressure when operating a centrifugal pump could be the result of pumping too much water for the available power and, in multistage pumps, pumping in the "volume" position instead of in the required "pressure" position. This problem can be checked by partially closing off all discharge valves until only a small flow is observed and then opening up the throttle until the desired pressure is reached, followed by slowly opening discharge valves and increasing the engine speed as necessary to maintain pressure until the desired capacity is obtained. An improperly adjusted or inoperative transfer valve can prevent the development of adequate pressure. Likewise, the pressure control system might be set too low or be defective.

Some possible causes of insufficient power are as follows:

- You might have failed to advance the throttle far enough or might be using the wrong transmission gear position.
- The engine might be in need of a tune-up.
- The grade of fuel might be improper for adequate combustion.
- Vaporization might be occurring in the fuel line.

Restriction in the intake arrangement is indicated if the pump speed is too high for the capacity and attained pressure levels. It could be the result of any one or a combination of the following conditions:

- A too-small suction hose
- A too-high altitude
- A too-high suction lift
- An incorrect strainer type
- An intake strainer that is clogged at the pump or at the end of the suction hose
- A collapsed or defective suction hose
- Aerated water or too warm water (greater than 90°F [32°C])
- Foreign material in the pump
- A pressure control device that is set too low or malfunctioning

An air leak in the suction hose connections or in the pump intake manifold also will result in excessive pump speed and eventually could cause loss of prime and complete cessation of flow.

Engine speed differences from the original pump test could be the result of any one or a combination of the following conditions:

- Operating the fire apparatus with the wrong transmission gear in use
- Stuck throttle control cable
- Restrictions in the intake arrangements
- Suction hose under an insufficient depth of water
- Air leak on the intake side of pump
- Changes in environmental conditions
- Pump and/or engine wear
- High gear lockup not functioning (automatic transmission)

Every effort should be made to correct any problem(s) that are found. Any portion of the performance tests that was deemed a failure should be redone to ensure that the problem(s) have been corrected.

Re-rating Fire Pumps

There are two conditions under which re-rating a pump on a fire apparatus should be considered. The first condition is when the apparatus is delivered or re-powered with an engine that is capable of supplying additional power beyond that needed by the pump, which warrants a larger capacity rating for the apparatus. This condition might require additional suction intakes or pump discharges to take advantage of that capacity. The fire apparatus manufacturer and/or pump manufacturer should be consulted as necessary to ensure that all components of the pumping system are adequate for the potential re-rating.

The second condition is when the environment in which the engine or pump was initially delivered has changed, such that the engine can no longer achieve its original performance. This situation can occur when an engine or pump passed the original pump rating test with little or no reserve power and the apparatus is now located at a higher elevation, or the natural wear within the engine has reduced its power output over time.

A pump should never be considered for re-rating if the engine is seriously worn or should undergo major restorative work. Likewise, the pump should not be considered for re-rating if the results of testing indicate that the pump has signs of wear or other problems. In these cases, there is a good chance that the pump will not pass the complete pump test.

If the AHJ wishes to re-rate the pump, the pump should be tested to the complete pumping test as specified in NFPA 1911, including having the test witnessed and certified by an accredited third-party testing organization. In such circumstances, problems with engine wear, pump wear, pump blockages, or other issues with the pump will worsen at an accelerated rate if they are not corrected, potentially resulting in catastrophic failure during an emergency incident.

It might be necessary, for a variety of reasons, to continue to use the pump on an apparatus that does not meet its original rating until the pump can be repaired. This operational decision needs to be made on a case-by-case basis, depending on the specific deficiency of the pump and the apparatus that is available to replace the deficient fire apparatus while repairs are being made. Re-rating the pump downward where such deficiencies are present merely creates a false sense of vehicle capability.

Safety Tip

Never conduct any performance tests on fire apparatus while you are alone; always have, at a minimum, one additional person with you. Preferably the crew assigned to the fire apparatus will participate in its testing. They will likely know the history of the fire apparatus and be familiar with any defects or deficiencies concerning that apparatus.

Safety Tip

- Whenever possible, park parallel to a sea wall, with the pump panel facing away from it if applicable.
- Chaffing blocks and a salvage cover should always be used to prevent damage to suction hoses.
- During testing, all personnel should wear the appropriate hearing protection. The noise level (decibels) produced by the apparatus is extremely significant during this process.
- Pitot gauges should be fixed or secured to the testing nozzles. Use of hand-held Pitot gauges is not recommended when the apparatus is used at high pressures.
- While taking nozzle Pitot readings, personnel are required to wear helmets and safety glasses.
- All appliances must be properly secured or affixed.
- All hose lines, whenever applicable, should be attached to discharges that are away from the pump panel.
- All fire apparatus must have wheel chocks and blocks in proper position during the test.

Wrap-Up

Chief Concepts

- Failure of a fire pump to deliver the required amount of water at a certain pressure impairs the suppression ability and the safety of fire fighters, possibly leading to injury and increased property damage.
- Engine-driven accessories should not be functionally disconnected or otherwise rendered inoperative during the tests.
- The first performance test is to check the governed engine speed.
- If the fire apparatus is equipped with an intake relief valve system or a combination intake/discharge system, an intake relief valve system test must be conducted to ensure that the system is operating in accordance with the manufacturer's specifications.
- A test of the pump shift indicators should verify that the pump shift indicators in the cab of the fire apparatus and on the driver/operator's panel indicate correct pump status when the pump is shifted from road mode to pump mode.
- Beginning with the 1991 edition of NFPA 1901, fire apparatus equipped with electronic or electric engine throttle controls were required to include an interlock system to prevent engine speed advancement, unless the chassis transmission is in neutral with the parking brake engaged; or unless the parking brake is engaged, the fire pump is engaged, and the chassis transmission is in pumping gear; or unless the fire apparatus is in the "Okay to pump" mode.
- Discharge pressure gauges can be checked quickly against test gauges for accuracy.
- Flow rates less than the rate when the fire apparatus was new or as established in previous testing indicate problems in the tank-to-pump line or tank pump.
- The vacuum test is basically a test of the priming system, including the tightness of the pump, plus its valves and fittings. It is not a test of the ability to maintain a vacuum while pumping water.
- At the beginning of the priming system test, attention should be paid to the ease with which the pump can develop a vacuum.
- In testing the pump, a change in any of three variable factors—pump speed, net pump pressure, and pump discharge rate—will cause a change in at least one of the other factors.
- The pressure control device, a relief valve, or pressure governor located on the pump panel should be tested at 150 psi, 90 psi, and 250 psi intervals. Care should be taken while performing the pressure control test to record the net pump pressure and net pressure rise readings.
- After conducting all of the performance tests, it is recommended to reduce the engine speed to idle.
  - This will allow the engine, pump, and transmission to cool down for approximately 10 minutes.
  - After the cool-down period, you can turn off all engine-driven accessories that were turned on during the testing period.
- There are two conditions under which re-rating a pump on a fire apparatus should be considered.
  - The first condition is when the apparatus is delivered or re-powered with an engine that is capable of supplying additional power beyond that needed by the pump, which warrants a larger capacity rating.
  - The second condition is when the environment in which the engine or pump was initially delivered has changed and the engine can no longer achieve its original performance—for example, when the engine or pump passed the original pump rating test with little or no reserve power and the apparatus is now located at a higher elevation, or if the natural wear within the engine has reduced the power output.

Hot Terms

ambient air temperature The temperature of the surrounding medium. It typically refers to the temperature of the air in which a structure is situated or a device operates.

authority having jurisdiction (AHJ) An organization, office, or individual responsible for enforcing the requirements of a code or standard, or for approving equipment, materials, an installation, or a procedure.

discharge pressure The water pressure on the discharge manifold of the fire pump at the point of gauge attachment.

failure A cessation of proper functioning or performance.

intake pressure The pressure on the intake passageway of the pump at the point of gauge attachment.

net pump pressure The sum of the discharge pressure and the suction lift converted to units of pounds per square inch (psi) or kilopascals (kPa) when pumping at draft, or the difference between the discharge pressure and the intake pressure when pumping at a hydrant or other source of water under positive pressure. [NFPA 1901]

performance tests Tests conducted after a fire apparatus has been put into service to determine if its performance meets predetermined specifications or standards.

rated capacity (water pump) The flow rate to which the pump manufacturer certifies compliance of the pump when it is new. [NFPA 1901]

service tests See *performance tests*.

Driver/Operator in Action

It's that time of year again: time for the annual performance testing for your fire apparatus. Just like the daily inspection that you perform every morning, these tests ensure that your fire apparatus is working properly. The safety of your entire crew depends on the safety of your fire apparatus, so you take annual testing very seriously. Before driving out to the testing grounds, you sit down to refresh your memory on a few points.

1. A test that is conducted after a fire apparatus has been put into service to determine if its performance meets predetermined specifications or standards is called a(n)
   A. acceptance test.
   B. Underwriters Laboratory test.
   C. annual performance test.
   D. governed no-load test.

2. What is the water pressure on the discharge manifold of the fire pump at the point of gauge attachment called?
   A. Total discharge pressure
   B. Discharge pressure
   C. Intake pressure
   D. Net pump pressure

3. True or false: If the pump has a rated capacity of 500 GPM (2000 L/min) or greater, the apparatus must be subjected to an overload test.
   A. True
   B. False

4. If the pump is a two-stage, parallel/series-type pump, the test at 100 percent capacity, 150 psi, should be run with the pump in which mode?
   A. Series mode
   B. Parallel mode
   C. Either series or parallel mode
   D. None of the above

5. When an apparatus is new, the pump must be able to develop a vacuum of ____ unless the altitude is greater than 2000' (610 m).
   A. 10 in. Hg
   B. 15 in. Hg
   C. 20 in. Hg
   D. 22 in. Hg

An Extract From: NFPA® 1002, *Fire Apparatus Driver/Operator Professional Qualifications*, 2009 Edition
© National Fire Protection Association

Appendix A material that involves NFPA:

Reproduced with permission from NFPA's 1002-09, *Fire Apparatus Driver/Operator Professional Qualifications*, Copyright © 2008, National Fire Protection Association. This reprinted material is not the complete and official position of the NFPA on the referenced subject, which is represented by the standard in its entirety.

Fire Service Pump Operator: Principles and Practices covers Chapters 4, 5, and 10 of NFPA 1002: *Fire Apparatus Driver/Operator Professional Qualifications*, 2009 Edition.

Chapter 4 General Requirements

4.1 General. Prior to operating fire department vehicles, the fire apparatus driver/operator shall meet the job performance requirements defined in Sections 4.2 and 4.3.

4.2 Preventive Maintenance.

4.2.1* Perform routine tests, inspections, and servicing functions on the systems and components specified in the following list, given a fire department vehicle, its manufacturer's specifications, and policies and procedures of the jurisdiction, so that the operational status of the vehicle is verified:

(1) Battery(ies)
(2) Braking system
(3) Coolant system
(4) Electrical system
(5) Fuel
(6) Hydraulic fluids
(7) Oil
(8) Tires
(9) Steering system
(10) Belts
(11) Tools, appliances, and equipment

(A) Requisite Knowledge. Manufacturer specifications and requirements, policies, and procedures of the jurisdiction.

(B) Requisite Skills. The ability to use hand tools, recognize system problems, and correct any deficiency noted according to policies and procedures.

4.2.2 Document the routine tests, inspections, and servicing functions, given maintenance and inspection forms, so that all items are checked for operation and deficiencies are reported.

(A) Requisite Knowledge. Departmental requirements for documenting maintenance performed and the importance of keeping accurate records.

(B) Requisite Skills. The ability to use tools and equipment and complete all related departmental forms.

4.3 Driving/Operating.

4.3.1* Operate a fire department vehicle, given a vehicle and a predetermined route on a public way that incorporates the maneuvers and features, specified in the following list, that the driver/operator is expected to encounter during normal operations, so that the vehicle is operated in compliance with all applicable state and local laws, departmental rules and regulations, and the requirements of NFPA 1500, Section 4.2:

(1) Four left turns and four right turns
(2) A straight section of urban business street or a two-lane rural road at least 1.6 km (1 mile) in length
(3) One through-intersection and two intersections where a stop has to be made
(4) One railroad crossing
(5) One curve, either left or right
(6) A section of limited-access highway that includes a conventional ramp entrance and exit and a section of road long enough to allow two lane changes
(7) A downgrade steep enough and long enough to require down-shifting and braking
(8) An upgrade steep enough and long enough to require gear changing to maintain speed
(9) One underpass or a low clearance or bridge

(A) Requisite Knowledge. The effects on vehicle control of liquid surge, braking reaction time, and load factors; effects of high center of gravity on roll-over potential, general steering reactions, speed, and centrifugal force; applicable laws and regulations; principles of skid avoidance, night driving, shifting, and gear patterns; negotiating intersections, railroad crossings, and bridges; weight and height limitations for both roads and bridges; identification and operation of automotive gauges; and operational limits.

(B) Requisite Skills. The ability to operate passenger restraint devices; maintain safe following distances; maintain control of the vehicle while accelerating, decelerating, and turning, given road, weather, and traffic conditions; operate under adverse environmental or driving surface conditions; and use automotive gauges and controls.

4.3.2* Back a vehicle from a roadway into restricted spaces on both the right and left sides of the vehicle, given a fire department vehicle, a spotter, and restricted spaces 3.7 m (12 ft) in width, requiring 90-degree right-hand and left-hand turns from the roadway, so that the vehicle is parked within the restricted areas without having to stop and pull forward and without striking obstructions.

(A) Requisite Knowledge. Vehicle dimensions, turning characteristics, spotter signaling, and principles of safe vehicle operation.

(B) Requisite Skills. The ability to use mirrors and judge vehicle clearance.

4.3.3* Maneuver a vehicle around obstructions on a roadway while moving forward and in reverse, given a fire department vehicle, a spotter for backing, and a roadway with obstructions, so that the vehicle is maneuvered through the obstructions without stopping to change the direction of travel and without striking the obstructions.

(A) Requisite Knowledge. Vehicle dimensions, turning characteristics, the effects of liquid surge, spotter signaling, and principles of safe vehicle operation.

(B) Requisite Skills. The ability to use mirrors and judge vehicle clearance.

4.3.4* Turn a fire department vehicle 180 degrees within a confined space, given a fire department vehicle, a spotter for backing up, and an area in which the vehicle cannot perform a U-turn without stopping and backing up, so that the vehicle is turned 180 degrees without striking obstructions within the given space.

(A) Requisite Knowledge. Vehicle dimensions, turning characteristics, the effects of liquid surge, spotter signaling, and principles of safe vehicle operation.

(B) Requisite Skills. The ability to use mirrors and judge vehicle clearance.

4.3.5* Maneuver a fire department vehicle in areas with restricted horizontal and vertical clearances, given a fire department vehicle and a course that requires the operator to move through areas of restricted horizontal and vertical clearances, so that the operator accurately judges the ability of the vehicle to pass through the openings and so that no obstructions are struck.

(A) Requisite Knowledge. Vehicle dimensions, turning characteristics, the effects of liquid surge, spotter signaling, and principles of safe vehicle operation.

(B) Requisite Skills. The ability to use mirrors and judge vehicle clearance.

4.3.6* Operate a vehicle using defensive driving techniques under emergency conditions, given a fire department vehicle and emergency conditions, so that control of the vehicle is maintained.

(A) Requisite Knowledge. The effects on vehicle control of liquid surge, braking reaction time, and load factors; the effects of high center of gravity on roll-over potential, general steering reactions, speed, and centrifugal force; applicable laws and regulations; principles of skid avoidance, night driving, shifting, gear patterns; and automatic braking systems in wet and dry conditions; negotiation of intersections, railroad crossings, and bridges; weight and height limitations for both roads and bridges; identification and operation of automotive gauges; and operational limits.

(B) Requisite Skills. The ability to operate passenger restraint devices; maintain safe following distances; maintain control of the vehicle while accelerating, decelerating, and turning, given road, weather, and traffic conditions; operate under adverse environmental or driving surface conditions; and use automotive gauges and controls.

4.3.7* Operate all fixed systems and equipment on the vehicle not specifically addressed elsewhere in this standard, given systems and equipment, manufacturer's specifications and instructions, and departmental policies and procedures for the systems and equipment, so that each system or piece of equipment is operated in accordance with the applicable instructions and policies.

(A) Requisite Knowledge. Manufacturer's specifications and operating procedures, and policies and procedures of the jurisdiction.

(B) Requisite Skills. The ability to deploy, energize, and monitor the system or equipment and to recognize and correct system problems.

Chapter 5 Apparatus Equipped with Fire Pump

5.1* **General.** The requirements of Fire Fighter 1 as specified in NFPA 1001 (or the requirements of Advanced Exterior Industrial Fire Brigade Member or Interior Structural Fire Brigade Member as specified in NFPA 1081) and the job performance requirements defined in Sections 5.1 and 5.2 shall be met prior to qualifying as a fire department driver/operator — pumper.

5.1.1 Perform the routine tests, inspections, and servicing functions specified in the following list in addition to those in 4.2.1, given a fire department pumper, its manufacturer's specifications, and policies and procedures of the jurisdiction, so that the operational status of the pumper is verified:

(1) Water tank and other extinguishing agent levels (if applicable)
(2) Pumping systems
(3) Foam systems

(A) Requisite Knowledge. Manufacturer's specifications and requirements, and policies and procedures of the jurisdiction.

(B) Requisite Skills. The ability to use hand tools, recognize system problems, and correct any deficiency noted according to policies and procedures.

5.2 Operations.

5.2.1 Produce effective hand or master streams, given the sources specified in the following list, so that the pump is engaged, all pressure control and vehicle safety devices are set, the rated flow of the nozzle is achieved and maintained, and the apparatus is continuously monitored for potential problems:
(1) Internal tank
(2)* Pressurized source
(3) Static source
(4) Transfer from internal tank to external source

(A) Requisite Knowledge. Hydraulic calculations for friction loss and flow using both written formulas and estimation methods, safe operation of the pump, problems related to small-diameter or dead-end mains, low-pressure and private water supply systems, hydrant coding systems, and reliability of static sources.

(B) Requisite Skills. The ability to position a fire department pumper to operate at a fire hydrant and at a static water source, power transfer from vehicle engine to pump, draft, operate pumper pressure control systems, operate the volume/pressure transfer valve (multistage pumps only), operate auxiliary cooling systems, make the transition between internal and external water sources, and assemble hose lines, nozzles, valves, and appliances.

5.2.2 Pump a supply line of 65 mm (2½ in.) or larger, given a relay pumping evolution the length and size of the line and the desired flow and intake pressure, so that the correct pressure and flow are provided to the next pumper in the relay.

(A) Requisite Knowledge. Hydraulic calculations for friction loss and flow using both written formulas and estimation methods, safe operation of the pump, problems related to small-diameter or dead-end mains, low-pressure and private water supply systems, hydrant coding systems, and reliability of static sources.

(B) Requisite Skills. The ability to position a fire department pumper to operate at a fire hydrant and at a static water source, power transfer from vehicle engine to pump, draft, operate pumper pressure control systems, operate the volume/pressure transfer valve (multistage pumps only), operate auxiliary cooling systems, make the transition between internal and external water sources, and assemble hose lines, nozzles, valves, and appliances.

5.2.3 Produce a foam fire stream, given foam-producing equipment, so that properly proportioned foam is provided.

(A) Requisite Knowledge. Proportioning rates and concentrations, equipment assembly procedures, foam system limitations, and manufacturer's specifications.

(B) Requisite Skills. The ability to operate foam proportioning equipment and connect foam stream equipment.

5.2.4 Supply water to fire sprinkler and standpipe systems, given specific system information and a fire department pumper, so that water is supplied to the system at the correct volume and pressure.

(A) Requisite Knowledge. Calculation of pump discharge pressure; hose layouts; location of fire department connection; alternative supply procedures if fire department connection is not usable; operating principles of sprinkler systems as defined in NFPA 13, NFPA 13D, and NFPA 13R; fire department operations in sprinklered properties as defined in NFPA 13E; and operating principles of standpipe systems as defined in NFPA 14.

(B) Requisite Skills. The ability to position a fire department pumper to operate at a fire hydrant and at a static water source, power transfer from vehicle engine to pump, draft, operate pumper pressure control systems, operate the volume/pressure transfer valve (multistage pumps only), operate auxiliary cooling systems, make the transition between internal and external water sources, and assemble hose line, nozzles, valves, and appliances.

Chapter 10 Mobile Water Supply Apparatus

10.1 General. The job performance requirements defined in Sections 10.1 and 10.2 shall be met prior to qualifying as a fire department driver/operator—mobile water supply apparatus.

10.1.1 Perform routine tests, inspections, and servicing functions specified in the following list, in addition to those specified in 4.2.1, given a fire department mobile water supply

apparatus, and policies and procedures of the jurisdiction, so that the operational readiness of the mobile water supply apparatus is verified:
(1) Water tank and other extinguishing agent levels (if applicable)
(2) Pumping system (if applicable)
(3) Rapid dump system (if applicable)
(4) Foam system (if applicable)

(A) Requisite Knowledge. Manufacturer's specifications and requirements, and policies and procedures of the jurisdiction.

(B) Requisite Skills. The ability to use hand tools, recognize system problems, and correct any deficiency noted according to policies and procedures.

10.2 Operations.

10.2.1* Maneuver and position a mobile water supply apparatus at a water shuttle fill site, given a fill site location and one or more supply hose, so that the apparatus is correctly positioned, supply hose are attached to the intake connections without having to stretch additional hose, and no objects are struck at the fill site.

(A) Requisite Knowledge. Local procedures for establishing a water shuttle fill site, method for marking the stopping position of the apparatus, and location of the water tank intakes on the apparatus.

(B) Requisite Skills. The ability to determine a correct position for the apparatus, maneuver apparatus into that position, and avoid obstacles to operations.

10.2.2* Maneuver and position a mobile water supply apparatus at a water shuttle dump site, given a dump site and a portable water tank, so that all of the water being discharged from the apparatus enters the portable tank and no objects are struck at the dump site.

(A) Requisite Knowledge. Local procedures for operating a water shuttle dump site and location of the water tank discharges on the apparatus.

(B) Requisite Skills. The ability to determine a correct position for the apparatus, maneuver apparatus into that position, avoid obstacles to operations, and operate the fire pump or rapid water dump system.

10.2.3* Establish a water shuttle dump site, given two or more portable water tanks, low-level strainers, water transfer equipment, fire hose, and a fire apparatus equipped with a fire pump, so that the tank being drafted from is kept full at all times, the tank being dumped into is emptied first, and the water is transferred efficiently from one tank to the next.

(A) Requisite Knowledge. Local procedures for establishing a water shuttle dump site and principles of water transfer between multiple portable water tanks.

(B) Requisite Skills. The ability to deploy portable water tanks, connect and operate water transfer equipment, and connect a strainer and suction hose to the fire pump.

APPENDIX B
NFPA® 1002 Correlation Guide

Chapter 4: General Requirements

Objectives	Corresponding Chapter	Corresponding Pages
4.1	1	1–360
4.2	6	128–148
4.2.1	6	130–148
4.2.1(A)	6	129–130
4.2.1(B)	6	130–148
4.2.2	6	130–148
4.2.2(A)	6	130–148
4.2.2(B)	6	130–148
4.3	7	155–185
4.3.1	8	193–214
4.3.1(A)	8	193–214
4.3.1(B)	7, 8	156–158, 166–168, 193–214
4.3.2	7	170–180
4.3.2(A)	7	170–185
4.3.2(B)	7	170–185
4.3.3	7	170–180
4.3.3(A)	7	170–185
4.3.3(B)	7	166–185
4.3.4	7	175, 177, 178
4.3.4(A)	7	166–185
4.3.4(B)	7	166–185
4.3.5	7	177, 179
4.3.5(A)	7	166–185
4.3.5(B)	7	166–185
4.3.6	8	194–199
4.3.6(A)	8	194–199
4.3.6(B)	8	194–199
4.3.7	9	220–250
4.3.7(A)	9	220–250
4.3.7(B)	9	237–250

Chapter 5: Apparatus Equipped with Fire Pump

Objectives	Corresponding Chapter	Corresponding Pages
5.1	1	1–360
5.1.1	6, 10, 11, 12	258–260, 298–323, 330–358
5.1.1(A)	6, 10, 12	258–260, 330–358
5.1.1(B)	6, 10, 12	257–292, 330–358
5.2	9	220–250
5.2.1	4, 9, 10	59–102, 220–250, 283–292
5.2.1(A)	4, 9, 10	63–102, 232–250, 283–292
5.2.1(B)	4, 9, 10	232–250, 283–292
5.2.2	10	275–283
5.2.2(A)	10	257–292
5.2.2(B)	10	275–292
5.2.3	11	298–323
5.2.3(A)	11	298–323
5.2.3(B)	11	298–323
5.2.4	9	232–250
5.2.4(A)	4, 9	60–102, 232–250
5.2.4(B)	9, 10	232–250, 275–292

Chapter 10: Mobile Water Supply Apparatus

Objectives	Corresponding Chapter	Corresponding Pages
10.1	10	257–292
10.1.1	10	258–263
10.1.1(A)	10	258–263
10.1.1(B)	10	258–263
10.2	10	257–292
10.2.1	10	275–292
10.2.1(A)	10	275–292
10.2.1(B)	10	275–292
10.2.2	10	275–292
10.2.2(A)	10	275–292
10.2.2(B)	10	275–292
10.2.3	10	283–292
10.2.3(A)	10	283–292
10.2.3(B)	10	283–292

Fire Protection Hydraulics and Water Supply (FESHE) Correlation Guide

APPENDIX C

Fire Protection Hydraulics and Water Supply

FESHE Objective	Corresponding Chapter	Corresponding Pages
Apply the application of mathematics and physics to movement of water in fire suppression activities.	4, 10	59–102, 257–285, 263–264
Identify the design principles of fire service pumping apparatus.	2	18–29
Analyze community fire flow demand criteria.	3	34–41
Demonstrate, through problem solving, a thorough understanding of the principles of forces that affect water at rest and, in motion.	4, 10	59–102, 257–285, 263–264
List and describe the various types of water distribution systems.	3	36–38
Discuss the various types of fire pumps.	2	18–29

Glossary

activity area Area of the incident scene where the work activity takes place; it may be stationary or may move as work progresses.

adapter Any device that allows fire hose couplings to be safely interconnected with couplings of different sizes, threads, or mating surfaces, or that allows fire hose couplings to be safely connected to other appliances.

adjustable-gallonage fog nozzle A nozzle that allows the driver/operator to select a desired flow from several settings.

advance warning area The section of highway where driver/operators are informed about an upcoming situation ahead.

aeration The process of introducing air into the foam solution, which expands and finishes the foam.

aerial device An aerial ladder, elevating platform, aerial ladder platform, or water tower that is designed to position personnel, handle materials, provide continuous egress, or discharge water. (NFPA 1901)

aerial fire apparatus A vehicle equipped with an aerial ladder, elevating platform, or water tower that is designed and equipped to support firefighting and rescue operations by positioning personnel, handling materials, providing continuous egress, or discharging water at positions elevated from the ground. (NFPA 1901)

aerial ladder A self-supporting, turntable-mounted, power-operated ladder of two or more sections permanently attached to a self-propelled automotive fire apparatus and designed to provide a continuous egress route from an elevated position to the ground. (NFPA 1904)

air pressure gauges Gauges that identify the air pressure stored in the tanks or reservoirs of an apparatus equipped with a pneumatic braking system.

ambient air temperature The temperature of the surrounding medium. It typically refers to the temperature of the air in which a structure is situated or a device operates.

ambulance A vehicle designed, equipped, and operated for the treatment and transport of ill and injured persons. (NFPA 450)

anti-siphon hole A very small hole in the fitting on top of the primer oil reservoir or in the top of the loop on the lubrication line leading to the priming pump.

apparatus inspection form A document that identifies who performed the inspection and when the fire apparatus inspection was performed, identifies any equipment that is damaged and/or repaired, and details other preventive maintenance procedures performed on the apparatus.

aqueous Pertaining to, related to, similar to, or dissolved in water.

aqueous film-forming foam (AFFF) A concentrated aqueous solution of fluorinated surfactant(s) and foam stabilizers that is capable of producing an aqueous fluorocarbon film on the surface of hydrocarbon fuels to suppress vaporization. (NFPA 11)

around-the-pump proportioning (AP) system A fire apparatus–mounted foam system that diverts a portion of the water pump's output from the discharge side of the pump and sends it through an eductor to discharge foam.

articulated booms An aerial device consisting of two or more folding boom sections whose extension and retraction modes are accomplished by adjusting the angle of the knuckle joints.

atmospheric pressure Pressure caused by the weight of the atmosphere; equal to 14.7 psi (101 kPa) at sea level.

attack hose (attack line) Hose designed to be used by trained fire fighters and fire brigade members to combat fires beyond the incipient stage. (NFPA 1961)

attack pumper An engine from which attack lines have been pulled.

authority having jurisdiction (AHJ) An organization, office, or individual responsible for enforcing the requirements of a code or standard, or for approving equipment, materials, an installation, or a procedure.

automatic-adjusting fog nozzle A nozzle that can deliver a wide range of water stream flows. It operates by means of an internal spring-loaded piston.

auxiliary appliance A standpipe and/or sprinkler system.

balanced-pressure system A fire apparatus–mounted foam system that uses a diaphragm-type pressure control valve to sense and balance the pressures in the foam concentrate and water lines to the proportioner.

ball valve A type of valve used on nozzles, gated wyes, and engine discharge gates. It consists of a ball with a hole in the middle of the ball.

bankshot (bankdown) method A method that applies the foam stream onto a nearby object, such as a wall, instead of directly at the fire.

batch mixing Pouring foam concentrate directly into the fire apparatus water tank, thereby mixing a large amount of foam at one time.

battery selector switch A switch used to disconnect all electrical power to the vehicle, thereby preventing discharge of the battery while the vehicle is not in use.

blind spots Areas around the fire apparatus that are not visible to the driver/operator.

booster hose (booster line) A non-collapsible hose that is used under positive pressure and that consists of an elastomeric or thermoplastic tube, a braided or spiraled reinforcement, and an outer protective cover. (NFPA 1962)

booster line A rigid hose that is ¾" (2 cm) or 1" (2.5 cm) in diameter. Such a hose delivers water at a rate of only 30 to 60 GPM (113 to 227 liters), but can do so at high pressures.

booster pump A water pump mounted on the fire apparatus in addition to a fire pump and used for firefighting either in conjunction with or independent of the fire pump.

brake fade Reduction in stopping power that can occur after repeated application of the brakes, especially in high-load or high-speed conditions.

braking distance The distance that the fire apparatus travels from the time the brakes are activated until the fire apparatus makes a complete stop.

Bresnan distributor nozzle A nozzle that can be placed in confined spaces. The nozzle spins, spreading water over a large area.

British thermal unit (Btu) The quantity of heat required to raise the temperature of one pound of water by 1°F at the pressure of 1 atmosphere and a temperature of 60°F. A British thermal unit is equal to 1055 joules, 1.055 kilojoules, and 252.15 calories. (NFPA 921)

bucket brigade An early effort at fire protection that used leather buckets filled with water to combat fires. In many communities, residents were required to place a bucket filled with water typically on their front steps at night in case of a fire in the community.

buffer space Lateral and/or longitudinal area that separates traffic flow from a work space or an unsafe area; it might also provide some recovery space for an errant vehicle.

burnback resistance The ability of a foam blanket to resist direct flame impingement.

butterfly valve A valve found on the large pump intake valve where the hard or soft suction hose connects to it.

bypass eductor A foam eductor that is mounted to the pump and can be used for either water application or foam application.

cavitation A condition caused by attempting to move water faster than it is being supplied to the pump.

cellar nozzle A nozzle used to fight fires in cellars and other inaccessible places.

centrifugal force The outward force that is exerted away from the center of rotation or the tendency for objects to be pulled outward when rotating around a center.

centrifugal pump A pump in which the pressure is developed principally by the action of centrifugal force. (NFPA 20)

chart A document summarizing the typical handline and master stream calculations for the hoses, nozzles, and devices specific to the fire apparatus. It lists the most commonly used or reasonably expected hose lays.

chassis The basic operating motor vehicle, including the engine, frame, and other essential structural and mechanical parts, but exclusive of the body and all appurtenances for the accommodation of driver, property, passengers, appliances, or equipment related to other than control. Common usage might, but need not, include a cab (or cowl). (NFPA 1901)

chemical foam Foam produced by mixing powders and water, where a chemical reaction of the materials creates the foam.

chemical wagon A small truck constructed with a large soda acid extinguisher and 50 feet (15 meters) or more of small hose (booster line size).

Class A foam Foam intended for use on Class A fires. (NFPA 1150)

Class B foam Foam intended for use on Class B fires. (NFPA 1901)

Code 1 response Response in a fire apparatus in which no emergency lights or sirens are activated.

Code 2 response Response in a fire apparatus in which only the emergency lights are activated; no audible devices are activated.

Code 3 response Response in a fire apparatus in which both the emergency lights and the sirens are activated.

coefficient (C) A numerical measure that is a constant for a specified hose diameter.

collapse zone An area encompassing a distance of 1½ times the height of a building, in which fire fighters and fire apparatus must not be located in case of a building collapse.

combustible liquid Any liquid that has a closed-cup flash point at or above 100°F (37.8°C) as determined by the test procedures and apparatus set forth in Section 4.4. Combustible liquids are classified according to Section 4.3. (NFPA 30)

command vehicle A vehicle that the chief uses to respond to the fire scene.

communications center A building or portion of a building that is specifically configured for the primary purpose of providing emergency communications services or public safety answering point (PSAP) services to one or more public safety agencies under the authority or authorities having jurisdiction. (NFPA 1221)

compressed-air foam system (CAFS) A foam system that combines compressed air and a foam solution to create firefighting foam.

concentration The percentage of foam concentrate contained in a foam solution. (NFPA 11)

condensed Q method Fire-ground method used to quickly calculate friction loss in 3" (76-mm) to 5" (127-mm) hose lines.

control zones A series of areas at hazardous materials incidents that are designated based on safety concerns and the degree of hazard present.

corner safe areas Areas outside a building where two walls intersect; these areas are less likely to receive any damage during a building collapse.

critical rate of flow The essential flow, measured in gallons per minute, that is needed to overcome the heat generated by fire.

critical speed Maximum speed that a fire apparatus can safely travel around a curve.

dependable lift The height that a column of water can be lifted in a quantity considered sufficient to provide reliable fire flow.

dip stick A graduated instrument for measuring the depth or amount of fluid in a container, such as the level of oil in a crankcase.

discharge header The piping and valves on the discharge side of the pump.

discharge pressure The water pressure on the discharge manifold of the fire pump at the point of gauge attachment.

discharge side The side of a pump where water is discharged from the pump.

dispatch To send out emergency response resources promptly to an address or incident location for a specific purpose. (NFPA 450)

distributors Relatively small-diameter underground pipes that deliver water to local users within a neighborhood.

double-acting piston pump A positive-displacement pump that discharges water on both the upward and downward strokes of the piston.

double-female adapter A hose adapter that is used to join two male hose couplings.

double-male adapter A hose adapter that is used to join two female hose couplings.

draft The pressure differential that causes water flow. (NFPA 54)

driver/operator A person having satisfactorily completed the requirements of driver/operator as specified in NFPA 1002, *Standard for Fire Apparatus Driver/Operator Professional Qualifications.* (NFPA 1521)

dry-barrel hydrant The most common type of hydrant; it has a control valve below the frost line between the footpiece and the barrel. A drain is located at the bottom of the barrel above the control valve seat for proper drainage after operation. (NFPA 25)

dry hydrant An arrangement of pipes that is permanently connected to a water source other than a piped, pressurized water supply system, provides a ready means of water supply for firefighting purposes, and utilizes the drafting (suction) capability of fire department pumpers. (NFPA 1144)

due regard The care exercised by a reasonably prudent person under the same circumstances.

dump line A small-diameter hose line that remains in the open position and flowing water during the entire drafting operation.

dump site A location where a fire apparatus can offload the water in its tank.

dump valve A large opening from the water tank of a mobile water supply apparatus for unloading purposes. (NFPA 1901)

duplex gauge A gauge at the pump panel that simultaneously monitors both the foam concentrate and water pressures.

eductor A device that uses the Venturi principle to siphon a liquid in a water stream. The pressure at the throat is below atmospheric pressure, allowing liquid at atmospheric pressure to flow into the water stream. (NFPA 11)

elevated water storage tower An above-ground water storage tank that is designed to maintain pressure on a water distribution system.

elevation gain The pressure gained when the nozzle is below the pump; it requires pressure to be subtracted from the discharge pressure.

elevation loss The pressure lost when the nozzle is above the pump; it requires pressure to be added to the discharge pressure to compensate for the loss.

elevation pressure The amount of pressure created by gravity.

elevation pressure (mathematical) In hydraulic calculations, this is the distance the nozzle is above or below the pump.

emitter A device that emits a visible flashing light at a specified frequency, thereby activating the receiver on a traffic signal.

exhaust primer A means of priming a centrifugal pump by using the fire apparatus's exhaust to create a vacuum and draw air and water from the pump.

extractor exhaust system A system used inside the fire apparatus bay that connects to the fire apparatus tailpipe and draws its exhaust outside the building.

failure A cessation of proper functioning or performance.

fill site A location where the fire apparatus can get water tanks filled.

film-forming fluoroprotein foam (FFFP) A protein-based foam concentrate incorporating fluorinated surfactants, which forms a foam capable of producing a vapor-suppressing, aqueous film on the surface of hydrocarbon fuels. This foam might show an acceptable level of compatibility to dry chemicals and might be suitable for use with those agents. (NFPA 412)

finished foam The homogeneous blanket of foam obtained by mixing water, foam concentrate, and air.

fire apparatus inspection An evaluation of the fire apparatus and its equipment that is intended to ensure its safe operation.

fire barrier A continuous membrane or a membrane with discontinuities created by protected openings with a specified fire protection rating, where such membrane is designed and constructed with a specified fire resistance rating to limit the spread of fire, that also restricts the movement of smoke. (NFPA 101)

fire department connection (FDC) A connection through which the fire department can pump supplemental water into the sprinkler system, standpipe, or other system, thereby furnishing water for fire extinguishment to supplement existing water supplies. (NFPA 25)

fire hook A tool used to pull down burning structures.

fire hose appliance A piece of hardware (excluding nozzles) generally intended for connection to fire hose to control or convey water. (NFPA 1965)

fire plug A valve installed to control water accessed from wooden pipes.

fire pump A water pump with a rated capacity of 250 GPM (1000 L/min) or greater at 150 psi (10 bar) net pump pressure that is mounted on an fire apparatus and used for firefighting. (NFPA 1901)

fire pump (fire apparatus) A device that provides for liquid flow and pressure and that is dedicated to fire protection. (NFPA 20)

fire-ground hydraulics Simpler fire-ground methods for performing hydraulic calculations.

fixed-gallonage fog nozzle A nozzle that delivers a set number of gallons per minute as per the nozzle's design, no matter what pressure is applied to the nozzle.

flammable liquid Any liquid that has a closed-cup flash point below 100°F (37.8°C), as determined by the test procedures and apparatus set forth in Section 4.4, and a Reid vapor pressure that does not exceed an absolute pressure of 40 psi (276 kPa) at 100°F (37.8°C), as determined by ASTM D 323, *Standard Test Method for Vapor Pressure of Petroleum Products (Reid Method)*. Flammable liquids are classified according to Section 4.3. (NFPA 30)

flow meter A device for measuring volumetric flow rates of gases and liquids. (NFPA 99)

flow pressure The amount of pressure created by moving water.

flow rate The volume of water moving through the nozzle measured in gallons per minute or liters per minute.

fluoroprotein foam (FP) A protein-based foam concentrate to which fluorochemical surfactants have been added. The resulting product has a measurable degree of compatibility with dry chemical extinguishing agents and an increase in tolerance to contamination by fuel. (NFPA 402)

foam A stable aggregation of small bubbles of lower density than oil or water, which exhibits a tenacity for covering horizontal surfaces. Air foam is made by mixing air into a water solution containing a foam concentrate, by means of suitably designed equipment. It flows freely over a burning liquid surface and forms a tough, air-excluding, continuous blanket that seals volatile combustible vapors from access to air. It resists disruption from wind and draft or heat and flame attack and is capable of resealing in case of mechanical rupture. Firefighting foams retain these properties for relatively long periods of time. Foams also are defined by expansion and are arbitrarily subdivided into three ranges of: (1) low-expansion foam—expansion up to a ratio of 20:1; (2) medium-expansion foam—expansion in the range of 20:1 to 200:1; and (3) high-expansion foam—expansion in the range of 200:1 to 1000:1. (NFPA 11)

foam concentrate (foam liquid) A concentrated liquid foaming agent as received from the manufacturer. (NFPA 11)

foam heat exchanger A device that uses water as its cooling source and prevents the foam concentrate from overheating.

foam proportioner A device or method to add foam concentrate to water so as to make foam solution. (NFPA 1901)

foam solution A homogeneous mixture of water and foam concentrate in the proper proportions. (NFPA 11)

foam tetrahedron A geometric shape used to illustrate the four elements needed to produce finished foam: foam concentrate, water, air, and mechanical aeration.

fog-stream nozzle A nozzle that is placed at the end of a fire hose and separates water into fine droplets to aid in heat absorption.

forward lay A method of laying a supply line where the line starts at the water source and ends at the attack pumper.

four-way hydrant valve A specialized type of valve that can be placed on a hydrant and is used in conjunction with a pumper to increase water pressure in relaying operations.

friction loss The result of the interaction between the hose or pipe and the flowing water, which leads to a reduction in flow pressure at the point of discharge.

front mount An apparatus pump that is permanently mounted to the front bumper of the apparatus and is directly connected to the motor.

fuel gauge A gauge that indicates the amount of fuel in the fire apparatus's fuel tank.

fuel pick-up The absorption of the burning fuel into the foam itself.

fuel resistance A foam's ability to minimize fuel pick-up.

gate valve A type of valve found on hydrants and sprinkler systems.

gated wye A valved device that splits a single hose into two separate hoses, allowing each hose to be turned on and off independently.

global positioning system (GPS) A satellite-based radio navigation system consisting of three segments: space, control, and user. (NFPA 414)

gravity-feed system A water distribution system that depends on gravity to provide the required pressure. The system storage is usually located at a higher elevation than the end users of the water.

hand lay A process in which the engine is positioned close to the fire scene and the driver/operator deploys the supply hose from the bed of the fire apparatus to the hydrant either alone or with very little assistance from other fire fighters.

hand pump A piston-driven, positive-displacement pump that is pushed up and down by fire fighters manning poles at the side of the pump.

handline nozzle A nozzle with a rated discharge of less than 350 GPM (1325 L/min). (NFPA 1964)

hard suction hose A hose used for drafting water from static supplies (e.g., lakes, rivers, wells). It can also be used for supplying pumpers from a hydrant if designed for that purpose. The hose contains a semi-rigid or rigid reinforcement designed to prevent collapse of the hose under vacuum. (NFPA 1963)

hardness The mineral content of the water; it usually consists of calcium and magnesium, but can also include iron, aluminum, and manganese.

heat of vaporization The energy required to transform a given quantity of a substance into a gas.

high-expansion foam Any foam with an expansion ratio in the range of 200:1 to approximately 1000:1. (NFPA 1145)

hose clamp A device used to compress a fire hose so as to stop water flow.

hose jacket A device used to stop a leak in a fire hose or to join hoses that have damaged couplings.

hydraulics Study of the characteristics and movement of water as they pertain to calculations for fire streams and fire-ground operations.

hydrocarbon A chemical substance consisting of only hydrogen and carbon atoms. (NFPA 36)

hydrolyzed Decomposition of a chemical compound by reaction with water.

ignition switch A switch that engages operational power to the chassis of a motor vehicle.

impeller A metal rotating component that transfers energy from the fire apparatus's motor to discharge the incoming water from a pump.

impeller vanes Sections that divide the impeller.

induction The use of an eductor to introduce a proportionate amount of foam concentrate into a stream of water flowing from a discharge.

initial attack fire apparatus Fire apparatus with a fire pump of at least 250 GPM (1000 L/min) capacity, water tank, and hose body, whose primary purpose is to initiate a fire suppression attack on structural, vehicular, or vegetation fires, and to support associated fire department operations. (NFPA 1901)

injection system A foam system installed on a fire apparatus that meters out foam by pumping or injecting it into the fire stream.

intake pressure The pressure on the intake passageway of the pump at the point of gauge attachment.

intake side The side of a pump where water enters the pump.

intermediate traffic incident A traffic incident that affects the lanes of travel for 30 minutes to 2 hours.

jet siphon Appliance that connects a 1½" or 1¾" hose to a hard-sided hose coupling and creates a pathway for flowing water that is being forced through the hose by Venturi forces.

job aid A tool, device, or system used to assist a person with executing specific tasks.

kinetic energy The energy possessed by an object as a result of its motion.

knockdown speed and flow The time required for a foam blanket to spread out across a fuel surface.

large-diameter hose (LDH) A hose of 3½ inches (90 mm) size or larger. (NFPA 1961)

Level I staging Initial staging of fire apparatus in which three or more units are dispatched to an emergency incident.

Level II staging Placement of all reserve resources in a central location until requested to the scene.

Glossary

lift The vertical height that water must be raised during a drafting operation, measured from the surface of a static source of water to the center line of the pump intake. (NFPA 1911)

liquid surge The force imposed upon a fire apparatus by the contents of a partially filled water or foam concentrate tank when the vehicle is accelerated, decelerated, or turned. (NFPA 1002)

low-expansion foam Any foam with an expansion ratio up to 20:1. (NFPA 1145)

low-volume nozzle Nozzle that flows 40 GPM (150 L/min) or less.

major traffic incident A traffic incident that involves a fatal crash, a multiple-vehicle incident, a hazardous materials incident on the highway, or other disaster.

manifold A large-diameter hose line that is split into four or five medium-size hose lines.

master stream A portable or fixed firefighting appliance supplied by either hose lines or fixed piping that has the capability of flowing in excess of 300 GPM (1140 L/min) of water or water-based extinguishing agent. (NFPA 600)

master stream device A large-capacity nozzle supplied by two or more hose lines of fixed piping that can flow 300 GPM (1135 L/min). These devices include deck guns and portable ground monitors.

master stream nozzle A nozzle with a rated discharge of 350 GPM (1325 L/min) or greater. (NFPA 1964)

mechanical foam Foam produced by physical agitation of a mixture of water, air, and a foaming agent.

mechanical generator An electrically powered fan used to aerate certain types of foams.

medium-diameter hose (MDH) Hose with a diameter of 2½″ (64 mm) or 3″ (76 mm).

medium-expansion foam Any foam with an expansion ratio in the range of 20:1 to 200:1. (NFPA 1145)

medium-size hose Hose that is 2½″ or 3″ in diameter.

metering device A device that controls the flow of foam concentrate into the eductor.

minor traffic incident A traffic incident that involves a minor crash and/or disabled vehicles.

miscellaneous equipment Portable tools and equipment carried on a fire apparatus, not including suction hose, fire hose, ground ladders, fixed power sources, hose reels, cord reels, breathing air systems, or other major equipment or components specified by the purchaser to be permanently mounted on the apparatus as received from the apparatus manufacturer. (NFPA 1901)

miscible Readily mixes with water.

mobile data terminal (MDT) A computer that is located on the fire apparatus.

mobile foam fire apparatus Fire apparatus with a permanently mounted fire pump, foam proportioning system, and foam concentrate tank(s) whose primary purpose is for use in the control and extinguishment of flammable and combustible liquid fires in storage tanks and other flammable liquid spills. (NFPA 1901)

mobile water supply apparatus A vehicle designed primarily for transporting (pickup, transporting, and delivering) water to fire emergency scenes to be applied by other vehicles or pumping equipment. Also known as a tanker or tender. (NFPA 1901)

motor vehicle accident (MVA) An incident that involves one vehicle colliding with another vehicle or another object and that may result in injury, property damage, and possibly death.

municipal water system A water distribution system that is designed to deliver potable water to end users for domestic, industrial, and fire protection purposes.

net pump discharge pressure (NPDP) Amount of pressure created by the pump after receiving pressure from a hydrant or another pump.

net pump pressure The sum of the discharge pressure and the suction lift converted to units of pounds per square inch (psi) or kilopascals (kPa) when pumping at draft, or the difference between the discharge pressure and the intake pressure when pumping at a hydrant or other source of water under positive pressure. (NFPA 1901)

normal operating pressure The observed static pressure in a water distribution system during a period of normal demand.

nozzle A constricting appliance attached to the end of a fire hose or monitor to increase the water velocity and form a stream. (NFPA 1965)

nozzle pressure (NP) Pressure required at the inlet of a nozzle to produce the desired water discharge characteristics. (NFPA 14)

nozzle shut-off A device that enables the fire fighter at the nozzle to start or stop the flow of water.

nurse tanker operation The process of supplying water to an attack engine directly by a supply engine.

oil pressure gauge A gauge that identifies the pressure of the lubricating oil in the fire apparatus's engine.

oleophobic Oil hating; having the ability to shed hydrocarbon liquids.

oxygenated To treat, combine, or infuse with oxygen.

parallel/volume mode Positioning of a two-stage pump in which each impeller takes water in from the same intake area and discharges it to the same discharge area. This mode is used when pumping water at the rated capacity of the pump.

parking brake The main brake that prevents the fire apparatus from moving even when it is turned off and there is no one operating it.

performance tests Tests conducted after a fire apparatus has been put into service to determine if its performance meets predetermined specifications or standards.

pick-up tube The tube from an in-line foam eductor that is placed in the foam bucket and draws foam concentrate into the eductor using the Venturi principle.

piercing nozzle A nozzle that can be driven through sheet metal or other material to deliver a water stream to that area.

piston intake valve (PIV) A large appliance that connects directly to the pump's intake and controls the amount of water that flows from a pressurized water source into the pump.

piston pump A positive-displacement pump that operates in an up-and-down action.

Pitot gauge A type of gauge that is used to measure the velocity pressure of water that is being discharged from an opening. It is used to determine the flow of water from a hydrant.

polar solvent A combustible liquid that mixes with the water contained in foam. Such solvents include non-petroleum-based fuels such as alcohol, lacquers, acetone, and acids.

polymer A naturally occurring or synthetic compound consisting of large molecules made up of a linked series of repeated simple monomers.

portable floating pump A small fire pump that is equipped with a flotation device and built-in strainer, capable of being carried by hand and flowing various quantities of water depending on its size.

portable master stream device A master stream device that may be removed from a fire apparatus, typically to be placed in service on the ground.

portable pump A type of pump that is typically carried by hand by two or more fire fighters to a water source and used to pump water from that source.

portable tank Any closed vessel having a liquid capacity in excess of 60 gal (227 L), but less than 1000 gal (3,785 L), and not intended for fixed installation. (NFPA 30)

positive-displacement pump A pump that is characterized by a method of producing flow by capturing a specific volume of fluid per pump revolution and reducing the fluid void by a mechanical means to displace the pumping fluid. (NFPA 20)

potential energy The energy that an object has stored up as a result of its position or condition. A raised weight and a coiled spring have potential energy.

power steering system A system for reducing the steering effort on vehicles in which an external power source assists in turning the vehicle's wheels.

power take-off unit A direct means of powering a pump with the fire apparatus's transmission and through a shaft directed to the gear case on the pump.

preincident plan A written document resulting from the gathering of general and detailed data to be used by responding personnel for determining the resources and actions necessary to mitigate anticipated emergencies at a specific facility. (NFPA 1620)

prepiped elevated master stream An aerial ladder with a fixed waterway attached to the underside of the ladder, with a water inlet at the base supplying a master stream device at the tip.

pressure-regulating valve (PRV) The type of valve found in a multistory building, which is designed to limit the pressure at a discharge so as to prevent excessive elevation pressures under both flowing (residual) and nonflowing (static) conditions.

preventive maintenance program A program designed to ensure that apparatus are capable of functioning as required and are maintained in working order.

primary feeder The largest-diameter pipes in a water distribution system, which carry the largest amounts of water.

prime the pump To expel all air from a pump.

primer oil reservoir A small tank that holds the lubricant for the priming pump.

priming The process of removing air from a fire pump and replacing it with water.

private water system A privately owned water system that operates separately from the municipal water system.

proportioned A combination of foam concentrate and water used to form a foam solution.

protein foam Foam created from a concentrate that consists primarily of products from a protein hydrolysate, plus stabilizing additives and inhibitors to protect against freezing, to prevent corrosion of equipment and containers, to resist bacterial decomposition, to control viscosity, and to otherwise ensure readiness for use under emergency conditions. These concentrates are diluted with water to form 3 percent to 6 percent solutions depending on the type. These concentrates are compatible with certain dry chemicals. (NFPA 11)

prove-out sequence A series of checks that an electrical system completes to ensure that all of the systems are functioning properly before the fire apparatus is started.

pump discharge pressure (PDP) Pressure measured at the pump discharge needed to overcome friction and elevation loss while maintaining the desired nozzle pressure and delivering an adequate fire stream.

pumper Fire apparatus with a permanently mounted fire pump of at least 750 GPM (3000 L/min) capacity, water tank, and hose body, whose primary purpose is to combat structural and associated fires. (NFPA 1901)

pumping element A rotating device such as a gear or vane encased in the pump casing of a rotary pump.

pump-mounted eductor An in-line eductor dedicated to the production of foam.

quint Fire apparatus with a permanently mounted fire pump, a water tank, a hose storage area, an aerial ladder or elevating platform with a permanently mounted waterway, and a complement of ground ladders. (NFPA 1901)

radiator cap The pressure cap that is screwed onto the top of the radiator, and through which coolant is typically added.

raindown method A method of applying foam that directs the stream into the air above the fire and allows it to gently fall onto the surface.

rated capacity (water pump) The flow rate to which the pump manufacturer certifies compliance of the pump when it is new. (NFPA 1901)

ratio controller A device required for each foam outlet to proportion the correct amount of foam concentrate into the water stream over a wide range of flows and with minimal pressure loss.

reaction distance The distance that the fire apparatus travels after the driver/operator recognizes the hazard, removes his or her foot from the accelerator, and applies the brakes.

receiver A device placed on or near a traffic signal to recognize a signal from the emitter on an emergency vehicle and preempt the normal cycle of the traffic light.

reducer A device that connects two hoses with different couplings or threads together.

relay pumping operation The process of moving water from a water source through hose to the place where it will be needed.

relay valve An appliance placed in a hose lay that allows an attack engine to connect into a relay pumper operation without interrupting the flow of water.

reservoir A water storage facility.

residual pressure The pressure that exists in the distribution system, measured at the residual hydrant at the time the flow readings are taken at the flow hydrant. (NFPA 291)

reverse lay A method of laying a supply line where the line starts at the attack pumper and ends at the water source.

rotary gear pump A positive-displacement pump that uses two gears encased inside a pump casing to move water under pressure.

rotary pump A positive-displacement pump that operates in a circular motion.

rotary vane pump A positive-displacement pump characterized by the use of a single rotor with vanes that move with pump rotation to create a void and displace liquid. (NFPA 20)

rotor A metal device that houses the vanes in a rotary vane pump.

secondary feeder A smaller-diameter pipe that connects the primary feeder to the distributors.

series/pressure mode Positioning of a two-stage pump in which the first impeller sends water into the second impeller's intake side and water is then discharged out the pump's discharge header, thereby creating more pressure with less flow.

service tests See *performance tests*.

shut-off valve Any valve that can be used to shut down water flow to a water user or system.

siamese connection A device that allows two hoses to be connected together and flow into a single hose.

single-acting piston pump A positive-displacement pump that discharges water only on the downward stroke of the piston.

single-stage pump A pump that has one impeller that takes the water in and discharges it out using only one impeller.

small-diameter hose (SDH) Hose with a diameter ranging from 1″ (25 mm) to 2″ (51 mm).

smooth-bore nozzle A nozzle that produces a straight stream that consists of a solid column of water.

smooth-bore tip A nozzle device that is a smooth tube and is used to deliver a solid column of water.

soft suction hose A large-diameter hose that is designed to be connected to the large port on a hydrant (steamer connection) and into the engine.

specific heat index The amount of heat energy required to raise the temperature of a substance. These values are determined experimentally and then made available in tabular form.

split hose bed A hose bed arranged such that supply line can be laid out, or such that two supply lines can be laid out.

split hose lay A scenario in which the attack pumper lays a supply line from an intersection to the fire, and then the supply engine lays a supply line from the hose left by the attack pumper to the water source.

split lay A scenario in which the attack engine lays a supply line from an intersection to the fire, and the supply engine lays a supply line from the hose left by the attack engine to the water source.

spotter A person who guides the driver/operator into the appropriate position while operating in a confined space or in reverse mode.

staging A specific function whereby resources are assembled in an area at or near the incident scene to await instructions or assignments. (NFPA 1561)

staging area A prearranged, strategically placed area, where support response personnel, vehicles, and other equipment can be held in an organized state of readiness for use during an emergency. (NFPA 424)

staging area manager The person responsible for maintaining the operations of the staging area.

standpipe The vertical portion of the system piping that delivers the water supply for hose connections, and sprinklers on combined systems, vertically from floor to floor in a multistory building. Alternatively, the horizontal portion of the system piping that delivers the water supply for two or more hose connections, and sprinklers on combined systems, on a single level of a building. (NFPA 14)

standpipe riser The vertical portion of the system piping within a building that delivers the water supply for fire hose connections, and sprinklers on combined systems, vertically from floor to floor in a multistory building.

standpipe system An arrangement of piping, valves, hose connections, and allied equipment with the hose connections located in such a manner that water can be discharged in streams or spray patterns through attached hose and nozzles, for the purpose of extinguishing a fire and so protecting designated buildings, structures, or property in addition to providing occupant protection as required. (NFPA 14)

starter switch The switch that engages the starter motor for cranking.

static pressure The pressure in a water pipe when there is no water flowing.

static water source The pressure that exists at a given point under normal distribution system conditions measured at the residual hydrant with no hydrants flowing. (NFPA 291)

stressors Conditions that create excessive physical and mental pressures on a person's body; a stimulus that causes stress.

supply hose (supply line) Hose designed for the purpose of moving water between a pressurized water source and a pump that is supplying attack lines. (NFPA 1961)

surface tension The elastic-like force at the surface of a liquid, which tends to minimize the surface area, causing drops to form. (NPFA 1150)

sweep (roll-on) method A method of applying foam that involves sweeping the stream just in front of the target.

tactical benchmarks Objectives that are required to be completed during the operational phase of an incident.

termination area The area where the normal flow of traffic resumes after a traffic incident.

theoretical hydraulics Scientific or more exact fire-ground calculations.

threaded hose coupling A type of coupling that requires a male fitting and a female fitting to be screwed together.

total pressure loss (TPL) Combination of friction loss, elevation loss, and appliance loss.

total stopping distance The distance that it takes for the driver/operator to recognize a hazard, process the need to stop the fire apparatus, apply the brakes, and then come to a complete stop.

tower ladders A device that provides fire fighters with a solid platform at the end of a boom.

traffic control The direction or management of vehicle traffic such that scene safety is maintained and rescue operations can proceed without interruption. (NFPA 1006)

traffic incident A natural disaster or other unplanned event that affects or impedes the normal flow of traffic.

traffic incident management area (TIMA) An area of highway where temporary traffic controls are imposed by authorized officials in response to an accident, natural disaster, hazardous materials spill, or other unplanned incident.

traffic signal preemption system A system that allows the normal operation of a traffic signal to be changed so as to assist emergency vehicles in responding to an emergency.

traffic space The portion of the highway where traffic is routed through the activity area of a traffic incident.

transfer case A gear box that transfers power, thereby enabling a fire apparatus's motor to operate a pump.

transfer valve An internal valve in a multistage pump that enables the user to change the mode of operation to either series/pressure or parallel/volume.

transition area The area where vehicles are redirected from their normal path and where lane changes and closures are made in a traffic incident.

triple-combination pumper A truck that carries water in a tank generally used for the booster line and commonly called the booster tank.

turbidity The amount of particulate matter suspended in water.

universal solvent A liquid substance capable of dissolving other substances, which does not result in the first substance changing its state when forming the solution.

vacuum Any pressure less than atmospheric pressure.

vane A small, moveable, self-adjusting element inside a rotary vane pump.

vehicle dynamics Vehicle construction and mechanical design characteristics that directly affect the handling, stability, maneuverability, functionality, and safeness of a vehicle.

vehicle intercom system A communication system that is permanently mounted inside the cab of the fire apparatus to allow fire fighters to communicate more effectively.

Venturi effect The creation of a low-pressure area in a chamber so as to allow air and water to be drawn in.

viscosity A measure of the resistance of a liquid to flow. (NFPA 1150)

voltage meter A device that registers the voltage of a battery system.

voltmeter A device that measures the voltage across a battery's terminals and gives an indication of the electrical condition of the battery.

volume The quantity of water flowing; usually measured in gallons (liters) per minute.

volunteer fire department A fire department in which volunteer emergency service personnel account for 85 percent or more of its department membership. (NFPA 1720)

volute The part of the pump casing that gradually decreases in area, thereby creating pressure on the discharge side of a pump.

water curtain nozzle A nozzle used to deliver a flat screen of water so as to form a protective sheet of water.

water hammer The surge of pressure caused when a high-velocity flow of water is abruptly shut off. The pressure exerted by the flowing water against the closed system can be seven or more times that of the static pressure. (NFPA 1962)

water main A generic term for any underground water pipe.

water pressure The application of force by one object against another. When water is forced through the distribution system, it creates water pressure.

water shuttle The process of moving water by using fire apparatus from a water source to a location where it can be used.

water supply A source of water that provides the flows (L/min) and pressures (bar) required by a water-based fire protection system. (NFPA 25)

water thief A device that has a 2½″ (64-mm) inlet and a 2½″ (64 mm) outlet in addition to two 1½″ (38 mm) outlets. It is used to supply many hoses from one source.

wet-barrel hydrant A hydrant used in areas that are not susceptible to freezing. The barrel of this type of hydrant is normally filled with water.

wye A device used to split a single hose into two separate lines.

Index

NOTE: Page numbers with italicized *f*, *t*, or *sd* indicate figures, tables, or skill drills respectively.

A

Activity areas, 209
Adapters
    characteristics, 225–226, 226*f*
    coming apart from LDH threaded connection, 96
    jet siphon, 290
    laying out hose with threaded couplings and, 234
    for relay pumping operations, 277
    for various thread sizes of hydrants, 3
Address of scene, 164, 164*f*, 199–200, 199*f*
Adjustable-gallonage fog nozzles, 230
Advance warning areas, 209
Aeration, in foam solutions, 299
Aerial devices
    fire apparatus with, 130*f*
    improvements in, 6, 6*f*
    inspection of, 145–146, 145*f*, 146*f*
    overhead obstructions and, 205, 205*f*
    preventive maintenance for, 129
Aerial fire apparatus, 25–27, 25*f*, 26*f*, 26*t*, 27*t*, 201
Aerial ladders, 25. *See also* Ladders
AFFF. *See* Aqueous film-forming foam
Aherns-Fox fire apparatus, 113, 113*f*
Air brakes. *See also* Brakes
    inspection of, 139–140, 139*f*
Air pressure gauges, 166–167
Air supply units, 6
Air-aspirating foam nozzles, 317
Albuquerque Fire Department, fire boxes for, 164
Alcohol-resistant aqueous film-forming foam (AR-AFFF), 302*t*, 303–304, 304*f*, 305
Alcohol-resistant film-forming fluoroprotein foams (AR-FFFP), 303
Alcohol-resistant foams, 299
Alley dock exercise, 212–213, 214*sd*
Alley lay, 235–236, 235*f*
Alternative automotive fuels, alcohol-resistant foams for incidents with, 304
Amber emergency lights, 208
Ambient air temperature, 331
Ambulances, positioning at fire scene, 206
American LaFrance, 25
Angulo, Raul A., 143
Anti-siphon hole, 258, 260, 260*f*
Apartment complex addresses, 199, 199*f*
Apparatus inspection form, 132*f*, 133, 146*sd*, 147*sd*
Appliances
    auxiliary, 205
    examples of, 73*f*
    friction losses and, calculation of, 72–75
    for high-flow devices, friction loss calculations due to, 73*t*
Approaching the scene, 199–200, 199*f*
Aqueous film-forming foam (AFFF), 298, 302*t*, 303, 303*f*
Aqueous substances, 35
AR-AFFF. *See* Alcohol-resistant aqueous film-forming foam
AR-FFFP, 303
Around-the-pump (AP) foam proportioning system, 309, 311, 312*sd*
Articulated booms, 6
Aspirating foam nozzles, 313, 317
Atmospheric pressure, 257, 258*f*
Attack hose/lines, 220, 221, 221*f*, 222
Attack line evolutions, 232
Attack pumpers, 233, 276, 276*f*, 279–280
Authority having jurisdiction (AHJ), 332
Automatic-adjusting fog nozzles, 230
Auxiliary appliances, 205

B

Backing procedures, 180–182, 181–182*f*, 183*f*, 184*sd*
Balanced-pressure foam proportioning systems, 311, 313
Ball, blow-up, for portable tank operations, 291
Ball valves, 227, 227*f*
Baltimore, Great Fire of (1904), 3
Bankshot (bankdown) method for applying foam, 321–322, 322*sd*
Barrel strainers, 267, 267*f*, 268
Batch mixing, of foam, 306, 306*f*, 307*sd*
Batteries, 137, 141, 141*f*, 183
Battery cables, safe removal of, 137, 138*f*
Battery selector switch, 166
Belts, inspection of, 137
Bernoulli's equation, 39
Bleeder valve, on attack pumper intake valve, 280, 280*f*
Blind spots, 158, 195. *See also* Backing procedures
Blue emergency lights, 208
Boat launching ramps, as drafting sites, 265
Booster hose/lines, 5, 220, 222
Booster pumps, 5–6, 5*f*
Booster tanks, 264
Bottoming out, chassis clearance and, 161
Brake fade, 194
Brake pedal test, 139
Brakes. *See also* Parking brake
    auxiliary, 195
    failure on large single axle rescue squad, 353
    faulty, near collision due to, 144
    inspection of, 139–140, 139*f*, 140*f*
    skidding and, 197

Index

Braking distance, 197
Breathing air system. *See also* Self-contained breathing apparatus
    aerial device, inspection of, 145, 146
Bresnan distributor nozzles, 232, 232f
Bridges
    as drafting sites, 265
    as height obstacles, 196
British thermal units (Btu), 60
Broken-stream nozzles, 61–62
Bucket brigades, 2–3
Bucking traffic, laws and policies on, 194
Buffer space, 209
Building collapse incidents, 213
Burnback resistance, 302
Burst hose jackets, 226, 226f
Butterfly valves, 227, 227f
Bypass eductors, 308

C

C. *See* Coefficients
Cab interior, 111, 111f, 137–139, 138f, 139f
Cab procedures, at fire ground, 236–237
Cable systems, aerial device, inspection of, 145
Calculated flow relays, 279
Calculators, hydraulic, 98
Camera, rear-mounted, backing up fire apparatus and, 182, 183f
Capacity test/150 psi test, 350, 351sd
Cascade units, 6
Cavitation, 273–274, 341
Cellar nozzles, 232, 232f
Centrifugal force, 196
Centrifugal pump(s)
    inspection of, 141, 142
    operational characteristics, 114, 116, 116f, 117f, 119–120
    priming, 110, 270
Certificate of Automotive Fire Apparatus Examination and Test (UL), 335f
Changeover operations, 245–247, 247f, 248sd
Charts, 97–98, 97f, 97t
Chassis, fire apparatus, 20, 161
Chemical foams, 299
Chemical wagons, 5
Clapper valve, 242, 245f
Class A foams, 301–302, 301–302f, 319, 320sd, 322
Class B foams
    applying, 319, 321–322, 321sd, 322sd, 323sd
    characteristics, 302–303
    compatibility of, 322
    types and properties of, 302t
Cleaning, 135, 162, 162f
Clear emergency lights, 208

Clearance, diminishing, exercise on, 177
Clothing. *See also* Turnout pants and boots
    reflective vests, 208
Code 1 response, 194
Code 2 response, 194
Code 3 response, 194
Coefficients (C)
    friction loss, 65, 65t
    for Siamese lines, 65t, 78, 79, 82, 83sd, 87t
    for standpipe risers, 91, 91t
Cold zones, at emergency medical scenes, 213
Collapse zone, apparatus positioning and, 203–204
Color-coding supply hose, 277
Colors, for fire hydrants, 46t
Combination load, 235
Combination nozzles, pressure ratings for, 61, 61f
Combustible and flammable liquid incidents, 298
Command vehicles, positioning at fire scene, 206
Communications center, 163
Communications system. *See also* Radio communication
    aerial device, inspection of, 145, 146, 146f
    for backing fire apparatus, 181, 182
    for filling tankers, 287
    for notifying on starting drafting water flow, 271
    for nurse tanker operations, 292
    vehicle intercom system, 158
Compartment doors, 166, 169, 169f, 180
Compressed-air foam system (CAFS), 314, 314f, 315sd, 318
Compression brakes, 195
Compressors, 6
Computer systems, battery cable removal and, 137, 138f
Concentration, of foam concentrate in foam solution, 302–303
Condensed Q method, 100–101, 101sd
Confined-space turnarounds, 175, 177, 178sd
Constant pressure relays, 279
Control valves, for hydrants, 285, 285f
Control zones, at emergency medical scenes, 213
Coolant level, inspection of, 136–137, 137f
Corner safe areas, apparatus positioning and, 204
Crew members. *See also* Spotters
    for apparatus inspections, 130, 130f
    backing procedures and, 180–182
    educating, 157–158
    fire apparatus public appearances and, 169
    hose loading operations in motion and, 169
    map reading by, 164
    monitoring fill site using a hydrant, 285
    notifying on starting drafting water flow, 271
    for performance tests, 358
    for relay pumping operations, 277
    responsibilities, 156

Crew members (*Cont.*)
   safe equipment handling and, 156–157
   trust and team building among, 158
Critical rate of flow, 60
Critical speed, for curves, 196
Curved roadways, driving on, 196, 196*f*

D

Daily Apparatus Checks, 143. *See also* Fire apparatus inspections
Daily engine inspection sheet, 132*f*, 160*f*, 258, 259*f*. *See also* Fire apparatus inspections
Daily fire apparatus inspections. *See* Fire apparatus inspections
Dalmatian dogs, 4
Dammed streams, as water source, 266, 266*f*
Dead-end water mains, 38
Defensive driving practices, 196–198
Dependable lift, 258
Dettman, Todd, 347
Diesel-powered fire apparatus, 9, 9*f*
Diminishing clearance exercise, 177, 179*sd*
Dip stick, 131, 136
Directional arrows, fire apparatus, traffic control and, 208
Dirt and grease, fire apparatus inspections and, 134
Discharge capacity of fire pump, effect of lift on, 258*t*
Discharge header, 116
Discharge pressure
   for drafting operations, 270, 270*f*, 271, 271*sd*, 272*sd*
   fire pump performance and, 349
   gauge meter test of, 338, 340*sd*
   for a standpipe, 90–91, 92*sd*, 93*sd*
Discharge side, of fire pumps, 113
Discharge valve(s)
   closed, operating foam system issues with, 311
   filling hose line from drafting operation and, 271–272, 273, 273*sd*
   gauge and flow meter test and, 338
   pump, inspection of, 141
   of standpipe risers, 89, 90*f*
Discharges, inspection of, 141
Dispatch, 163–164
Distance from vehicle ahead, 197
Distributors, 38
Double-acting piston pumps, 113
Double-female adapters, 226, 226*f*, 234
Double-male adapters, 226, 226*f*, 234
Drafting water. *See also* Static water sources
   complications during, 273–275
   establishing pumping operations, 266–268, 267–268*f*, 269*sd*
   mechanics of, 257–258
   operational considerations for site, 266
   positioning fire apparatus for, 268, 269*sd*, 270

   priming the pump for, 270
   producing the water flow, 271–273
   in rural areas, 257
   selecting site for, 263–264, 264*f*
   in shallow water, 270
   special accessibility considerations, 265–266, 265*f*, 266*f*
   starting, and producing discharge pressures for, 270–271
Driver/operators. *See also* Training exercises
   backing procedures for, 180–182, 181–182*f*, 184*sd*
   duties on scene, 249, 249–250*sd*
   educating crew members, 157–158
   emergency vehicle laws and, 193–194
   emergency vehicle response, 163–164, 163*f*, 164*f*, 166, 167*sd*
   fire apparatus and equipment functions and limitations, 161–162
   fire apparatus and equipment inspections, 162, 162*f*
   getting underway, 169–170, 169*f*, 180
   leading by example, 159, 161
   maintaining safe work environment, 158–159
   protecting, at the scene, 9
   returning to the station, 180
   roles and responsibilities, 2, 155–156
   safety role of, 10, 10*f*, 156–157, 162–163, 168–169
   selection, 11–12
   shutting down the fire apparatus, 183–184, 183*f*, 185*sd*
   starting the fire apparatus, 166–168, 168*f*, 169*f*, 171–173*sd*
   teams of experienced and inexperienced, 60
   trust and team building by, 158
   visibility for, 9, 10*f*
   Voice of Experience on adapting to the situation by, 347
   Voice of Experience on emergency response by, 165
   Voice of Experience on staying calm, 115
Driving exercises. *See also* Driving to fire ground
   confined-space turnarounds, 175, 177, 178*sd*
   diminishing clearance exercise, 177, 178*sd*
   importance of, 170, 175
   serpentine maneuvers, 175, 176*sd*, 177
Driving tankers, safety considerations for, 287
Driving to fire ground
   approaching the scene, 199–200, 199*f*
   emergency vehicle laws, 193–194
   intersections, 198
   safe driving practices, 194–198, 195*f*, 196*f*
Dry hydrants, 49, 51*f*, 265, 265*f*
Dry standpipe systems, 241
Dry-barrel fire hydrants, 41, 41*f*
Dual air brake system warning light and buzzer test, 140
Due regard, 161, 177, 194
Dump lines, 270–271, 281
Dump sites
   offloading tankers for, 289–290

as static water source, 287
traffic flow within, 291, 291f
using portable tanks at, 289, 289f
for water shuttle, 283
Dump valves, 289, 289f, 290
Dunn, Vincent, 204
Duplex gauge, 313, 313f

E

Eductors, 306, 308, 308f, 309, 310sd
Electromagnetic retarders, 195
Electronic Fire Commander (EFC), 102
Electronic pump controllers, 102
Elevated master streams, 83, 87, 90sd, 91sd. *See also* Prepiped elevated master streams
Elevated water storage towers, 37, 38f
Elevated water streams, 5
Elevation gain, 70
Elevation loss calculations, 70–71
Elevation pressure (EP), 39, 70–71, 71sd, 72sd
Emergence response, 162–163
Emergency communications, 158. *See also* Communications system
Emergency lights, 198, 206, 208, 236
Emergency medical care, seat belts and, 168
Emergency scene. *See also* Motor vehicle accidents
medical, positioning at, 212–213
with no specific address, 199–200, 200f
securing, on interstate under construction, 118–119
Emergency shutdown feature, 184
Emergency vehicle response
dispatch, 163–164
maps, 164, 164f
360-degree inspection for, 166, 167sd, 180
Emergency vehicle technicians, 162, 162f
Emergency warning equipment inspection, 139
Emitter, for traffic signal preemption, 198
Engine compartment inspection, 136–137, 136f, 137f
Engine oil level inspection, 131, 136, 136f
Engine speed test, no-load governed, 334–335
Engine warm-up, 169–170
Engines' and ladders' positioning. *See also* Fire apparatus positioning
exposures and, 203–204
fire conditions and, 204
rescue potential and, 203
setup position for, 203
water supply and, 204
EP. *See* Elevation pressure
Equipment. *See also* Fire equipment
for fire pump performance tests, 333–334f

Exhaust brakes, 195
Exhaust extractor system, 134, 170, 183
Exhaust primer, 114
Exhaust system inspection, 135
Exposure protection, of apparatus at fire scenes, 203
Exterior fire apparatus inspections, 134–136
Exterior functional control switches, inspection of, 138–139
Extractor exhaust system, 134, 170, 183

F

Failure, performance, 331
Fatal injuries to U.S. fire fighters, causes (2005) of, 206, 206t
FDCs. *See* Fire department connections
Female connections, 242. *See also* Double-female adapters
50 percent test, 352, 354
Fill sites, 283, 285–286. *See also* Drafting water
Filling tankers, 286–287
Film-forming fluoroprotein foam (FFFP), 302–303, 302t
Finished foam, 299, 300
Fire apparatus. *See also* Driving to fire ground; Emergency vehicle response; Fire apparatus inspections; Fire apparatus positioning; Fire equipment
aerial, 25–27, 25f, 26f, 26t, 27t
with aerial device, 130f
Aherns-Fox, 113, 113f
auxiliary cooling system for, 249sd
backing procedures, 180–182
changeover operations and, 246–247, 247f, 248sd
congestion, staging and, 202
distance to hydrant from, 240
driving with turnout pants and boots, 170
emergency shutdown feature, 184
emergency type and handling of, 162–163
engine speed, drafting water and, 272, 273
engine temperature during drafting operations, 273
evolution of, 2–6, 3–6f, 9, 9f
functions and limitations, 161–162
initial attack, 21, 24, 24t
mobile foam, 28–29, 29f, 29t
mobile water supply, 24–25, 24f, 25t
modern, 10–11
in motion, hose loading and, 168–169
parking, emergency vehicle laws on, 194
performance test requirements, 331
positioning at fire scene, 200–201
positioning for drafting water, 267
preventive maintenance for, 129
pumpers, 20–21, 20f, 20, 21t
purchasing requirements, 18–19, 19f
quint, 27, 27f, 27t, 28t
safety considerations, 9–10, 10f

Fire apparatus (Cont.)
    shutting down, 183–184, 183f, 185sd
    special service, 27–28, 28f, 28t
    specialized, 205–206
    starting, 166–168, 168f
    static water source accessibility for, 264–265, 285–286
    stopping considerations, 197
    traffic control devices on, 210
    water on, 19–20
    with water tank, 20
Fire apparatus inspections
    aerial devices, 145–146, 145f, 146f
    brakes, 139–140
    cab interior, 137–139
    completing forms for, 148
    daily engine inspection sheet, 132f
    driver/operators and, 162, 162f
    engine compartment, 136–137, 136f, 137f
    exterior, 134–136
    fire apparatus sections, 134–142
    general tools and equipment, 140–141, 140f
    organization of process, 133
    positioning for, 133f
    priming system, 258
    procedures for, 130–131
    process for, 133–134
    pumps, 141–142, 141f, 142f, 145
    review previous report on, 133, 146sd, 147sd
    safety considerations, 148, 148f
    in sections, 144–146
    steps ins, 146–147sd
    360-degree, 166, 167sd
    tires, 159
    Voice of Experience on, 143
    weekly, monthly, or other period inspection items, 148
Fire apparatus positioning. *See also* Engines' and ladders' positioning
    alley dock exercise, 212–213, 214sd
    collapse zone and, 203–204
    at emergency medical scenes, 212–213
    engines and ladders, 203–205, 205f
    evaluating on approach, 200–201
    at intersections or on highways, 206, 208
    at motor vehicle accidents, 210, 212
    near railroad crossings, 212
    other apparatus, 205–206
    staging, 201–202, 201f, 202f
    traffic incidents and, 208–210
    Voice of Experience on, 207
    wearing reflective vests, 208, 209
Fire barrier, Class A foam as, 302

Fire boxes, for Albuquerque Fire Department, 164
Fire crew. *See* Crew members
Fire department(s)
    on aerial apparatus inspections, 145
    on apparatus inspections, 130–131, 133–134
    emergency vehicle rules and regulations of, 193–194
    volunteer, preventive maintenance programs, 129–130
Fire department connections (FDCs), 200f. *See also* Master stream device
    connecting supply hose lines to, 243, 246sd
    locating, 242, 243f
    preincident plans on, 205
    Siamese connections of, 78
    standpipe operations and, 87–91
Fire department engines. *See also* Fire apparatus
    connecting to a water supply, 236
Fire equipment. *See also* Equipment
    on aerial fire apparatus, 26t, 27t
    evolution of, 2–6, 9
    functions and limitations, 161–162
    for initial attack fire apparatus, 24, 24t
    inspection of, 140–141, 140f, 162, 162f
    on mobile foam fire apparatus, 29t
    for mobile water supply apparatus, 25, 25t
    modern, 10–11
    for pumpers, 21, 21t
    purchasing requirements, safety and, 157
    on the quint, 27t, 28t
    safe retrieving and restoring of, 156–157, 157f
    safe storage of, 148, 148f
    secured, getting underway and, 169, 169f
    on special service fire apparatus, 28t
Fire extinguishers, 159, 298, 299, 301f
Fire hooks, 3
Fire hose. *See* Hoses
Fire hose appliances, 223. *See also* Siamese connections; Valves
    adapters, 3, 96, 225–226, 226f, 234, 277
    hose clamps, 227, 227f
    hose jackets, 226, 226f
    reducers, 226, 226f
    water thiefs, 225, 225f
    wyes, 223, 225, 225f
Fire hose evolutions, 232–236, 233–235f
Fire hydrants
    adapters for varying thread sizes of, 3
    charging line from, 234
    colors for, 46t
    connecting Storz coupling soft suction hose to a pump from, 223, 224sd
    control valves for, 285, 285f
    distance to fire apparatus from, 240

dry, 49, 51*f*, 265, 265*f*
as fill site(s), 285
four-way valves for, 227, 233, 233*f*
inspecting and maintaining, 44, 46, 46*f*, 47*f*
locations, 42, 44, 239
operating source pumpers from, 279
operation of, 41–42, 43*sd*
positioning fire apparatus for soft suction hose to, 240
screw-type hydrant valve for, 285*f*
shutting down, 42, 44*sd*
testing, 46–47, 47*f*, 48*sd*, 49
wet- or dry-barrel, 41, 41*f*, 42*f*
Fire officers, 134, 156
Fire plugs, 3. *See also* Fire hydrants
Fire pumps. *See also* Centrifugal pump(s); Hand pumps; Portable pumps; Primer/priming pump; Priming; *under* Pump
capacity ratings, 258
connecting Storz coupling soft suction hose from hydrant to, 223, 224*sd*
of different capacities, test data for, 348*t*
discharging foam solution and, 309
disengaging, 249–250*sd*
drafting preparation, 271
engaging, at fire ground, 236–237, 238*sd*
engine control interlock test, 336, 338, 339*sd*
first mechanized, 4
flushing/refilling after drafting operations, 264
gauge and flow meter test, 338, 341
hose suction size, number of suction lines, and lift for, 348, 349*t*
inspection of, 141–142, 141*f*, 142*f*, 145
interlock system, 338*f*
lift effect on discharge capacity of, 258*t*
municipal water distribution, 37
NFPA 20 Standard on, 111–112
NFPA 1901 definition of, 19
NFPA 1901 requirements for, 20
performance test form, 356, 356*f*
performance tests, 348–350, 349*f*, 349*t*, 350*f*, 351*sd*, 352, 354
power supplies for, 120–122, 120*f*, 121*f*
preventive maintenance, 129, 129*f*
priming two-stage, 270
rated capacity, 330
re-rating, 357–358
shift controls, 336*f*
shift indicator test, 335–336*sd*, 337*sd*
speed reading for, 348, 349*f*
temperature during drafting operations of, 273
test requirements, 344, 348, 348*t*
types, 113–114, 113–114*f*, 116–117, 116–117*f*, 119–120
vacuum test of, 260–261, 262*sd*
water conditions and maintenance of, 36

Fire service hydraulic calculations, 93–95
Fire station
backing into, 180–182, 181–182*f*, 184*sd*
pulling out of, 170, 180
returning to, 180
Fire-ground hydraulics, 60, 93–95. *See also* Mathematics
Fires
early protection in American cities, 3–4, 3–4*f*
One Meridian Plaza, Philadelphia (1991), 92
petroleum-fueled, 299, 302
vehicle, 212, 212*f*
Fixed straight-stream nozzles, 4
Fixed-gallonage fog nozzles, 230
FL. *See* Friction loss
Flashlights, backing up fire apparatus and, 181
Floating pumps, 266
portable, 286
Floating strainers, 268, 268*f*, 286
Flow meters (flow minders), 101–102, 338, 341, 341*f*, 342*sd*
Flow pressure, 39, 47, 47*f*, 49. *See also* Nozzle pressure; Pump-discharge-pressure
Flow rate, 62, 64*sd*
Fluid levels, inspection of, 131, 136, 136*f*, 137*f*
Fluoroprotein foam (FP), 298, 302–303, 302*t*
Foam. *See also* Foam proportioning systems
applying, 319, 320*sd*, 321–322, 321*sd*, 322*sd*
characteristics, 298, 300–301
classifications, 301–304, 301–304*f*, 302*t*
compatibility, 322
expansion rates, 305
foam concentrates, 305, 305*t*
history of, 298–299
method of operation, 299, 300*f*
nozzles for, 317
overview of, 299
storage supplies for, 317, 319, 319*f*
tetrahedron components, 299–300, 300*f*
Voice of Experience on efficient use of, 316
Foam blanket, 300*f*, 301, 301*f*, 302, 302*f*
Foam concentrate pump, 311, 313
Foam concentrates, 299, 300, 304, 305, 305*t*, 306
Foam heat exchanger, 313
Foam proportioning systems
around-the-pump (AP), 309, 311, 312*sd*
balanced-pressure, 311, 313
batch mixing, 306, 306*f*, 307*sd*
compressed-air foam, 314, 314*f*, 315*sd*
eductors, 308, 308*f*
injection, 313–314
for mechanical foam production, 298
metering devices, 308–309, 308*f*

Foam proportioning systems (*Cont.*)
   operating in-line eductors, 309, 310*sd*
   premixing, 306, 308
Foam solution, 299
Foam supply tank inspection, 141
Foam tetrahedron, 299–300, 300*f*
Fog distributor nozzles, 61–62, 62*f*
Fog nozzles, 61–62, 61*f*, 302, 303*f*, 317
Fog-stream nozzles, 228, 230, 230–231*sd*, 230*f*
Forward (hose) lay, 233, 233*f*, 277, 277*f*
Four-way hydrant valves, 227, 233, 233*f*
FP (fluoroprotein foam), 298, 302–303, 302*t*
Friction loss (FL)
   in appliances, calculations of, 72–75, 75*sd*
   calculating, 65
   calculators for, 98
   coefficients, standard or metric, 65*t*
   condensed Q calculation, 100–101
   GPM flowing calculation, 100
   hand method calculations, 98–99, 98*sd*, 99*sd*
   in-line gauges for testing, 77*sd*
   in multiple hose lines of different lengths and sizes, 66–67, 68*sd*, 69*sd*, 70
   Pitot gauges for measurement of, 76*sd*
   relay pumping calculations, 278–279
   in relay pumping operations, 277
   in single hose lines, 66*sd*, 67*sd*
   standard equation for, 63–64
   in standpipe operations, 91
   subtract 10 calculation, 100
   total pressure loss and, 76–77
   water flow and, 39
   in wyed hose lines, 77, 78*sd*
Front mount pumps, 120–121, 120*f*
Fuel cap inspection, 136
Fuel gauge, checking during starting procedure, 166
Fuel level
   inspection of, 139
   for power tools, 141
Fuel pick-up, foam and, 301
Fuel resistance, of foam, 300–301

G

Garrity, Charles, 115
Gasoline-powered fire apparatus, 5, 5*f*
Gate valves, 227, 227*f*
Gated ball valves, 241
Gated wyes, 73*f*, 223, 225, 225*f*. *See also* Wyes
Global positioning system (GPS), 164
GPM (gallons per minute) flowing method, 100, 100*sd*

Gravity-feed water distribution systems, 37, 37*f*
Ground ladders. *See also* Ladders
   on fire apparatus, 24, 25–26, 27, 28
   straight, for drafting operations, 268, 268*f*

H

Hand lays, 239, 241, 242*sd*
Hand method, for calculating FL and PDP, 98–99, 100*sd*
Hand pumps, 3, 4, 4*f*, 112
Hand signals, for backing fire apparatus, 181, 181–182*f*
Handline nozzles, 228
Handline selection, 93
Hard suction hose, 223, 223*f*, 267, 268, 270, 290
Hardness, of water, 36
Hayes, Daniel, 25
Hazardous materials incidents, 213
Hazards, recognizing potential, 200
Headlights, fire apparatus, 170, 180
Hearing protection, vacuum test of priming system and, 344
Heat of vaporization, 36
Heavy rescue vehicles, 6
Height restrictions, for fire apparatus, 161
High-expansion foam generators, 317
High-expansion foams, 304, 304*f*, 305
High-pressure auxiliary pumps, 112
High-rise structures, 70–71, 70*f*, 92
Highways. *See also under* Traffic
   positioning fire apparatus on, 206, 208, 210, 212
Hills. *See also* Slopes
   driving up or down, 194
   positioning at fire scene on, 204
Hook and ladder, 4–5, 4*f*
Hook and ladder truck, 26*f*
Horses, for crew transportation, 3*f*
Hose carts, 3
Hose clamps, 227, 227*f*
Hose jackets, 226, 226*f*
Hoses. *See also* Fire hose appliances; Fire hose evolutions; Friction loss; Large-diameter hose; Medium-diameter hose
   access to, positioning at fire scene and, 201
   evolution of, 4, 5
   for filling tankers, 286–287
   functions of, 220
   hand methods for calculating FL, 98–99, 100*sd*
   for initial attack fire apparatus, 24
   in-line pressure gauges for testing, 74*f*
   loading, while fire apparatus is in motion, 168–169
   for mobile water supply apparatus, 25
   for pumpers, 21
   for quints, 28*t*

size of, 220, 221f
suction, 20
Hot zones, at emergency medical scenes, 213
Hydrants. *See* Fire hydrants
Hydraulic calculators, 98
Hydraulic system, aerial device, inspection of, 145
Hydraulics/hydraulic calculations. *See also* Mathematics
 definition of, 60
 Voice of Experience on, 86
Hydrocarbon, combustible fuels, 302
Hydrolyzed keratin protein, 302

I

Ignition switch, 166, 183
Impeller, centrifugal pump, 116, 116f
Impeller vanes, 116, 116f
Incident commander (IC)
 cavitations and, 274
 Class B foam application and, 319
 drafting operations and, 263
 on duration of drafting operations, 272–273
 on emergency scene positioning, 213
 on fire apparatus placement, 203, 204, 205
 on fire scene positioning, 200–201
 on relay pumping operations, 275, 276, 282, 283
 on staging, 201–202
 on tanker operations, 292
 traffic incidents and, 209
 on water shuttle operations, 287, 292
 water supply monitoring and, 291
Incident Management System (IMS), 261, 263, 292
Incident space, 209
Incontrol (electronic pump controller), 102
Induction, 308
Initial attack fire apparatus, 21, 24, 24t
Injectors/injection systems, 306, 313–314
In-line eductors, 308, 308f, 309, 310sd
In-line pressure gauges, 73–75, 74f, 77sd, 90
Inspections. *See also* Fire apparatus inspections
 fire equipment, 140–141, 140f
 work-safe environment and, 159
Insurance Services Office (ISO), 330
Intake pressure, 273, 349
Intake relief valve system test, 335
Intake side, of fire pumps, 113
Intakes, 141, 142, 274
Interior functional control switches, inspection of, 138
Intermediate traffic incidents, 209
Intersections, 198, 206, 208, 210
Inventory, fire apparatus, 162

J

Jet dump/dumping, 290
Jet siphon, 290, 290f
Job aids for inspections, 159, 160f

K

Kinetic energy, 39
Knockdown speed and flow, of foam, 300

L

L (length of hose), friction loss and, 65, 66–67, 68sd, 69sd, 70
Ladder companies, setup position for, 203
Ladder hook truck, 4f
Ladder wagons pulled by horses, 4–5
Ladders. *See also* Ground ladders; Tower ladders
 aerial, 5, 5f, 25
 for drafting operations, 268, 268f, 286
 early use in fire protection, 3
 fire apparatus positioning and, 201
 for initial attack fire apparatus, 24
 for pumpers, 21
Large-diameter hose (LDH)
 characteristics, 220
 to establish a fill site using a hydrant, 285
 hand lays using, 239, 239f
 for relay pumping operations, 276–277
 as supply hose, 204, 221
Leaks, checking for, 141, 141f
Leather fire buckets, 2–3, 3f
Left turns, blind spots and, 195
Level I staging, 202, 202f
Level II staging, 202, 202f, 206
Lift, height of, from water source, 265
Lights, fire apparatus. *See also* Emergency lights
 inspection of, 130, 139, 139f
Lines. *See* Hoses
Liquid surge, driving fire apparatus and, 197
Long pipe nozzles, 4
Low-expansion foams, 305
Low-level strainers, 267, 267f, 268, 291
Low-volume nozzles, 228

M

Major traffic incidents, 209
Manifolds, 72, 73f, 277–278, 281, 281f
Manual on Uniform Traffic Devices for Streets and Highways (MUTCD), 208–210, 212
Manual water drain valve, for air brake system, 140, 140f
Manufacturers of fire apparatus, 18–19, 19f, 130
Maps, 164, 164f. *See also* Address of scene

Master stream devices. *See also* Fire department connections
   medium-diameter hose for, 220, 222f
   nozzles for, 228
   Siamese connections for, 225
   smooth-bore nozzles for, 229
Master stream foam nozzles, 317
Master stream nozzles, 228
Master streams
   definition of, 73
   elevated, 83, 87, 90sd, 91sd
   portable devices for, 78
   prepiped elevated, 83, 87, 87f, 88sd, 89sd
   smooth-bore nozzles for, 61, 62, 62f
Mathematics
   appliance loss, 72–75, 73f, 75sd
   charts, 97–98, 97f, 97t
   chunking calculations for understanding, 75
   condensed Q calculation, 101sd
   elevated master streams, 83, 87, 90sd, 91sd
   elevation pressure, 70–71, 71sd, 72sd
   fire service hydraulic calculations, 93–95
   friction loss equation, 64–65
   friction loss in Siamese hose lines, 78–79, 82–83, 82sd, 83sd, 84sd, 85sd
   friction loss in single hose lines, 65, 66sd, 67sd
   friction loss in wyed hose lines, 77, 78sd, 79sd, 80sd, 81sd
   frictions loss in multiple hose lines of different sizes and lengths, 66–67, 68sd, 69sd, 70
   hand method, 98–99, 99sd, 100sd
   hydraulic calculators, 98
   introduction to, 59–60
   net pump discharge pressure, 93
   nozzle flow, 62–63, 63t
   nozzle pressures, 61–62, 61f, 62f
   PDP for standpipe systems, 94sd, 95sd
   prepiped elevated master streams, 83, 87, 88sd, 89sd
   preplanning discharge pressure for a standpipe, 92sd, 93sd
   pressure-regulating valves, 92
   pump-discharge-pressure, 60–61
   quantity of water available estimates, 263–264
   standpipe systems, 87–91, 92sd, 93sd
   subtract 10 method, 100, 100sd
   total pressure loss, 76–77
   Voice of Experience on, 86
McGrail, David, 97
MDTs (mobile data terminals), 163, 163f
Mechanical foams, 299
Mechanical generators, 317, 317f
Medium-diameter hose (MDH), 220, 221f, 276–277, 285, 286–287, 290
Medium-expansion foam generators, 317

Medium-expansions foams, 305
Metric measurements
   coefficients for multiple Siamese lines, 87t
   coefficients for standpipe risers, 91t
   elevation pressure (loss and gain), 72sd
   friction loss coefficients, 65t
   friction loss in multiple hose lines of different sizes and lengths, 69sd
   friction loss in single hose lines, 67sd
   preplanning discharge pressure for a standpipe, 93sd
   of pump-discharge-pressure in prepiped elevated master streams, 89sd
   of pump-discharge-pressure in wye scenario with equal lines, 79sd
   of pump-discharge-pressure in wye scenario with unequal lines, 81sd
   of Siamese hose FL by coefficient method, 83sd
   of Siamese hose FL by percentage method, 85sd
   of Siamese hose FL by split flow method, 82sd
   smooth-bore handline nozzle flow, 64sd
   for smooth-bore nozzles flows, 63t
   of total pressure loss, 76
   of vacuum to water column and pressure, 257t
Mid-ship pump panels, for viewing scene, 9, 10f
Minor traffic incidents, 209
Mirrors, fire apparatus, 139, 166, 168f, 181f
Miscellaneous equipment. *See* Fire equipment
Miscible fuels, 302
Mnemonics, crew training using, 159
Mobile data terminals (MDTs), 163, 163f
Mobile foam fire apparatus, 28–29, 29f, 29t
Mobile water supply apparatus, 24–25, 24f, 25t, 34–35, 35f
Moisture removal system, for air brakes, 140, 140f
Monthly fire apparatus inspections, 148
Motor vehicle accidents (MVA), 159, 210, 212. *See also* Emergency scene
Motorists, reducing vision impairment of, 206
Moving water sources, estimating quantity of, 263–264
Multiple-connection standpipes, 245f
Multistage pumps, 116, 117, 117f, 119–120
Multistory structures. *See* High-rise structures
Municipal water systems
   characteristics and sources for, 36–37
   distribution system, 37–38, 37f, 38f
   as source for water, 34, 34f
   treatment facilities, 37, 37f

N

National Fallen Firefighters Foundation (NFFF), 177
National Incident for Occupational Safety and Health (NIOSH), 180

National Institute of Standards and Technology (NIST), 298
Near Miss report(s)
  apparatus lurches forward during training exercise, 8
  brake failure on large single axle rescue squad, 353
  broken coupling from overpressurized supply line, 23
  downed power lines are energized, 211
  fire fighter falls off back step while laying supply line, 288
  fire stream delays and compromised safety, 174–175
  LDH adapter came apart when charged with water, 96
  loss of water during fire attack, 45
  near collision due to faulty brakes, 144
  problem with compressed-air foam unit, 318
  seatbelts averted fatality en route to medical emergency, 244–245
  securing scene on interstate under construction, 118–119
Net pump discharge pressure (NPDP), 93
Net pump pressure, 349
NFPA 20 (*Standard for the Installation of Stationary Pumps for Fire Protection*), 111–112
NFPA 291 (*Recommended Practice for Fire Flow Testing and Marking of Hydrants*), 46
NFPA 1002 (*Standard for Fire Apparatus Driver/Operator Professional Qualifications*), 2, 11, 175, 182, 212
NFPA 1221 (*Standard for the Installation, Maintenance, and Use of Emergency Services Communications Systems*), 163
NFPA 1403 (*Standard on Live Fire Training Evolutions*), 319
NFPA 1500 (*Standard on Fire Department Occupational Safety and Health Program*), 9, 168
NFPA 1561 (*Standard on Emergency Services Incident Management System*), 201
NFPA 1620 (*Recommended Practice for Pre-incident Planning*), 205
NFPA 1901 (*Standards for Automotive Fire Apparatus*)
  on aerial fire apparatus, 25–27, 26t, 27t
  on engine compartment maintenance, 136
  on fire apparatus equipped with water tanks, 20
  on fire pumps, 19, 260–261
  on initial attack fire apparatus, 21, 24, 24t
  on manufacturer documentation, 130–131
  on mobile foam fire apparatus, 28–29, 29f, 29t
  on mobile water supply apparatus, 24–25, 25t
  on parking apparatus on slopes, 204
  on parking brake during starting procedure, 166
  on pumpers, 21
  on quints, 27, 27t, 28t
  on seat belt warning devices, 9
  on seat belts, 168
  on special service fire apparatus, 27–28, 28f, 28t
  on vehicle height, 196
  on water delivery rates, 20
NFPA 1911 (*Standard for the Inspection, Maintenance, Testing, and Retirement of In-Service Automotive Fire Apparatus*), 145, 330–331, 356, 356f
NFPA 1931 (*Standard for Manufacturer's Design of Fire Department Ground Ladders*), 24
Nighttime
  driving during, 196
  identifying address/location during, 199–200
Nonmoving water sources, 263. See also Static water sources
Normal operating pressure, of water, 39
Notepads, for emergency vehicle response, 164
Nothing showing, apparatus positioning when, 200
Nozzle pressure (NP), 61–62, 61f, 62f. See also Pump-discharge-pressure
Nozzle shut-offs, 228
Nozzles
  description and types of, 228, 232, 232f
  determining flow rate for, 62–63
  fog-stream, 230, 230–231sd, 230f
  for hand pumps or leather hoses, 4
  hose diameter and size of, 63
  for initial attack fire apparatus, 24
  maintenance and inspection, 232
  for mobile water supply apparatus hoses, 25
  for performance tests, 333, 334, 334f
  for pumpers, 21
  for quints, 28t
  smooth-bore, 228–230sd, 228f
NP (nozzle pressure), 61–62, 61f, 62f. See also Pump-discharge-pressure
$NPDP_{draft}$, 93
$NPDP_{pps}$, 93
Nurse tanker operations, 266, 292

O

Oil pressure gauge, 168
Oleophobic quality, of foam, 301
1½″ attack hose, 222
1¾″ attack hose, 222
100 percent test, 350
On-scene operations, 162–163
Operating manuals, 11, 11f, 159, 161
Overhead obstructions, 204–205
Overload test/165 psi test, 350, 352
Overpasses, as height obstacles, 196
Oxygenated fuel additives, 302

P

Paddlewheel flow meters, 101
Paper maps, 164f
Parallel/volume mode, of multistage pumps, 117, 117f, 119

Parked fire apparatus, 201
Parking, 194, 206, 208f, 236, 236f
Parking brake. *See also* Brakes
   dual air brakes and, 140
   pump throttle and, 236
   shutting down the fire apparatus, 183
   spring brake test, 140
   starting the fire apparatus and, 166
   testing, 139
   traffic signal preemption systems and, 198
Pascal's law, 39
PASS mnemonic, 159
Passing other vehicles, 194–195
PDP. *See* Pump-discharge-pressure
Percentage method (calculation), 78, 79, 83, 84sd, 85sd
Performance tests
   capacity/150 psi, 350
   environmental requirements for, 331
   equipment requirements, 333–334, 333–334f
   final results of, 356–357
   fire apparatus requirements for, 331
   fire pump, 348–350, 349f, 349t, 350f
   flow meter, 338, 341, 342sd
   gauge meter, 338, 340sd
   intake relief valve system, 335
   no-load governed engine speed, 334–335, 335f, 336sd
   overload/165 psi, 350, 352
   pressure control, 354
   priming system, 344, 345sd
   problem solving, 357
   procedures after, 356
   pump engine control interlock, 336, 338, 339sd
   pump shift indicator, 335–336, 337sd
   pumping test requirements, 344, 348
   safety monitoring with, 330
   safety procedures during, 358
   site for, 331–333, 332f
   tank-to-pump flow, 341, 342sd
   250 psi (50 percent), 352
   200 psi (70 percent), 352
   vacuum, 341, 344, 345sd
Periodic fire apparatus inspections, 148
Personal protective equipment (PPE), 131, 134, 161, 168, 283
Petroleum-fueled fires, 299, 302
Pick-up tubes, 317
Piercing nozzles, 232, 232f
Piston intake valve (PIV), 142
Piston pumps, 112–114, 113–114f, 120
Pitot gauge(s)
   for fire hydrant testing, 47, 49, 49f, 50sd
   manual or threaded, 74f
   for testing appliance loss, 73–75, 76sd
   and tubes, for performance tests, 334, 334f
Pitot pressure, 47
Polar solvents, 302
Police
   monitoring fill site using a hydrant, 285
   traffic control at MVAs and, 210
Polymer, in AR-FFFP, 303
Pony hose, 239, 239f. *See also* Large-diameter hose
Poole, Todd, 22
Portable floating pumps, 286
Portable master stream devices, 78
Portable pumps
   for accessibility to static water sources, 51, 51f, 265–266, 266f, 286
   positive-displacement pumps as, 112
   for use in remote areas, 120, 120f
Portable tanks
   for drafting operations, 286
   dump sites and, 287
   multiple, use of, 290, 290f
   source pumper considerations with, 290–291
   using, 289, 289f
Positive-displacement pumps, 112–114, 113–114f
Potential energy, 39
Power lines, downed, 211
Power steering system fluid inspection, 137
Power steering system inspection, 135
Power take-off units, 121, 121f
Power tools, inspection of, 141
Preconnect lines, marking pressures for, 71
Preincident plans, 102, 205, 242. *See also* Preplanning
Premixing foam solutions, 306, 308
Prepiped elevated master streams, 83, 87, 87f, 88sd, 89sd
Preplanning, 92sd, 93sd, 265, 283, 287. *See also* Preincident plans
Pressure control test, 354, 355sd, 356
Pressure governor, 246, 247f, 354, 354f, 355sd, 356
Pressure relief valve
   characteristics, 246, 247, 247f
   performance test, 354, 355sd, 356
   setting, 142
Pressure-regulating valves (PRV), 92
Preventive maintenance program, 129, 159
Primary feeders, 38
Prime the pump, 110. *See also* Priming
Primer oil reservoir, 258, 260f
Primer valve, 260
Primer/priming pump
   inspection of, 141, 142, 142f
   lubrication of, 258, 260f
   positive-displacement pumps as, 112

rotary pumps as, 113, 114
vacuum test, 341, 344, 345sd
Priming, 258, 270, 271sd, 272sd, 274–275. *See also* Prime the pump
Priming system test, 341, 344, 345sd
Private water systems, 34, 34f
Procedures. *See* Standard operating procedures
Proportioned amounts
　of foam concentrate to water, 305, 305t
　of mechanical foam components, 299
Protective parking, 206, 208f
Protein foam, 298, 299f, 302, 302t, 303f
Prove-out sequence, 166
Public response to emergency vehicles, 198
Pump panel, 111, 111f
Pump shift control switch, 236–237, 237f
Pump Test Pit Design Recommendations, 332
Pump test pits, 332–333, 333f
Pump transmission
　engaging, for drafting operations, 267
　switching from road transmission to, 236–237, 237f
Pump-and-gravity-feed water distribution system, 38
Pump-discharge-pressure (PDP). *See also* Friction loss
　calculating, 60–61
　condensed Q method in relay operation for, 101
　hand method calculations, 98–99, 99sd
　hydraulic calculations of, 59
　in standpipe operations, 91, 94sd
　in wye scenario with equal lines, 78sd
Pumpers, 20–21, 20f, 21t
　exterior of, 110
Pumping element, 113
Pump-mounted eductors, 308
Pumps. *See* Fire pumps

Q

Q (quantity of water). *See also* Water supply
　friction loss and, 65
Quarter-turn hydrant valve, 285
Quint fire apparatus, 27, 27f, 27t, 28t

R

Radiator cap, safe removal of, 131, 137
Radio communication, 181, 202, 236, 292. *See also* Communications system
Railroad crossings, 198, 212
Raindown method for applying foam, 322, 323sd
Rated capacity, of water pumps, 330
Ratio controller, 311
Reaction distance, 197
Readiness, fire apparatus, 130, 130f, 183–184, 183f, 185sd
Receiver, for traffic signal preemption, 198
Red (stop) lights, emergency vehicle laws on, 194
Red emergency lights, 208
Reducers, 226, 226f, 277
Reflective vests, ANSI-approved, 208, 209
Relay pumpers, 276, 277f, 280–281
Relay pumping operations, 275f
　calculating friction loss for, 278–279
　components, 275–276
　condensed Q method in, 101
　equipment for, 276–277
　joining an existing relay pumping operation, 281–282
　operating the attack pumper, 279–280
　operating the relay pumper, 280–281
　operating the source pumper, 279
　personnel for, 277
　preparing for, 277–278
　pressure fluctuations in, 282
　safety for, 283
　shutting down, 282–283
　types, 279
　for uninterrupted water supply, 266
　water relay delivery options, 281
Relay valve, 281, 281f, 282f
Repairs, fire departments on identification vs. performance of, 130
Rescue potential, positioning at fire scene with, 203
Reservoirs, 37
Residential structures, floor spacing in, 70, 70f
Residual pressure, of water, 39, 49
Reverse (hose) lay, 233, 234–235, 234f, 277, 278f
Rider positioning, assigned, 157–158
Right turns, blind spots and, 195
Road surface condition, stopping and, 197
Roll-on method for applying foam, 319, 321, 321sd
Rotary gear pumps, 113–114, 114f
Rotary pumps, 113–114, 114f
Rotary vane pumps, 114, 114f
Rotors, in rotary pumps, 114
Rural areas, finding address of fire in, 199
Rural water supplies, 49

S

Safe driving practices, 194–198, 195f, 196f
Safety. *See also* Near Miss report(s)
　adding fuel to power tools, 141
　alcohol-resistant foams for alternative fuel incidents, 304
　appropriate use of foam concentrates, 304
　charging line from hydrant, 234
　with compressed-air foam systems, 314
　as driver/operator priority, 156–157
　due regard, 161, 177, 194

Safety (*Cont.*)
   during fire apparatus inspection, 131, 131f, 133, 134
   of fire apparatus operation, 148
   fire apparatus public appearances and, 169
   fire fighter stance for forward hose lay, 233
   fire stream delays and, 174–175
   getting underway and, 169–170
   handline selection, 93
   hose loading operations in motion, 169
   intake or discharge outlets on pump panel, use of, 70
   no casual chat in the cab, 158
   during performance testing, 358
   personnel for performance tests, 358
   with portable tanks and pumping operations, 286
   positioning fire apparatus for drafting water, 267
   protecting driver/operator at the scene, 9
   for relay pumping operations, 283
   seat belt issues, 168–169
   standards on, 9
   teaching, 157
   traffic incidents at fire scenes, 210
   two crew members on a tanker, 291
   vacuum test of priming system and, 344
   Voice of Experience on, 7
   water hammer and, 39
   water tank level during foam operations, 311, 311f
   wheel chocks to prevent fire apparatus rolling, 61
   work environment and, 158–159
Safety and Survival on the Fireground (Dunn), 204
Safety equipment, training and education on, 159
Safety system of aerial devices, inspection of, 145–146
Salvage companies, 6
Salvage covers, for portable tanks, 289
Savannah, Ga., first fully motorized fire department in, 5
School bus loading/unloading, emergency vehicle laws on, 194
Screw-type hydrant valve, 285, 285f
Seagraves, first aerial truck by, 25
Seat belts
   driver/operator as first to buckle, 156, 156f
   fire apparatus public appearances and, 169
   getting underway and, 180
   Near Miss vehicle event and wearing of, 244–245
   NFPA 1901 on warning devices for, 9
   safety issues, 168–169
   SCBA belts and straps vs., 138, 138f
Seats, fire apparatus, positioning for starting, 166
Secondary feeders, of municipal water systems, 38
Self rescue training, 174–175
Self-contained breathing apparatus (SCBA). *See also* Breathing air system; Fire equipment
   as hazard in cab, 157, 168, 170
   inspection of, 137–138, 140–141
   seat belt accommodation for, 156
   seat belts vs. belts and straps of, 138, 138f
Series/pressure mode, of multistage pump, 117, 117f, 119
Serpentine maneuvers, 175, 176sd
Service manuals, 159. *See also* Operating manuals
Service testing, 330. *See also* Performance tests
70 percent test, 352
Shut-off valves, for water mains, 38, 38f
Siamese connections, 73f
   description of, 225, 225f
   determining friction loss in, 78–79, 82–83, 82–85sd
   in fire department connections, 242, 245f
   standard and metric coefficients for, 65t, 87t
Single-acting piston pumps, 112–113, 113f
Single-stage pumps, 116–117, 117f
Sirens, 177, 193, 194, 198
Skids/skidding, 197–198
Slide rule hydraulic calculators, 98
Slides and rollers, aerial device, inspection of, 145, 146f
Slopes. *See also* Hills
   positioning at fire scene on, 204
Small-diameter hose (SDH), 220, 221f
Smooth-bore tips (nozzles), 228f
   characteristics, 228
   for compressed-air foam systems, 317
   distributor type, 63f
   flow calculations, 64sd
   handline, 61, 61f, 62, 64
   inspecting, 228–230sd
   master stream, 61, 62, 62f
   for performance tests, 334
   standard or metric flows by size or type, 63t
Soda acid extinguishers, 5
Soft suction hose, 222–223, 223f, 224sd, 239, 239f
Solid-tip nozzles, 61, 61f
Songer, Daryl, 86
Source pumpers, 275–276, 275f, 279, 290–291
Special service (speicalized) fire apparatus, 27–28, 28f, 28t, 205–206
Specific heat index, 35
Speed limits, emergency vehicle laws on, 194
Speed of fire apparatus, stopping and, 197
Split (hose) lay, 235–236, 235f, 277, 278f
Split flow method (calculation), 78–79, 82, 82sd
Split hose bed, 235
Spotters, 9, 10f, 180–182, 180–182f, 184sd
Spring brake test, 140, 140f
Spring probe flow meters, 101
Sprinkler systems, 205, 241–243, 243f
Stabilizing system, aerial device, inspection of, 145

Staging, 201–202, 201f, 202f
Staging area, 202
Staging area manager, 202
Standard(s). *See under* NFPA
Standard measurements
  coefficients for multiple Siamese lines, 87t
  coefficients for standpipe risers, 91t
  elevation pressure (loss and gain), 71sd
  friction loss coefficients, 65t
  friction loss in multiple hose lines of different sizes and lengths, 68sd
  of friction loss in Siamese hose by coefficient method, 83sd
  of friction loss in Siamese hose lines by percentage method, 84sd
  of friction loss in Siamese lines by split flow method, 82sd
  friction loss in single hose lines, 66sd
  hand method, 99sd, 100sd
  preplanning discharge pressure for a standpipe, 92sd
  of pump-discharge-pressure in prepiped elevated master streams, 88sd
  of pump-discharge-pressure in wye scenario with equal lines, 78sd
  of pump-discharge-pressure in wye scenario with unequal lines, 80sd
  smooth-bore handline nozzle flow, 64sd
  for smooth-bore nozzles flows, 63t
  of total pressure loss, 76
  of vacuum to water column and pressure, 257t
Standard operating procedures (SOPs)
  for backing fire apparatus, 180–182
  for emergency communications, 158
  following and promoting compliance with, 159, 161
  for positioning at fire scene, 200
  safety role of, 10
  for work-safe environments, 156
Standpipe(s), 87
Standpipe risers, discharge valve of, 89, 90f
Standpipe systems
  characteristics and calculations for, 87–91
  connecting supply hose lines to, 241–243
  connection to, 242f
  discharge pressure calculations for, 92sd, 93sd
  preincident plans on, 205
  pressure-regulating valves for, 92
  pump-discharge-pressure calculations for, 94sd, 95sd
Starter switch, 166
State laws, on emergency vehicles, 193–194
Static pressure, of water, 39, 47, 48sd, 49
Static water sources. *See also* Drafting water
  accessibility to, 264–265
  determining reliability of, 263
  establishing a fill site from, 285–286
  establishing dump site operations from, 287
  evaluating quality of, 264, 264f
  as fill sites, 283
  operating source pumpers at, 279
  pump maintenance and conditions of, 36
  in rural areas, 34, 49, 49f, 51
Steam-powered pumps (steamers), 4, 4f, 113
Steering, during skids, 197–198
Steering wheel, free play in, 135, 135f
Stipp, Bill, 40
Stop signs, emergency vehicle laws on, 194
Storz connections, 222, 223, 224sd
Straight roads, driving on, 195, 195f
Strainers
  debris in, 273
  for drafting operations, 267–268, 267–268f, 286
  pump intake, inspection of, 142, 142f
Stressors, 158
Subtract 10 method (calculation), 100, 100sd
Suction hoses, 20
Supply hose/lines
  bleeding, in relay pumper operations, 280, 280f, 282
  color-coding, 277
  fire fighter falls off back step while laying, 288
  functions of, 220, 221f
  testing, 221
Supply line evolutions
  changeover operations, 245–247
  forward hose lay, 233, 233f
  hand lays of, 239, 241, 242sd
  objectives, 232, 233
  reverse hose lay, 234–235, 234f
  split hose lay, 235–236, 235f
  to standpipe and sprinkler systems, 241–243, 242f, 243f, 245f, 246sd
Surface conditions, fire apparatus positioning and, 204
Surface tension, 35, 299–300
Suspension system inspection, 135
Sweep method for applying foam, 319, 321, 321sd
Synthetic detergent foams, 304, 304f
Synthetic foams, 303

T

Tactical benchmarks, 158
Tankers, 283, 285–287, 291. *See also* Nurse tanker operations; Water tanks
Tank-to-pump flow test, 341, 342sd
Team building, 158
Temperature
  engine and pump, during drafting operations, 273
  weight of water and, 36

Tenders, 24–25, 24f. *See also* Tankers
Termination areas, 209
Terrorism incidents, 213
Testing gauges. *See also* Pitot gauge(s)
    fire apparatus's gauges vs., 333, 333f
Theoretical hydraulics, 60, 93–95
Threaded hose couplings, 222
360-degree inspection, 166, 167sd, 180
Three-second rule, 197
Three-stage pumps, 120
Tiller (aerial fire apparatus type), 26f, 169
Tires, inspection of, 134–135, 159
Tools
    for fire apparatus inspections, 130, 131f
    functions and limitations, 161
    inspection of, 140–141, 140f
Total pressure loss (TPL), 76–77
Total stopping distance, 197
Tower ladders, 6, 6f
Traffic
    direction of, emergency vehicle laws on, 194
    never trusting, 206
    into and out of dump sites, 291, 291f
Traffic control, 208–210, 291
Traffic incident management area (TIMA), 209
Traffic incidents, 208–210, 209f, 210
Traffic signal preemption systems, 198
Traffic space, 209
Training exercises. *See also* Driving exercises
    apparatus lurches forward during, 8
    for driver/operators, 11
    for filling tankers, 287
    fire apparatus driving, 179sd
    on foam application, 319
    with foam concentrates and foam systems, 298
    for joining an existing relay pumping operation, 282
    mnemonics for, 159
    self rescue, 174–175
    for tiller device, 169
Transfer case, 121–122, 121f
Transfer valve, 145, 246, 247f, 349, 350f
Transition areas, 209
Transmission. *See also* Pump transmission
    switching from "road" to "pump," 236–237, 237f
Transmission fluid inspection, 137
Transmission retarders, 9, 195
Trench collapse incidents, 213
Triple-combination pumpers, 6, 6f
Trust, team building and, 158
Turbidity, of water, 36
Turnaround, confined-space, 175, 177, 178sd
Turnout pants and boots, 170, 208
250 psi test, 352, 354
200 psi test, 352
2½" attack hose, 222
Two-stage pumps, 117, 117f, 119–120, 270

U

Underwriters Laboratory (UL), 119, 330, 335f
Underwriters Laboratory (UL) plate, 335–336, 336f
Universal solvent, 35
U-turns, 195–196

V

Vacra, Jesse, 165
Vacuum
    for drafting operations, 257–258, 257t, 272, 273
    for dump line, 270
    priming a pump using, 114
    whirlpools while drafting water and, 273
Vacuum leaks, 261, 274
Vacuum test, of fire pumps, 260–261, 262sd, 341, 344, 345sd
Valves. *See also specific types*
    fire hose appliance, 227, 227f
Vanes, in rotary pumps, 114
Vehicle dynamics, 161
Vehicle fires, 212, 212f
Vehicle intercom system, 158
Velocity, of moving water sources, 264
Venturi effect/forces, 114, 290, 308, 311
Violent scenes, response to, 164, 212
Viscosity, of protein foams, 302
Voices of Experience
    Angulo on fire apparatus inspections, 143
    Dettman on adapting to the situation, 347
    Garrity on staying calm in emergencies, 115
    Poole on uninterrupted water supply, 22
    Songer on hydraulic calculations, 86
    Stipp on water sources, 40
    Vacra on driving narrow city streets, 165
    Warriner on wind conditions at fire scene, 207
    Washington on water mains for water supply, 284
    Westhoff on efficiency in foam applications, 316
    Willis on securing water source, 240
    Winkler on safety components, 7
Voltage meter inspection, 137
Voltmeter, 167–168, 168f
Volume, as measure of water flow or quantity, 38, 39
Volume pumps, piston pumps and, 120
Volute, 116

W

Warm zones, at emergency medical scenes, 213
Warriner, Mike, 207
Washtubs, drafting water and, 286
Water. See also Foam; Water shuttle; Water sources; Water supply
   chemical properties of, 35, 35f
   elevated streams for, 5
   fire hydrant locations, 42, 44
   fire hydrant operation, 41–42, 43sd
   fire hydrant testing, 46–47, 47f, 48sd, 49
   fire hydrant types, 41, 41f, 42f
   flow and pressure, 38–39, 41
   harmful characteristics of, 36
   inspecting and maintaining fire hydrants, 44, 46, 46f, 46t, 47f
   municipal systems for, 36–38, 37f, 38f
   NFPA 1901 on rate of delivery of, 20
   physical properties of, 35–36, 36f
   rural supplies of, 49, 51, 51f
   securing source for, 239, 241
   shutting down a hydrant, 42, 44sd
   Voice of Experience on reliable supply of, 40
   Voice of Experience on uninterrupted supply of, 22
Water curtain nozzles, 232, 232f
Water flow monitor, 73f
Water hammer, 39, 41
Water mains, 37, 38
Water pressure, 38–39
Water shuttle
   establishing dump site operations for, 287
   filling tankers for, 286–287
   in Incident Management System, 292
   offloading tankers for dump sites, 289–290
   safety for, 287
   traffic flow within dump sites, 291
   for uninterrupted water supply, 266, 283, 285–286
   using multiple portable tanks, 290
   using portable tanks, 289
Water sources. See also Static water sources
   characteristics, 34–35, 35f
   for fire pump performance tests, 331–333
   moving, 263–264
Water supply. See also Drafting water
   characteristics, 34–35, 34f

   efficiency in establishing, 42
   filling tankers for, 286–287
   management in the Incident Management System, 261, 263
   positioning at fire scene and, 204
   relay pumping operations for, 275–283
   secondary or supplemental, thinking ahead for, 63
   static or dynamic, NPDP from, 93
   uninterrupted, 275
   Voice of Experience on, 240, 284
   water shuttle for, 283, 285–287, 289–291
Water supply officers, 60, 275, 291
Water supply tank inspection, 141
Water tanks, 20, 20, 311, 311f. See also Nurse tanker operations; Tankers
Water thiefs, 225, 225f
Water treatment facilities, 37, 37f
Water-miscible fuels, 302
Weekly fire apparatus inspections, 148
Weight restrictions, for fire apparatus, 161
Westhoff, Roger, 316
Wet standpipe systems, 241
Wet vs. dry fire pumps, 141
Wet-barrel fire hydrants, 41, 42f
Wheel chocks
   aerial device stabilizing system inspection and, 145
   air brake inspection and, 139, 139f
   parking at scene procedures, 237, 238sd
   for parking on slopes, 204
   when working around fire apparatus, 260
Wheels or rims, inspection of, 135
Whirlpools, vacuum for drafting water and, 273
Willis, Brent, 240
Wind conditions, 204, 205f, 207
Windows, inspection of, 139
Windshield wipers, inspection of, 139
Winkler, Dave, 7
Wivell, Abraham, 25
Work environment, safe, 156, 158–159
Working fire apparatus, 201. See also Fire apparatus positioning
Writing utensils, for emergency vehicle response, 164
Wyed hoses, 77, 78sd, 79sd, 80sd, 81sd
Wyes, 223, 225, 225f. See also Gated wyes

Photo Credits

Cloud of smoke © Greg Henry/ShutterStock, Inc.

Chapter 1
Opener © National Library of Medicine; **1-1, 1-2** Courtesy of the FASNY Fire Museum of Firefighting, New York; **1-3** © Trinity Mirror/Mirrorpix/Alamy Images; **1-4, 1-5, 1-6** Courtesy of the FASNY Fire Museum of Fire Fighting, New York; **1-7** © National Library of Medicine; **1-8** Courtesy of Phillip S. "Sean" Barrett; **1-9** Courtesy of the FASNY Fire Museum of Fire Fighting, New York; **1-10** © Jack Dagley Photography/ShutterStock, Inc.; **1-11** © GTibbetts/ShutterStock, Inc.; **1-12** Courtesy of the FASNY Fire Museum of Fire Fighting, New York; **1-15** © Peter Willott, St. Augustine Record/AP Photos

Chapter 2
Opener © Mike Brake/ShutterStock, Inc.; **2-2** © JoLin/ShutterStock, Inc.; **2-3** © Peter Kim/ShutterStock, Inc.; **2-4** Courtesy of Troy S. Huizar/Ferrara Fire Apparatus Inc.; **2-5** © Peter Kim/ShutterStock, Inc.; **2-7** © SVLumagraphica/ShutterStock, Inc.; **2-8** Photo supplied by PBCFR; **2-10** Courtesy of Roy Robichaux, ConocoPhillips Alliance Refinery

Chapter 3
3-2 © Dan Myers; **3-5** © Paul Glendell/Alamy Images; **3-10** Courtesy of American AVK Company; **3-11** Courtesy of Captain David Jackson, Saginaw Township Fire Department; **3-16** © Will Powers/AP Photos; **3-18** Courtesy of Best Fire Defense LLC

Chapter 4
4-7 © Howard Sandler/ShutterStock, Inc.; **4-8** © haveseen/ShutterStock, Inc.; **4-9A** Courtesy of Akron Brass Company; **4-9C** Courtesy of Akron Brass Company; **4-11** © Kurt Hegre, The Fresno Bee/AP Photos

Chapter 5
5-1 © Keith Muratori/ShutterStock, Inc.; **5-2A, 5-2B** Courtesy of Jim Hylton; **5-6** Courtesy of Britton Crosby; **5-7** Courtesy of Pentair Water; **5-15** © FirePhoto/Alamy Images; **5-16** Courtesy of Har-Rob Fire Apparatus, Inc.

Chapter 6
6-1 © Bryan Eastham/ShutterStock, Inc.; **6-2** © Mike Voss/Alamy Images; **6-3** © Rorem/Dreamstime.com

Chapter 8
Opener © Zoran Milich/age fotostock; **8-1** © AlanHaynes.com/Alamy Images; **8-2** © iofoto/ShutterStock, Inc.; **8-3** © jean schweitzer/ShutterStock, Inc.; **8-4** © PeterG/ShutterStock, Inc.; **8-5** © Jim Parkin/ShutterStock, Inc.; **8-6** © Vladislav Gurfinkel/Dreamstime.com; **8-10, 8-11** © Adam Alberti, NJFirePictures.com; **8-14** © Aaron Kohr/ShutterStock, Inc.; **8-15** © Keith Muratori/ShutterStock, Inc.

Chapter 9
Opener © Thinkstock/age fotostock; **9-4** © Samuel Acosta/ShutterStock, Inc.; **9-8, 9-9, 9-10** Courtesy of Akron Brass Company; **9-12, 9-13** © 2003, Berta A. Daniels; **9-15** Courtesy of Akron Brass Company; **9-18** Courtesy of Flamefighter Corporation; **9-19B** Courtesy of Akron Brass Company; **9-20** Courtesy of POK of North America, Inc.; **9-22** Courtesy of Jim Hylton; **9-31A** © Artdirection/Dreamstime.com; **9-31B** © Eugene Feygin/Dreamstime.com; **9-32** © Vladislav Gurfinkel/ShutterStock, Inc.

Chapter 10
10-11, 10-12, 10-13 Courtesy of Kochek Co., Inc.; **10-17** © Steve Redick

Chapter 11
Opener © Uladzimir Chaberkus/Dreamstime.com; **11-1** © Bortel Pavel/ShutterStock, Inc.; **11-4** © Lksstock/Dreamstime.com; **11-16** Courtesy of Williams Fire & Hazard Control, Inc.; **11-18** © Dale A. Stork/ShutterStock, Inc.

Chapter 12
Opener Courtesy of Underwriters Laboratories, Inc.; **12-3** Courtesy of Class 1

Unless otherwise indicated, all photographs and illustrations are under copyright of Jones and Bartlett Publishers, LLC, courtesy of Maryland Institute for Emergency Medical Services Systems, photographed by Glen E. Ellman, or have been provided by the authors.